The Neural Crest and Neural Crest Cells
in Vertebrate Development and Evolution

Brian K. Hall

The Neural Crest and Neural Crest Cells in Vertebrate Development and Evolution

 Springer

Brian K. Hall
Department of Biology
Dalhousie University
Halifax, Nova Scotia
Canada B3H4J1
bkh@dal.ca

and

Centre for Science and Society
School of Life Sciences
Arizona State University
Tempe, Arizona
85287-3301

ISBN 978-0-387-09845-6   e-ISBN 978-0-387-09846-3
DOI 10.1007/978-0-387-09846-3

Library of Congress Control Number: 2008941099

springer.com

Preface

Knowledge of the development and evolution of the neural crest sheds light on many of the oldest questions in developmental and evolutionary biology. What is the role of germ layers in early embryogenesis? How does the nervous system develop? How does the vertebrate head arise developmentally, and how did it arise evolutionarily? How did the vertebrate dorsal nervous system, heart, skeleton, teeth (and the neural crest itself) originate? How do growth factors and *Hox* genes direct cell differentiation and embryonic patterning? What goes wrong if development is misdirected by mutations, or if embryos are exposed to exogenous agents such as drugs, alcohol, or excess vitamin A (retinoic acid)?

Twenty years ago, I was instrumental in organizing the publication of a facsimile reprint of the classic monograph by Sven Hörstadius, *The Neural Crest: Its properties and derivatives in the light of experimental research*, originally published in 1950. Included with the reprint was an analysis of subsequent studies on the neural crest and its derivatives. A decade later, the first edition of this book was published (Hall, 1999a). The explosion of interest in and knowledge of the neural crest over the past decade prompted me to write this second edition.

As in my 1988 overview of the reprinting of '*Hörstadius*'—as his book is known to many—and as in the first edition of this book, I take a broad approach in dealing with the discovery, embryological and evolutionary origins, migration, differentiation and cellular derivatives of the neural crest. Cells from the neural crest are associated with many developmental abnormalities, many of which have their origins in a defective neural crest (NC) or in defective neural crest cells (NCCs). The book would be incomplete without discussing neurocristopathies—those tumors and syndromes involving NCCs or those birth defects in which NCCs play a role.

The book is organized into three parts.

Part I (Discovery and Origins) begins with a chapter devoted to the discovery of the neural crest and the impact of that discovery on entrenched notions of germ-layer specificity and the germ-layer theory, a theory that placed a straitjacket around embryology and evolution for almost a century. Primary and secondary neurulation and the neural crest as the fourth germ layer are introduced in this chapter.

In Chapter 2, I discuss the embryological origins of the neural crest, including the identification of future NCCs in gastrula-stage embryos; molecular and cellular

markers of future NCCs; neural and neural crest induction; and rostrocaudal patterning of the developing neural tube and neural crest.

Chapter 3 takes NCCs out of the neural tube with discussions of:

- the delamination of NCCs from the neural tube as mesenchymal cells (a process
  requiring the transformation of epithelial to mesenchymal cells, usually written
  in the text as EMT or epithelial —> mesenchymal transformation),
- NCC migration and the nature of the extracellular matrices (ECM) through
  which or along which they migrate, and
- the differentiative potential of NCCs.

Chapter 4 is devoted to the evolutionary origins of the neural crest through an
analysis of fossils and of cell types, genes, and gene networks in extant cephalochordates (amphioxus) and in urochordates (chiefly ascidians) in an effort to answer
the question 'Is there any evidence of precursors of the neural crest in urochordates
or in cephalochordates?' The second aim of Chapter 4 is to examine the origin of
neural and skeletal tissues of neural crest origin in the first vertebrates (i.e., chordates with a head), and the origin of the jaws in the transition from jawless to jawed
vertebrates.

Part II (Neural-Crest Derivatives) presents an analysis of our knowledge of the
cell types into which NCCs differentiate. The organization of this part differs from
the first edition in which the chapters were organized by major groups of vertebrates,
each of which included a discussion of similar cell types—neural, pigment, and
skeletal cells. In this edition, I have organized each of the four chapters around major
class of cells and the tissues and organs they form or to which they contribute:

- pigment cells and color patterns (Chapter 5);
- neurons and the nervous system (Chapter 6);
- cartilage, bone, and skeletal systems (Chapter 7); and
- dentine-forming cells and teeth, and the smooth muscle, septa and valves of the
  heart (Chapter 8).

These chapters cover:

- **trunk neural crest cells** (TNCCs)—Chapter 5;
- the **vagal and sacral neural crest** (VNC, SNC), peripheral nervous system
  (spinal and cranial ganglia), autonomic and parasympathetic nervous systems
  (sympathetic and parasympathetic ganglia, enteric ganglia, adrenal chromaffin
  cells), Schwann and glial cells, and Rohon–Béard neurons—Chapter 6;
- **cranial neural crest cells** (CNCC), chondroblasts and osteoblasts, mesenchyme,
  the skeletogenic (chondrogenic) neural crest, and epithelial–mesenchymal
  interactions—Chapter 7;
- the odontogenic neural crest, odontoblasts (dentine-forming cells), tooth formation, and the **cardiac neural crest (CarNC)**, the heart, and development of
  valves, septa and the aortic arches—Chapter 8.

Part III consists of two chapters, Chapter 9 devoted to tumors of neural-crest origin (neurocristopathies), Chapter 10 to a reconsideration of NCC development in the context of birth defects.

Chapter 9 includes discussions of neuroblastomas, neoplasia, and examples of syndromes based in defective NC or NCCs, the two major examples being APUDomas and DiGeorge syndrome.

Chapter 10 broadens the scope to birth defects (often but not always involving the neural tube to which NCCs contribute) or which are induced by a teratogen—vitamin A and craniofacial defects in this case. Mutations affecting NCCs are discussed as is the ability of NCCs to compensate for lost cells, a developmental property known as regulation and a discussion that brings us full circle to the differing potentials of subpopulations of NCCs and whether any NCCs persist as stem cells in embryos or adults.

To avoid interrupting the flow of the text, I have placed most references and some supporting statements in numbered notes, which are gathered at the end of each chapter, and which serve as an annotated bibliography through which access to the literature may be obtained. I have not included all of the literature published before 1999, much of which is in the first edition (Hall, 1999a*). Otherwise, I have surveyed the literature to early 2008. References marked with an * are significant reviews or analyses. Occasionally, I use footnotes $^{(\otimes)}$ for general points that apply throughout. Similarly, boxes are used for items of general interest, biographies, or interesting case studies. †signifies an extinct taxon. Gene names are italicized and capitalized (*Shh*), proteins are in plain text and capitalized (Shh). Human genes and proteins are capitalized (*SHH*, SHH). As a shorthand expression for a transformation or interaction I use the symbol —>. The text is extensively illustrated and there is a detailed index. A list of abbreviations is provided. From that list, the following are abbreviations for regions of the neural crest (NC) or for populations of neural crest cells (NCCs).

NC	**NCCs**
CarNC — cardiac neural crest	CarNCCs — cardiac neural crest cells
CNC — cranial neural crest	CNCCs — cranial neural crest cells
NC — neural crest	NCCs — neural crest cells
SNC — sacral neural crest	SNCCs — sacral neural crest cells
TNC — trunk neural crest	TNCCs — trunk neural crest cells
VN — vagal neural crest	VNCCs — vagal neural crest cells

I am grateful to the following experts who provided invaluable comment on individual chapters—chapter reviewed are shown in parenthesis—Marianne Bronner-Fraser (2), Carol Erickson (3), Daniel Meulemans (4), Lennart Olsson (5), Ryan Kerney (7), and Gerhard Schlosser (6, Part). Ryan Kerney and Jennifer Quinn provided helpful comments on stem cells, Cory Bishop comments on ascidians. Tim Fedak prepared 20 of the new figures. Many thanks, Tim. June Hall edited the manuscript for style and comprehensibility in her inimitable way. Many thanks, June. Individuals who provided figures are acknowledged in the appropriate figure legend. Figure 7.10 is modified from Del Pino and Medina (1998), published by UBC Press, Leioa, Vizcaya, Spain. Financial support for my research program on the neural crest and its derivatives from the Natural Sciences and Engineering Research Council (NSERC) of Canada is gratefully acknowledged.

Halifax and Tempe   Brian K. Hall

Contents

Abbreviations

A

Adam13   a member of the Adam family of membrane-anchoring metalloprotease (named from *a d*isintegrin *a*nd *m*etalloprotease domain)

Amphi   prefix for genes in amphioxus (*Branchiostoma* spp.), for example, *AmphiOtx*

APUDomas   tumors that share *amine precursor uptake and decarboxylation*

B

BALB/c   an inbred strain of mice that develops numerous tumors in later life

Bdnf   brain-derived neurotrophic factor

bHLH   basic helix–loop–helix transcription factors

Bmp   bone morphogenetic protein family of genes and their product

BmpR   bone morphogenetic protein receptor, for example, BmpR1

C

CarNC   cardiac neural crest

CarNCCs   cardiac neural crest cells

Cbfa1   Core binding factor alpha 1; see Runx2

Cdh   Cadherin, a family of 20 Ca^{++}-binding transmembrane proteins that function in cell adhesion

ch   *congenital hydrocephalus* mutant mice

CHD7   chromodomain helicase DNA binding protein 7

Chn   *Chinless* mutant zebrafish

Ci   prefix for genes in the sea vase, *Ciona intestinalis* (an ascidian)

cls   *colorless* mutant in zebrafish

CNC   cranial neural crest

CNCCs   cranial neural crest cells

CNS   central nervous system of vertebrates

Col2α1   the gene for the procollagen type II alpha 1 chain.

Con   *Chameleon* mutant zebrafish

CraBP   cellular retinoic acid binding protein

CRKL   *v-crk sarcoma virus CT10 oncogene homolog [avian]-like*

D

DiI	1,1′-dioctadecyl-3,3,3′,3′-tetramethyl indocarbocyanine perchlorate
Dil	the *dilute* (*dil*) allele in the budgerigar, a mutation in melanocytes
Disp1	the protein dispatched (DISP1), a regulator of *Shh*
Dll	a family of genes Delta proteins, which are type-1 cytokine receptor family protein
Dlx	*distalless* gene family in vertebrates, for example, *Dlx1*
DOPA	3, 4-dihydroxyphenylalanine, a catecholamine precursor
DRG	dorsal root ganglia
Dsh	Disheveled gene and protein product
D–V	dorsoventral axis/polarity of embryonic regions/organ rudiments (e.g., pharyngeal arches, limb bud), organs (e.g., limbs) or organism. Sometimes referred to in the literature as medio-lateral or proximo-distal polarity.

E

ECM	extracellular matrix
EDAR	Ectodysplasin 1, anhidrotic receptor
Edn3	endothelin-3, a mitogenic peptide
Egf	epithelial growth factor genes and their proteins
eIF-4AIII	Eukaryotic translation initiation factor 4AIII
EMT	epithelial-mesenchymal transformation (sometimes shown as epithelial —> mesenchymal transformation)
En	*Engrailed* gene family, for example, *En1*
EphA, EphB	eph receptor tyrosine kinases A and B, members of large subfamilies of receptor protein-tyrosine kinases consists of receptors related to Eph, a receptor expressed in an *erythropoietin-p*roducing human *h*epatocellular carcinoma cell line.

F

Far	*First- arch* murine craniofacial mutation
Fgf	fibroblast growth factor gene and protein family
Fox	*forkhead transcription factor* binding element
Frzb	*Frizzled-related protein precursor*, a secreted antagonist of *Wnt* signaling
FUDR	5-fluoro-2′-deoxyuridine (blocks DNA synthesis)

G

Gdnf	glial-cell-line-derived neurotrophic factor
GFP	green fluorescent protein
GnRH neurons	gonadotropin-releasing hormone (GnRH) neurons; see also LhRH neurons

H

Hh	*Hedgehog* gene family, for example, *Sonic hedgehog* (*Shh*)
H.H.	Hamilton–Hamburger stage of chick embryonic development
HMG	high-mobility group proteins
Hnf3β	*hepatocyte nuclear factor3β*
HNK-1	a cell surface carbohydrate (known as CD-57 in immunology) used as a marker for NCCs
Hox	homeotic gene classes in vertebrate, for example, *Hoxd10*
Hr	prefix for genes in the western northern Pacific ascidian *Halocynthia roretzi*

I

Igf	insulin-like growth factor genes and products
Insm1	*insulinoma-associated 1* gene

J

Jag1	gene for the transmembrane protein Jagged1, which functions via the Notch pathway

K

kDa	kilo daltons
Krox20	a gene encoding a zinc-finger transcription factor

L

La-N-5	human neuroblastoma cell line
LhRH neurons	luteinizing hormone releasing neurons; see also GnRH neurons
Lp	the *Loop-tail* mutant mouse, in which hindbrain and spinal cord fail to close and so NCCs fail to migrate
LRD	lysinated rhodamine dextran

M

MAPK	ras/mitogen-activated protein kinase, a signaling pathway involved in the phosphorylation of target molecules such as transcription factors and other kinases in cell membrane, cytoplasm, and nucleus
Mash1	*Mouse achaete-scute homologue 1* gene
Mdkb	midkine-b (a heparin-binding growth factor)
MEN1	*Multiple endocrine neoplasia type 1* in humans—characterized by endocrine neoplasia of the parathyroids, pituitary, and pancreas
MIF	Macrophage inhibitory factor
Mitf	*microphthalmia-associated transcription factor*
Mmp	matrix metalloprotease family
Msh	melanocyte stimulating hormone
Msx	homeobox genes (e.g., *Msx1*, *Msx2*) of vertebrates of the *Drosophila msh* (melanocyte-stimulating hormone) family
MTN	*Mesencephalic Trigeminal Nucleus*
M.W.	molecular weight

N

NC	neural crest
N-CAM	neural cell adhesion molecule
NCCs	neural crest cells
NF1	the *neurofibromin1* gene responsible for von Recklinghausen neurofibromatosis (type 1 neurofibromatosis)
NF1	von Recklinghausen neurofibromatosis (type 1 neurofibromatosis)
NF2	bilateral acoustic neurofibromatosis
Ngf	nerve growth factor family of genes and their products
Nkx	family of transcription factors that function downstream of *Shh*
Nof	*No-fin* mutant in zebrafish, which lacks pectoral fins and gill cartilages
Nrp1, Nrp2	neuropilin1, neuropilin2 (co-receptors for semaphorins)
Nt3	neurotrophin3, a nerve growth factor

O

Oca2	*oculocutaneous albinism2* gene
Osf2	osteoblast-stimulating factor2; see Runx2
Otx	*orthodenticle* family of genes in vertebrates, for example, *Otx1*

P

p27	cell cycle inhibitor
Pax	a family of nine mammalian genes containing a paired-type homeodomain as a DNA-binding motif; for example, *Pax1*, *Pax9*
P–D	proximo-distal axis/polarity of embryonic regions/organ rudiments (e.g., pharyngeal arches, limb bud), organs (e.g., limbs) or organism
Pdgf	platelet-derived growth factor genes and gene products
PdgfR	platelet-derived growth factor receptor genes and gene products
Pitx2	Pituitary homeobox gene-2 in mouse related to *bicoid* in *Drosophila*. Also known as *Ptx*.
PNA	peanut agglutinin lectin
PRE	pigmented retinal epithelium
Ptch	Patched, a binding protein for hedgehog gene products
PTEN	tumour-suppressor gene, *p*hosphatase and *ten*sin homologue
Ptx	*see Pitx*

Q

QCPN	quail non-chicken perinuclear antigen

R

r	rhombomeres, a segment of the hindbrain in vertebrates
RA	retinoic acid, a biologically active form of vitamin A

Raldh2	a gene for retinaldehyde dehydrogenase, required for synthesis of retinoic acid
RaLP	a member of the Src family of tyrosine kinase substrates
R–B	neurons Rohon–Béard neurons
Rhob	a low-molecular-weight GTPase in the Ras protein family
RTK	receptor tyrosine kinase
Runx2	*runt-related transcriptional factor-2*; older names are *cbfa-1* and *osf2*

S

SEM	scanning electron microscopy
Shh	*Sonic hedgehog* gene
Six2	*sine oculis*-related homeobox 2
Snail1, 2	zinc-finger transcription factor-encoding genes, orthologues of *Drosophila Snail homologue 1* and *Snail homologue*
SNC	sacral neural crest
SNCCs	sacral neural crest cells
sof	*short fin*, a zebrafish mutant
Sox	multigene families that encode transcription factors with high-mobility group DNA-binding domains, the acronym coming from coming from *sex*-determining region homeobo*x*
S phase	the phase of cell division during which DNA is synthesized
Suc	the gene *sucker* in zebrafish, *Danio rerio*, which disrupts Endothelin-1. Also known as *endothelin-1* (*edn1*)

T

T-box	a family of genes encoding transcription factors, for example, *Brachyury, Tbx1, Tbx5*
Tbx,	a class of genes within the T-box family of transcription factors, for example, *Tbx6*
Tcf/Lef	T-cell specific/lymphoid enhancer binding factor (transcription factors)
TCOF1	*Treacher-Collins Franceschetti syndrome 1* gene coding for nucleolar phosphoprotein Treacle
TEM	transmission electron microscopy
Tgfß	transforming growth factor beta genes and their products
TgfßR	transforming growth factor beta receptors
Timp	tissue inhibitor of metalloprotease
TNC	trunk neural crest
TNCCs	trunk neural crest cells
Trk	a family of tyrosine kinases receptors for neurotropins
Tsp	thrombospondins (family of five glycoproteins involved in cell migration and proliferation)

V

Vegf	vascular endothelial growth factor gene and protein
VER	ventral ectodermal (epithelial) ridge on developing tail buds
Vgr1	older name for *Bmp6*
VMA	vanillinemandelic acid (4-hydroxy-3-methoxymandelic acid, a metabolite of catecholamine
v-myc	a proto-oncogene from retrovirus-associated DNA sequences originally isolated from an avian myelocytomatosis virus
VNC	vagal neural crest
VNCCs	vagal neural crest cells
VNT	ventral neural tube
vt	the *vestigial tail* tailless mouse mutant

W

Wnt	a large gene family orthologous to *wingless* in *Drosophila* that produce secreted molecules involved in intercellular signaling. *Wnt* is a combination of *Drosophila* w*ingless* and *int1* for *Wnt1*, the first vertebrate (mouse) family member discovered

Part I
Discovery and Origins

The topic of this book is the neural crest—its embryological and evolutionary origins; the multitude of cells, tissues and organs that develop either directly from neural crest cells (NCCs) or under their influence; the ways and means by which those cells migrate and differentiate; and abnormalities resulting from the involvement of the neural crest or of NCCs in tumors, deficiencies or defects.

In Chapter 1, I provide a brief overview of the major phases of investigation into the neural crest and the major players involved, discuss how the origin of the neural crest in vertebrate embryos is intertwined with the development of the nervous system, discuss the impact on the germ-layer theory of the discovery of the neural crest and of secondary neurulation, and present evidence of the neural crest as the fourth germ layer in vertebrates alongside ectoderm, mesoderm, and endoderm.

Chapters 2 and 3 are devoted to the embryological origins, delamination, migration, and potentiality of NCCs and Chapter 4 to the evolutionary origins of the neural crest and NCCs.

Chapter 1
Discovery

The **neural crest** (NC) has long fascinated developmental biologists and, increasingly over the past decades, evolutionary and evolutionary-developmental biologists.

The neural crest is the name given to the fold of neural ectoderm at the junction between neural and epidermal ectoderm in neurula-stage vertebrate embryos (Fig. 1.1). In this sense, neural crest is a morphological term akin to neural tube or limb bud. The neural crest consists of **neural crest cells** (NCCs), a special population(s) of cells that give rise to an astonishing number of cell types and to an equally astonishing number of tissues and organs (Table 1.1).

Major classes of NCCs can be grouped in various ways. Figure 1.2 details the relationships of human NCCs as determined using standard histological criteria and in a cladistic analysis.

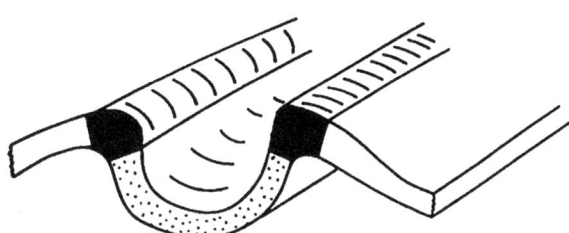

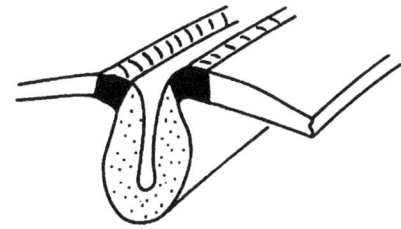

Fig. 1.1 Localization of the neural crest (*black*) at open neural plate (*above*) and closing neural fold (*below*) stages as seen in an avian embryo. Neural crest is located at the boundary between neural ectoderm (*stippled*) and epidermal ectoderm

B.K. Hall, *The Neural Crest and Neural Crest Cells in Vertebrate Development and Evolution*, DOI 10.1007/978-0-387-09846-3_1,
© Springer Science+Business Media, LLC 2009

Table 1.1 A list of the cell types derived from the neural crest and of the tissues and organs that are entirely neural crest or that contain cells derived from the neural crest

Cell types

Sensory neurons	Cholinergic neurons
Adrenergic neurons	Rohon–Béard cells
Satellite cells	Schwann cells
Glial cells	Chromaffin cells
Parafollicular cells	Calcitonin-producing (C) cells
Melanocytes	Chondroblasts, chondrocytes
Osteoblasts, osteocytes	Odontoblasts
Fibroblasts (mesenchyme)	Cardiac mesenchyme
Striated myoblasts	Smooth myoblasts
Mesenchymal cells	Angioblasts
Merkel cells	

Tissues or organs

Spinal ganglia	Parasympathetic nervous system
Sympathetic nervous system	Peripheral nervous system
Thyroid and parathyroid glands	Ultimobranchial body
Adrenal gland	Craniofacial skeleton
Teeth	Dentine
Connective tissue	Adipose tissue[a]
Smooth muscles	Striated muscles
Cardiac septa, valves, aortic arches	Dermis
Eye	Cornea
Endothelia	Blood vessels
Heart	Dorsal fin
Brain	Connective tissue of thyroid, parathyroid, thymus, pituitary, and lacrymal glands

[a] A recent investigation using both murine and Japanese quail embryos has demonstrated the origin of a population of adipocytes (fat cells) from cranial neural crest (Billon *et al.*, 2007). Adipocytes have long been thought to arise from cells within mesodermal lineages; indeed, there is a vast literature on the origin of adipocytes, chondro- and osteoblasts from clonal cell lines derived from periosteal or bone marrow cells (Hall, 2005b*). Billon and colleagues demonstrated that cultured mouse neural crest cells form adipocytes, that cultures of quail NCC can be induced to differentiate as adipocytes, and that a subset of adipocytes associated with the developing ear—confirmed by visualization of perilipin, a lipid marker—in mouse embryos arise from the neural crest during normal development. In another analysis, Takashima *et al.* (2007) demonstrated that mesenchymal stem cells capable of producing adipocytes, chondro- and osteoblasts arise from Sox1[+]-neuroepithelial cells (i.e., NCCs) as well as from paraxial mesoderm.

NCC contribution may be direct (providing cells) or indirect (providing a necessary, often inductive, environment in which other cells develop). The enormous range of cell types produced provides an important source of evidence of the NC as a germ layer, bringing the number of germ layers to four: ectoderm, endoderm, mesoderm, and neural crest.

This chapter provides a brief overview of the major phases of investigation into the neural crest and the major players involved, discusses how the origin of the neural crest relates to the origin of the nervous system in vertebrate embryos, discusses the

Fig. 1.2 Two classification schemes for human neural crest cells showing the different relationships obtained in a cladistic analysis (cladistic) and in an analysis using standard histological criteria (artificial). Of five cell-type categories (1–5) shown, only the skeletal cells (chondrocyte, osteocyte) separate out in both schemes, although chromaffin cells of the adrenal glands and parafollicular cells of the thyroid glands cluster as a subgroup. Note that neural crest-derived neurons (sensory, sympathetic) do not cluster as an individual (monophyletic) group, reflecting their origins from different NCC populations (see Chapter 6). See Vickaryous and Hall (2006) for further analysis and for the 19 characters used in the cladistic analysis. From Vickaryous and Hall (2006)

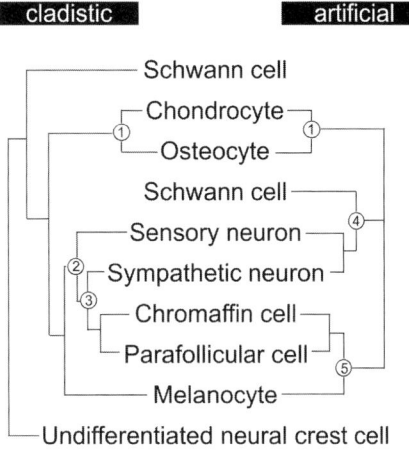

Cell type categories
1. Non-polarized extracellular matrix-secreting
2. Endocrine
3. Biogenic amine-handling
4. Nervous tissue
5. Epithelial tissue

impact on the germ-layer theory of the discovery of the neural crest and of secondary neurulation, and presents evidence of the neural crest as the fourth germ layer.

This chapter begins with the story of how this amazing embryonic region was discovered.

Zwischenstrang

Wilhelm His (1831–1904), professor of anatomy and physiology in Basel, Switzerland, was the first to provide a causal explanation for embryonic development, basing his explanation on mechanics (embryos as rubber tubes), developmental physiology, and predetermined organ-forming germinal regions, each of which contained cells with specified fates.

In 1868, His identified a band of cells sandwiched between the developing neural tube and the future epidermal ectoderm as the source of spinal and cranial ganglia in chicken embryos. He named this band *Zwischenstrang*, the intermediate cord.[1] In 1874, His included *Zwischenstrang*—the **neural crest** as we now know it—as one of the organ-forming germinal regions.

As far as can be determined, the term 'neural crest' was first used in a paper on the development of the olfactory organ published in 1879. The author, Arthur Milnes Marshall (1852–1893), was a professor of zoology at Owens College in Manchester, where he remained until his death at age 41 in a fall from Scafell Pike, England's highest peak.[2] In a paper on the development of the cranial nerves in chicken embryos, Marshall (1878) used the term **neural ridge** for the cells that give rise to cranial and spinal ganglia. Realizing that this term was less descriptive than was desirable, a year later he replaced neural ridge with neural crest. As told in his own words,

> I take this opportunity to make a slight alteration in the nomenclature adopted in my former paper. I have there suggested the term neural ridge for the longitudinal ridge of cells which grows out from the reentering angle between the external epiblast and the neural canal, and from which the nerves, whether cranial or spinal arise. Since this ridge appears before closure of the neural canal is effected, there are manifestly two neural ridges, one on either side; but I have also applied the same term, neural ridge, to the single outgrowth formed by the fusion of the neural ridges of the two sides after complete closure of the neural canal is effected, and after the external epiblast has become completely separated from the neural canal. I propose in future to speak of this single median outgrowth as the neural crest, limiting the term neural ridge to the former acceptation.
>
> (Marshall, 1879, p. 305, n. 2)

A Brief Overview of the Past 120 Years

1890–1950s

His and Marshall independently identified the NC as the origin of cranial and spinal ganglia and neurons, an origin that was easy for others to accept because of the relationship of these cell types and of the neural crest to the dorsal neural tube, the source of the dorsal nervous system.

In the 1890s, however, Julia Platt claimed that the cartilages of the craniofacial and pharyngeal arch skeletons[⊕] and the dentine-forming cells (odontoblasts) of the teeth of the mudpuppy, *Necturus maculosus*, arose from the ectoderm adjacent to the neural tube. Although supported by several contemporary researchers, Platt's conclusion was not accepted. In fact, her proposal of an ectodermal origin of the pharyngeal arch skeletons raised major controversies.

Why?

Because her conclusions ran completely counter to the entrenched **germ-layer theory**, according to which skeletal tissues arose from mesoderm, not ectoderm (see the section on germ-layer theory below).

Because an NC origin for skeletal tissues was so contentious, there was a 40-year gap between Platt's papers and independent studies in the 1920s and 1930s by Stone, Raven, and Holtfreter demonstrating the NC as a major source of mesenchyme, connective tissue, and cartilage (see Chapter 7).[3] Even more detailed reports by Sven Hörstadius, Sven Sellman, and Gavin de Beer were published in the 1940s. de Beer (1947) also thought it probable that NCCs differentiated into the osteoblasts of

---

[⊕] Three terms—pharyngeal, visceral, and branchial—are often used interchangeably for the multiple arches that arise from the pharynx in vertebrates. I use the term *pharyngeal arches*, except when referring to gills in urodeles or fish, when gill arch seems more appropriate.

dermal bones in *Ambystoma*, although the evidence was less convincing than was his evidence of the NC origin of cartilage and teeth.[4] Nowadays, not only has the skeletogenic capability of the cranial neural crest (CNC) been documented in all classes of vertebrates (see Chapter 7) but also the NC and its cells occupy a central position in studies of vertebrate development and evolution (see Chapters 2 and 4).

Despite these studies on the skeletogenic neural crest but because of the entrenched germ-layer theory, the focus of interest until the 1940s and 1950s was the NC as a source of pigment cells (chromatophores) and neural elements, such as spinal ganglia (see Chapters 5 and 6). Amphibian embryos were the embryos of choice.◊

Standing as a milestone on the road to understanding the NC is *The Neural Crest: Its properties and derivatives in the light of experimental research*, a monograph by Sven Hörstadius (Box 1.1). Published in 1950, 82 years after the discovery of the neural crest, it was reprinted in 1969 and again in 1988. 'Hörstadius,' as the monograph is known, was based on a series of lectures delivered during 1947 at the University of London at the invitation of Professor (later Sir) Gavin de Beer, then head of the Department of Embryology at University College and later director of the British Museum (Natural History). As noted above, de Beer had just completed his extensive experimental study of the NC origin of craniofacial cartilages and dentine in the Mexican axolotl and had provided suggestive evidence of a NC contribution to the splenial, a membrane bone of the skull.

Box 1.1 Sven Otto Hörstadius (1898–1996)

Born in Stockholm on 18 February 1898, Hörstadius began his academic career at Stockholm University, from which he graduated in 1930 and where he was appointed first lecturer and then associate professor of zoology. Early marked for recognition, in 1936, Hörstadius was awarded the Prix Albert Brachet by the Belgian Academy of Science. From 1938 to 1942 he directed the Department of Developmental Physiology and Genetics at the Wenner-Gren Institute of Experimental Biology. In 1942, Hörstadius became professor of zoology at the University of Uppsala, a position he occupied for 22 years, retiring in 1964 as Professor Emeritus.

A pioneering embryologist, brilliant lecturer, and an expert ornithologist, Hörstadius had a reputation for producing some of the best and among the earliest close-up photographs of difficult-to-photograph birds. The numerous

---

◊ Although studies on anuran (frogs, toads) and urodele (newt, salamander) embryos are often discussed together, bear in mind that relationships between these two groups of amphibians are not fully resolved. Indeed, use of the term amphibian at all is contentious; in current terminology, amphibian would imply a monophyletic group, which amphibians clearly are not (Hall and Hallgrímsson, 2008).

honors he received reflect his standing in the European scientific circles and the breadth of his interests and accomplishments.[a]

Our knowledge of the most fundamental aspects of echinoderm development derives from his studies (Hörstadius, 1928, 1939). Hörstadius was the first to demonstrate a fundamental feature of life now taken as a given, namely, *nuclear control of the species-specific characteristics of organisms*. By enucleating an egg from one species of sea urchin and fertilizing it with the sperm from another, Hörstadius created the first chimeric sea urchin embryos. The characteristics of the resulting embryos were those of the species providing the nucleus, not the species providing the cytoplasm. Hörstadius' experimental studies culminated in a book, *Experimental Embryology of Echinoderms*, published in 1973.

Based on the work undertaken with Sven Sellman, Hörstadius published two large papers devoted to an extensive experimental analysis of the development of the neural crest (NC)-derived cartilaginous skeleton of *Ambystoma* (see Chapter 7). This experimental verification of Platt's observations on the mudpuppy ran counter to the dogma enshrined in the germ-layer theory, according to which skeletal tissues develop from mesoderm and from no other germ layer. According to one assessment of Hörstadius' experimental work on the NC, 'This is the first time that the phenomenon of complex and additive inductive action by different structures has been demonstrated under experimental conditions' (*Nature* 1950, 169, p. 821).[b] A symposium to mark the centennial in 1998 of his birth, to honor his contributions to the developmental biology of echinoderms and the NC, and to chart future directions for developmental biology, was organized by Carl-Olof Jacobson and Lennart Olsson at the Wenner-Gren Research Institute in Stockholm. A handsome publication based on the symposium, the only volume devoted to studies on echinoderm embryology and the NC, was published in 2000 under the editorship of Olsson and Jacobson.

---

[a] Hörstadius was a member of the Royal Swedish Academy of Sciences and the Academia Pontifica (Vatican); Fellow of the Royal Society of London, the Royal Institution of Great Britain and the Societé Zoologique de France; and held honorary doctorates from the Université de Paris and Cambridge University. Active in scientific administration at the international level, Hörstadius was Secretary-General and organizer of the Xth International Ornithological Congress in Uppsala in 1950 and edited the proceedings, to which he contributed a paper on Swedish ornithology (Hörstadius, 1951). A founding member of the Council of The World Wildlife Fund, Hörstadius was also chairman of the European Section of the International Council for Bird Preservation, president of the Swedish Ornithological Society, president of the International Union of Biological Sciences (1953–1958) and president of the International Council of Scientific Unions (1962–1963).

[b] See Ebendal (1995) and O. Jacobson (2000) for details of his scientific work and Olsson (2000) for a bibliography.

1960s–1970s

The 1960s ushered in investigations of mechanisms of NCC migration and a move away from amphibian and toward avian embryos as the organisms of choice. The floodgate was opened by the seminal studies of Jim Weston (1963) and Mac Johnston (1966) on migrating trunk and cranial neural crest cells (TNCCs, CNCCs)[⊕] in chicken embryos, by Pierre Chibon (1964) with his studies on the skeletogenic NC in the Spanish ribbed newt, by the discovery and exploitation of the quail nuclear marker by Nicole Le Douarin (1974*), and by an influential review by Weston (1970) on the migration and differentiation of NCCs.

Detailed maps of the fate of NCCs appeared during the 1970s. The microenvironment encountered by these cells was revealed as a major determinant of their migration, differentiation, and morphogenesis in normal embryos and in embryos with abnormalities resulting from mutations or the consequences of exposure to teratogens, such as alcohol or drugs (see Chapters 9 and 10).

Syndromes[◊] involving one or more cell types derived from the NC were recognized as separate and identifiable entities and classified as **neurocristopathies** (see Chapter 9). Monoclonal antibodies against individual populations or types of NCCs were developed in the 1980s and used to analyze cell determination, specification, lineage, and multipotentiality (see Table 1.2 for definitions of terms such as multi- and pluripotentiality). Even mammalian embryos, which are difficult to study, began to yield the secrets of their NCCs to skilled and persistent experimental embryologists.[5]

1980s to the 21st Century

Homeotic transformations, and a code of *Hox* genes that patterns the major axis of most (and all bilaterally symmetrical) animal embryos, were discovered and analyzed in some detail during the 1980s and 1990s (Box 1.2). The NC, which had earlier been shown to be divisible into **cranial** and **trunk** regions, was further subdivided following the recognition of:

Table 1.2 Terms used in the text for the potentiality of individual cells

Totipotent	Able to generate all cell types—stem cells
Pluripotent	Able to generate cell types from all four germ layers—stem cells
Multipotent	Able to generate cell types within a germ layer—germ layer-restricted cells
Bipotential	Able to generate two cell types—typical of many embryonic cells
Unipotent	Able to generate a single cell type—committed progenitor cells

---

⊕ CNCC migrate from the developing brain, TNCC from the developing spinal cord.

◊ A syndrome is a group of symptoms that collectively characterize an abnormal condition or disease state.

- subpopulations in the hindbrain associated with specific **rhombomeres** (r)[6] and segmental patterns of expression of *Hox* genes;
- a **vagal** and a **sacral neural crest** in the neck and caudal body, respectively, from which arise the enteric ganglia and the neurons of the parasympathetic nervous system of the intestine and blood vessels (see Chapter 6); and
- a **cardiac neural crest** that contributes cells to the valves, septa, and major vessels of the heart (see Chapter 8).

Box 1.2 *Hox* genes

Recognition of the importance of *Hox* genes marks a fascinating episode in the history of the search for relationships between animals, the origin of body plans associated with individual phyla and, with respect to our topic, the origin of chordates and vertebrates.

Vertebrate homeobox (Hox) genes with sequence homology to such gene complexes as *Ultrabithorax* and *Antennapedia* in the fruit fly *Drosophila* are orthologs of a series of transcription factors organized as homeobox clusters throughout the animal kingdom. As in *Drosophila*, the order of *Hox* genes within a cluster is paralleled by an anterior–posterior sequence of gene expression. Conservation of the roles of these genes in vertebrates and in *Drosophila* is demonstrated by research showing that, for example, after being transfected into *Drosophila*, the mouse *Hoxb6* gene elicits leg formation in the place of antennae.

Considerable information now is available on the evolution of the genes leading to the *Ultrabithorax* and *Antennapedia* gene complexes. The scenario (outlined below[a]) is that a single protoHox gene (□) duplicated to produce a ProtoHox cluster of four genes (▣) arranged in an anterior–posterior (A–P) sequence,

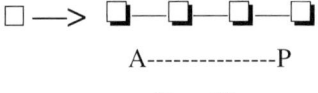

A--------------P

ProtoHox

Duplication of this ProtoHox cluster produced two clusters of four genes each, four Hox and four ParaHox.

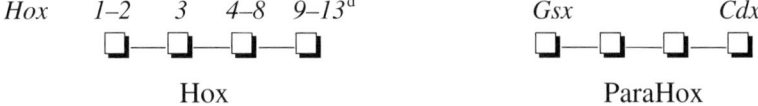

Hox   *1–2*   *3*   *4–8*   *9–13*[d]   *Gsx*   *Cdx*

Hox   ParaHox

Further tandem duplications of three of the Hox cluster at the origin of the bilaterally symmetrical animals (the Bilateria) produced the Hox clusters found in bilaterians; Hox cluster 3 remained as a single gene in both Hox and ParaHox clusters.[b]

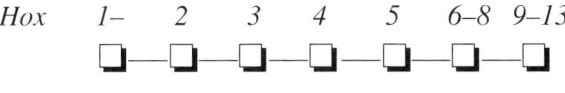

Hox   *1–*   *2*   *3*   *4*   *5*   *6–8 9–13*

Bilaterian Hox cluster

The number of Hox clusters varies among vertebrates: four clusters of 39 genes in mice, three clusters in lampreys, and up to seven in teleost fish. "Duplication of the genome" at the origin of the chordates is the most likely current explanation for the four clusters; duplication sets up the possibility of future structural and functional divergence and *specialization of function among copies of the genes*.

Four possibilities, which are not mutually exclusive, could explain evolutionary changes in gene function:

- **Two involve change in gene number**, either (i) the number of Hox gene clusters or (ii) the number of genes per cluster. Duplication of Hox clusters *before* the teleosts arose—perhaps associated with duplication of large portions of chromosomes or entire chromosomes or genomes— would have taken the number from four to eight in teleosts. This, coupled with subsequent loss of one cluster, would explain the seven clusters in zebrafish.
- **The other two possibilities involve altered function,** either (iii) modification of individual *Hox* genes through regulatory or other changes or (iv) increasing the complexity of interaction between gene networks, either of which could come about by alteration in the upstream and/or downstream regulation of a Hox gene(s).[c]

  The *patterning role* carried out by *Hox* genes is demonstrated by studies in which knocking out or knocking in a *Hox* gene to eliminate or enhance its function in mice is followed by the transformation of skull, vertebral, or other features into a more anterior element in the sequence. Such a transformation is known as **homeotic**, a term introduced into biology by William Bateson in the early 20th century. In the tadpoles of some species of frogs, an amputated tail can be made to transform homeotically into the duplicated posterior portion of the body, including hindlimbs and a pelvic girdle, rather than regenerating a tail. Developmental and evolutionary transformations of middle ear ossicles are discussed in Boxes 7.1 and 10.2. Meckel's cartilage and the ossicles are duplicated following *Hoxa2* knockout, essentially because the second pharyngeal arch fails to form. Instead, a second set of first-arch elements forms more anteriorly than the normal position of the first arch; that is, there is a homeotic transformation of second arch to first-arch structures (see Box 10.2).

---

[a] Based on the scheme outlined in Shimeld (2008).

[b] A 14th Hox gene (*Hox14*) has been identified in amphioxus, a lamprey, the coelacanth and a shark. It is not expressed in the CNS, somitic mesoderm, or fin buds of the Japanese lamprey, *Lethenteron japonicum*, but rather is expressed in association with the hindgut. The decoupling of *Hox14* from the collinear arrangement of *Hox1–Hox13* is thought by Kuraku and colleagues (2008) to have facilitated its loss in tetrapods and teleost fish.

[c] See Holland and Graham (1995), Peterson and Davidson (2000), Shimeld and Holland (2000), and Carroll *et al.* (2005) for literature on the major roles of *Hox* genes in bilaterian evolution. Modes of evolutionary change in gene activity at levels other than structural changes in gene sequences have emerged in recent years. Changes in *cis*-regulation and changes in noncoding RNA molecules such as MicroRNAs (MiRNAs) are two changes under active investigation. See Hall and Hallgrímsson (2008) for the evolution of gene regulation, Eberhart *et al.* (2008) for MiRNA regulation of the *PdgfRa* gene and cleft palate in zebrafish, Heimberg *et al.* (2008) for an overview of MiRNAs, and Miller *et al.* (2007) for parallel evolution of *cis*-regulation of the gene for c-kit ligand in the evolution of pigmentation in marine and freshwater species of threespine sticklebacks and in humans (see Chapter 5).

The major derivatives of each of these regions are outlined in Table 1.3. Relationships between these derivatives are shown in Fig. 1.2.

Mapping the NC became ever more fine grained as comparative studies were undertaken, some within specific phylogenetic frameworks to test explicit evolutionary hypotheses. Knowledge of the molecular basis of the migration, differentiation, and death (apoptosis) of NCCs advanced considerably.

Last but by no means least in this brief overview is the role that studies of the NC have played in forging a new synthesis between developmental and evolutionary biology. The NC is a **vertebrate synapomorphy**, that is, a character shared by all vertebrates to the exclusion of all other animals. Evolutionary biologists sought the origin of the vertebrate head in the unique properties of the NC, placing the NC at centerstage in the vertebrate evolutionary drama. Indeed, as discussed in Chapter 4, the vertebrate head is a 'new head' added to the rostral end of a more ancient invertebrate head.[7] The vertebrate heart is also new, for it, too, contains NC derivatives, the cardiac neural crest (CarNC; see Chapter 8).

Neural Crest and Germ-Layer Theory

The bulk of the discussion in de Beer's 1947 paper evaluated the significance of his own results concerning the NC origin of pharyngeal cartilages to germ-layer theory, which had placed evolutionary studies of embryonic development in a straitjacket for almost a century. The germ-layer theory makes three claims:

(1) Early embryos are arranged into equivalent (homologous) layers: ectoderm and endoderm in diploblastic animals, such as sponges and coelenterates; ectoderm,

Table 1.3 Derivatives of the neural crest in relation to the four major regions of the crest

Cellular Derivative (Other Contributions)

Cranial neural crest (CNC, CNCCs)

Mesenchyme (Induction of the thymus and parathyroid glands)
Connective tissue (including muscle sheaths)
Cartilage
Bone
Dentine (odontoblasts)
Parafollicular cell (ultimobranchial bodies) of the thyroid gland
Cornea
Sclera
Ciliary muscle and muscles for eye attachment
Inner ear (with otic placode)
Sensory ganglia of cranial nerves V, VI, IX, and X

Vagal and sacral neural crest

Neurons of parasympathetic nervous system of alimentary canal
Neurons of parasympathetic nervous system of blood vessels
Enteric ganglia

Trunk neural crest (TNC, TNCCs)

Pigment Merkel cells
Dorsal root ganglia
Neurons and ganglia of the sympathetic nervous system
Chromaffin cells of the adrenal medulla
Epinephrine-producing cells of the adrenal gland

Cardiac neural crest (CarNC)

Connective tissue associated with the great vessels of the heart
Aorticopulmonary septum of the heart
Smooth muscles of the great arteries
Ganglia (celiac, superior and inferior mesenteric, and aortical renal)

mesoderm, and endoderm in triploblastic animals, such as sea urchins, flies, fish,[8] and humans.

(2) Embryos (and so larvae and adults) form by differentiation from these germ layers.

(3) Homologous structures in different animals arise from the same germ layers.

The germ-layer theory exerted a profound influence on those claiming a NC—that is, an *ectodermal*—origin for tissues such as mesenchyme and cartilage, traditionally believed (indeed 'known') to arise from mesoderm. Hörstadius (1950, p. 7) commented on the 'violent controversy' that followed the assertion of a NC origin for mesenchyme by Platt and others such as Brauer, Dohrn, Goronowitsch, Lundborg, Kastschenko, and Kupffer and the opposition to such a heretical idea by

Buchs, Corning, Holmdahl, Minot, and Rabl—a veritable who's who of comparative morphologists of the day.

A brief discussion of germ layers and the germ-layer theory follows.[9]

Germ-Layer Theory

Germ-layer theory had its origin in the early 19th century.

In the course of a pioneering study on the development of chicken embryos undertaken for his MD thesis, Christian Heinrich Pander (1817) recognized that the blastoderm is organized into the three germ layers we now know as ectoderm, mesoderm, and endoderm. Pander referred to an upper 'serous', a lower 'mucous', and a rather ill-defined middle 'vessel' layer and coined the terms *Kembla* (germ layer) and *Keimhaut* (blastoderm). Eleven years later, Karl von Baer extended Pander's findings by demonstrating that all vertebrate embryos are built on this three-layered plan.

On the basis of his studies of the organization of coelenterates, Thomas Huxley came to the important conclusion that the outer and inner layers of **adult coelenterate**s are homologous with the outer and inner layers of **vertebrate embryos** (Huxley, 1849). At one fell swoop, Huxley extended the concept of germ layers from vertebrates to invertebrates, from embryos to adults, and from ontogeny to phylogeny. No biologist could ignore germ layers.

In 1853, George J. Allman coined the terms ectoderm and endoderm for the outer and inner layers of the hydroid *Cordylophora*. In his encyclopedic treatment of animal development published between 1850 and 1855, the Polish embryologist-cum-physiologist Robert Remak identified Pander's rather indistinct middle layer as a distinctive germ layer, which Huxley (1871) named mesoderm. Remak provided the first histological descriptions of each germ layer.

Against the background that not all animals develop from three-layered embryos, in the 1870s the influential English zoologist Sir Edwin Ray Lankester expanded the germ-layer concept from ontogeny into systematics by dividing the animal kingdom into three grades based on numbers of germ layers:

- *Homoblastica*—single-celled organisms
- *Diploblastica*—sponges and coelenterates
- *Triploblastica*—the remainder of the animal kingdom

Lankester's scheme stood for 125 years until evidence was assembled that vertebrates are tetrablastic and not triploblastic, the NC constituting a fourth germ layer (Fig. 1.3)—the topic of the penultimate section of this chapter.[10]

Not all embraced the germ-layer theory. Adam Sedgwick, Chair of Zoology at Cambridge, rejected the germ-layer and cell theories entirely, claiming that one could not even state what the cell theory is—'it is a phantom,' he said, and if extended to the germ layers it is the 'layer phantom' (Sedgwick, 1894, p. 95). For many, the problem was the apparent origin of multiple cell types from a single layer or region of the embryo.

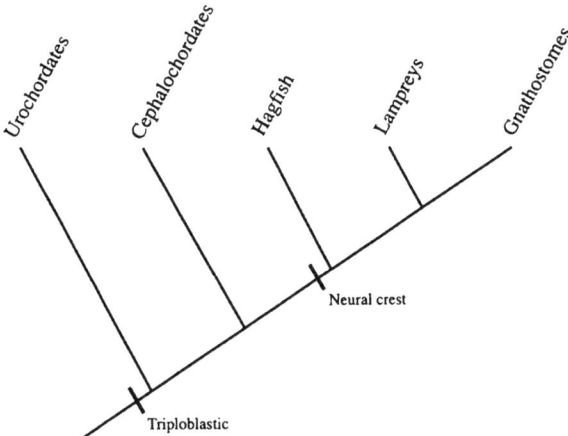

Fig. 1.3 A phylogeny of extant chordates (urochordates, cephalochordates), jawless vertebrates (hagfish, lampreys), and jawed vertebrates (gnathostomes). Urochordates and cephalochordates are triploblastic, having ectoderm, endoderm, and mesoderm as germ layers. The fourth germ layer, the neural crest, arose in the common ancestor of jawless and jawed vertebrates. This phylogeny shows cephalochordates as the sister group to vertebrates, with urochordates as a more basal group

Multiple Tissues from Single Layers

Sedgwick's criticisms arose, in part, from the nature and origins of mesenchyme and peripheral nerve trunks and of the NC itself.

The term **mesenchyme** was coined in 1882 by Oscar and Richard Hertwig (brothers and brilliant zoologists) for those cells that leave the mesodermal germ layer during formation of the coelom and form elements of connective tissue or blood. The term is now used for meshworks of cells **irrespective** of their germ layer of origin; NCCs migrating as mesenchymal cells in a skate embryo can be seen in Fig. 7.11.[11] A century ago, however, an ectodermal NC producing head mesenchyme created major problems for the entrenched germ-layer theory. So too did the observation that mesenchyme, nerves, muscles, connective and vascular tissues all develop from a single layer in the vertebrate head. This was not what was demanded by a rigid germ-layer theory in which:

- ectoderm formed nerves and epidermis;
- mesoderm formed muscle, mesenchyme, connective and vascular tissues; and
- endoderm formed the alimentary canal.

A diversity of cell and tissue types, however, **do** arise from single germ layers. As Kölliker (1884) demonstrated 125 years ago, epithelial, neuronal, and pigmented cells all arise from ectoderm and, under experimental conditions, structures can form from a different germ layer than the one from which they arose embryonically. de Beer (1947) cited asexual reproduction, regeneration, and adventitious (ectopic) differentiation as further situations in which structures develop from a different germ layer than the one from which the original structure formed. Consequently, the germ-layer theory speaks neither to the full developmental potential of individual germ layers nor to determination or cell fate within a germ layer.

De Beer carefully pointed out that the germ-layer theory is a morphological concept that does not speak to developmental potencies or to cell fate, concluding

> that there is no invariable correlation between the germ layers and either the presumptive organ-forming regions or the formed structures. It follows that the germ layers are not determinants of differentiation in development, but embryonic structures, which resemble one another closely in different forms although they may contain materials differing in origin and fate. *The germ-layer theory in its classical form must therefore be abandoned.*
>
> (1947, p. 377, emphasis added)

In the original form of the germ-layer theory—thus also in its strictest application—only those structures that develop from equivalent layers were regarded as homologous. Again, according to de Beer, the 'problem' with fitting an NC origin of cartilage into the germ-layer theory is largely due to a misconception of the theory of homology and the misapplication of homology based on adult structures to homology based on developmental origin and developmental processes— the 'attempt to provide an embryological criterion of homology' (p. 393). This has important consequences, for we now know that homologous structures need not arise from the same embryonic area or, indeed, by the same developmental processes.[12]

Despite his denigration of germ-layer theory, de Beer was reluctant to abandon it entirely, as can be seen in the conclusion to his 1947 paper, a nice example of 'the dogged attempt of the human mind to cling to a fixed idea' (Oppenheimer, 1940, p. 1):

> There is just sufficient constancy in the origins and fates of the materials of which the germ layers are composed to endow the ghost of the germ-layer theory with provisional, descriptive, and limited didactic value, in systematizing the description of the results of the chief course of events in the development of many different kinds of animals; provided that it be remembered that such systematization is without bearing on the question of the causal determination of the origin of the structures of an adult organism.
>
> (p. 394)

Sedgwick's rejection of and de Beer's concerns over germ-layer theory highlight a fundamental feature of vertebrate development, discussed in the following two sections and that is either unknown or underappreciated by most students of development and indeed by most biologists.

Heads and Tails

Of the three claims of the germ-layer theory,

(1) that early embryos are arranged into equivalent layers is always true;
(2) that embryos form by differentiation from these germ layers is not true for the caudal development of vertebrate embryos (see below); and

(3) that homologous structures in different animals arise from the same germ layers need not be true.

Our consideration of these issues begins with Holmdahl (1928), who distinguished two phases of vertebrate embryonic development: (1) **primary development** for the laying down of germ layers and (2) **secondary development** for the development of the caudal region of the embryo without the specification and segregation of germ layers. Holmdahl's second phase is predicated on a **lack** of germ-layer involvement in the caudal region and tail buds of vertebrate embryos. Why? Because the mechanisms of neurulation operate differently in the head and body than they do in the tail. Indeed, the opposite ends of the same embryo develop by fundamentally different developmental mechanisms, **primary and secondary neurulation.**

Secondary Neurulation and Tail Buds

The cranial region of vertebrate embryos arises by primary neurulation through delamination and migration of the germ layers. The caudal end arises by secondary induction and the transformation of epithelial cells into a mesenchymal tail bud.[13] Secondary neurulation is characteristic of all vertebrates studied—lampreys, fish, amphibians, birds, and mammals (including humans)—a finding consistent with secondary neurulation being an ancient developmental process inherited from the common vertebrate ancestor.[14]

Kölliker (1884) was aware of the problem raised for the germ-layer theory by his finding that the most caudal part of the nervous system arises from mesoderm, not ectoderm. Using the method for vital dye staining of cells published by Vogt in 1925, Bijtel confirmed Kölliker's finding by demonstrating that tail somites in amphibian embryos arise from the medullary (neural) plate, that is, that 'mesodermal' cells arise from ectoderm. Studies using ^{3}H-thymidine-labeled grafts and quail/chicken chimeras extended these findings to birds, in which neural, muscular, vascular, and skeletal tissues arise from a common tail bud mesenchyme.

Muscle, cartilage, neuroepithelium, and pigment, all differentiate in culture from what appears to be homogeneous tail bud mesenchyme. Because these tissue derivatives represent multiple germ layers, Griffith *et al.* (1992) concluded that the tail bud consists of three unseparated germ layers, an interpretation that assumes that germ layers have been specified in this region, which we now know they have not (Fig. 1.4). Kanki and Ho (1997) showed that pluripotent cells (Table 1.2) within zebrafish tail buds contribute to caudal trunk tissue that lies rostral$^\oplus$ to the anus as well as to the tail bud itself. Davis and Kirschner (2000) used photoactivation of fluorescence labeling of groups of cells to show that the tail bud in *Xenopus*

$^\oplus$ Throughout the text, I endeavor to be consistent in using rostral and caudal rather than anterior and posterior if referring to the organization of axial embryonic structures, such as the neural tube, NC, or somites.

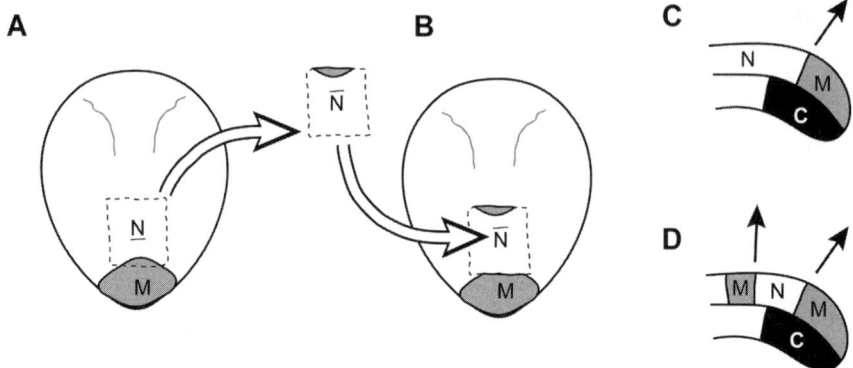

Fig. 1.4 Autonomous development of tail buds in *Xenopus* as seen after rotation and reinsertion of the posterior neural plate and associated mesoderm (N) (*arrows* connecting **A** and **B**) elicits an extra tail. (**C**) Interactions between genes in region C and regions N and M in unoperated embryos have already specified a tail bud, which is shown by the *arrow*. (**D**) Additional tail buds after transplantation, the *arrows* showing the orientation of the two tail buds. Adapted from Beck and Slack (1998)

contributes cells to the neural tube, notochord, somites (muscle cells), and other structures. They concluded that the tail bud represents a region in which germ-layer decisions are delayed. As the tail bud is of mixed origin, caudal NC may not be entirely ectodermal in origin (Fig. 1.4). This is a far cry from germ layers laying the foundation for the more rostral embryo.

With primary neurulation from germ layers and secondary neurulation from a tail bud blastema, fundamental distinctions exist between the origins of cranial and caudal NCCs. Indeed, we can think of primary and secondary NCCs and the NC as having a dual origin: cranial (primary) NC from neurectoderm; caudal (secondary) NC from a tail bud. An additional line of evidence is the origin of the tail bud by embryonic induction and not from germ-layer delamination.[15]

Induction of Tail Buds

Epithelial–mesenchymal interactions, which are secondary inductions, initiate or regulate differentiation, growth, and/or morphogenesis of most organs in vertebrate embryos (see Fig. 3.12 and the discussion in Box 3.4).

Limb outgrowth is controlled through epithelial–mesenchymal signaling involving an apical ectodermal ridge (AER). The tail bud is also surmounted by a ridge known as the ventral ectodermal ridge (VER). Is the VER the tail bud equivalent of the AER?

Removing tail ectoderm almost completely eliminates tail growth in embryonic chickens just as removing the AER eliminates limb growth. Limb and tail induction in avian embryos share common mechanisms, as shown by experiments in which limb bud mesenchyme grafted beneath tail ectoderm induces tail ectoderm to form an ectodermal ridge that regulates limb outgrowth, chondrogenesis, and skeletal formation.[16]

Mutations that decrease or eliminate the AER slow or stop limb development. Similarly, the VER is missing in *vestigial tail* (*vt*) and *Brachyury* (*T*) mutant mice in which tails fail to form. In *repeated epilation* (*Er*) mutant embryos, the VER and the AER are abnormal, as are tail and limb development. Reduced proliferation in the ventral part of the tail accounts for rostral defects in *curly tail* mutant mice.[17]

As with the development of the limbs, heart, kidney, and other organs, tail development occurs by secondary induction and not by primary development from a germ layer (which occurs before epithelial–mesenchymal interactions are initiated), reinforcing the fundamental similarity between the development of the most caudal region of early vertebrate embryos, on the one hand, and organ systems arising later in development, on the other.

Neural Crest as the Fourth Germ Layer

The NC meets all the criteria used to define and identify a germ layer.

Ectoderm and endoderm are **primary germ layers**; they were the first to appear in animal evolution, are the earliest to form embryonically, are present in the unfertilized egg, and are determined by cytoplasmic factors deposited by the mother into the egg by a process known as maternal cytoplasmic control. Mesoderm is a **secondary germ layer**; it is not preformed in the vertebrate egg but arises by activation of zygotic genes following inductive interactions between ectoderm and endoderm.[18]

Like mesoderm, the NC arises early in development and gives rise to divergent cell and tissue types. Like mesoderm (and as discussed in Chapter 2), the NC arises by secondary induction from a primary germ layer. The NC therefore meets the criteria of a secondary germ layer.

As the fourth germ layer, the NC is confined to vertebrates, which are therefore tetrablastic and not triploblastic.[19] Indeed, possession of an NC is a vertebrate synapomorphy (Fig. 1.3). A cladistic analysis of 411 human cell types undertaken by Vickaryous and Hall (2006) and illustrated in Fig. 1.2 and an analysis of tissue-specific genes and gene programs using a bioinformatics approach by Martinez-Morales *et al.* (2007) provide independent support for lineage relationships among NCCs from developmental, molecular, and evolutionary points of view (see the section on Comparative Genomics and Bioinformatics in Chapter 4 for elaboration). Clinicians and medical geneticists acknowledge the NC as a germ layer by recognizing neurocristopathies, in which the common link among affected tissues and organs is their origin from the NC (see Chapter 9).

Just as the evolution of mesoderm allowed triploblastic organisms to form new, often novel body parts, so the evolution of the NC allowed vertebrates to form new, often novel body parts (Hall, 2005a). To quote from a recent analysis of how new features appear in evolution,

The evolutionary versatility of the neural crest justifies Hall's claim that it is an emergent, fourth germ layer. Along with duplication of the whole genome in the early vertebrates, and further duplication and differentiation of genes that regulate development and physiology, the neural crest provided powerful experimental tools for emergent evolution.

(Reid, 2007, p. 216)

The three germ layers recognized for the past almost 180 years can be replaced by four germ layers, two being primary (ectoderm and endoderm) and two secondary (mesoderm and NC). This is a far cry from the days when to claim that mesenchyme arose from ectoderm was biological heresy and professional suicide, as Julia Platt found (see Box 7.2).

Neural Crest as Inhibitor

A little-studied aspect of NCCs, one that may be of considerable importance for signaling and integration among and between tissues during development, is the inhibition by NCCs of other cell types.

NCCs inhibit induction of the heart, kidney, and lens, apparently by producing diffusible inhibitory factors. Although intimations of such inhibition first appeared in the early 1960s, the concept that NCCs inhibit induction, indeed, the general concept of inhibition as part and parcel of induction, is underappreciated, even now.

Woellwarth (1961) demonstrated that NC inhibits lens induction in the common newt, *Lissotriton vulgaris*. Forebrain, or forebrain plus neural folds containing NC, was removed from embryos that were then reared at 14°C, the optimal temperature for lens induction. Lenses formed in 26.5% of embryos from which forebrain was removed but in 95% of embryos lacking forebrain and NC. Henry and Grainger (1987) confirmed this inhibitory action. Their claim is that by coming into contact with surface ectoderm, NC-derived mesenchyme suppresses the lens field except where the optic cup contacts the overlying ectoderm. This contact prevents mesenchyme from invading the lens field. Effectively, NC-derived mesenchyme positions the lens over the center of the developing optic cup.

NCCs inhibit heart development in California newt embryos, with CNCCs suppressing precardiac mesoderm to a greater extent than do TNCCs. Other studies have demonstrated that premigratory NCCs and mesenchyme derived from NC suppress kidney formation in newt and *Xenopus*[⊕] embryos.

Inhibitory or delaying actions on kidney development may not be limited to amphibians. Kidney development is aberrant in *congenital hydrocephalus* (*ch*) mutant mice, which die at birth. These mutants display bulging cerebral hemispheres from early in development (because they retain fluid within the brain), multiple

---

[⊕] Throughout the book, I use common names for most species. The list following Chapter 10 contains an alphabetical list of common names for organisms discussed in the book. A second list contains species discussed, arranged by major groups of organisms. Because it is in such common usage I use the generic name *Xenopus* when referring to *Xenopus laevis*, the South African clawed-toed frog (the African clawed frog). Other species of *Xenopus* are referred to in full.

skeletal defects, fewer than normal celiac ganglia, and additional mesonephric kidney tubules. The latter defect was attributed to NCCs inhibiting kidney development during the ninth or early tenth days of gestation, a time when preganglionic crest cells are migrating along mesonephric mesoderm.[20]

With this story of the discovery of the NC and of NCCs as background we turn to Chapter 2 to the identification and origin of the NC during embryonic development, to Chapter 3 to the delamination, migration, and differentiative potential of NCCs, and to Chapter 4 to the evolutionary origin(s) of the NC.

Notes

1. Wilhelm His was professor of anatomy at Leipzig from 1872 until his death in 1904. A man of many accomplishments, he invented a microtome with a system that allowed section thickness to be controlled when cutting thin serial sections of animal and plant material; discovered that nerve fibers arise from single nerve cells; and coined the term neuroblasts for those cells. His 1895 treatise, *Nomina Anatomica*, introduced order into anatomical terminology. He was a vigorous and, if Ernst Haeckel's ferocious response is any guide, an effective opponent of the Haeckelian biogenetic law that ontogeny recapitulates phylogeny.
2. Marshall's scientific reputation rests on his work on cranial nerves, olfactory organs, and head cavities; his textbook of frog embryology (perhaps the first devoted to the development of a single species); and to a lesser extent, on his later work on corals. See Kuratani *et al.* (2000) for an analysis of head cavities in sturgeon embryos.
3. See Landacre (1921), Stone (1926, 1929), Raven (1931, 1936), and the literature and discussion in Hall (1999a*) for the NC origin of head mesenchyme and craniofacial cartilages.
4. See Platt (1893, 1894, 1897), Rodaway and Patient (2001), Hörstadius and Sellman (1941, 1946), de Beer (1947), and Chapter 7 for seminal studies on the contribution of the NC to the skeleton.
5. See Bolande (1974) and Morriss-Kay and Tan (1987) for early studies with mammalian embryos.
6. We have known for over 120 years that the hindbrain is organized into segments, which Orr (1887) named rhombomeres on the basis of studies on lizard brain development.
7 For major reviews of the NC to the end of the 20th century, see Landacre (1921), Holmdahl (1928), Hörstadius (1950*), Weston (1970*), Le Douarin (1982*), Maderson (1987), Hall and Hörstadius (1988*), Le Douarin *et al.* (1992), Hall (1999a*), and Le Douarin and Kalcheim (1999*).
8. Use of the term 'fish' does not imply that fish form a natural group of vertebrates or that the various groups of fish are more closely related to one another than to other vertebrates. See Hall and Hallgrímsson (2008) for discussions of terminology for groups of vertebrates and invertebrates.
9. Oppenheimer (1940) and Hall (1997, 1999a*) contain detailed discussion of the discovery of germ layers, their naming, germ-layer theory, and references to early studies.
10. See Lankester (1873, 1877) for the tripartite subdivision of the animal kingdom on the basis of germ layers and Hall (1997, 1999a*) for NC as the fourth germ layer and for evaluations of Lankester's work.
11. See Hall (1997*, 2005b*) for discussions of the use of the term mesenchyme.
12. For discussion of the development of homologous structures from different embryonic regions or by different developmental processes, see de Beer (1971), Hall (1994, 1995, 1999a,b*, 2003a*), Stone and Hall (2004), and Janvier (2007).
13. For current understanding of epithelial–mesenchymal transformations, see the papers in the special issue of *Acta Anatomica* edited by Newgreen (1995*) and see Savagner (2001*), Kang and Svoboda (2005*), and Morales *et al.* (2005*).

14. See Schoenwolf and Nichols (1984), Schoenwolf *et al.* (1985*), and Le Douarin *et al.* (1996) for secondary neurulation in avian embryos; Beck and Slack (1998, 1999) and Davis and Kirschner (2000) for *Xenopus*; Wilson and Wyatt (1988) and Hall (1997, 2000b) for murine embryos; Müller and O'Rahilly (2004) and O'Rahilly and Müller (2006*, 2007*) for human embryos; and Hall (1997, 1999a*, 2000b,c, 2007b) and Handrigan (2003) for further discussions of phases of development and secondary neurulation.

15. See Schoenwolf and Nichols (1984) and Schoenwolf *et al.* (1985) for the dual origins of the NC and Griffith *et al.* (1992) and Hall (2000b) for primary and secondary body development. In studies using chicken embryos, Weston *et al.* (2004) found that PdgfRα-positive mesenchymal cells arose from nonneural ectoderm adjacent to the neural tube. Taking PdgfRα as a NC marker, they regard these as NC cells that arise from beside the neural tube, a 'metablast' in their terminology.

16 See Grüneberg (1956), Hall (2000b,c, 2005b*, 2007a), and Handrigan (2003) for limb and tail bud development.

17. See Grüneberg (1956), Johnson (1986), and Hall (2005b*) for mutations affecting mouse tail-development and Wilson and Wyatt (1988) for closure of the caudal neuropore in *vL* mutant mice and for suggesting that defective secondary neural crest may contribute to spina bifida in these mutants.

18. For recent overviews of the origin of the mesoderm in development and evolution, see Sasai and De Robertis (1997), Martindale *et al.* (2004), and Putnam *et al.* (2007).

19. See Hall (1997*, 1999b*, 2000d, 2005b), Opitz and Clark (2000), Carstens (2004), Stone and Hall (2004), and Vickaryous and Hall (2006) for discussions of the evidence for the NC as the fourth germ layer.

20. See Hall (1997*, 1999a*) for literature and further discussions of inhibitory actions of NCCs.

Chapter 2
Embryological Origins and the Identification of Neural Crest Cells

Neural Crest

This chapter seeks to answer a twofold question: When and by what mechanisms do the neural crest (NC) and neural crest cells (NCCs) arise during embryonic development?

In one sense, the embryological origin of the NC is self-evident. The NC is the apex of the neural folds of neurula-stage embryos (Fig. 2.1). The very name neural crest—like the crest of a mountain—is indicative of this location. But NCCs are not the only derivatives of the neural folds; neural and epidermal ectoderms arise from the neural folds. Furthermore, placodal ectoderm (the development and derivatives of which are discussed in Chapter 6) arises from the lateral neural folds or from ectoderm immediately lateral to the neural folds. Indeed, it can be difficult, if not impossible, to label NCCs in the neural folds without labeling placodal ectoderm.

The NC can also be defined as a region that lies at or forms the border between the neural and epidermal ectoderm (Fig. 2.1) or as the region of the embryo from which NCCs arise.

Before Neurulation

Although most evident in neurula-stage embryos, the intimate association between the four presumptive areas—neural crest, neural, epidermal, and placodal ectoderm—does not arise at neurulation: **Specification of the NC begins during gastrulation**, although without special methods, however, neither NC nor other ectodermal cell types can be identified before neurulation.

Three different methods—vital staining, extirpation, and cell labeling—show that in early amphibian blastulae the future NC lies at the border between presumptive epidermal and neural ectoderm (Fig. 2.1). Grafting ^{3}H-thymidine-labeled regions of chicken epiblasts into unlabeled epiblasts similarly reveals presumptive NC at the epidermal–neural ectodermal border at the blastula stage of embryonic development (Fig. 2.2), although neither epidermal nor neural markers are expressed until after the onset of neurulation.[1]

B.K. Hall, *The Neural Crest and Neural Crest Cells in Vertebrate Development and Evolution*, DOI 10.1007/978-0-387-09846-3_2, © Springer Science+Business Media, LLC 2009

Fig. 2.1 Fate map of a late blastula urodele to show the location of future neural crest (NC) at the boundary between epidermal (Ee) and neural (Ne) ectoderm. Modified from Hörstadius (1950)

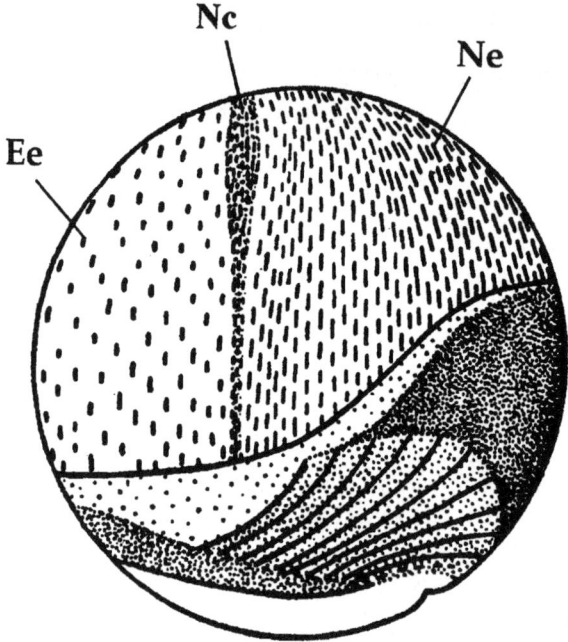

Fig. 2.2 Fate map of the epiblast of a chicken embryo showing the location of future neural crest (NC) at the boundary between epidermal (Ee) and neural (Ne) ectoderm. Based on data from Rosenquist (1981) and Garcia-Martinez *et al.* (1993)

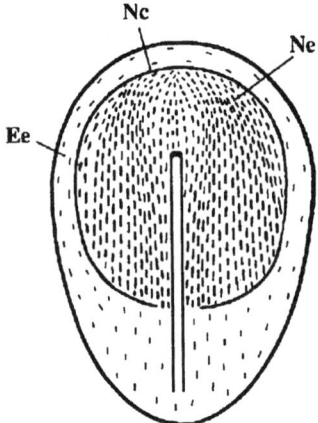

Fate maps of the epiblast of murine embryos provide insights into the location of prospective NC before neurulation. Using clonal analysis, Lawson and colleagues (1991) derived a sufficiently detailed fate map of the mouse NC that they could compare it with fate maps of the epiblast in chicken and urodele embryos. Research from Patrick Tam's laboratory provided an insightful comparison into what they termed the 'striking homology' between the fate maps of representative fish, amphibian, avian, and mammalian embryos.[2] The congruence of these fate maps includes the location of the presumptive neural crest at the border between neural and epidermal ectoderm. One important finding in mouse embryos, consistent with what is known from secondary neurulation (see Chapter 1), is that clonal descendants within the epiblast are not confined to single germ layers and that germ layers are not fully segregated until gastrulation.

Cruz and colleagues (1996) used DiI injection to map the fate of the epiblast in the Australian marsupial 'mouse', *Sminthopsis macroura*. Although they demonstrated that neurectoderm gives rise to epidermal and neural ectoderm, they did not map the NC. Indeed, as noted in Chapter 1, only a few individuals have investigated the NC during marsupial embryonic development, although Hill and Watson, in studies published in 1958 but begun in 1911, documented NCCs and their contribution to cranial mesenchyme and ganglia in a number of Australian marsupials[3] and in American opossums (*Didelphis* spp.); see Box 7.5 for available information on marsupial NCCs.

Establishing the Epidermal–Neural Border

A long-standing interpretation of studies directed at determining the origin of the NC has been that NC arises at the **border** of neural and epidermal ectoderm precisely, because this is where neuralizing and epidermalizing influences meet, the combined action of these influences generating the NC.

In perhaps the first study to raise this interpretation, undertaken using the common European salamander and the alpine (Arctic) newt, Rollhäuser-ter Horst (1980) replaced future NC of neurula-stage embryos with future epidermal ectoderm from gastrula-stage embryos. The grafted ectoderm formed neural folds that, according to the interpretation, responded to the combined neuralizing induction of the notochord and epidermalizing induction of the lateral mesoderm and differentiated into NC.

Do we know which events determine that cells at the presumptive epidermal–neural border in such early embryos will form NC? Are NCCs induced or do they self-differentiate? If induced, is their induction separate from, part of, or subsequent to (and/or dependent upon) neural induction?

Some of the answers to these questions come from analyses of the origin of placodal ectoderm, some from the origin of Rohon–Béard neurons, both discussed in Chapter 6. In these studies, as in those outlined below, cellular and/or molecular markers are essential in tracing the origin of the NC and NCCs. Early studies, discussed below, used particular cell types as markers. More recently, and as discussed in the following section, molecular markers have been used almost exclusively to follow the initiation of NCCs.

Although not without their problems (Box 2.1), early studies in which neural folds were isolated and transplanted provided important information on the origin of the NC. Cell/tissue types such as mesenchyme, pigment cells, and cartilage known to arise from NCCs were used as markers to indicate that NC had been induced, although mesenchyme, which arises from NC and from mesoderm, is not a reliable

Box 2.1 Isolating, extirpating, or grafting premigratory NCCs

Attempts are often made to isolate NCCs from neural folds before the cells have delaminated and migration has begun.

Because neural folds contain neural, epidermal, and perhaps placodal ectoderm in addition to NCCs, it can be difficult to isolate NC from the neural folds without including other cell types, and knowing whether you are isolating (or grafting) neural folds or NC is important. Unless neural folds are isolated carefully, a graft of a neural fold may contain neural and epidermal ectoderm (and perhaps placodal ectoderm; see below) in addition to NC. Drawing conclusions about intrinsic patterning of NCCs or about NC and/or placodal origins or particular cell types can thus be problematic. If epithelial ectoderm is included in the grafts, patterning that appears intrinsic to NCCs may, in fact, be imposed by the epithelial ectoderm. On the other hand, in situations in which ectoderm is required to induce NCCs, grafting NC alone will not reveal the differentiative potential of the grafted NCCs.

Extirpating NC is not totally satisfactory either. Some NCCs may be left behind; others may have delaminated and begun to migrate before the extirpation. Adjacent cells—neural ectoderm, or NC rostral or caudal to the region extirpated (or from the contralateral side if NC is removed from only one side)—may replace the extirpated NC through *regulation*, a topic discussed in Chapter 10. The NCCs removed may normally have played a role in inducing non-NC cells. Absence of a tissue or cell type after NC extirpation, therefore, is not unequivocal proof of NC origin.

All these caution means that the results of studies using extirpation have to be interpreted with caution; replacing extirpated NCCs with a similar population of labeled cells from another embryo provides an essential marker to follow the fate of the transplanted cells. Nevertheless, before such labeling methods were discovered, important (and usually correct) conclusions about NCCs were made.

marker for NC origin if mesoderm is also present. (A recent analysis by Blentic *et al.* (2008) demonstrates that *Fgf* signaling from pharyngeal arch epithelium is required (but not sufficient) for CNCCs to be directed into differentiating as mesenchyme; see Box 3.4)

However, the differentiation of pigment cells can be an excellent marker for the differentiation of a NC phenotype. As the retina and small populations of dopamine-producing neurons in the substantia nigra in the midbrain are the only other sources of pigment cells in vertebrates, differentiation of pigment cells is often a sufficient marker of NC origin (see Chapter 5). Cartilage also can be a marker of NCCs. Cartilage (and neural tissue) was evoked from early gastrula ectoderm of the European common frog using concanavalin A as the evoking agent. As the starting tissue was the embryonic ectoderm, and as no mesoderm was included, the cartilage that formed was presumed to be NC in origin and the gastrula ectoderm to have produced NCCs.

In a different approach, NC is induced and NC derivatives differentiate in lateral epiblast ectoderm in Japanese quail embryos into which a chicken Hensen's node (the site of the future notochord) is grafted. Chondrocytes form and can be positively identified as NC in origin because they express the Japanese quail nuclear marker; that is, they have been induced from the host epiblast by Hensen's node.[⊕] With the development of further markers, Bronner-Fraser and her colleagues used a similar approach to identify HNK-1-positive or *Snail2*-expressing cells (see below) in association with grafted neural plates.[4]

Arising as they do at the border between future neural and epidermal ectoderm (see Fig. 2.1), NCCs could be epidermal or neuroectodermal. Because they arise from the apical region of the neural folds but more especially because they produce neurons and ganglia and because some lineages can give rise to NCCs *and* CNS neurons, NCCs are regarded as derivatives of neural and not epidermal ectoderm; this is why we call it the NC and not the epidermal crest. In support of this designation, NCCs do not appear when epidermal–ectodermal derivatives arise in the absence of neural derivatives. The experimental induction of neural tissue, however, is accompanied by NC formation, although derivatives of the NC such as pigment cells and mesenchyme *can* arise in the absence of neural derivatives. The molecular markers outlined below provide further evidence of the neural lineage connection.

NCC Markers and Specification of the NC

Markers of NC are more than convenient labels allowing us to identify NC or NCCs. Many play a role in the formation of the NC and/or in the delamination of NCCs. It is as markers, as active players, and, in some cases, as providing evidence of the connection of the NC to the neural lineage, that they are discussed below. More cellular aspects of NCC delamination are discussed in Chapter 3.

In an insightful series of papers published over the past 5 years, Daniel Meulemans, Marianne Bronner-Fraser, and their colleagues have addressed genes associated with the NC in the context of what they term a 'Neural crest gene regulatory network', with three levels of action:

(i) **Inductive** (interactive) **signals** (Bmps, Wnts, Fgfs, and Notch/Delta) that establish the neural plate border (see Fig. 2.9) and upregulate **transcription factors** in the *Msx*, *Pax*, and *Zic* families at the neural–epidermal border.

(ii) In turn, and after NC induction, these transcription factors **upregulate genes** of the *Snail*, *SoxE*, *FoxD3*, and other gene families that are specific to NCCs and

---

[⊕] In addition to being the site of the future notochord and, therefore, a major player in the induction of neural ectoderm and NC, Hensen's node in tetrapods and its homolog, Kupffer's vesicle, in fish (see Box 9.1) imposes rostrocaudal patterning onto the NC during primary neurulation.

that are expressed before **NCC epithelial —> mesenchymal transformation and migration** (see Fig. 2.7 for *FoxD3*).

(iii) These transcription factors activate **downstream effector genes** associated with the **migration and differentiative potency of NCCs.**

This is an ancient network[5]—upstream elements (components of (i) and (ii)) are present in the North American sea lamprey—but, as you might expect, downstream or distal elements under (iii) show greater differences between agnathans and gnathostomes.[⊕]

HNK-1 and Pax7[◇]

HNK-1, an antibody against a cell surface sulfoglucuronyl glycolipid labels avian premigratory and *some* postmigratory NCCs, labels odd-numbered rhombomeres in the hindbrain, but does not label NCCs that are more fully differentiated (Fig. 2.3). Nor is HNK-1 required for NCC delamination in chicken embryos.[6] HNK-1 also identifies NCCs in embryonic lampreys, fish, birds, and mammals, but not in amphibians, while HNK-1-positive cells associated with the neural tubes in ascidian embryos provide one class of evidence that ascidians possess precursors of NCCs (see Chapter 4). The retinoid X receptor-γ nuclear receptor gene, which is expressed in migrating chicken NCCs as they enter the somites—and at Hamilton–Hamburger (H.H.) stages 24–27 in the peripheral nervous system, dorsal root, and cranial ganglia—may be an earlier marker than HNK-1 for migrating avian TNCCs.

One has to be cautious in using HNK-1 as the sole marker for NCCs, however; the antigen was generated by immunizing a mouse with extracts of human natural killer cells—hence, HNK1—but is present on the surfaces of many cell types. HNK-1-positive cells are present in the avian embryonic gut before it is colonized by NCCs and mesenchymal cells of mesodermal origin can be labeled with HNK-1. During gastrulation in chicken embryos, HNK-1 and *Snail1* (see below) are regulated by *Pax7*, suggesting that *Pax7* could be used as an early marker for NC-fated cells; *Pax7* is broadly expressed in cranial and trunk NCCs in zebrafish (see Chapter 4).[7]

[⊕] Living jawless vertebrates (agnathans) were formerly included in the cyclostomes, a group comprised of lampreys (petromyzontids), hagfish (myxinoids), and various groups of extinct jawless vertebrates. Cyclostomes, however, are not a natural (monophyletic) group. Researchers have grappled with whether lampreys and hagfish represent a monophyletic group of vertebrates with a common ancestor, or whether they represent two separate lines of jawless vertebrates (Fig. 1.3; and see Figs. 4.3 and 4.4).

[◇] Pax (Paired box) genes, of which there are nine arranged in four groups, are transcription factors linked on the basis of a shared Paired domain. A partial or complete homeodomain also may be present. Each of the nine *Pax* genes acts within a specific tissue. Eight of the nine are discussed in this book. The only one not discussed, *Pax4*, functions in the β cells of the islets of Langerhans in the pancreas. Vertebrates have multiple copies of *Pax* genes resulting from gene duplication (see Box 1.2). Where vertebrates have a single gene, amphioxus has a single copy of the orthologous gene: *Pax 3* and *Pax7* in vertebrates, *Pax 3/7* in amphioxus (*AmphiPax3/7*); *Pax 1* and *Pax9* in vertebrates, *AmphiPax1/9* in amphioxus.

Fig. 2.3 These two fluorescent micrographs of adjacent thin sections through the trunk of an H.H. stage 18 chicken embryo show the comparative distribution of antibodies against the cell adhesion molecule Cad2 [N-cadherin] (**A**), and HNK-1 (**B**) HNK-1 is expressed strongly in migrating neural crest cells, which appear *white* in **B**. Cad2 is expressed in the lumen of the neural tube (NT), notochord (N) and myotome (M), but not in neural crest cells. Reproduced from Akitaya and Bronner-Fraser (1992), Copyright © (1992), from a figure kindly supplied by Marianne Bronner-Fraser. Reprinted by permission of Wiley-Liss Inc., a subsidiary of John Wiley & Sons, Inc.

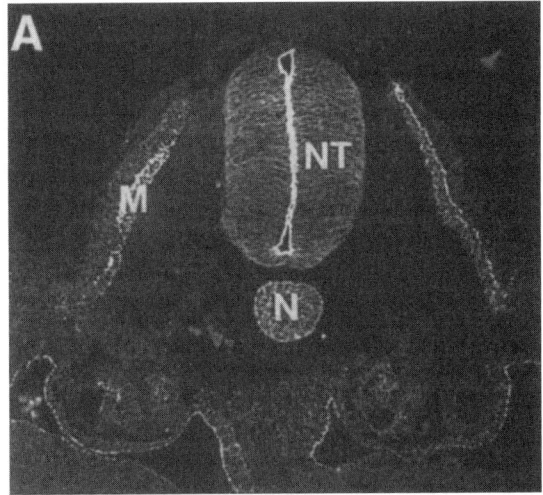

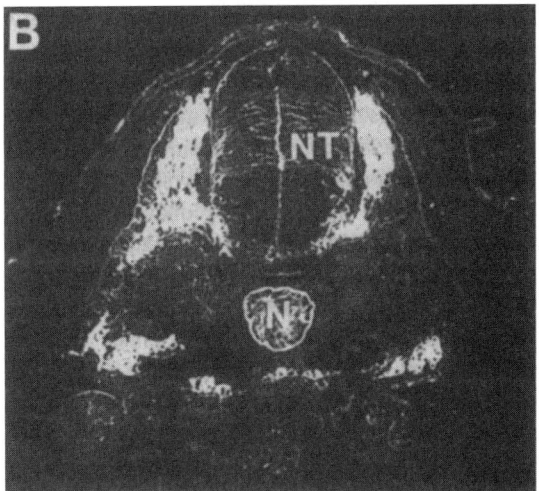

Snail-2, Bmp4, and Cadherins

The zinc-finger transcription factor-encoding gene *Snail2* —previously known as *Slug*⊕—has been used to great advantage as a marker for pre- and postmigratory NCCs.

⊕ The Human Genome Organization (HUGO) Nomenclature Committee has approved a new terminology for the genes previously known as *Snail* (now *Snail1*) and *Slug* (now *Snail2*). *Snail 1* and *Snail 2* are orthologs of the *Drosophila* genes *Snail homologue 1* and *Snail homologue 2*, respectively.

Xsnail2 (where *X* stands for *Xenopus*) is expressed within the NC in neurula-stage embryos and has been used as a marker for NC in studies in which NC is induced in *Xenopus*. *XSnail2* is downstream of *XSnail1*. As it induces NCC markers, *XSnail2* could play a role in NC induction, a role investigated by overexpressing mutant constructs in *Xenopus*. Early inhibition of *XSnail2* blocks the formation of NC; later inhibition prevents the migration of NCCs.[8] A recent analysis of NCC formation showed that the basic helix–loop–helix (bHLH) transcriptional repressor gene, *Xhairy2*, is localized in presumptive NC before expression of *Snail2* or *FoxD3* (see below and Fig. 2.4). *Xhairy2* appears to maintain presumptive NCCs as proliferative and nondifferentiating.[9]

As introduced in the previous section and discussed more fully in Chapter 3, migrating NCCs express cell adhesion molecules, such as N-CAM (neural cell adhesion molecule), Cad2 (N-cadherin), and Cad6B (Figs. 2.3 and 2.4), molecules that are regulated by *Snail2*. Cadherins are regulated by the genes *Snail2* and *Bmp4*, the latter a member of the TGFβ family of secreted factors (see Chapter 3). The binding of *Snail2* to regulatory sites for *Snail2* on *Cad6B* represents the first demonstration of a direct target of Snail2. Downregulation of Cad6B is triggered by Bmp4, which acts via an Adam10-dependent mechanism to cleave Cad2 into soluble fragments within the cytoplasm (Fig. 2.5, and see Chapter 3).[10]

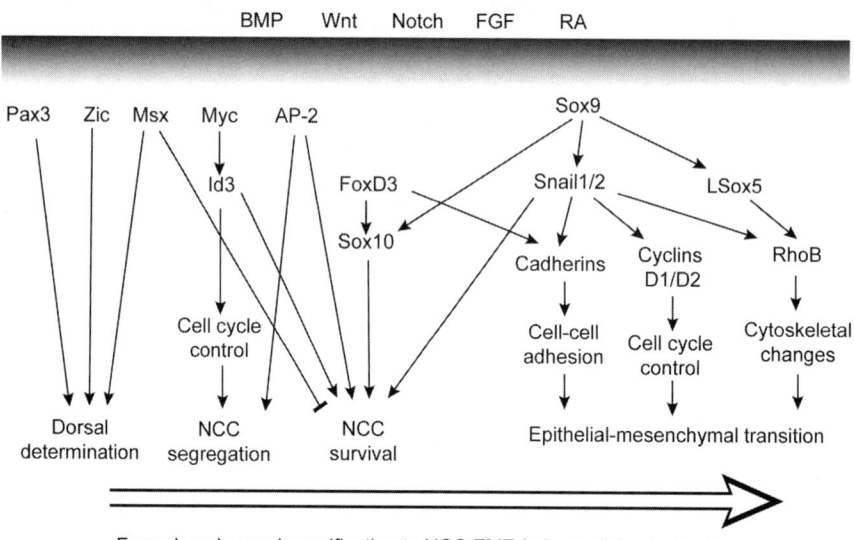

Fig. 2.4 Major genes and pathways known to regulate the early development of NCCs, shown as four steps: determination of the dorsal neural tube (dorsal determination), segregation and survival of NCCs, and the epithelial —> mesenchymal transformation that allows delamination. Bmp, Wnt, Notch, FGF and retinoic acid (RA) are involved at all stages. Adapted from Morales *et al.* (2005)

Fig. 2.5 A summary of the genetic cascade involved in epithelial —> mesenchymal transformation and the delamination of NCCs. High levels of Noggin in mesoderm adjacent to the neural tube blocks Bmp4, Wnt1, and cyclin-D1 to prevent delamination. Expression of Cad2 (N-cadherin) in the dorsal neural tube (*left*) along with Adam10 also block cyclin1, preventing delamination. Inhibition of *Noggin* transcription initiates delamination by activating Cyclin-D1 via the canonical *Wnt* pathway (Bmp4 —> Wnt1 —> Cyclin-D1) and by cleavage of Cad2 to CTF1 and 2. Adapted from Shoval *et al.* (2007)

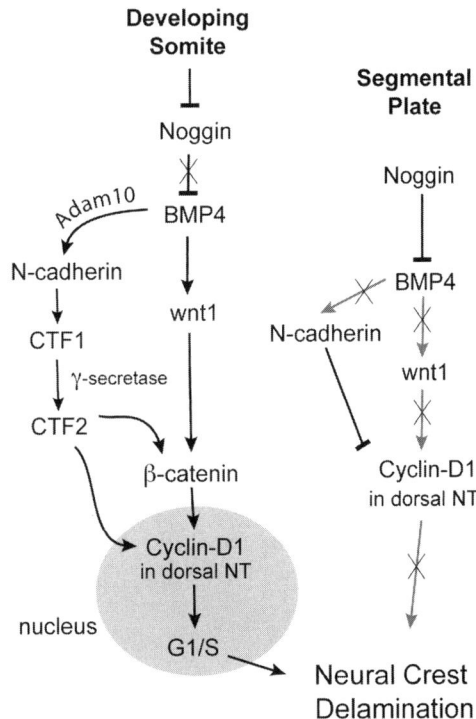

Sox Genes

Sox genes are transcription factors that produce high-mobility group (HMG) proteins with many and varied functions. The name Sox is an acronym for **S** ry HMG-**box** transcription factors.

Sox genes are organized into 10 families, SoxA–SoxJ, which are related on the basis of similarity in the sequence of their DNA-binding HMG domain; all share the DNA motif (A/T)(A/T)CAA(A/T)G. Because all Sox genes are activated following interactions with partner molecules, they can exert different roles at different stages in the initiation, differentiation, and/or maintenance of the *same cell type*. Consequently, as important regulators of NCC initiation, development, and maintenance (Fig. 2.4), Sox genes appear over and over again on the pages ahead.

The SoxE subfamily is united on the basis of a shared C-terminal transcriptional activation domain. An important group of three SoxE genes (*Sox8*, *Sox9*, and *Sox10*) involved in NCCs was revealed in 1998 from studies with mice, in which defects in NC-derived ganglia of the colon were traced to a mutation in *Sox10*. The wide-ranging action of *Sox9* in NC and non-NC tissues is seen in *Campomelic dysplasia*, a human condition characterized by craniofacial defects, sex reversal, and malformed endochondral bones, resulting from a mutation in one allele of *Sox9*.

Sox10: *Sox10* is a major player in the four major processes responsible for the development of NCCs, processes that underlie the development of many cell types:

- initiation of the neural crest;
- maintenance of the multipotency of NCCs;
- specifying NCCs into particular lineage fates; and
- initiating the differentiation of specified cells.[11]

Sox10 is expressed prominently in premigratory NCCs along the entire neural axis. Were it not for the fact that it is rapidly downregulated in the earliest stages of the differentiation of many NCCs, *Sox10* would be a good pan-NCC marker. The one exception is glial cells, in which expression of *Sox10* continues in embryos and adults (see Chapter 6).

Involvement of *Sox10* in NC formation is evident in the requirement for the expression of *Sox10* to activate expression of the NCC marker gene *Snail2* as early as blastula or gastrula stages of development. In *Xenopus* and in chicken embryos, *Sox9* induces *Sox10* expression (Fig. 2.4). Consequently, separating the actions of these two members of the SoxE subfamily is difficult, especially when different taxa are compared; NCCs are reduced in number in *Sox10*-mutant *Xenopus* and zebrafish but are present in normal numbers in *Sox10*-mutant mice.

Once formed, NCCs are maintained in a multipotent state by *Sox10* under the regulation of Bmp2 and Tgfβ; see Fig. 7.7 for *Sox10* as a marker for ectopic expansion of NCC in mouse embryos.

Sox9: *Sox9* appears in several contexts through the book as an important regulator of various aspects of NCC development (Fig. 2.4, and see Fig. 4.10). Recent analysis implicates a mediator coactivator complex in the interaction between *Sox9* and transcriptional regulation via RNA polymerase II (Rau *et al.*, 2006).

Taxon-specific differences are evident in the role of *Sox9*. For example,

(i) The induction of NCCs in *Xenopus* is dependent on *Wnt* signaling, which in turn is dependent on *Sox9*.

(ii) *Sox9* is involved in suppressing the death and so maintaining the survival of NCCs in zebrafish (Fig. 2.4). Zebrafish have two orthologs of *Sox9*, *Sox9a* and *Sox9b*, which function together as the single *Sox9* gene functions in tetrapods.

(iii) *Sox9* regulates the expression of *FoxD3* in mice, but not in zebrafish or *Xenopus* (Fig. 2.4).

(iv) *Sox9* is required to induce the otic placode in *Xenopus* but not in mice.

Sox8: The role of *Sox8* is less well understood than are those of *Sox9* and *Sox10*, in part because of surprising differences in apparent function between taxa, and in part because of a combination of overlapping and nonoverlapping functions between the three genes.

Sox8-deficient mice show weight loss but no defects that can be traced back to the NC or to NCCs. This is not because NCCs are unaffected. Rather, it is because of functional redundancy with *Sox9* and *Sox10*. Because it functions upstream of *Sox9* and *Sox10* in mouse embryos, *Sox8* can modify *Sox10* function in *Sox10*-mutant mice (Hong and Saint-Jeannet, 2005).

Taxon-specific differences were highlighted in a recent study of the expression and function of *Sox8* in *Xenopus* (Fig. 2.6 [Color Plate 1]). The chief differences from previous studies using chicken and mouse embryos are:

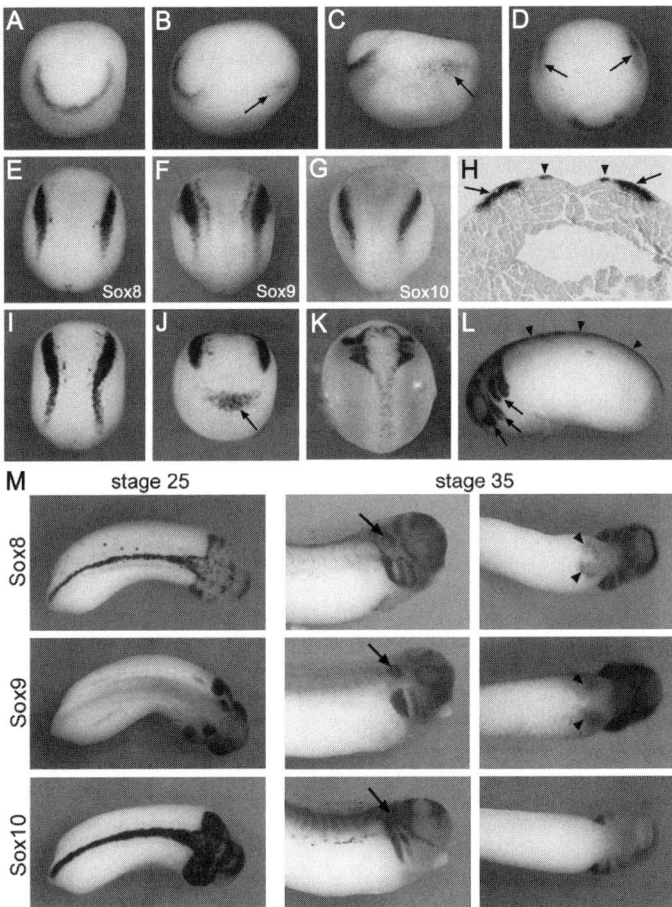

Fig. 2.6 Expression of *Sox8*, *Sox9*, and *Sox10* in NCCs and in NCC derivatives in *Xenopus* embryos. *Sox8* is expressed around the blastopore in gastrulae (**A**, **B**, **C** and **D**) and then lateral to the neural plate (*arrows* in **B**, **C**, and **D**). In slightly later embryos, *Sox8* (**E**), *Sox9* (**F**), and *Sox10* (**G**) are expressed in the neural folds, the site of the future NC. Panel (**H**) shows *Sox 8* expression in both medial (*arrowheads*) and lateral (*arrows*) NC, shown here in a transverse histological section. (**I** and **J**) slightly later stage in neurulation showing the extent of expression of *Sox8* in the NC and expression in the future cement gland (*arrow* in **J**). With closure of the neural tube — shown dorsally in **K** and laterally in **L** — Sox8 is expressed in migrating CNCCs (*arrows* in **K**) and in premigratory TNCCs (*arrowheads* in **L**). The nine panels in (**M**) compare expression of the three *Sox* genes at two tail bud stages (25 and 35). Note co-expression in CNCCs but down-regulation of *Sox9* in TNCCs. The three genes are expressed in the otic vesicle (*arrows*). *Sox8* and *Sox9* (but not *Sox10*) are expressed in the primordium of the pancreas (*arrowheads*). Figure kindly provided by Jean-Pierre Saint-Jeannet (see Color Plate 1)

- *Sox8* is expressed early in *Xenopus* embryos, as early as the mid-gastrula stage, and so is the earliest marker of future NC known.
- Expression of *Sox8* in the NC precedes *Sox9* in *Xenopus* and follows *Sox9* and *Sox10* in chicken and mouse embryos, but *Sox8* is not expressed in the NC of zebrafish embryos.
- The earlier expression of *Sox8* in *Xenopus* embryos has the consequence that for a short time, coinciding with when NCCs are specified, *Sox9* and *Sox10* are not available to compensate for any loss of *Sox8*.
- The timing of the induction of the NC is delayed in *Xenopus* embryos in which *Sox8* is knocked down using a *morpholino*, an effect that can be rescued with restoration of *Sox8* expression.
- In *Xenopus*, *Sox8* regulates the onset of expression but not the maintenance of the marker genes for the NC, *Snail*, and the winged-helix transcription factor, *FoxD3*.
- Migration, but not the proliferation of NCCs, is delayed in *Sox8*-deficient *Xenopus* embryos, resulting in major defects in several NCC lineages, severe loss or reduction of the craniofacial skeleton and dorsal root ganglia in all embryos, and reduction of pigmentation in two-thirds of treated embryos.

LSox5: *LSox5* is the long form of *Sox5*; the functions of *Sox5* have not been elucidated, although it seems more associated with glial cells than with NC- or placode-derived neurons. *LSox5* was first isolated in a screen of chicken embryos, where it is expressed in premigratory and migratory cranial and more caudal TNCCs. Regulated by *Sox9* (see Fig. 4.10), *LSox5*, *Snail1*, and *Snail2* initiate migration by acting through RhoB, a low-molecular-weight GTPase in the Ras protein family (Fig. 2.4). Overexpressing *Snail2* by gain of function in chicken embryos enhances RhoB expression and increases the number of NCCs (HNK-1-positive cells) that form in the neural tube.[12]

A role in specifying NCCs, revealed after misexpressing *LSox5* in the dorsal neural tube, elicited additional and ectopic NCCs beside the dorsal neural tube. Expression of *LSox5* alone, however, is *not* sufficient to generate NCCs; active *Sox9* is required to generate a full complement of NCC markers and functions (Hong and Saint-Jeannet, 2005*, and Fig. 2.4). *LSox5* acts cooperatively with *Sox6* and *Sox9* to promote chondrogenesis.

Wnt genes

Wnt genes have emerged as important signaling molecules in development, in no small part because they signal through several transduction pathways. The major pathway, the one by which Wnts exert their effects on NCCs, is through stabilization and regulation of the transcriptional role of β-catenin in **the canonical Wnt pathway** (Fig. 2.5). Much remains to be discovered, and Wnt signaling pathways are understood in considerably greater detail than are outlined in Fig. 2.5. Furthermore, cross-regulation between Wnt and Notch signaling pathways ('Wntch signaling') and roles for Wnt in the specification of cell fate in bipotential cells are emerging, for which see Hayward *et al.* (2008).

The Canonical Wnt *Pathway*: The phrase 'canonical Wnt pathway' refers to a cascade initiated by Wnt proteins binding to their cell surface receptors (members of the Frizzled family), resulting in the activation of proteins in the Disheveled (Dsh) protein family that form part of the Wnt receptor complex in cell membranes. Further downstream changes culminate in regulation of the amount of β-catenin reaching the nucleus (Fig. 2.5). β-Catenin interacts with transcription factors of the T-cell specific/lymphoid enhancer binding factor (Tcf/Lef) family, which upregulate specific gene expression.

Frizzled genes, which encode Frizzled Wnt receptors, are upregulated in NCCs and in condensing mesenchyme. The protein Kermit interacts with the C-terminus of *Frizzled3* (*Xfz3*) in *Xenopus*; NCC induction is blocked in *Xenopus* if Kermit is knocked out, and expression of *Xfz3* is required for *XWnt1* to be expressed and NC to be formed.[13]

The NonCanonical Wnt *Pathway*: Noncanonical (planar cell polarity or Wnt–protein kinase C–Ca^{++}) *Wnt* signaling is independent of β-catenin, but acts through domains on Dsh proteins to phosphorylate regulatory sites of JNK proteins, which are the products of *mitogen-activated protein kinase* (*MAPK*) genes. Pescadillo, a nuclear protein regulated by the noncanonical Wnt pathway, plays a role in CNCC migration in *Xenopus*; loss of function of Pescadillo leads to cranial cartilage defects.

Although the canonical andnoncanonical pathways are separate, individual *Wnt* genes can operate in tandem to regulate cell specification. For example, when operating via the canonical Wnt pathway, *Wnt1* inhibits the induction of NCCs in chicken embryos. When operating via the noncanonical Wnt pathway, *Wnt6* induces NCCs through specification of the neural plate border (Fig. 2.7 [Color Plate 2]); *Wnt6* can operate through both canonical and noncanonical pathways.[14]

Wnt *Expression and Function*: Of the 21 genes in the *Wnt* family, at least 10 are expressed in 8–9.5-day-old mouse embryos (Table 2.1), three with sharp boundaries of expression in the forebrain immediately before the onset of CNCC migration.[15]

Wnt1 is involved in the determination of the midbrain–hindbrain boundary—an important organizing center (see below and Box 3.3)—and in patterning the midbrain. Given that *Wnt1* is expressed in the dorsal neural tube throughout most of the body axis, *Wnt* expression cannot be used as a marker for specific populations of NCCs; Wnt1-cre mice were generated to take advantage of the finding that *Wnt* is a marker for all NC derivatives. Indeed, using Wnt-cre as a marker system in mice it was demonstrated that conditionally knocking out *Wnt* results in loss of NC derivatives, while constitutive activation of *Wnt* directs most NCCs into a neuronal cell fate.[16]

Wnt-signaling also plays a role in regulating the proliferation of NCCs. Double mouse mutants (*Wnt1$^-$/Wnt3a$^-$*) display defective NC and deficient dorsal neural tubes. The stapes and hyoid bones—both derivatives of hindbrain NC—are missing and thyroid cartilages abnormal. Mice lacking either *Wnt1* or *Wnt3a* form reduced numbers of TNCCs, resulting in reduced numbers and inhibited differentiation of

Table 2.1 Ages and Theiler stages of mouse development in relation to NCC origins[a]

Day of gestation	Theiler stage[b]	Morphological stage	Somite numbers	Neural tube development
8	12	Late primitive streak	1–8	Open neural plate with neural groove and neural folds
			(3–4)[c]	NCCs delaminate from the midbrain and rostral portion of the hindbrain
			(5–7)[c]	NCCs delaminating from all levels of the brain
8.5	13	Rotation of embryo	8–12	Initial elevation of neural folds
9	14	Anterior neuropore	13–20	Elevation, convergence, and fusion of the neural folds to form the hollow neural tube; formation and closure of anterior neuropore
			(16)[c]	Delamination of CNCCs complete
9.5	15	Forelimb buds appears	21–29	Migrating CNCCs; formation of posterior neuropore
10	16	Hindlimb buds appear	21–30	Neural tube completely fused; closure of posterior neuropore; ganglia of cranial nerves as condensations
11	18	lens vesicle detaching ectoderm	36–42	Regions of the brain are distinct. Neural tube is fused from

[a] Given as in a typical inbred strain.
[b] As described in Theiler (1972) on the basis of whole and sectioned embryos.
[c] As determined by Nichols (1987) using transmission electron microscopy.

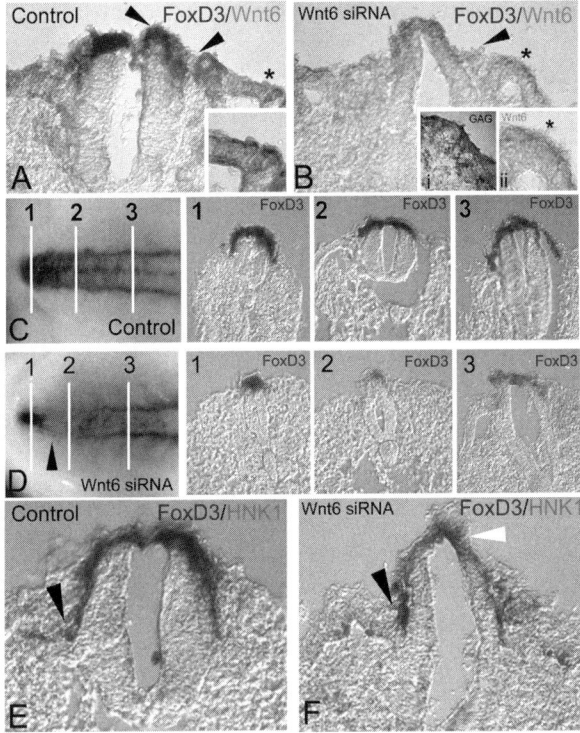

Fig. 2.7 *Wnt6* and NC induction depicted in cross-sections of the neural tubes of H.H. stage 18 (3-day) chicken embryos using FoxD3 protein and HNK-1 as NC markers. (**A**) Wnt6 (*brown*) and FoxD3 (*blue*) expression in neural ectoderm (*arrowheads*) in a control embryo. The area marked * is shown in the insert. (**B**) Reduced Wnt6 (*brown*) expression and absence of FoxD3 (*blue*) expression in a Wnt6 siRNA-treated embryo. (**C** and **D**) Reduced expression of FoxD3 (*blue*) at three rostrocaudal levels of the dorsal neural tube in a Wnt6 siRNA-treated embryo (**D**) when compared with control (**C**). (**E**) FoxD3 (*blue*) expression in the dorsal neural tube of a control embryo (**E**) is reduced significantly in Wnt6 siRNA-treated embryo (**D**, *white arrowhead*). Figure kindly supplied by Imelda McGonnell (see Color Plate 2)

melanocytes. The transcription factors Ap2α and Ap2γ,[⊕] discussed in Chapters 6 and 7, are regulated by *Wnt* genes and, at least in zebrafish, regulate the expression of *Snail2* to play a role in NC induction. In skeletogenic NCCs, *Hoxa2* is a target of *Ap2α,* which in turn is regulated by (and can substitute for) Bmp in NC induction.[17]

[⊕] Ap2 is a family of four transcription factors (Ap2α, Ap2ß, Ap2γ, and Ap2γ∂) that share conserved DNA binding and dimerization domains. Ap2α plays a critical role in NC induction, and in NCC initiation and maintenance. Ap2α is expressed in amphioxus, so its role in neural tube development preceded the origin of the NC and the vertebrates (Meulemans and Bronner-Fraser, 2002).

Specification of Ectoderm as Neural or Epidermal

Does the association between NC and neural tissues mean that the NC, like neural ectoderm, arises during or in association with neural induction, or is the NC set aside as a determined layer earlier in development? Given that the NC arises at the border between neural and epidermal ectoderm, we need to take a brief look at neural induction (a topic worthy of a book in its own right) and at how neural and epidermal ectoderm are specified.

According to the classic interpretation of the associations between notochord, neural ectoderm, and NC proposed by Raven and Kloos in 1945, the neural tube is induced by notochord, and NC is induced by lateral mesoderm (Fig. 2.8). The argument went as follows: the presumptive notochord contains more inducers than the lateral mesoderm. The notochord therefore induces neural structures and NC, while the lateral roof induces NC alone (Fig. 2.8). This interpretation rests on the assumption of a lower threshold for induction of NC than for induction of neural tissue, and on a graded distribution of neuralizing inducer with the mesoderm, with the highest concentration in dorsal mesoderm (notochord). Below is an outline of how ectoderm is dorsalized, and neural and epidermal ectoderm are specified, as essential background to discussing the induction of the NC itself. Members of the Bmp family of growth factors play major roles at several stages during neural induction, as outlined below.[18]

(1) **Dorsalization of ectoderm**: Following interactions during gastrulation, ectoderm is dorsalized by *Bmp7* from the mesoderm, and ventralized/caudalized by *Hox* genes; in *Xenopus* and chicken embryos, overexpressing *Bmp7* in ventrolateral mesoderm dorsalizes the neural tube and promotes expansion of neural ectoderm.

(2) **Neural or epidermal ectoderm**: Neural induction flows from interactions between axial mesoderm (presumptive notochord) and the overlying dorsalized ectoderm, establishing the location of the future dorsal nervous system.

Initially, Bmp4 is distributed throughout the neural ectoderm in a gradient that is highest rostrally and decreases caudally. The gradient is established by a reciprocal gradient of the Bmp4 inhibitor, Noggin, a secreted polypeptide. Induction and rostrocaudal patterning of the nervous system both involve cascades of signals that suppress Bmp4 (a growth factor that plays a key role in

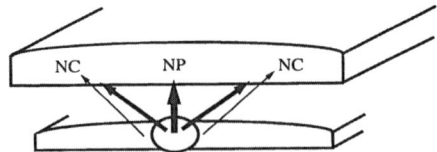

Fig. 2.8 A model of induction of neural plate (NP) and neural crest (NC), based on differential strength of induction (shown by the *thickness* of the *arrows*). Notochord (*circle*) is a stronger inducer than is lateral mesoderm. See text for details

NC induction; see below). Bmp4 and Bmp7 must both be inhibited if ectoderm is to become neural. Receptor mediation is part of the mechanism; injecting a dominant negative Bmp4 receptor into *Xenopus* animal cap ectoderm neuralizes the ectoderm, a neural fate that can be reversed after injecting *Bmp4* mRNA. Activin, another growth factor in the Tgfβ superfamily, inhibits neuralization but does not induce an epidermal cell fate.[19]

Bmp2 also ventralizes the embryonic dorsoventral axis and mesoderm and nervous system and is involved in later organogenesis; in *Xenopus*, neurula-stage embryos, zygotic transcripts of Bmp2 are expressed in the NC, olfactory placodes, pineal gland, and heart primordia.

(3) **Fore- and hindbrain**: Depending on the species, Chordin (a protein involved in the determination of the dorsoventral body axis; see Box 4.1), Noggin, and/or follistatin (a protein that binds to the growth factor, activin, and inhibits Bmp7) bind to Bmp4 to prevent Bmp4–receptor interactions, and so specify the most rostral neural ectoderm associated with fore- and hindbrain. *Noggin* is then downregulated in gradient fashion within the ectoderm, effectively setting the earliest stage when NCCs can delaminate from the neural tube (see Chapter 3).

NC Induction

Bmps, Wnts, and Fgfs

Three major classes of genes have emerged as involved in NC induction at the neural–epidermal border in different vertebrates: **Bmps, Wnts,** and **Msx genes** (Fig. 2.9). In a recent review emphasizing the functions of Bmps and Wnts in the NC, Raible and Ragland (2005*) discuss several models by which these gene families and their products interact to initiate NCC formation. Each is a two-step model, involving specification of NC by:

(1) sequential activity of ectoderm-derived *Bmp* and *Wnt* at the border of the neural tube, with Bmp conferring competence on the neural plate to respond to Wnts (Fig. 2.9);
(2) combined activity by which Notch signaling in the dorsal neural tube modulates the activity of Bmp and so induces NC (Fig. 2.9); and/or
(3) interaction between different signaling pathways such as *Fgf* —> homeobox, (msh-like 1 gene) *Msx1* and *Wnt* —> *Pax3* to induce NC.

All three modes of specification, or combinations of two or more modes, may operate in a single species. The pathway utilized may vary from species to species, or one or more pathways may act as a backup for a third and most usually used pathway. Although termed 'modes' and 'pathways', there may well be considerable conservation in signaling across the vertebrates. At least one Bmp and one Wnt gene signal high up in the cascade. Differences between species may be as 'simple' as which

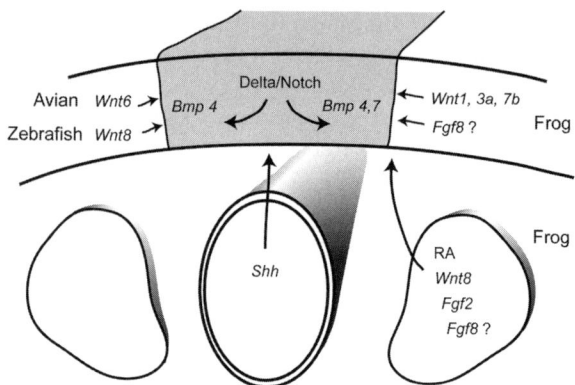

Fig. 2.9 A comparison of the signaling molecules involved in induction of NC at the neural ecto-derm (*gray*)–epidermal ectoderm border of avian, zebrafish, and frog embryos as seen in a cross-section with the neural plate (*gray*) and lateral epidermal ectoderm *above* and the notochord and somitic mesoderm *below*. For the genes activated by these signaling molecules, see the text. In all three taxa, *sonic hedgehog* (*Shh*) from the notochord activates *Bmp4* and *Bmp7* in the neu-ral plate via the Delta–Notch pathway to establish the neural–epidermal ectoderm border. Addi-tional signals from the epidermal ectoderm differ by taxa; *Wnt6* in chicken embryos, *Wnt8* in zebrafish, and *Wnts1, 3a* and *7* in frog embryos (shown on the *left* and *right*, respectively). Addi-tional signals required to generate the border in *Xenopus* include *Wnt8*, *Fgf2*, and retinoic acid (RA) from the somitic mesoderm, and *Fgf8*; whether the *Fgf8* is somitic or epidermal ectoder-mal in origin is uncertain (shown as *Fgf8*?). Arrows show signaling to the border, except for *Shh*, which signals to the neural plate. Data from various sources. Presentation adapted from Jones and Trainor (2005)

Bmp or which Wnt paralogs is used or precisely when in the cascade the Bmp or Wnt is activated. Nonetheless, because differences occur, the evidence for *Xenopus*, chicken, and mouse embryos are discussed separately.

Xenopus

Once neural ectoderm is induced by notochord (see above), induction of NC occurs at the epidermal/neural ectodermal border. Depending on the species, this may complete the induction, or epithelial ectodermal signaling may continue to be required for the induction/specification of particular types of NCCs. Lateral meso-derm (Fig. 2.10) is involved in NC induction in *Xenopus*, acting in concert with axial mesoderm. Lateral (paraxial) mesoderm evokes at least four NC markers from *Xeno-pus* neural ectoderm—*Snail2, FoxD3, Zic3*, and *Sox9*—axial mesoderm evokes only a subset, while in the absence of mesoderm, *Fgf8* upregulates all but *Snail2* in the list above. The implication is that *Fgf8* from lateral mesoderm is a necessary and may be a sufficient signal to evoke NC in *Xenopus*.[20]

Although the role of mesoderm was thought not to extend beyond the initial induction of neural ectoderm by the notochord in *Xenopus* (Fig. 2.10), paraxial

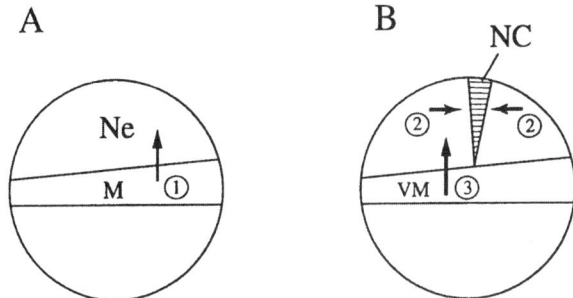

Fig. 2.10 The sequence of the major steps in the induction of the neural crest, as seen in early amphibian embryos. (**A**) Step 1: the mesodermal (M) induces neural ectoderm (Ne). (**B**) Step 2: neural and epidermal ectodermal induce neural crest (NC) at the neural–epidermal boundary. Step 3: ventral mesoderm (VM) may play a role in epidermal ectodermal induction of neural crest

mesoderm may participate in inducing the NC by regulating a gradient of Bmp associated with induction of NC and inhibition of epidermal differentiation; future epidermal ectoderm dorsalized by Noggin produces melanophores, even though no notochord is present; chicken neural ectoderm associated with mesoderm forms melanocytes but not neurons. In tissue recombination experiments, and on the basis of the induction of high levels of expression of *Snail2* and the differentiation of melanophores, lateral mesoderm was found to be a more potent inducer of NC than is notochord. Studies with whole embryos support this conclusion: removing lateral mesoderm reduces markers of NC induction/differentiation; removing notochord has no effect.[21]

One of the Wnt genes, *Xwnt7b*, expressed in future epidermal ectoderm, plays an active role in the induction of *Xenopus* NCCs (Fig. 2.9). The evidence is based on (i) the finding that *Xwnt7b* induces the NC markers *Xsnail2* and *Xtwist* (which encodes a bHLH transcription factor) in epidermal ectoderm cotreated with Noggin and in neuralized ectoderm *in vitro* and (ii) that exogenous *Xwnt7b* enhances the expression of *Xtwist in vivo*.

A role for *Wnt8* in NC induction in *Xenopus* (and in zebrafish) has also been demonstrated through inhibition of Bmp4 at gastrulation, a stage of development when Noggin does not inhibit Bmp4 (Baker *et al.*, 1999). *Wnt8* from somitic mesoderm is involved is establishing the border at which NC arises in *Xenopus* and in zebrafish (Fig. 2.9). *AmphiWnt8* is also expressed in the paraxial mesoderm in amphioxus, and although functional studies have not been performed, the patterns of expression are consistent with the possibility that *AmphiWnt8* may play a role in neural induction, upregulating *Pax3/7* and *Msx* at the border between epidermal and neural ectoderm (see Chapter 4).

Fgf2 from somitic mesoderm and *Fgf8* from mesoderm or epidermal ectoderm (Fig. 2.9) are involved in NC induction in *Xenopus*; neural differentiation declines and melanophore differentiation increases if gastrula ectoderm from increasingly older embryos is exposed to Fgf2, a finding that is consistent with altered competence$^{\oplus}$ of the ectoderm with age and with a progressive shift from neural to NC induction.

Chicken Embryos

Fgf2 appears to play a role in neural and NC induction in avian embryos. Overexpressing Fgf2—achieved by Rodríguez-Gallardo *et al.* (1997) by placing Fgf-soaked beads within the primitive streak—induces ectopic neural cells from epidermal ectoderm; whether NC also arose in these ectopic neural cells was not reported.

Juxtaposing neural and nonneural ectoderm from H.H. stage 4–10 avian embryos elicits NC; juxtaposing the same tissues from older (H.H. stage 8–10) embryos dorsalizes the ectoderm, conclusions based on upregulation of such dorsal and NC markers as *Wnt1*, *Wnt3a*, and *Snail2*. This result is consistent with a two-step model for induction of the NC in chicken embryos, involving *Wnt6* and *Fgf* as major upstream regulators (Fig. 2.9). *Snail2*, which by activating Rhob can enhance NCCs production, may be part of the second step (del Barrio and Nieto, 2002).

Mouse Embryos: In mice, *Snail2* is not required for NC or mesoderm formation.

Snail2 is not expressed in murine premigratory NCCs but is expressed in migrating NCCs. *Snail2* alone does not provide a sufficient signal to induce NC in *Xenopus,* indicating modulation of NCC inducing pathways between different vertebrates. As neither *Snail1* nor *Snail2* is involved in NC induction or NCC delamination, a two-step model involving *Snail* genes cannot apply to mice, although *Snail1* does play a role in establishing the left–right symmetry of murine embryos. Nevertheless, murine and chicken *Snail2* are functionally equivalent in that *Snail2* from either species can function in chicken hindbrain.[22]

A Role for Notch in NCC Induction

Members of the Notch family of transmembrane domain proteins—Notch1–Notch4 in mammals—are receptors for two families of transmembrane ligands, Jagged (Jagged1, Jagged2) and delta-like (Delta-like1, 3, and 4). Alagille syndrome, which includes heart and facial defects, results either from *absence of the gene JAG1* (5–7% of individuals) or spontaneous mutation(s) in *JAG1* (perhaps as many as 50% of cases).

$^{\oplus}$ Competence is the term used in developmental biology for the ability of a group of embryonic cells or an embryonic region to respond to inductive signals. Competence is gained and lost progressively during development (see this chapter and Box 2.2). Loss of competence is the proximate explanation for the loss of the lateral line and cement glands in the direct-developing Puerto Rican frog, *Eleutherodactylus coqui* (see this chapter and Box 2.3), for the inability of *premature death* (*p*) mutant Mexican axolotls to form NC-derived cartilages (see Chapter 7), for the loss of teeth in birds (see Box 3.5), and for the variability in regulative ability in subpopulations of cells along the body axis or in different species (see Chapter 10).

Receptor–ligand interaction enables signal transduction between adjacent cells and regulation of gene expression by activating bHLH repressors of transcription. After receptor binding, the Notch intracellular domain is cleaved off and transported to the nucleus, where it binds to a C-repeat binding factor (CBF) in nonmammalian vertebrates and to the conserved DNA-binding protein RBP-Jk in mammals.

Notch signaling plays a critical role in NC specification by facilitating **lateral induction** (Box 2.2) in gastrulae, determining the ectodermal domain from which NC will arise. In chicken and amphibian embryos, Notch is involved in determination of the ectodermal NC domain for CNC, reflecting a conserved role for Notch in generating boundaries—for example, at wing margins in *Drosophila*, and at limb bud margins in vertebrates—and in refining boundaries, for example, somite boundaries.

Box 2.2 Lateral induction

Lateral (homoiogenetic, horizontal, planar) induction is the spreading of an induced state by cells that were induced following interaction with an inductor derived from an adjacent cell layer, the latter sometimes known as vertical induction (Figs. 2.6 and 2.11). Neural induction proceeds laterally in *Xenopus*; that is, additional neural tissue is induced from already induced neural tissues through a signal traveling along the ectoderm, rather than from continuous vertical induction from the notochord below (Fig. 2.11).

Lateral induction has been demonstrated by the neuralization of ectoderm transplanted adjacent to the neural tube or placed in culture with neural tube, or by replacing early future neural plate ectoderm of Mexican axolotls or alpine newts with uninduced gastrula ectoderm (Servetnick and Grainger, 1991). There is loss of ectodermal competence, lateral spread of neural induction along the ectoderm, and placode formation in association with weak competence at the boundary. In chicken embryos, trunk but not cranial neural ectoderm is induced laterally, cranial neural tube requiring contact with the invaginating Hensen's node (Box 9.1). Hindbrain from zebrafish can induce ventral epidermis to become NC, a finding that is inconsistent with lateral induction in this species.

Some studies raise the issue of whether induction of NC always follows the induction of neural ectoderm. Although Mitani and Okamoto (1991) claimed evidence for separate inductions of neural tube and NC in a microculture assay of *Xenopus* early gastrula cells, close range and/or lateral inductions cannot be ruled out in such an experimental approach. Mitano and Okamoto used antibody markers for neurons, melanophores, and epidermal cells, but not NC markers. Using the genes *Snail1, Snail2,* and *Noggin* as NC markers, Mayor and colleagues (1995) claimed that NC was induced independently of the neural plate. *Noggin* is an important inducer of rostral neural tissues and associated structures such as the cement gland (Box 2.3), but does not induce hindbrain or spinal cord; induction of postotic and TNC is controlled by genes other than *Noggin*.[a]

----

[a] See Lamb *et al.* (1993) for *Noggin* as an inducer of rostral neural structures and Holtfreter (1968) and Nieuwkoop *et al.* (1985) for older studies on lateral induction affecting the NC.

Box 2.3 Cement glands

Cement glands located on the ventral surface of the head of anuran tadpoles (see panel J in Fig. 2.6) are used for attachment during feeding.[a] In *Xenopus*, the cement glands arise following a series of interactions initiated during the induction of rostral neural and epidermal ectoderm. Disrupting any of these interactions blocks cement gland formation. Induction is evidenced by differential expression of epidermal and nonepidermal keratins and by the expression of an antibody against tyrosine hydroxylase associated with the glands (Drysdale and Elinson, 1993).

A gene involved in cement gland induction, *XOtx2*, the *Xenopu* s ortholog of the *Drosophila* gap gene *Orthodenticle*, is expressed in rostral neurectoderm during gastrulation. Ectopic expression of *XOtx2* is a sufficient signal to induce an extra cement gland.

Dlx is another gene expressed in *Xenopus* cement glands. In Puerto Rican coqui, which lack cement glands, *Dlx* is expressed in a region of ectoderm that corresponds to the ectodermal region from which cement glands arise in *Xenopus*. Fang and Elinson (1996) used cross-species transplantation and tissue recombinations to investigate the potential developmental mechanisms responsible for the loss of the cement glands in coqui. They found that coqui cranial tissues can induce cement glands from *Xenopus* ectoderm, but that coqui ectoderm cannot respond to inductive signals from *Xenopus*; competence of coqui ectoderm to respond to induction is modified without modifying the inductive signal. Therefore, loss of competence, not loss of induction, leads to loss of cement glands in coqui.

Loss of ectodermal competence is also responsible for the loss of balancers in some amphibians, for loss of limbs in avian mutants such as *limbless*, and for loss of teeth in birds.[b]

An important series of messages lies in these examples of the ways in which cell and tissue interactions are modified in association with the loss of structures during evolution:

(1)   An organ may be lost without the loss of the entire developmental system that produces that organ.

---

(2) Loss of organs is often mediated through modification (not loss) of inductive interactions.

(3) Modification of competence is the usual means by which inductive interactions are altered.

(4) Inductive signaling can persist even if competence to respond is lost.

(5) Provided that competence can be restored, the potential exists for the organ to reappear.

---

[a] In their description of larval cement glands in 20 species of frogs, Nokhbatolfoghahai and Downie (2005) documented five patterns—not necessarily restricted to families—and three species that lacked cement glands, two of which bore traces of the glands as evaginations.
[b] See Maclean and Hall (1987) and Hall (1987, 1999a*, 2005b*) for examples of loss of ectodermal competence.

After NC is specified Notch plays a further critical role, specifying the fates of NCCs through **lateral inhibition**; Notch limits the number of cells that adopt a primary fate, holding them in reserve for a second fate. The cells held in reserve express a high level of the Notch ligand Delta1. Adoption of the second fate requires activation of Notch signaling, which, depending on species, may or may not involve maintaining cells in a proliferative state, and may or may not always act by inhibiting the expression of genes required for cells to adopt a neuronal fate.

In chicken embryos *Notch* signaling is modulated by *Lunatic fringe* (*Lfng*), which encodes for a glycosyltransferases that modifies *Notch* and its ligands. *Lfng* is expressed in the neural tube except along the dorsal midline. The border between expression and nonexpression therefore marks the site of future NC formation. In the presence of excess *Lfng*, Nellemans and colleagues (2001) found a 68% increase in CNCCs as a result of enhanced proliferation of existing NCCs; *Lfng* upregulated *Delta1* leading to the redistribution of *Notch1* and enhanced development of CNCCs.

In *Xenopus*, Notch signaling is required for the expression of *Xsnail2*, which is induced by *Xmsx1*, which in turn is induced by *Bmp4* under Notch control

$$Notch \longrightarrow Bmp4 \longrightarrow Xmsx1 \longrightarrow Xsnail2$$

Similarly, in chicken embryos, *Bmp4* (and therefore Notch) is required for expression of *Snail2* as CNCCs are specified; epidermal expression of the Notch ligand, Delta1 is required to upregulate *Bmp4* and to induce NC in chicken embryos (Endo *et al.*, 2002).

Notch signaling appears to play a minor role in induction of the NC in fish and mammals. Knocking out *Delta1* in mice or *DeltaA* or *Notch1a* in zebrafish does not eliminate NC. It may reduce the numbers of NCCs that form, although redundancy with other pathways may obscure the primary effects in these knockouts.

A Role for Bmps in NCC Induction and Beyond

As just discussed, as members of the TGF-β family of secreted factors Bmps are regulated by Notch signaling. Bmps also are regulated in the extracellular environment by binding proteins, Noggin and Chordin being the two most well studied. At the transcriptional level, *Bmps* are regulated by **Smad proteins**[⊗] translocated to the nucleus.

In *Xenopus*, Smad7 inhibits Bmp4-mediated induction of mesoderm, thereby activating a default neural induction pathway. Smad4 is not required for NCC migration in mouse embryos but is required for the correct patterning of the epithelium of the first pharyngeal arch, which, in turn, patterns the craniofacial skeletal elements (see Chapter 7). Smad4 is required for the development of tooth buds beyond the dental lamina stage, and for development of the NCC contribution to the cardiac outflow tract (see Chapter 8). In all three situations, Smad4 mediates epithelial–mesenchymal interactions (see Chapters 7 and 8).[24]

Bmps play multiple roles in development, so it is perhaps not surprising that they play multiple roles in NCCs in chicken, *Xenopus*, and zebrafish embryos, including:

- induction of the neural crest;
- epithelial → mesenchymal transformation and migration of NCCs (*Noggin* is downregulated in gradient fashion within the ectoderm, setting the earliest time for NCC delamination from the neural tube);
- specification of some types of NCCs, especially cells of the autonomic nervous system; and
- regulation of those mesenchymal NCCs that form craniofacial skeletal and heart structures via *Smad4*.[25]

After NC induction and specification, NCCs reuse Bmps at different times, in different places, and in different ways. Bmp2 is expressed in distinct fields in facial epithelia and in NC-derived mesenchyme but not in somatic or prechordal mesoderm.

As discussed earlier, acquiring or maintaining a dorsal fate involves interaction between neural ectoderm and signals from the epidermal ectoderm. Genes preferentially expressed in the dorsal neural tube (from which NCCs arise) initially have a uniform distribution throughout the neural tube but are inhibited ventrally by genes, such as *Shh*, although cells of the ventral neural tube can be switched to NCCs if they are grafted into the migration pathway taken by NCCs (see Box 6.3).

[⊗] *Smads (Small Mothers Against Decapentaplegic)* are three classes of transcription factors. Their most common role is to modulate the action of ligands of members of the Tgfβ family of growth factors, with which they complex before entering the nucleus to transcribe gene activity. *Smad1*, *Smad4*, and *Smad7*, each play roles in the NC or in NCCs.

As discussed when evaluating neural induction, a gradient of Bmp expression along the neural tube, at its highest rostrally and lowest caudally, is suggested from analyses of mutant zebrafish that lack *Bmp2b*, and in which the NC fails to form. The role of such a gradient, whether it is counteracted by an inverse gradient of the Bmp inhibitor, Noggin, and whether it is required to induce NC in all taxa remain active areas of research, as evidenced by studies in which *Bmp2* has been shown to be required for NCC migration but not for NC induction in mouse embryos, even though *Bmp2* is required for induction in other vertebrates; blocking *Bmp2* or *Bmp4* in murine embryos with Noggin depletes CNCCs, resulting in small pharyngeal arches and inhibition of chondrogenesis (because *Bmp2* is required for NCC migration) but does not block NC induction.[26]

Bmp4 is expressed initially in the lateral neural plate and subsequently in the dorsal neural tube and midline ectoderm. It now appears that *Shh* mediates regionalization of the medial portion of the neural plate—the region from which neurons and NC arise—while Bmp from the adjacent epidermal ectoderm regionalizes the lateral neural plate, from which placodes arise (Figs. 2.11 and 6.4; and see Box 10.1). *Bmp4* and *Bmp7* are expressed in epidermal ectoderm adjacent to the border with neural tube (Fig. 2.12); either one can substitute for ectoderm to promote NCC delamination and activate NC markers such as *Snail1*. The Bmp4 expressed at the edges of the neural plate and in the dorsal neural tube also signals to paraxial mesoderm; grafting Bmp4-producing cells into paraxial mesoderm induces *Msx1* and *Msx2* expression (see the following section) and is associated with ectopic cartilage formation. Similarly, suppressing Bmp inhibits its ventralizing action in dorsal locations.[27]

Zic3 and Zic5

The pair-rule family of homeobox genes in *Drosophila* is responsible for the subdivision of the embryonic body into regions. *Zic3*, an ortholog of the *Drosophila* pair-rule gene *odd-paired*, encodes a zinc-finger transcription factor that promotes NC over neural differentiation. *Zic3*, which is blocked by Bmp4, is expressed in

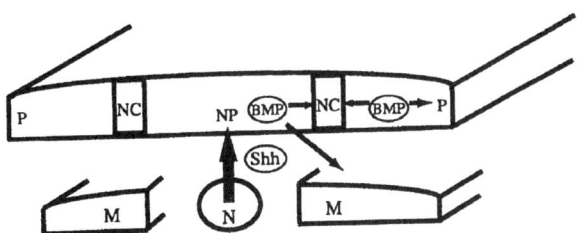

Fig. 2.11 The role of sonic hedgehog (Shh) and Bmp4 and Bmp7 in the induction of the neural crest. Shh in the notochord (N) induces neural ectoderm from the neural plate (NP). Bmp in the neural plate and epidermal ectoderm induces neural crest (NC) at the neural–epidermal ectoderm boundary. Epidermal ectodermal BmpP diffuses laterally to induce placodal ectoderm (P). Neural plate Bmp diffuses to the mesoderm (M) to induce somitic mesoderm

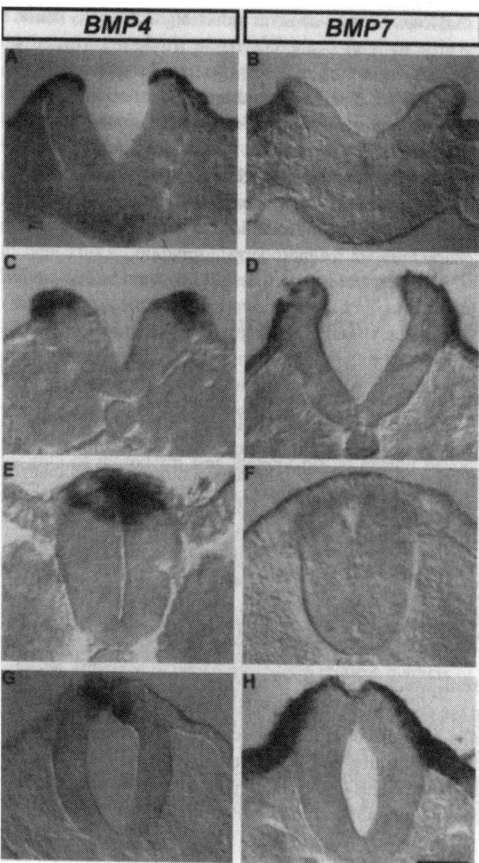

Fig. 2.12 Expression of Bmp4 and Bmp7 as seen in cross-sections through the developing neural folds, dorsal neural tube, and epidermal ectoderm in a chicken embryo of H.H. stage 10. At the level of the open neural folds (**A**, **B**, **C**, and **D**), Bmp4 is expressed in the neural folds and in the epidermal ectoderm flanking the neural folds (**A** and **C**), while Bmp7 is only expressed in epidermal ectoderm (**B** and **D**). At the level of the closed neural tube (**E**, **F**, **G**, and **H**), Bmp4 is concentrated in the dorsal midline of the neural tube (**E** and **G**), while Bmp7 is concentrated in the epidermal ectoderm, especially in the region of the future forebrain (**H**). Bar = 80 μm (**A**, **B**, **C**, and **D**); 100 μm (**E**, **F**, **G**, and **H**). Reproduced from Liem *et al.* (1995) from a figure kindly provided by Karel Liem and with the permission of the publisher, Copyright © Cell Press

neural ectoderm and NC, appearing first in the neural plate at gastrulation (Fig. 2.13). *Zic3* is one of the earliest genes so far identified as involved in neural ectoderm, induction, and/or proliferation of NC. Overexpressing *Zic3* induces NCC markers in animal cap explants and expands the population of NCCs (Fig. 2.13), both of which can be induced by Bmp4 or Bmp7.[28]

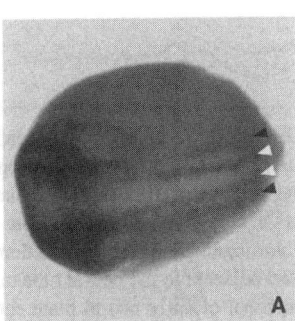
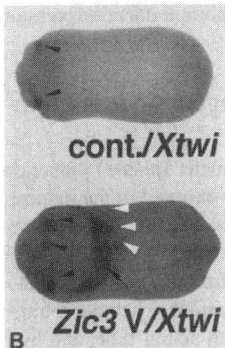

Fig. 2.13 Expression of *Zic3* in the South African clawed-toed frog *Xenopus laevis*. (**A**) Expression in a stage-16 neurula (anterior to the *left*) is in the lateral edges of the neural plate (*white arrowheads*) and in the neural crest (*black arrowheads*). (**B**) A control embryo (cont.) and an embryo injected with *Zic3* mRNA at the eight-cell stage (*Zic3*). *Xtwist* (*Xtwi*) is used as a marker for neural crest cells; in the cephalochordate *Branchiostoma belcheri*, *twist* is expressed in mesoderm and pharyngeal endoderm. In control *Xenopus* embryos, *Xtwist* is confined to CNCCs (*black arrowheads*). In the embryo in which *Zic3* was overexpressed, *Xtwist* visualizes an expanded cephalic neural crest (*black arrowheads*, *arrow*) and ectopic clusters of pigment cells (*white arrowheads*). Reproduced from Nakata *et al.* (1997) from a figure kindly supplied by Jun Aruga. Copyright © (1997) National Academy of Sciences, U.S.A.

A second pair-rule gene, *Zic5*, also is expressed in *Xenopus* NC. Overexpression enhances NC markers (with corresponding loss of epidermal markers) and induces NC in animal cap ectoderm. A dominant-negative construct blocks NC formation *in vivo*. While *Zic3* primarily functions rostrally, *Zic5* evokes more caudal NC, converting cells from epidermal to NC (Nakata *et al.*, 2000).

Msx Genes and Specification of NCCs

Once NC is induced by Notch and Bmp4, Msx genes are required to upregulate *Snail2* to specify populations of cells at the border as NCCs:

$$\text{Notch} \longrightarrow \textit{Bmp4} \longrightarrow \textit{Xmsx1} \longrightarrow \textit{Xsnail2}$$

Grafting Bmp4 into paraxial mesoderm induces *Msx1* and *Msx2* expression and associated formation of ectopic (presumably NC) cartilage.

Msx genes are homeobox-containing genes. A code of 13 homeobox-containing (*Hox*) genes patterns cranial and pharyngeal regions of vertebrate embryos (see Box 1.2). A 14th Hox gene, *Hox14*, has been identified in some vertebrates (note b in Box 1.2). Links between Bmp and *Hox* genes in the induction of NC are being uncovered; in *Xenopus, Msx1* mediates the role of Bmp4 in inhibiting epidermal and neural ectodermal induction. In concert with *Msx* genes, Bmp2 and Bmp4 regulate apoptosis of NCCs, with Bmp4 eliciting apoptosis, a topic discussed in

Chapter 10. Because *Msx1* induces apoptosis, while *Snail2* inhibits apoptosis; the balance between the two genes—coupled with the regulation of transcription caspase enzymes and other genes—generates discrete boundaries with NC-forming territories, and therefore facilitates formation of the NC.[29]

Three *Msx* genes have been characterized from mouse embryos (Fig. 2.14).[⊕]*Msx1* and *2* have similar patterns of expression in early mouse embryos, initially in the dorsal neural tube and in migrating NCCs, then in pharyngeal arches, facial processes, tooth germs, hair buds, and limb buds. *Msx3* is confined to the dorsal neural tube. In embryos with 5–8 pairs of somites, *Msx3* is expressed segmentally in the hindbrain in all rhombomeres except r3 and r5, from which lower numbers of NCCs emerge (Fig. 2.15). By the 18-somite stage, expression is no longer segmental but is uniform within dorsal hindbrain and dorsal rostral spinal cord (Figs. 2.14 and 2.15 [Color Plate 3]). As for the other Msx genes,

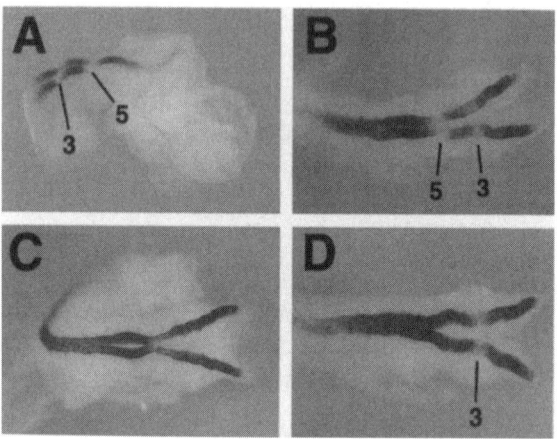

Fig. 2.14 Expression of *Msx3* in 8–9-day-old mouse embryos. (**A**) *Msx3* is expressed strongly in rhombomeres 1, 2, and 4 and in the spinal cord, and weakly in r3 in this 7-somite embryo seen in lateral view with anterior to the *left* and r3 and r5 identified. (**B**) This 10-somite embryo, seen in dorsal view with anterior to the *right*, shows weak expression of *Msx3* in r3 and lack of expression in r5. (**C**) There is uniform expression of *Msx3* throughout the hindbrain and spinal cord in this 18-somite embryo seen in dorsal view (anterior to the *right*). (**D**) The gap in expression in r5 seen in the normal embryos (**B**) is not seen in this 10-somite embryo carrying the *Kreisler* (*Krmlkr*) mutation. Kreisler codes for a transcription factor that regulates rhombomere segment identity through *Hox* genes. Indeed, r5 may not have developed in this embryo. Reprinted from a figure kindly provided by Paul Sharpe from *Mechanisms of Develop*, Volume 55, Shimeld *et al.* (1996). Copyright © (1996) with permission from Elsevier Science (see Color Plate 3)

[⊕] Zebrafish have at least five *Msx* genes (*MsxA–MsxE*), although these are not orthologous to *Msx1* and *2* of amphibians, birds, and mammals, a finding that is consistent with separate gene duplications in fish and with potentially different functions of *Msx* genes in fish and tetrapods.

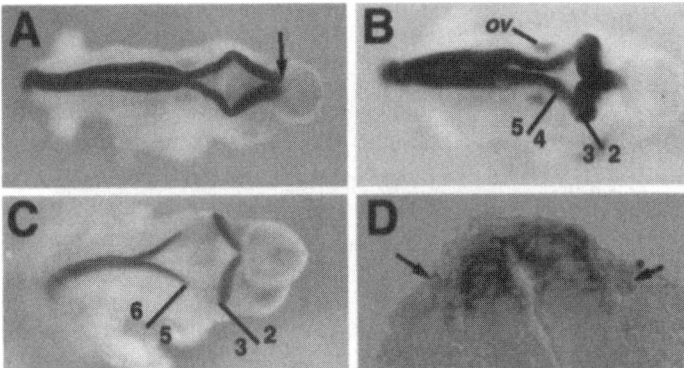

Fig. 2.15 Expression of *Msx3* in 9.5–11.5-day-old mouse embryos seen in dorsal view with anterior to the *right* (**A**, **B**, and **C**) and in histological cross-section (**D**). (**A**) Expression is strong in hindbrain and spinal cord at 9.5 days. The arrow marks the hindbrain–midbrain boundary, expression being negative in the midbrain. (**B**) Expression is similar at 10.5 days of gestation. 2, 3, 4, and 5, rhombomeres 2, 3, 4 and 5; OV, the otic vesicle, which displays nonspecific trapping of the antibody. (**C**) At 11.5 days of gestation, expression is restricted dorsally and is absent from rhombomeres 3–5. (**D**) A transverse section of the neural tube of an embryo of 9.5 days of gestation shows *Msx3* expression in the dorsal neural tube and in NCCs adjacent to the neural tube. Reprinted from a figure kindly provided by Paul Sharpe from *Mech Develop*, Volume 55, Shimeld *et al.* (1996). Copyright © (1996) with permission from Elsevier Science (see Color Plate 3)

Msx3 can be upregulated by Bmp4 and expression (which is normally restricted dorsally) extended ectopically into the ventral neural tube.[30]

Establishing Cranial and Trunk Neural Crest

Hensen's node is the site of future notochord in chicken embryos; the role of Kupffer's vesicle, the teleost homolog of Hensen's node, is discussed in Box 9.1. As discussed earlier, neural ectoderm is induced by notochord during primary neurulation as notochordal cells invaginate beneath the ectoderm and extend caudally, visible externally as the primitive streak. During this caudal extension, Hensen's node imposes rostrocaudal patterning onto the future NC, in all likelihood with the same signals (FGFs, Wnt, and retinoic acid) that posteriorize the neural tube as a whole. A consequence of rostrocaudal patterning is the regionalization of the NC into cranial and trunk, which can broadly be equated with NCCs arising from the brain and from the spinal cord, respectively.

Evidence is accumulating to indicate that CNC and TNC may be able to produce similar cell types, provided they are exposed to the appropriate signals, which they can be when maintained *in vitro*, but which they are not *in vivo* or *in ovo*. Nevertheless, differentiation of NC-derived cell types is regionalized *in vivo*, and it is important to know how that restriction in potential occurs, because the mechanisms that pattern the NC into distinct regions along the rostrocaudal axis appear to differ between chicken and mouse embryos are discussed separately.

Chicken Embryos

In the section on establishing the epidermal–neural border earlier in this chapter, we saw that an ectopic ectodermal–neural border can be established and NC induced at that border when Hensen's node from a chicken embryo is grafted into the lateral (normally ectodermal) epiblast of a Japanese quail embryo. Notochord induces ectoderm to become neural plate (and suppresses epidermal ectodermal fate). Through subsequent signaling (Fig. 2.11) NC forms at the neural–epidermal border.

Fate mapping and lineage analysis of Hensen's node in H.H. stage 4 chicken embryos demonstrate that the node consists of presumptive notochord, endoderm, and somitic mesoderm, and that the progeny of individual cells within the node can contribute to all three regions. Fate is restricted by Hensen's node during a narrow window between H.H. stages 4 and 6 (18–25 h of incubation). This conclusion comes from co-culturing NCCs with Hensen's node and observing modification of NCC fate, a determining factor being the age of the embryo from which Hensen's node is derived:

- Nodes from young (H.H. stage 4) embryos respecify TNCCs as cranial; cranial markers, such as fibronectin and actin, are upregulated, and the trunk marker, melanin, is downregulated.
- This ability is lost from Hensen's node by H.H. stage 6, which corresponds with the timing of neural induction and regionalization by Hensen's node *in ovo*.
- Nodes from H.H. stages 2–4 induce rostral and caudal nervous system, while nodes from H.H. stages 5 and 6 induce only caudal nervous system. In part, this reflects declining competence of the epiblast at H.H. stage 4.[31]

One of the molecules involved in fate determination is Bmp6, a protein localized in the posterior marginal zone of the epiblast. Additional primitive streaks can be induced in ectopic locations in the epiblast following localized injection of Bmp6, which changes the fate of epiblast cells from epidermal to neural. Once ectodermal is fated to be neural, regression of Hensen's node and the accompanying induction of notochord and neural ectoderm impart rostral–caudal patterning onto the NC.[32]

Tgfβ has been identified as a candidate molecule establishing rostral identity through a mechanism that may involve regulating NCC–substrate adhesion. Although cranial and trunk crest have similar amounts of Tgfβ messenger RNA (mRNA), cranial crest is more sensitive to exogenous Tgfβ. Immortalized Hensen's node cells secrete a Tgfβ-dependent factor that enhances cranial but suppresses TNC. Furthermore, treating TNC with Tgfβ enhances CNC markers: 400 picomolar Tgfβ decreases the number of melanocytes (TNCCs) that form, while increasing the number of fibronectin-positive (cranial) cells; blocking Tgfβ downregulates cranial and upregulates trunk markers in CNCCs. Such a mechanism, tied to NC induction as outlined above, would impose rostral–caudal patterning onto the NC during primary neurulation.

Mouse Embryos

Quinlan and colleagues (1995) mapped the neurectodermal fate of epiblast cells at the egg-cylinder stage of mouse development and demonstrated that neural primordia exhibit cranio-caudal patterning before neurulation.

An early regionalization of future cranial and TNC in murine embryos is suggested by a ^{3}H-thymidine-labeling study, in which rostral or caudal ectoderm from embryos labeled immediately before the onset of NCC migration (late primitive streak stage, 8 days of gestation; Table 2.1) was inserted into the equivalent position in unlabeled embryos, which were then maintained in whole embryo culture for 3 days. Rostral ectoderm formed cranial neuroepithelium, while caudal ectoderm formed trunk neuroepithelium, indicating that segregation into future cranial and neural tissue (and NC?) occurs before the incorporation of future neural ectoderm into the neural tube (assuming that NC is patterned at the same time as the neural ectoderm, as suggested by the 1995 study). Also recall that the most caudal NCCs in mice consist of two cell populations, one derived from neurectoderm (primary neural crest) and the other from the tail bud (secondary neural crest).[33] This fundamental, but little appreciated process of secondary neurulation, and the knowledge that the tail region does not develop directly from primary germ layers, were discussed in Chapter 1.

Ectoderm from the Most Rostral Neural Tube

NCCs and cells of the central nervous system are closely related, indeed so closely related that they share a common lineage: central neurons and NC derivatives can arise from the same cloned cells. Nevertheless, not all the cells in the neural folds form neurons or even NC.[34]

Intracelomic grafting of future rostral neural ectoderm from H.H. stage 4–5 chicken embryos demonstrates that much of the neural tube arises from the medial and not the lateral region of the neural folds. (The origin of placodes from lateral neural folds is discussed in Chapter 6.) However, NC does not arise from neural folds in the region of the future forebrain. Surprisingly, this most rostral 'neural' ectoderm forms facial ectoderm (Fig. 2.16). Quail/chicken chimeras demonstrate that the facial ectoderm of chicken embryos arises from the neural folds of the forebrain and is patterned into regions or **ectomeres**, the epidermal ectodermal equivalent of the neuromeres in the hindbrain, discussed in the following section. The superficial ectoderm of the roof of the mouth, of the olfactory cavities, and of the head and face in avian embryos, all arise from the neural folds and neural plate of the prosencephalon; that is, from cells immediately rostral to the most rostral NC (Fig. 2.16). While the fate of prosencephalic neural crest is committed early (certainly by H.H. stages 10–14), the fate of the more caudal mesencephalic–metencephalic neural crest is not set until later.[35]

In this connection, the two studies by Couly and Le Douarin (1985, 1987) on the existence of ectomeres may be especially important. Neural ectoderm, NC, and

Placodes   Ectoderm

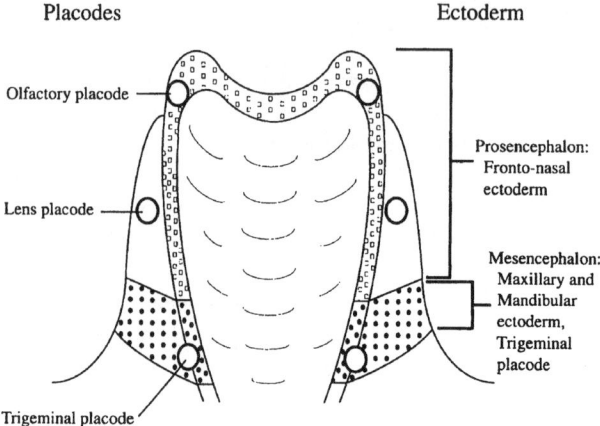

Olfactory placode

Prosencephalon:
Fronto-nasal
ectoderm

Lens placode

Mesencephalon:
Maxillary and
Mandibular
ectoderm,
Trigeminal
placode

Trigeminal placode

Fig. 2.16 Diagrammatic representations of the origin of the most rostral placodes and of the craniofacial ectoderm in embryonic chickens as seen from the dorsal surface. Placodes such as the olfactory may arise from the prosencephalic neural folds (*open squares*), from ectoderm adjacent to the neural folds (lens), or from neural folds and adjacent ectoderm (trigeminal). Ectoderm from the prosencephalic neural folds gives rise to the ectoderm of the frontonasal processes but not to neural crest or neural ectoderm. Ectoderm from the rostral mesencephalon and adjacent ectoderm (*black circles*) gives rise to ectoderm of the maxillary and mandibular processes and to the trigeminal placode. The mesencephalic contribution to the trigeminal placode includes neural crest. Based on data from Couly and Le Douarin (1985, 1987, 1990) and Dupin *et al.* (1993)

facial ectoderm are set aside during the primary embryonic induction by a code of *Hox* genes. Alternatively, although not necessarily to the exclusion of early determination of some cell lineages, restriction may occur during NCC migration; facial ectoderm and pharyngeal endoderm play important roles in eliciting differentiation from lineages of NCCs at different stages of migration (see Box 3.4).

Rostrocaudal Patterning of CNC

The pharyngeal region of vertebrate embryos forms by coordinated interactions between NCCs, pharyngeal arches, and the developing brain under the direction of a code based on overlapping expression boundaries of *Hox* genes (see Box 1.2). Alternate (odd-numbered) rhombomeres (neuromeres) in mice have characteristic boundaries of *Hox* gene expression. Other gene products are also expressed segmentally; odd-numbered rhombomeres of the hindbrain of chicken embryos bind to HNK-1 and express *Msx2* and Bmp (see Chapter 10).

Hox Genes

Using knowledge of expression boundaries in the rhombomeres of embryonic chicken hindbrains, Paul Hunt and colleagues demonstrated that the rostral limits of expression of *Hoxb1–Hoxb4* coincide with particular rhombomere boundaries. As with cranial-nerve patterning, these expression patterns are intrinsic to each rhombomere, provided that the rhombomeres develop in their normal position. Consequently, they are maintained if rhombomeres are transplanted to another site along the neural axis, being driven by the *Hox* code they carry with them (Figs. 2.17 and 2.18 [Color Plate 4]).[36] For example:

- *Hoxb1* is expressed only in r4, even if r4 is allowed to form more rostrally within the neural tube (Fig. 2.17).
- *Hox3a* has its rostral boundary of expression at the border between r4 and r5, an autonomous expression boundary that is reflected in neural tube and NC and retained if r4 and r5 are transplanted.[37]

Vielle-Grosjean et al. (1997) described a *Hox* code with a high degree of conservatism of *Hox1–Hox4* in the hindbrain and pharyngeal arches of human embryos. Differential downregulation of individual *Hox* genes in different human tissues occurs later in development.

NCCs migrating from chicken hindbrains express a combination of *Hox* genes appropriate to their rhombomere of origin and carry this combination to the pharyngeal arches. Pharyngeal-arch ectoderm expresses the same combination of *Hox* genes as does NC-derived mesenchyme of that arch; similar expression boundaries are detected in early mouse embryos in surface ectoderm, cranial ganglia, migrating NCCs, and in the mesenchyme of the pharyngeal arches. Consequently, a **fundamental unity of mechanism patterning this region** of the embryo is the common pattern of *Hox* gene expression in the hindbrain, NC, and the pharyngeal arches into which NCCs migrate.

A similar pattern is seen in zebrafish, in which the transcription factor *Krox20* is expressed in r3, r5, and in migrating NCCs. This is a relic of an ancient segmentation; *AmphiKrox* is expressed with a one-somite periodicity in the neural tube of the cephalochordate, amphioxus (Jackman and Kimmel, 2002). Interestingly, expression of *Krox20* in the pharyngeal NC in chicken embryos does not equate

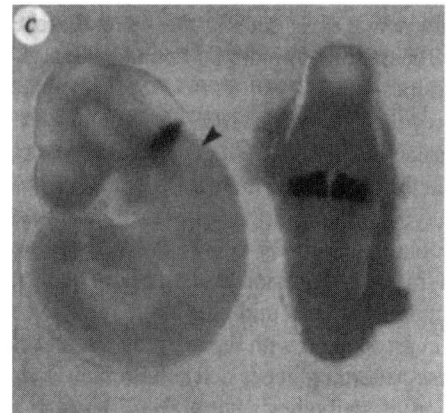

Fig. 2.17 Expression of *Hoxb1* (visualized with β-galactosidase) is restricted to r4 in 9.5-day-old mouse embryos. The *arrow* marks the position of the otic vesicle. Reproduced in *black* and *white* from the colored original in Guthrie *et al.* (1992) from a figure kindly supplied by Andrew Lumsden. Reprinted with permission from *Nature* 356:157–159. Copyright © (1992), Macmillan Magazines Limited

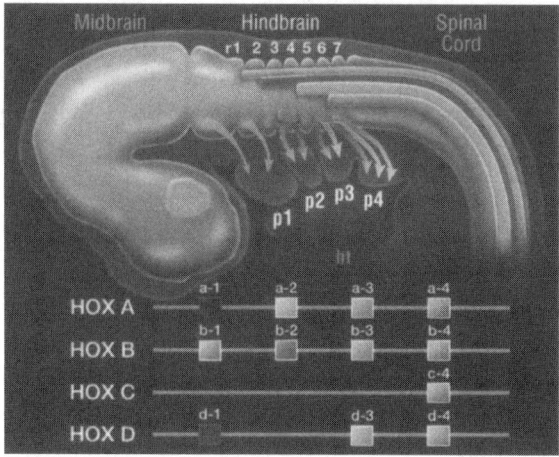

Fig. 2.18 *Hox*-gene expression in the rhombomeres of the hindbrain (r1–r7) and in the pharyngeal arches (p1–p4) is shown in this reconstruction of a mouse embryo of 9.5 days of gestation. *Colored bars* in the neural tube and *colored arrows* in migrating neural crest cells represent expression domains of *HoxA–HoxD*, which are also shown in the panel at the *bottom*. Some genes, such as *Hoxa2*, are expressed in the hindbrain but not in migrating neural crest cells. Reproduced from Manley and Capecchi (1995), with the permission of Company of Biologists Ltd. (see Color Plate 4)

with segmentation of the hindbrain, highlighting the existence of different postmigration patterning mechanisms in different species. A *cis*-acting enhancer element 26 kB upstream of *Krox20* in conserved among chickens, mice, and humans and can rescue the pattern of *Krox20* expression in transgenic mice. Localization of *Krox20* is to r5 in these embryos because of two conserved *Krox20* binding sites in the enhancer and a conserved binding site for Sox10, which, in concert, direct expression to the r5 NC.[38]

Hox group 3 paralogous genes (*Hoxa3*, *Hoxb3* and *Hoxd3*) act in a combinatorial fashion in patterning neurectoderm, and in patterning NC and mesoderm-derived mesenchyme, although it appears from studies by Manley and Capecchi (1997) that the identity of specific *Hox* genes may be less critical than the number of genes functioning in a region or developmental field (see Box 10.1).

A Role for Mesoderm

As we have seen, neural ectoderm, NC, and pharyngeal-arch endoderm share boundaries of expression of the same *Hox* genes. Initially thought to reflect transfer of the *Hox* code from

neural ectoderm   —>   neural crest   —>   pharyngeal-arch endoderm

The similar expression boundaries may not reflect a straightforward *Hox* code transfer; for example, separate enhancer elements are present in *Hox* gene clusters in the neural tube and in NCCs.

A study by Frohman and colleagues (1990) based on the expression pattern of the murine homeobox-containing gene *Hoxb1* suggests that the primary pattern lies with head mesoderm, not within rhombomeres of the hindbrain. According to this scenario, the *Hox* code arises in mesoderm, is transferred to rhombomeres of the hindbrain, then to the pharyngeal arch mesenchyme (via migrating NCCs), and finally to pharyngeal endoderm and superficial ectoderm:

> mesoderm —> hindbrain —> arch mesenchyme —> arch
> endoderm and superficial ectoderm

This study raises an important point. Is the primary rostrocaudal regionalization of neural ectoderm and NC derived from neural induction, or is it secondarily imposed upon the neural tube from mesoderm?

The cranial nerves of avian embryos are patterned by dorsoventral signals that are both cranial mesodermal and rhombomeric in origin. A role for paraxial mesoderm is supported by studies in which rhombomeres were transplanted more rostrally or more caudally along the neural axis than their normal locations, and the resulting altered patterns of *Hox* gene expression analyzed. The boundary of *Hox* gene expression is controlled, in part, by paraxial mesoderm, in part by signals from the neural epithelium itself, with the constraint that posterior properties and posterior *Hox* genes overrule anterior properties and anterior genes. Transplanting rhombomeres from caudal —> rostral alters neither the pattern of *Hox* genes expressed nor the fate of the cells. Transplantation from rostral —> caudal modifies *Hox* code and cell fate to that appropriate to the new location.[39]

The Midbrain–Hindbrain Boundary

The midbrain–hindbrain boundary (the isthmus), a major organizing center in vertebrate embryos, is regulated by members of at least five gene families: *Otx2*, *Wnt1*, *Fgf8*, the *Engrailed* genes *En2*, *En5*, and En8, and three Pax genes, *Pax2*, *Pax5*, and *Pax8*. Figure 4.8 shows the relationships between these *Pax* genes in vertebrates. Mutations in these patterning genes can delete mid- and hindbrain.

Through secretion of *Fgf8*, the isthmus functions as a developmental organizer and patterns the midbrain; *Fgf8* grafted into the caudal diencephalon can induce an ectopic midbrain. Studies in mouse embryos show that *Fgf8* expressed in the isthmus plays a regulatory role in the specification of first- and second-arch NCCs; second-arch craniofacial structures form if NCCs express *Hoxa2* (see Boxes 7.1 and 10.3) and *Fgf8* downregulates *Hoxa2* in first arch crest (Trainor et al., 2002).

En 2 is upregulated during neural induction in a region-specific manner. *XEn2* is expressed at the boundary between mid- and hindbrain and in the mandibular arches, optic tectum, and anterior pituitary. Although *En2* is normally restricted to the boundary between mid- and hindbrain in avian embryos it can be induced ectopically in association with repatterning neural ectoderm. Injecting antibodies against *Pax2* into zebrafish embryos leads to malformations of the midbrain–hindbrain boundary, downregulation of *Pax2* transcripts in the caudal midbrain, and alterations of *Wnt1* and *En2*, two genes regulated by *Pax2*.

Of evolutionary interest is the expression of *En2* at the midbrain–hindbrain boundary in Japanese lamprey embryos. In the lamprey, however, *Engrailed* is expressed not in the NC but in one muscle of the mandibular arch, the velothyroideus (see Chapter 4).[40]

Dlx Genes and Dorsoventral Patterning of CNC

Another genetic cascade governing pharyngeal-arch development and specification is the differential dorsoventral expression of *Distal-less* (*Dlx*) genes[⊗] in mouse pharyngeal endoderm, expression domains that pattern the endoderm dorsoventrally. Expression boundaries of *Hox* and *Dlx* pattern the pharyngeal arches in orthogonal rostrocaudal and dorsoventral[◊] directions. *Dlx* is a marker for the forebrain and patterns the forebrain and the rostral craniofacial skeleton, which develops from the first and second pharyngeal arches; specific arch deficiencies occur in embryos carrying *Dlx* mutations. Additional roles for *Dlx* genes are discussed in Chapters 4 and 7.[41]

Notes

1. See Hörstadius (1950, pp. 4–6) for early studies on the origin of the NC, mostly based on analysis of amphibian embryos. Garcia-Martinez *et al.* (1993) and Basch *et al.* (2006) mapped the future NC in blastula-stage chicken embryos. Schoenwolf and Alvarez (1991) used quail/chicken chimeras to establish the timing of determination of neural–epidermal cell fate. See Colas and Schoenwolf (2001) for an overview of the cellular and molecular bases of neurulation.
2. See Quinlan *et al.* (1995), Tam and Quinlan (1996) and Tam and Selwood (1996) for these studies.
3. Hill and Watson (1958) studied the northern quoll ('native cat'), *Dasyurus hallucatus*, bandicoots (*Perameles* spp.), kangaroos and rock wallabies (*Macropus* spp., *Petrogale* spp.), and American opossums (*Didelphis* spp.).
4. For induction of NC in birds and for the combined use of quail/chicken chimeras and HNK-1 or *Snail2*, see Dickinson *et al.* (1995), Bronner-Fraser (1995), and Selleck and Bronner-Fraser

---

[⊗] *Distal-less* (*Dlx*) genes are a family of seven homeodomain transcription factors, the vertebrate ortholog of the gene *distal-less* (*Dll*) in *Drosophila*. With respect to the NC and NCCs, *Dlx* genes play important roles in forebrain and craniofacial development. The roles of five *Dlx* genes (*Dlx1-4* and *Dlx7*) are discussed in the text. Amphioxus expresses the invertebrate ortholog, *Dll*, in the neural tube (see Chapter 4).

[◊] The terms dorsoventral, mediolateral, and proximodistal can be used interchangeably for this second axis that extends from the dorsal midline.

(1995*). For overviews and recent studies on NC induction see Baker and Bronner-Fraser (1997a), Barembaum and Bronner-Fraser (2005), Basch *et al.* (2006), Correia *et al.* (2007) and Schmidt *et al.* (2007).

5. See Meulemans and Bronner-Fraser (2004) and Sauka-Spengler *et al.* (2007) for the first and latest of these studies, and see Davidson and Erwin (2006) for gene regulatory networks and the evolution of major organismal features. Marianne Bronner-Fraser has generously provided access to a manuscript which was in press at the time of this being written (April, 2008).

6. See Le Douarin and Kalcheim (1999*) and Tucker (2004*) for summaries of the characterization of the HNK-1 antigen. See Kuratani (1991) for HNK-1 expression in alternate rhombomeres, and Bronner-Fraser (1987) for perturbation studies with HNK-1 antibody. Information on HNK-1 labeling in different groups may be found in the relevant sections of Chapters 5, 6, 7, and 8.

7. See Luider *et al.* (1992) for HNK-1-positive cells in the embryonic gut, and Basch *et al.* (2006) and Lacosta *et al.* (2007) for *Pax7*.

8. Molecular markers often have to be used in conjunction with cellular markers to avoid ambiguity or false-positive results. Because *Xsnail2* is expressed in NC-derived and in mesoderm-derived mesenchyme, patterns of expression must be interpreted with caution.

9. Nieto *et al.* (1994) and Duband *et al.* (1995) demonstrated the involvement of *Snail2* at the onset of NCC migration. Sechrist *et al.* (1995) used *Snail2* to monitor regulation of the NC; Mancilla and Mayor (1996), Mayor *et al.* (1997), La Bonne and Bronner-Fraser (2000) and Aybar *et al.* (2003) analyzed the role of *XSnail2*; and Nagatomo and Hashimoto, 2007) studied *Xhairy2*.

10. For a comprehensive overview of the Tgfβ superfamily and their receptors in the context of mouse craniofacial development, including tables of family members and their receptors, see Dudas and Kaartinen (2005). See Taneyhill *et al.* (2007) and Coles *et al.* (2007) for Cad6B as a *Snail2* target, and for overviews.

11. See Hong and Saint-Jeannet (2005*), Kelsh (2006*), and Saint-Jeannet (2006*) for analyses of and access to the primary literature on *Sox10*.

12. See Morales *et al.* (2007) for Sox5, and del Barrio and Nieto (2002) for *Snail2* and RhoB.

13. See Logan and Nusse (2004*) and Hayward *et al.* (2008*) for *Wnt* signaling in development, Raible and Ragland (2005*) for *Wnt* signaling in NCCs, C. Tan *et al.* (2001) for *Kermit*, Baranski *et al.* (2000) for *Frizzled* upregulation in avian NCCs, and Deardorff *et al.* (2001) for *Xfz3*.

14. See Gessert *et al.* (2007) for Pescadillo and Schmidt *et al.* (2007) for canonical and non-canonical signaling pathways operating in tandem.

15. See Northcutt (1995) and Smith Fernandez *et al.* (1998) for discussions of whether the vertebrate forebrain is segmented.

16. See McMahon and Bradley (1990) and Ikeya *et al.* (1997) and H. Y. Lee *et al.* (2004) for the studies on *Wnt* genes, and Urbánek *et al.* (1997*) for gene families that establish the midbrain–hindbrain boundary.

17. See Dunn *et al.* (2000) for Wnts and melanocytes, and Luo *et al.* (2003) for Hoxa2 and Ap2α.

18. For cascades of signals in embryonic induction, see Nieuwkoop *et al.* (1985), Gurdon (1992), Hall (2005b*) and Gilbert (2006). See Hemmati-Brivanlou and Melton (1997), Sasai and de Robertis (1997), Weinstein and Hemmati-Brivanlou (1997), Marchant *et al.* (1998) and Hurtado and de Robertis (2007) for the role of Bmps in neural induction.

19. For development of the NC at the border between neural and epidermal ectoderm, see Bronner-Fraser (1995), Selleck and Bronner-Fraser (1995*) and Graveson *et al.* (1997). See Nieuwkoop and Albers (1990) for altered ectodermal responsiveness, Xu *et al.* (1995) for the Bmp4 receptor, and Wilson and Hemmati-Brivanlou (1995) for activin.

20. See Selleck and Bronner-Fraser (1995) for melanocyte induction, Marchant *et al.* (1998) for the Bmp gradient, Barembaum and Bronner-Fraser (2005*) for an overview of early steps in NC induction, Sasai *et al.* (2001) for *FoxD3*, and Monsoro-Burq *et al.* (2003) for *Fgf8* and lateral mesoderm.

21. La Bonne and Bronner-Fraser (1998) found that although *Snail2* is a marker for NC, *XSnail2* alone is not a sufficient signal to induce NC in *Xenopus*, although downregulation of *Xsnail1* and *Xsnail2* inhibits NCC migration, resulting in deficiencies, especially in the rostral craniofacial skeleton (Carl *et al.,* 1999).

22. See Dickinson *et al.* (1995) and del Barrio and Nieto (2002) for studies with chicken embryos, La Bonne and Bronner-Fraser (1998) for those with *Xenopus*, Jiang *et al.* (1998) for mice, Murray and Gridley (2006) for *Snail* genes in mice.

23. See Cornell and Eisen (2005*) and Jones and Trainor (2005*) for reviews of Notch–Delta signaling.

24. See Bhushan *et al.* (1998) for Smad7, and Ko *et al.* (2007) and Nie *et al.* (2008) for Smad4.

25. See Raible and Ragland (2005*) and Nie *et al.* (2006*) for overviews of the roles of Bmp in NCC and craniofacial development, and Nie *et al.* (2008) for *Smad4* signaling.

26. See Marchant *et al.* (1998) and Barth *et al.* (1999) for initial studies, Kanzler *et al.* (2000) and Correia *et al.* (2007) for *Bmp2* and *Bmp4* in murine NCC migration but not induction, and see Jones and Trainor (2005) and Trainor (2005) for recent summaries. At the gastrula stage in *Xenopus*, Noggin does not inhibit *Bmp4*, although *Wnt8* does (J. C. Baker *et al.,* 1999). A gradient of *Noggin* in the neuroepithelium of chicken embryos inactivates *Bmp4* in response to signals from the paraxial mesoderm, and so coordinates the timing of NCC delamination (Sela-Donenfeld and Kalcheim, 2000).

27. For involvement of Bmp4 and Bmp7 in neural or NC induction in *Xenopus*, chicken, mouse, and zebrafish, see Dickinson *et al.* (1995), Liem *et al.* (1995), Selleck and Bronner-Fraser (1995*), and the literature discussed by Hall (1999a*), Le Douarin and Kalcheim (1999*), Nie *et al.* (2006), and Saint-Jeannet (2006). For overviews of molecular control of neural induction, see Sasai and de Robertis (1997) and Hurtado and de Robertis (2007). See Watanabe and Le Douarin (1996) for Bmp4 and chondrification of paraxial mesoderm, and Korade and Frank (1996) for NC from ventral neural tube cells.

28. See Nakata *et al.* (1997) and Suzuki *et al.* (1997) for *Zic3* and *Msx1* and their relationship to Bmps and NC induction, and Woo and Fraser (1998) for the zebrafish study.

29. See Bennett *et al.* (1995) for the distribution of Bmp in murine orofacial tissues, Tribulo *et al.* (2004) for *Msx1, Snail2,* and apoptosis, and Hall and Ekanayake (1991) and Chapter 7 for other growth factors that regulate differentiation of NCCs.

30. See Ekker *et al.* (1997) and Shimeld *et al.* (1996) for *Msx* genes in zebrafish and mice.

31. Similarly, the ability of specific rostrocaudal regions of the neural ectoderm to induce lens formation and the ability of specific regions of the epidermal ectoderm to respond to those inductions are determined during neural induction and primary axis formation.

32. See Selleck and Stern (1991) for the fate map and lineage analysis of Hensen's node, and Shah *et al.* (1997) for the studies with chicken Bmp6 (formerly known as Vg1). In a different context, prostate cancers secrete Bmp6, which, in turn, induces further Bmp6, which plays a role in the invasiveness and metastasis of the cancer (Dai *et al.,* 2005).

33. See Beddington and Robertson (1998) for the ^{3}H-thymidine study. Thomas and Beddington (1996) determined that anterior primitive endoderm may be responsible for initial patterning of the rostral neural plate (especially the prosencephalon) in murine embryos. See Schoenwolf and Nichols (1984) and Schoenwolf *et al.* (1985) for caudal NC and tail bud formation in mice and avian embryos. See Chapter 1 and Hall (1997, 1999a*) for secondary neurulation.

34. See Bronner-Fraser and Fraser (1997*), Fraser and Bronner-Fraser (1991), Selleck and Bronner-Fraser (1995*) and Mujtaba *et al.* (1998) for evidence for the common lineage of central neurons and NCCs.

35. See Couly and Le Douarin (1990) for these studies on patterning of neural fold ectoderm, Fernández-Garre *et al.* (2002) for fate mapping of the neural plate at H. H. stage 4, and Alvarado-Mallart (1993) for the timing of commitment of the mesencephalic–metencephalic neuroepithelium, a commitment that requires at least two steps.

36. See Kuratani (1991) for HNK-1 expression in alternate rhombomeres, and Hunt and Krumlauf (1992*), Kessel (1992*), and Krumlauf (1993*) for the *Hox* code.

37. See Saldivar *et al.* (1996) for cell autonomous rather than environmentally regulated expression of *Hox3a*, and Guthrie *et al.* (1992), Kuratani and Eichele (1993) and Wilkinson (1995) for intrinsic patterning of cranial nerves and rhombomere *Hox* gene expression patterns.

38. See Nieto *et al.* (1995) for *Krox20* expression in zebrafish, and Ghislain *et al.* (2003) for the enhancer elements that restrict expression to r5 in tetrapods.

39. See Kuratani and Aizawa (1995) for the transplantation studies, Kuratani and Aizawa (1995) for cranial nerve patterning, and Carstens (2004) for importance of understanding neuromeric organization for pediatric surgery.

40. See Hemmati-Brivanlou *et al.* (1990) for expression of *Engrailed*, Simeone *et al.* (1992) for nested expression domains of *Otx1* and *Otx2* in the murine rostral hindbrain, and Rhinn *et al.* (1998) for *Otx2* in murine pharyngeal endoderm and its role in specification of fore- and midbrain territories. For gene families that establish the midbrain–hindbrain boundary, see Urbánek *et al.* (1997*).

41. See Qiu *et al.* (1997*) for the dorsoventral *Dlx* code. An issue (# 5) of Volume 235 of the *J Dev Dyn* (2006) is devoted to craniofacial development and patterning.

Chapter 3
Delamination, Migration, and Potential

Although other embryonic cells migrate—primordial germ cells from their site of origin to the embryonic gonadal ridges, neuromast precursors from the lateral line (see Box 6.1), sclerotomal cells from the somites to surround the notochord and spinal cord—the extraordinary migratory ability of neural crest cells (NCCs) sets them apart from all other embryonic cells. Loss of cell-to-cell attachments, basal translocation of cytoplasmic contents, movement of cells to a fenestrated basal lamina, penetration through the basal lamina—all of which are summed up in the terms and processes *epithelial to mesenchymal transformation* and *delamination* — adoption of a mesenchymal phenotype, and the *migration* of NCCs along epithelial basal laminae or through extracellular matrices, are the major topics of this chapter. Another is how NCCs stop migrating once they reach their final site.[1]

Evidence of the following three conclusions regarding the genetic control of NCC delamination were made in Chapters 1 and 2:

- Members of the *Wnt* gene family are expressed in the cranial neural tube of 8–9.5-day-old mouse embryos before NCC delamination.
- HNK-1, a marker of migrating NCCs is required for NCC delamination.
- The earliest time when NCCs can delaminate is set by the timing of the downregulation of *Noggin*, an event that allows *Bmp4* and *Bmp7* to upregulate the expression of *Snail1* and *Snail2*, neither of which are required for NCCs to delaminate in mouse embryos.

Delamination

In embryology, the term **delamination** is used for the splitting of the blastoderm into two cellular layers. In this chapter, as throughout the text, the term delamination is used for the separating of NCC from the neural tube. For delamination to occur, epithelial cells in the dorsal neural tube lose their epithelial organization, transform to a mesenchymal phenotype, and leave the neural tube as mesenchymal cells. Genes involved in the specification of NCCs that can delaminate from the neural tube were discussed in Chapter 2. In the present chapter, the emphasis is on the cellular events associated with delamination and migration.

B.K. Hall, *The Neural Crest and Neural Crest Cells in Vertebrate Development and Evolution*, DOI 10.1007/978-0-387-09846-3_3,
© Springer Science+Business Media, LLC 2009

As discussed in Chapter 1, all germ layers are initially epithelial, as indeed are all blastula-stage embryos. With a few exceptions, endoderm and ectoderm and their derivatives remain epithelial. However, mesoderm and neural crest, the third and fourth germ layers, delaminate as mesenchymal cells after undergoing an epithelial —> mesenchymal transformation.[2] NCCs delaminate from open neural plates in amphibian and rodent embryos and from closed neural tubes in avian embryos. Delamination differs in many teleost fish in which neurulation is by cavitation of a neural keel and not by invagination of a flat neural plate (Box 3.1).

Box 3.1 The neural keel

Neurulation is often thought to always occur by the **invagination** of a flat neural plate that rolls up into a tube (Fig. 3.1). In some groups, however, the neural tube arises by the **cavitation** of a solid neural keel, seen, for example, in teleost fish (bowfin, sea bass, zebrafish) and in the European brook lamprey.[a]

Although textbooks describe a neural keel as typical of neurulation in teleosts, ganoid fish, and cyclostomes, neurulation by cavitation may only appear typical because so few species have been studied. There is, for example, no solid neural keel in the convict cichlid, *Cichlasoma nigrofasciatum*, in which involution occurs in tightly opposed neural folds with a narrow neural groove. Neurulation by involution rather than by cavitation in this species was only revealed through high-resolution microscopical analysis, a technique that should be applied to other species.

The mode of neurulation can also affect NCC delamination. In sharks—or, at least in those studied—NCCs delaminate from a solid neural keel along the neural tube except for the midbrain, where cells delaminate from the dorsal midline, appearing to accumulate above the neural tube before beginning to migrate, as Balfour depicted almost 140 years ago (Fig. 3.2).

---

[a] See Schmitz *et al.* (1993) and Papan and Campos-Ortega (1994) for neurulation in *Danio rerio*. See Schilling and Kimmel (1994) for studies on zebrafish.

The patterns of NCC delamination and migration, first documented in the 1920s by Bartelmez and Adelmann, have now been mapped in much greater detail for mice and rats. NCC delamination in human embryos before and after fusion of the neural folds has been described on the basis of the interpretation of serially sectioned embryos (see below), although few studies have been undertaken on nonhuman primates. One exception is a study in long-tailed macaques of NCC delamination from the unfused neural folds of rostral and preotic regions of the hindbrain after the neural folds fuse. N-CAM labeled a sufficient number of migrating NCCs that populations could be readily visualized (Fig. 3.3).[3]

Fig. 3.1 Neurulation by invagination of a neural plate (*left*) is compared with neurulation by cavitation of a neural keel at three successive stages. The *black circle* represents the neural cavity

NEURAL PLATE   NEURAL KEEL

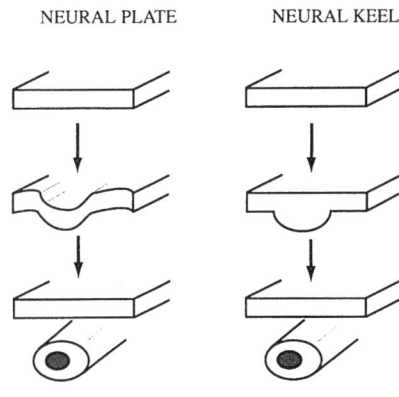

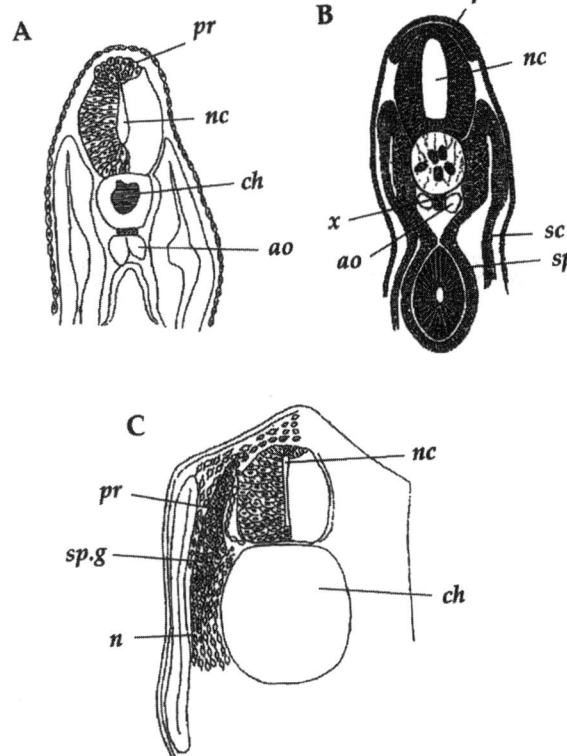

Fig. 3.2 The neural crest (**A**) and successive stages in development of the roots and ganglia of the spinal nerves (**B** and **C**) in a shark embryo. (**A**) Origin of the neural crest (pr) by proliferation from the dorsal surface of the neural tube. (**B**) Initial formation of the root of the spinal nerve (pr). (**C**) An older embryo to show the position of the posterior root (pr) and spinal ganglion (sp.g) of the spinal nerve (n). Other abbreviations are: ch, notochord; ao, aorta; nc, neural canal; sc, somatic mesoderm; sp, splanchnic mesoderm; x, subnotochordal rod. Modified from Balfour (1881)

Fig. 3.3 Cross-sections
(**A**, **B**, and **C**) and a coronal
section (**D**) through the
hindbrain of embryonic
long-tailed monkeys, *Macaca
fascicularis* to show
delamination and migration
of NCCs (*arrows* in **A** and **B**)
from the unfused neural folds
of the rostral hindbrain
(**A**) and from the dorsal
neural tube in the preotic
region of the hindbrain (**B**).
In the postotic hindbrain
(**C**) delaminated NCCs can
be seen dorsal to the neural
tube and in connection with
NCCs within the neural tube
(*arrow*). In the coronal
section in (**D**) three
populations of NCCs can be
seen emigrating from three
rhombomeric regions of the
hindbrain. Modified from
Peterson et al. (1996)

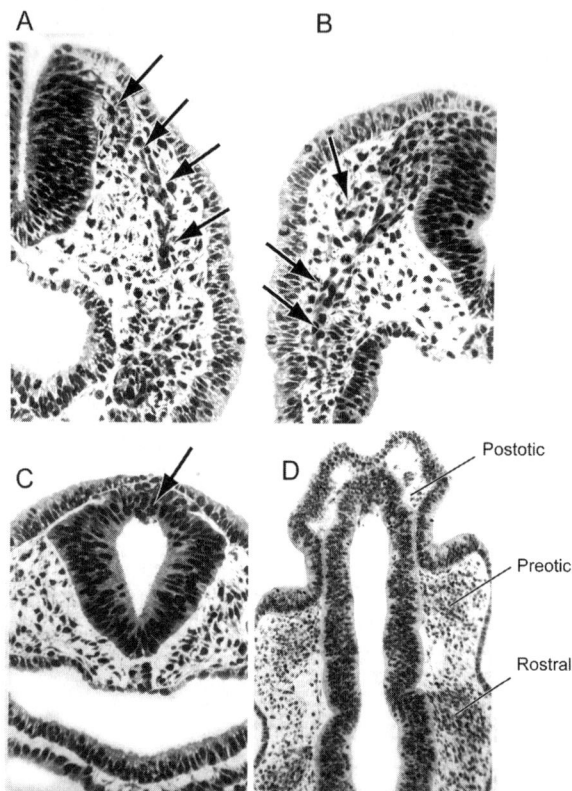

The development and application of scanning electron microscopy in the 1970s
provided previously unavailable details of neural tube and neural fold formation in a
number of mammalian embryos, including humans.[4] In 1979, O'Rahilly and Gardner integrated an analysis of 16–37-day-old (Carnegie stages 9–11) human embryos
from the Carnegie collection with other published reports to identify three populations of CNCCs (rostral, facial, and postotic) migrating between 20 and 24 days
of gestation. Subsequently, in a series of elegant and important studies detailing
the development of the human brain, O'Rahilly and Müller documented the timing
of NCC delamination and the patterns of migration between Carnegie stages 9–13
(Table 3.1).[5]

Cellular Changes Driving Delamination

Two early classic studies on mouse embryos are used to introduce cellular changes
associated with NCC delamination from cranial and trunk neural crest.

Using a combination of scanning and transmission electron microscopy, light
microscopy and alcian blue staining to analyze wild-type C57BL/J6 inbred mice

Table 3.1 Ages and Carnegie stages of human development when NCCs arise[a]

Carnegie stage	Age (days)[b]	Neural tube development
9	20	Major divisions of the brain evident; first appearance of NCCs in the mesencephalon. Open neural plate with neural groove and neural folds
10	22	No NCCs delaminate from the future forebrain; prominent populations of mesencephalic and rhombencephalic crest cells (trigeminal, facial, and postotic)
11	24	Delamination and migration of NCCs from open and closed regions of the neural tube, appearance of NCC-free zones in r1, r3, and r5
12	26	Maximal extent of optic neural crest; cardiac NC identified as migrating from r6 and r7. Closure of caudal neuropore; onset of secondary neurulation
13	28	Delamination of TNCCs; initiation of placodes; neural tube fully closed

[a] Compiled from data in O'Rahilly and Gardner (1979) and in O'Rahilly and Müller (2006*, 2007*).
[b] Note that while Carnegie stages are based on morphological criteria and so are quite precise, age is more variable.

and embryos carrying the *Patch* mutation (for more on *Patch*, see Chapters 6 and 8), Erickson and Weston (1983) demonstrated that the first delaminating TNCCs appear at 8.5-days gestation with lamellipodial extended to the dorsal neural tube and detached trailing processes. Basal lamina associated with such cells is discontinuous, but as migration proceeds the basal lamina becomes continuous (is reformed) except immediately associated with a narrow region of discontinuous basal lamina where delamination continues. Separation of these cells from the neural tube and reformation of the basal lamina on the neural tube are contemporaneous processes. In *Patch* embryos, on the other hand, spaces in the ECM not seen in wild-type embryos are spatially associated with altered patterns of TNCC migration.

Utilizing staining with toluidine blue after cetylpyridinium chloride fixation, Nichols (1987) differentiated early delaminating NCCs from the adjacent neuroepithelium and ectoderm in mouse embryos and visualized NCCs delaminating from the midbrain and rostral portion of the hindbrain in 3- to 4-somite stage embryos. By 5- to 7-somite stages, NCCs are moving away from the ectoderm at fore-, mid-, and hindbrain levels. Delamination continues until the 16-somite stage (Table 3.1), facilitated by discontinuities in what was previously a continuous basal lamina underlying the neuroepithelium. Unlike the CNCCs discussed above, these cells do not immediately break through the *basement membrane*⊕ of the neuroepithelium. Rather, they accumulate above it to lie beneath the basal surface of the ectoderm.

⊕ The term *basement membrane* refers to the ECM deposited on the basal surface of epithelial cells and visible in the light microscope. It consists of several subcomponents visible only with transmission electron microscopy, including the basal lamina and lamina lucida.

Nichols described the following sequence of intracellular changes in mouse dorsal brain:

- NCCs elongate and reposition their organelles to the basal region of the epithelium.
- Apical cell-to-cell contacts are lost, leading to the development of a stratified epithelium with basal free elongated cells.
- Processes from the basal cells penetrate the basal lamina, which is then degraded or disrupted, allowing the basal cells to break free of the neuroepithelium.
- Finally, the apical cells form a new basal lamina.

It has been assumed that escaping NCCs create focal breaks in the basal lamina, perhaps by selective enzymatic degradation, although Nichols' study shows this may not be so. Studies on NCC delamination in the long-tailed macaque show that gaps in the basement membrane at stages 11 and 12 (16–21 somites) are repaired by stage 13.[6]

We might not expect these patterns seen in delaminating mammalian CNC to hold for mammalian TNC or, indeed, to hold for other vertebrates. The cellular events associated with delamination and epithelial —> mesenchymal transformation—loss of cell-to-cell adhesions, dissolution of the basal lamina, development of leading and trailing edge polarity and initiation of migration—are more conserved, especially in the role that cadherins play.

Cadherins

Cadherins (Cad) are members of a family of some 20 transmembrane Ca^{++}-dependent cell-adhesion proteins that share Ca^{++}-binding domains (cadherin repeats). Cad2 (N-cadherin) possesses five extracellular cadherin-binding domains, a transmembrane, and an intercellular β-catenin-binding domain.

While NC, neural plate, neural tube, and all differentiated neuronal cells express neural cell adhesion molecule (N-CAM),[⊗] only cells in the dorsal neural tube (NCC-precursors) express Cad2 and Cad6B. Both classes of cell adhesion molecules are downregulated in migrating NCCs (see Fig. 2.3). Retention or overexpression of *XCad11* delays migration of NCCs in *Xenopus*. As a further indication of the link between NC and neural ectoderm, epidermal ectoderm does not express N-CAM.[7]

Expression of endogenous galactoside-binding lectins increases during NCC migration and decreases with cell adhesion, coincident with expression of N-CAM and cadherins. Expression and/or downregulation of cell adhesion molecules are

---

[⊗] N-CAM, a large cell-surface glycoprotein, is, like the cadherins, involved in cell-to-cell adhesion. Unlike cadherins, which comprise a large family of proteins, N-CAM is a single molecule with immunoglobulin-like and fibronectin type III domains. N-CAM function is derived from its ability to form 27 alternatively spliced mRNAs and even more isoforms.

required for the sequence of events controlling the EMTs that initiate mesenchymal cell migration, whether NC or mesodermal in origin; mesenchymal cells delaminate from somites as the sclerotome develops.

At least four processes are required to trigger epithelial —> mesenchymal transformations:

(i) decreasing cell adhesion,
(ii) reduction in expression of N-CAM,
(iii) loss of Cad2, and
(iv) modulation of expression of **integrins**, a family of transmembrane receptors that bind fibronectin, laminin, type I collagen, and other components of the ECM encountered by migrating mesenchymal cells (see Chapter 3).

At the cellular level during epithelial —> mesenchymal transformation there is downregulation of adherent junctions and desmosomes, massive rearrangement of the cytoskeleton, and remodeling of actin, much of which is under the control of members of the Rho family of GTPases or *Ras–Rho* interactions and required to initiate epithelial —> mesenchymal transformation.[8]

Overexpressing *Cad2* or *Cad7* in the neural tubes of chick embryos prevents many NCCs from delaminating, prevents melanocytes from migrating along the normal dorsolateral migration pathway, and leads to an accumulation of melanocytes and melanocyte precursors within the neural tube. Cad2 prevents epithelial —> mesenchymal transformation via both cell adhesion-dependent and Wnt-signaling pathways; levels of Cad2 decrease in the dorsal neural tube leading to reduction in *Bmp4* via *Adam10* signaling, and NCC delamination. Similarly, exposing premigratory neural crest to vitamin A (retinoic acid) enhances the retention of N-CAM. Consequently, crest cells cannot initiate the epithelial —> mesenchymal transformation. Rather, they accumulate at the neuroepithelium, a failure that leads to craniofacial anomalies, as discussed in Chapters 9 and 10.[9]

Cad6B regulates the timing of NCC epithelial —> mesenchymal transformation, which is disrupted when *Cad6B* is overexpressed and premature when *Cad6B* is knocked down. Recent studies demonstrate that expression of *Cad6B* is depressed in embryonic chick NCCs 30 min after injecting a morpholino antisense oligonucleotide against *Snail2* and that the regulatory region of the chicken *Cad6B* gene contains three pairs of clustered E boxes that are binding sites for *Snail2*. The binding of *Snail2* to these sites represents the first demonstration of a direct target of *Snail2*.[10]

Bmp4 counteracts the action of Cad2 on NCC migration in avian embryos. Cad2 inhibits delamination (which marks the onset of migration) in at least two ways: by a cell-adhesion-dependent mechanism and by repressing canonical *Wnt* signaling (see under *Wnt* genes below). *Cad2* is downregulated in the neural tube as NCCs emerge, a downregulation triggered by Bmp4, which acts via an Adam10-dependent mechanism to cleave Cad2 proteins into soluble fragments within the cytoplasm (see Fig. 2.5). Overexpression is followed by translocation of these Cad2 fragments to the nucleus, where, by stimulating the transcription of cyclin D1 (see Fig. 2.5), the inhibition of epithelial —> mesenchymal transformation elicited by full-length Cad2 molecules is overcome, and NCC delamination is initiated (Shoval *et al.,* 2007).

Extracellular Spaces and Delamination

NCCs, especially TNCCs, require cell-free extracellular spaces through which to migrate. In comparison with the forces exerted by fibroblasts, NCCs exert relatively weak traction forces on the substrate, an attribute that allows them to migrate on or through what appear to be relatively weak extracellular matrices.

Hyaluronan, also known as hyaluronic acid, is an alternating polymer of glucuronic acid and *N*-acetylglucosamine, joined by β1–3 linkage. Early in development, hyaluronan is involved in the elevation and closure of the neural folds, the ECM around the neural folds being rich in hyaluronan. Subsequently, by virtue of diverse binding domains—hyaluronan interacts with at least 30 genes or their products (see below)—hyaluronan is thought to be involved in NCC delamination from the neural tube and (especially in the trunk) in opening up spaces through which TNCCs migrate. It is said 'thought to be involved' because, although NCC from Mexican axolotl embryos preferentially bind to hyaluronan over binding to fibronectin *in vitro*, the distribution of hyaluronan *in vivo* is not consistent with hyaluronan playing a role during normal NCC development. On the other hand, knocking out *hyaluronan synthase-2* (*Has2*), one of a family of three genes involved in the synthesis of hyaluronan in mammals, results in loss of the NC-derived endocardial cushion of the heart, a defect attributed to inhibition of migration of CarNCCs. However, as a similar phenotype is seen after knocking out the ECM protein versican, versican–hyaluronan interactions, or versican itself may be regulating the EMT.[11]

Inhibitors of protein kinases can initiate the epithelial —> mesenchymal transformation required for NCC delamination. NCCs also produce proteases such as plasminogen activator that help create the spaces through which they migrate. NCCs from murine embryos synthesize urokinase- and tissue-type plasminogen activators from as early as 8.5 days of gestation, although a small population of NCCs that fail to migrate and remain attached to the neuroepithelium during normal development, do not produce plasminogen activator.[12]

Metalloproteinases are emerging as regulators of both the transformation of NCCs from epithelial to mesenchymal and the migration of NCCs; for the latter see below. Matrix metalloprotease-2 (Mmp2) is expressed in NCCs as they detach from the neural tube in chicken embryos. Expression disappears as migration begins. Mmps are inhibited by inhibitors in a family known as tissue inhibitors of metalloprotease or Timps. Morpholino-knockdown of *Mmp2* or addition of exogenous Timp2 inhibits EMT but not NCC migration, consistent with a role for Mmp2 in the former but not the latter process (Duong and Erickson, 2004). Recent studies also have identified another class of proteins, the semaphorins, with primary roles in delamination and migration (see below).

Migration

Early descriptions of migration were based on analyses of serially sectioned whole embryos in which investigators 'saw' TNCCs move away from their superficial association within the developing dorsal neural tube to take up more peripheral positions as sensory ganglia. Such cytological analyses, and the colonization of increasingly more distant regions as development ensued, provided eloquent evidence of the massive cell movements associated with the relocation of NCCs.

To take mammalian embryos as an example, in cytological studies that extended from the 1920s to the 1960s, Bartelmez and Holmdahl saw cells apparently migrating away from the neuroepithelium at the neural-plate stage of human and other mammalian embryos, a situation paralleling that seen in embryos of other vertebrates. Bartelmez reported NCCs migrating away from the forebrain on either side of the optic primordium of eight-somite stage (Carnegie stage 20; 22nd day of gestation) human embryos, and traced these cells in older embryos into the trigeminal ganglion and as mesenchyme migrating around the developing pharynx to contribute to the mandibular and hyoid arches. Bartelmez also described NCCs in the neural folds of the fore-, mid-, and hindbrain regions of four- to five-somite-stage rat embryos. Migration, which begins cranially, gradually extends caudally as in birds and amphibians; migration away from the forebrain begins at the five-somite stage and is maximal at the six- to seven-somite stage, when spurs of mesenchyme project into the cranial mesoderm and the developing mandibular arches. Some 40 years after his first study, Bartelmez (1962) was still providing information on the NC in rats, pigs, the tenrec *Hemicentetes* (an insectivore), primates, and humans.[13]

Pathways of CNCC Migration

By grafting radioisotopically labeled cells in place of the NC in unlabeled host chicken embryos—as was undertaken in the pioneering studies by Weston and Johnston in the 1960s—Johnston determined that CNCCs migrated in streams to colonize the maxillary, mandibular, and median and lateral nasal processes. These cells went on to form connective tissue, cartilage, and bone (see Chapter 7 for further details).

As is now well recognized, CNCCs in all the vertebrates examined migrate in **three distinct streams from discrete rhombomeres of the hindbrain**:

- a most rostral stream from r1 and r2 migrates to and produces the first (mandibular) arch from which the jaws develop;
- a more caudal stream from r4 migrates to and produces the second (hyoid) arch (Fig. 3.4); while
- the most caudal stream from r6 migrates to and gives rise to more caudal pharyngeal arches—those that produce the gills in amphibians and fish.

Fig. 3.4 Migration of NCCs into the second pharyngeal (branchial) arch (ba2) in chicken embryos under normal conditions is from r4 to r5 (**A**). In the absence of neuropilin1 (Nrp1), NCCs migrate normally but do not invade ba2 (**B**). Reducing the size of the NCC population in r4 results in much more directed migration and filling up of ba2 (**C**) Modified from McLennan and Kulesa (2007)

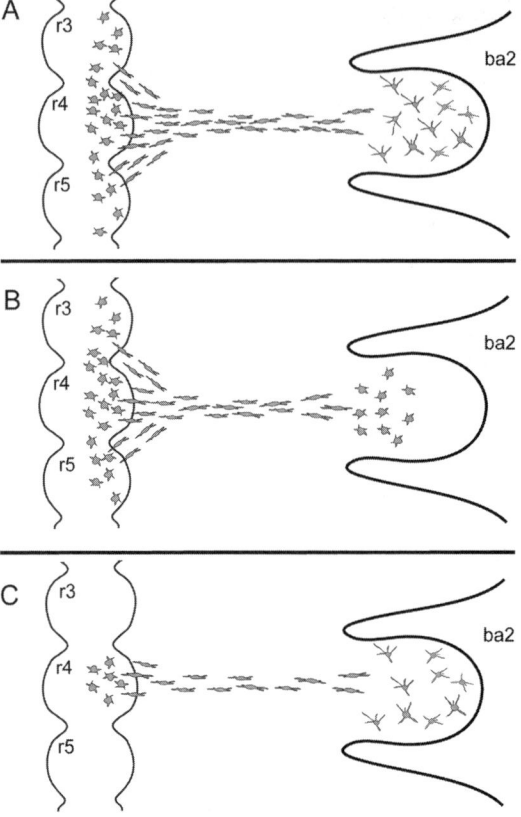

Equally well recognized is that r3 and r5 contribute far fewer CNCCs than the other rhombomeres.

These pathways of CNCC migration and populations of CNCCs are conserved across the vertebrates; initial studies on populations of migrating NCCs in the Australian lungfish, *Neoceratodus forsteri*, identified mandibular, hyoid, and branchial streams, although delamination is delayed in lungfish in comparison with timing in amphibians, especially when hyoid and branchial streams are compared (Falck *et al.,* 2000).

Conservation of pathways and populations is seen even in direct-developing frogs, which have lost the tadpole stage from their life history and, therefore, do not form larval skeletons. CNCC migration in amphibians with tadpoles (indirect developers), and in the direct-developing Puerto Rican frog, is in rostral, rostral otic, and caudal otic streams typical of other vertebrates studied; patterns of early migration are not altered with loss of the tadpole stage. As discussed in Chapter 5, the only difference in coqui is some enhancement of the mandibular stream.

Pathways of TNCC Migration

Beginning with Weston's (1963) study in which he grafted radioisotope-labeled NC into unlabeled chicken embryos, followed in the next decade by other studies of embryonic chickens, mice, and frogs using HNK-1 or DiI, two streams of migratory dorsolateral and ventral populations of TNCCs, each with predictable pathways of migration, were identified. TNCCs migrate:

- superficially, along the dorsal trunk epidermal ectoderm;
- more medially, along the lateral edges of the somites;
- between the somites;
- along the dermatome/myotome boundary in the somites themselves (subsequently shown to migrate through the rostral half of each somite) and into somitomeres, the equivalent of somites in the head; and
- between spinal cord and somites in the most medial population.

These migrating cells, whose major pathways are shown in Fig. 3.5, go on to form pigment cells, spinal ganglia, and sympathetic neurons. As discussed later in this chapter, by exposing cells to different environments, differing pathways of migration can influence the final fate of NCCs.[14]

Migration of TNCCs was investigated by Sadaghiani and Thiébaud (1987) in a study in which lysinated fluorescein dextran was injected into eggs of *X. laevis* and then labeled neural folds removed and transplanted into the equivalent positions in the neural folds of embryos of *X. borealis*. Migration is along three pathways: a ventral pathway around the notochord and neural tube; a lateral pathway under the ectoderm; and a dorsal route across the caudal two-thirds of each somite and into the dorsal fin of the tadpole.

Are the migratory routes taken by NCCs an intrinsic property of or imposed by the extracellular environments through which NCCs migrate? Avian and mammalian NCCs (and the axons of cranial nerves) migrate only through the dorsorostral sclerotome, specifically avoiding ventrocaudal sclerotome and perinotochordal mesenchyme. The rostral sclerotome is rich in tenascin and cytotactin, molecules that facilitate intrasclerotomal migration. The surfaces of caudal sclerotomal cells contain glycoconjugates that can be visualized by binding to peanut agglutinin lectin (PNA), inhibit NCC migration through the rostral halves of each sclerotome, and cause growth cones of neurons of dorsal root ganglia to collapse.

Transplanting NCCs to more rostral or caudal regions within the neural folds to follow their pathways of migration shows that, for the most part, directionality of migration is not intrinsic. NCCs grafted into a different position along the neural tube do not seek out their original path. Rather, they migrate along a path typical of their new location; rostral midbrain CNCCs, which would normally migrate rostral to the eye, migrate caudal to the eye if grafted more caudally along the neural tube.[15]

TNCCs emerging from a single region such as adjacent to a single somite contribute to structures both more rostral and more caudal than the region of the neural tube from which they emerged. In embryonic chickens:

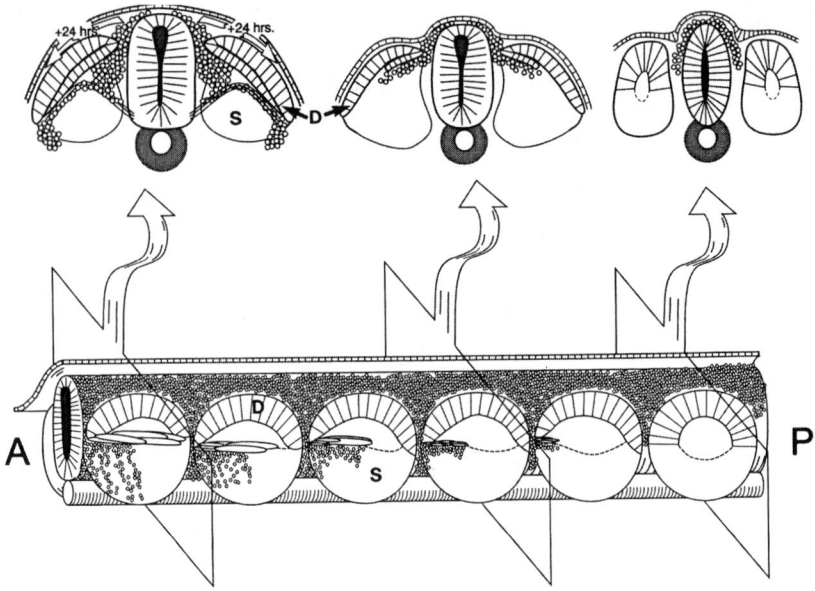

Fig. 3.5 The migratory pathways followed by TNCCs are shown in cross-sections of three levels of the neural tube (*top*) and in longitudinal section (*bottom*). A is anterior and P posterior. The progressive migration of cells between the neural tube and dermatome (D) and through the rostral and dorsal portions of the sclerotome (S) can be seen. Open arrows (*top left*) show migration of cells beneath the surface ectoderm. Reproduced from Erickson and Perris (1993) from a figure kindly supplied by Carol Erickson. Reprinted by permission of Academic Press, Inc.

- precursors of dorsal root ganglia from TNC adjacent to three somites contribute to single dorsal root ganglion;
- neurons that make up an individual ganglion migrate two somite lengths rostrally and three lengths caudally from their points of emergence, which means that NCCs adjacent to six somites contribute to a single ganglion; while
- melanocyte precursors migrate extensively along the rostrocaudal embryonic axis;
- A recent study identified a population of embryonic chicken CNCCs that delaminate and then cross the dorsal midline—previously only known to occur during regulation to replace extirpated NCCs—to colonize the dorsal root ganglion on the other side, contributing almost half the population of sensory neurons involved in pain reception in the ganglion.[16]

In an especially informative study, Erickson and Goins (1995) labeled cells with fluoro-gold, grafted them into the migration pathway taken by NCCs, and followed their migration with high-resolution microscopy. Migration along the dorsolateral pathway is restricted to cells that are *already specified as melanocytes*; other NCCs are excluded from that pathway. Migration *per se*, however, does not evoke the differentiation of this subpopulation, which requires specific environmental signals to differentiate, one of which is the mitogenic peptide endothelin-3 (Edn3), promotes differentiation of NCCs into melanocytes in mice, but initiates proliferation and formation of glial–melanocyte precursors in chicken embryos; see Box 3.2, and Harris and Erickson (2007) for further development of what one might call an 'early restriction–later specification model' and how it generally applies to subpopulations of TNCCs.

Box 3.2 Endothelins

The three known endothelins (Edn1, Edn2, and Edn3), each a 21-amino-acid peptide, are potent mitogens that act by binding to one of the two G-protein-coupled receptors, EdnRA and EdnRB.[a]

Edn1, which is expressed in pharyngeal epithelia and mesodermal mesenchyme, upregulates several classes of transcription factors in craniofacial mesenchyme, including *goosecoid*, *Dlx-2*, *Dlx-3*, *Hand2* (*dHand*) *eHand*, and *Barx1*. *Edn1* also regulates *Msx1* expression through the basic helix–loop–helix transcription factor *Hand2*.

In zebrafish, *Edn1* from pharyngeal ectoderm functions through the receptors Ednra1 and Ednra2 to regulate lower jaw skeletal precursors and jaw joint formation. Expression is in pharyngeal endoderm, pharyngeal-arch mesodermal mesoderm, but not in arch NC-derived mesenchyme. *Edn1* creates the environment in which arch NCCs and mesodermal mesenchyme are specified as ventral chondrogenic and ventral myogenic, respectively. Knocking out the endothelin1 receptor in mice produces severe anomalies of NC-derived craniofacial and cardiac structures.[b]

In mice, mutations in the gene for Edn3 or in its receptor, EdnRB, result in the phenotypic mutants *Lethal Spotted* and *Piebald Lethal*, respectively. Both mutants have megacolon—lack of enteric ganglia in the terminal 2–3 mm of the bowel (see Chapter 9).[c] Mutation in EDNRB, if they co-occur with mutation in the *RET* proto-oncogene, result in Hirschsprung disease, one of the two major forms of aganglionic megacolon in humans discussed in Chapter 9.

---

[a] See Lahav *et al.* (1996) and Thomas *et al.* (1998) for endothelin3 and melanocyte proliferation and differentiation.
[b] For genes regulated by *Edn1*, for the distribution of *Edn1* and for the knockout study, see Thomas *et al.* (1998), Clouthier *et al.* (1998, 2000), Miller *et al.* (2000), and Clouthier and Schilling (2004). See Nair *et al.* (2006) for action via the receptors Ednra1 and Ednra2.
[c] See Rothman *et al.* (1993) and Puffenberger *et al.* (1994) for *Piebald Lethal* and *Lethal Spotted.*

Contributions from more than one region of NC may not be required for particular structures to form normally: enteric ganglia in chicken embryos normally arise from VNC adjacent to somites 3–6. NCC from adjacent to somite 3 but not from adjacent to somite 1 can form the entire enteric nervous system under conditions where the VNC is removed and replaced with 'one somite length' of NC (Barlow *et al.,* 2008). Regional specificity always has to be taken into account when assessing NCC potential.

Migration into Dorsal Fins and Tails

Recall from Chapter 1 that the tail develops from a mesenchymal tail bud that contains cells that are not segregated into germ layers, which is a very different situation from more rostral regions of the embryos. This chapter discusses tail mesenchyme in two situations—the dorsal fins of amphibian tadpoles and those of teleost larvae/adults.

Tadpole Dorsal Fin Mesenchyme: Studies undertaken by DuShane (1935) demonstrated that TNCCs induce the tadpole median dorsal fin. Bilateral extirpation of TNCC resulted in larvae that lacked the dorsal fin and pigment cells in areas that would have been populated by the extirpated NCCs (Fig. 3.6). Pigment cells were known to arise from NC (see Chapter 5). The presumption was that NCCs also provided the fin mesenchyme.

As discussed above in the context of migration pathways of TNCCs, one approach to labeling cells is to use two species (*Xenopus laevis* and *X. borealis* in that case) whose cells can be distinguished from one another because of a naturally occurring marker. Sadaghiani and Thiébaud (1987) were able to track mesenchymal TNCCs into the median dorsal fin mesenchyme. It turns out that in *Xenopus* and in the Mexican axolotl, and in what was thought to contrast with other amphibians and fish, TNCCs provide only a minor contribution to the mesenchyme of the dorsal fin. The bulk of the fin mesenchyme comes from mesoderm (dermomyotome), a derivatives of the somites whose origin predates the last common ancestor of all vertebrates (Devoto *et al.*, 2006).

Collazo *et al.* (1993) injected vital dyes into groups or single TNCCs in *Xenopus* embryos and confirmed that *some* of the mesenchyme of the medial, unpaired fins

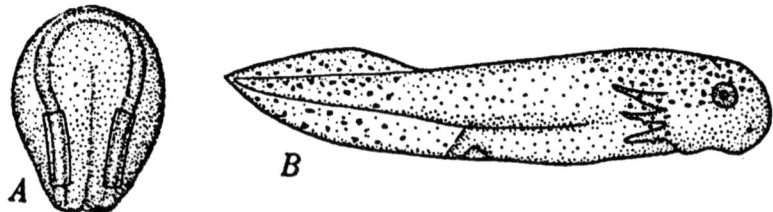

Fig. 3.6 DuShane (1935) demonstrated involvement of the TNC in amphibian dorsal fin formation. Bilateral extirpation of TNC (*boxed areas* in the dorsal view of a neurula in (**A**) resulted in larvae lacking the dorsal fin and pigment cells in the areas that would have been populated by the extirpated neural crest cells (**B**). Reproduced from Hörstadius (1950) with the permission of Dagmar Hörstadius Ågren

is NC in origin. They also showed that fin mesenchyme, pigment cells, spinal ganglia, adrenal medullary cells, and cells of the pronephric kidney arise as progeny of single TNCCs. Most clones produced fin mesenchyme, indicating that the majority of trunk cells are multipotential for mesenchyme and the other NC derivatives.

The combined use of cell labeling and grafting enabled Epperlein and colleagues to conclude that most of the fin mesenchyme in the Mexican axolotl was mesodermal in origin. The contribution of somitic mesoderm to mesenchyme of the median fin in the Mexican axolotl (and as seen below in another study using shark and lamprey embryos) was confirmed in a recent report of the first transgenic axolotl, and the transplantation of cells from GFP transgenic to unlabeled embryos.[17]

Therefore, mesenchyme of the dorsal fin of larval anurans and axolotls is primarily of mesodermal origin, with only a small contribution from NCCs.

Teleost Fin Mesenchyme: Distinct populations of migrating TNCCs can be visualized in embryonic teleost fish (Fig. 3.7).

Sadaghiani and Hirata and their colleagues studied NCC migration in three species (platyfish, the swordtail, and Japanese medaka) using a variety of techniques including SEM, reactivity with HNK-1, and immunohistochemistry; HNK-1 labels migrating cells in all three species. As shown using HNK-1, the dorsal fin of swordtail embryos is populated by mesenchymal cells derived from TNC. The components of the ECM along the pathway taken by migrating NCCs in platyfish

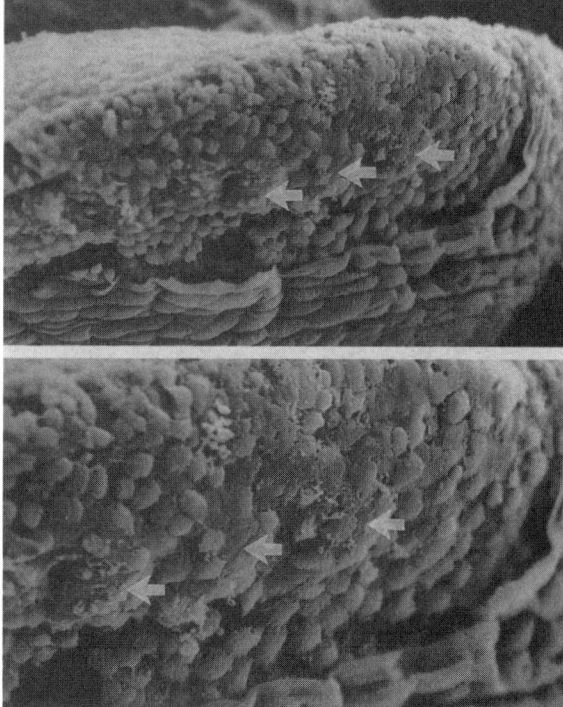

Fig. 3.7 Scanning electron micrographs at lower (*top*) and higher (*bottom*) magnifications of the hindbrain region of a 16-somite-stage (17-h) embryo of the zebrafish viewed from the dorsal surface. Ectoderm was partially removed to reveal streams of migrating neural crest cells (*arrows*). Figure kindly supplied by Janet Vaglia

include chondroitin sulfate and fibronectin, the roles of which in directing NCC migration are discussed later in this chapter.[18]

Considerable mesenchyme arises from the TNC of fish embryos. In zebrafish, this mesenchyme comes only from the medial portion of the neural keel (see Box 3.1). The lateral portion may form placodes, as it does in amphibians. We know from the studies in which TNCCs of zebrafish embryos were labeled with DiI that mesenchyme of the tail fin in regions where fin rays later will form is TNC in origin.

TNCC migration has been studied in *Spadetail* mutant zebrafish, in which the somites are partly deleted and tail development is disrupted because lateral mesoderm fails to form. *Spadetail* represents a mutation in a T-box gene that is the ortholog of zebrafish *VegT*, a gene involved in regulating the Nodal-signaling pathway. *Snail2* is downregulated in the TNCCs of *Spadetail* embryos. NCCs in the medial migration pathway in wild-type embryos are contact inhibited by somitic cells. Contact inhibition (and therefore somite-based repulsion of migrating NCCs) fails to occur in *Spadetail* mutants, resulting in diminished tail development.[19]

Although tail fins are not paired fins, recent analyses indicate that median and paired fins share genetic pathways under the control of *Fgf* signaling.

Although sharks are not teleosts, Freitas *et al.* (2006) demonstrated that much of the mesenchyme of the median fins of the lesser-spotted catshark originates in the somitic mesoderm (mesenchyme of the paired fins arising from lateral plate mesoderm). Because they also found that the median fins of the North American sea lamprey receive somitic mesodermal cells (and share expression of *Hoxd* and *Tbx18* genes with the catshark). Freitas and colleagues conclude that fins originated in a somitic mesoderm program that was co-opted by lateral plate mesoderm when paired fins arose. The role of the NC in fin evolution remains to be elucidated.[20]

Molecular Control of NCC Migration

Extracellular Matrices, Cell Surface Ligands, and Receptors

As demonstrated by analyses of the extracellular matrices through which they migrate, and by examining the migratory behavior of cells cultured on or in various ECM products, NCCs make use of both extracellular matrix and cell-to-cell signaling molecules when migrating. Whether this is true for the population of NCCs that remain behind in the neural tube to form the mesencephalic trigeminal nucleus (Box 3.3) remains to be determined.

The ECM exists as a fibrillar meshwork that is structurally altered immediately before NCCs enter it and modified biochemically by the transit of NCCs through it. The matrix is rich in the glycosaminoglycans hyaluronan and chondroitin sulfate, and also contains chondroitin sulfate proteoglycan (aggrecan, versican), type I and type II collagens, tenascin, laminin, and fibronectin. Basement membranes and basal laminae of epithelia also constitute ECMs used by migrating NCCs. TNCCs only migrate through rostral portions of epithelial somites because of the differential distribution of molecules in the somites. Developing somites also synthesize and release products such as thrombospondins into the ECM and to which migrating NCCs respond.

Cranial and trunk cells use different mechanisms of attachment to these extracellular matrices. TNCCs attach to fibronectin, laminin, collagen type I, and collagen type IV. CNCCs do not attach to the two collagens, although they do attach to basal laminae. N-CAM labeling of migrating hindbrain NCCs in long-tailed macaque revealed that NCCs migrate under the ectoderm and into pharyngeal arches 1, 2, and 3, using laminin and collagen type IV of the basement membrane as substrata. Initial migration of rat NCCs is also mediated via laminin and collagen type IV, which remain attached to NCCs as they migrate. Although basement membranes and basal laminae are epithelial extracellular matrices, some of their constituents are not epithelial in origin; mesenchyme contributes important structural components such as fibronectin.[21]

By decreasing their adhesion to molecules that enhance NCC migration, components of extracellular matrices effectively direct the speed of migration. The rate of migration increases when NCCs are placed into the interstices of hydrated collagen gels containing chondroitin sulfate, chondroitin sulfate proteoglycan (aggrecan), or hyaluronan.[22]

To summarize the functions of these molecules:

- Aggrecan and versican inhibit migration.
- Fibronectin, laminin, collagen types I and IV, and thrombospondins permit migration.
- Collagen types II, V, and IX, small proteoglycans and molecules that bind PNA—which are expressed in regions from which NCCs are excluded—inhibit or deflect migrating cells, often forming a barrier to migration[23]

Several classes of ligands and cell surface receptors have emerged over the past decade as important players in all aspects of NCCs migration:

- EphB receptors both permit and inhibit migration depending on NCC population and temporal and spatial embryonic context.
- Ephrins and endothelins can either inhibit migration or act as attractants for migrating NCCs, again in both NCC subpopulation and context-dependent ways.
- Semaphorins and their receptors (neuropilins and plexins) regulate proliferation and pathways of migration of NCCs.

In the sections that follow, the roles of extracellular matrix and cell surface molecules are discussed with respect to whether they act as factors permitting, inhibiting, and/or attracting or forming a barrier to migrating NCCs.

Permitting Migration

Cell surface and ECM products, such as fibronectin, laminin, and entactin, play important roles in NCC migration in avian and amphibian embryos, as they do in other situations of active cell migration. Laminin provides a scaffold for migrating cells during primitive streak formation in avian embryos; antibodies that bind laminin alter migration to such an extent that no normal primitive streak forms. Indeed, no embryonic axis forms (Zagris and Chung, 1990). Antibodies generated against fibronectin, laminin, and entactin have been used to analyze the composition of the cell surface and pericellular matrix of the neural tubes of rat and mouse embryos. All three are present coincident with NCC delamination. Although glycosaminoglycans modify and perhaps even inhibit NCC migration, fibronectin is the major player in NCC migration.

Fibronectin

Fibronectin is a complex, large (400,000 daltons mw) glycoprotein. Structurally and functionally distinct cell-, collagen-, heparin-, and hyaluronan-binding domains make up some two-thirds of the molecule, the cell-binding domain (120,000 daltons) being the largest.

Fibronectin occurs in especially high concentrations in epithelial basement membranes, including those of the superficial dorsal ectoderm and somites along which NCCs migrate. Localization of NCCs to the rostral half of each somite at the boundary between dermatome (future dermal connective tissue) and myotome (future muscle), as shown in Fig. 3.5, may be facilitated because migrating NCCs use fibronectin in the somitic basal lamina as a substratum. Migration therefore requires that somitic mesoderm be at a stage of differentiation equivalent to the sclerotome; NCCs do not migrate through segmental plate mesoderm.[24]

Fibronectin binds cells and other molecules. When presented with a choice of ECM products, NCCs preferentially bind to fibronectin-coated substrates. Such direct evidence of the binding of fibronectin to NCCs, coupled with the co-localization of fibronectin and migrating NCCs in vivo, and the expression of fibronectin receptors on NCCs, implicates fibronectin in guiding NCCs in vivo.[25]

As might be expected, NCCs attach to the cell-binding domain of fibronectin. This need not have been the case, for fibronectin could, in theory, attach to components of the pericellular matrix via the collagen-binding or hyaluronan-binding domains. Latex beads can be coated with cell-, collagen-, or heparin-binding portions of fibronectin and implanted into the pathway taken by migrating avian NCCs in ovo to assess which portions of fibronectin inhibit bead translocation. Only beads coated with the cell-binding domain fail to translocate, further supporting the role of the cell-binding domain in mediating crest cell migration. NCCs fail to migrate on fibronectin to which an antibody raised against the cell-binding domain has been bound. They also fail to migrate on a synthetic decapeptide with the amino acid sequence Arg-Gly-Asp-Ser-Pro-Ala-Ser-Ser-Lys-Pro—the recognition site for the cell-binding domain. Antibodies against fibronectin can also be used as immunoselective agents to isolate subpopulations of NCCs.[26]

Fig. 3.8 The numbers of neurons in the mesencephalic trigeminal nucleus (MTN) of the Peking duck (*Anas platyrhynchos*) declines during embryonic development to a low at 16 days of gestation that is maintained into adult life. Based on data in Petrosino *et al.* (2003)

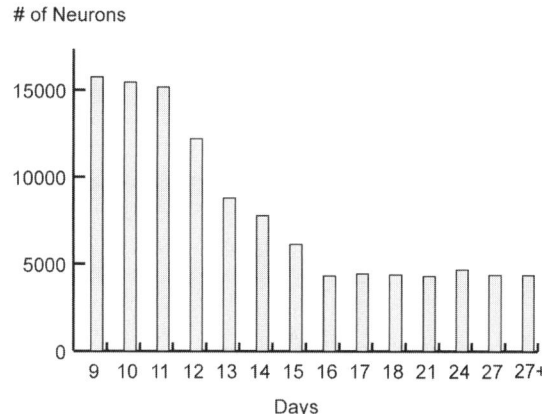

Proteoglycan Complexes

Initiation of NCC migration, onset of the segregation of paraxial mesoderm, and initiation of chondrogenesis in *Xenopus* are all associated with production and deposition of the ECM protein cytotactin and cytotactin-binding proteoglycan (Williamson *et al.* 1991), implicating ECM control over onset of migration *and* differentiation.

In Mexican axolotl presumably in other amphibians also, large proteoglycan complexes within the ectoderm influence NCC migration. Extracellular matrix products from Mexican axolotl and avian embryos have been adsorbed onto Nuclepore filters and the filters implanted *in vivo* (Fig. 3.9) or used as substrates on which NCCs are cultured. ECM associated with the superficial ectoderm preferentially

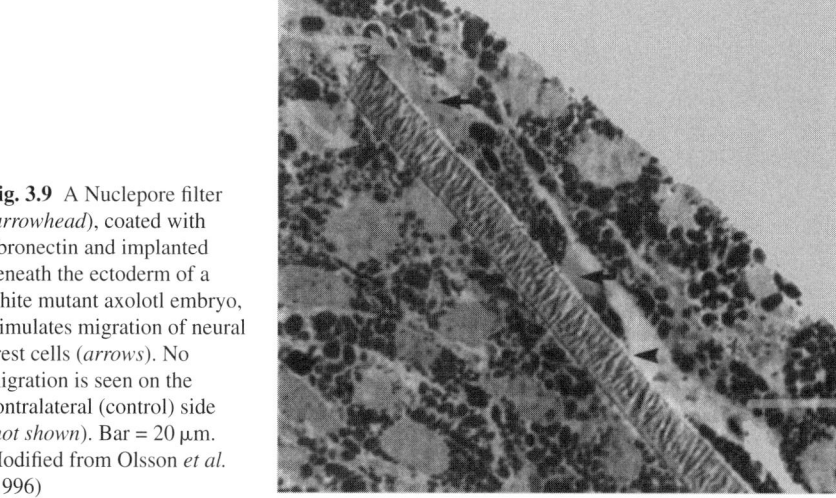

Fig. 3.9 A Nuclepore filter (*arrowhead*), coated with fibronectin and implanted beneath the ectoderm of a white mutant axolotl embryo, stimulates migration of neural crest cells (*arrows*). No migration is seen on the contralateral (control) side (*not shown*). Bar = 20 µm. Modified from Olsson *et al.* (1996)

promotes migration of the superficial population of TNCCs. The basal lamina of the dorsal epidermal ectoderm, along which superficial cells migrate, promotes adherence between NCCs and basal laminae so that the superficial cells effectively migrate as a sheet. In the pericellular matrix immediately surrounding NCCs, enzymes such as galactosyltransferases bind to sugar residues such as *N*-acetylglucosamine in the ECM and on basal laminae. Such binding, which affects cell adhesion and de-adhesion, and therefore cell motility, could control the movement of NCCs *in vivo* (Brauer and Markwald, 1987).

Neural Crest Cells Contribute to Extracellular Matrices to Permit Migration

As already noted, NCCs produce proteases and plasminogen activator to create a path through which they migrate. Some matrix components are synthesized by the NCCs themselves; ascorbic acid, which is released from migrating chicken TNCCs in amounts as high as 1.5 µg/mg protein, increases collagen synthesis by somites or muscles by as much as 2.5–6 times (Tucker, 2004*). Avian NCCs synthesize tenascin (a member of a family of extracellular cell-adhesion proteins) to create an environment through which they migrate. Tenascin binds to chondroitin sulfate proteoglycan but not to fibronectin.

Novel classes of adhesion (and de-adhesion) molecules have been localized in NCCs at different phases of migration. One is the Adam family of membrane-anchoring metalloproteases, Adam being an acronym for **a d** isintegrin **a**nd **m**etalloprotease domain. The disintegrin domain functions in adhesion, the metalloprotease domain in de(anti)adhesion. *Adam13* has been cloned and mRNA and protein localized to CNC and somitic mesoderm during *Xenopus* embryogenesis (Fig. 3.10). Overexpressing *Adam13* in *Xenopus* elicits massive invasion of NCCs into the surrounding tissues; inhibiting *Adam13* inhibits NCC motility (Alfandari *et al.*, 1997, 2001*).

Thrombospondins

Thrombospondins are a family of five large glycoproteins (Tsp1–5). Tsp1 and Tsp2 are homodimers, composed of three identical 120 kDa subunits linked by disulfide bonds. Tsp1, which inhibits the proliferation of angioblasts and so is antiangiogenic, plays a permissive role on NCC migration.

Tsp1 is highly expressed in basement membranes of embryonic mouse epithelia and in association with peripheral neuronal outgrowth. Expression is especially high adjacent to the neural tube late in the ninth day in regions where NCCs are migrating, and continues to be expressed on NCCs as they initiate ganglion formation; Tsp1 functions, at least in part, by an intracellular signal regulated by integrins (O'Shea and Dixit, 1988).

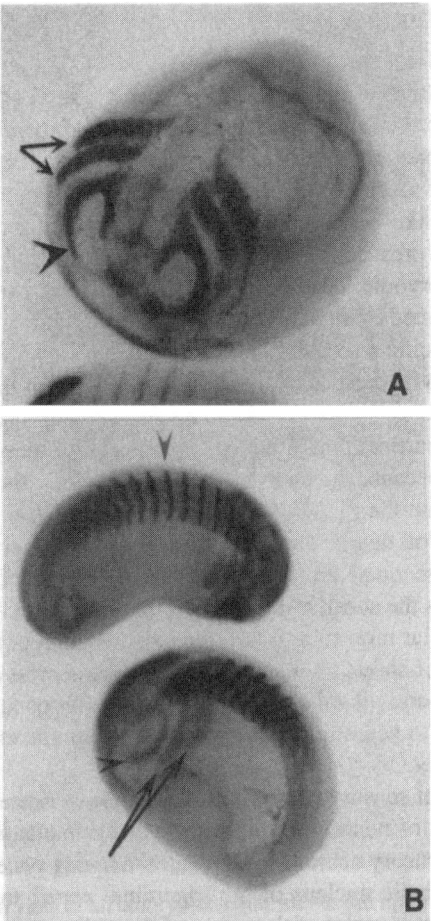

Fig. 3.10 (**A**) *Adam13* mRNA is expressed in streams of migrating neural crest cells (*arrows*) in *Xenopus* embryos. The most rostral stream in this stage-20 embryo subdivides into pre- and postoptic streams around the optic vesicle (*arrowhead*). (**B**) Lateral views of an early tailbud (stage 22) embryo (*top*) and a late neurula (*below*) to show expression of Adam13 protein in somitic mesoderm (*arrowhead* in the stage-22 embryo) and in migrating neural crest cells (*arrows*) in the late neurula (*arrows*). Reproduced with permission from Alfandari *et al.* (1997) from a figure kindly supplied by Douglas DeSimone

In avian embryos, Tsp1 is expressed in the ECM associated with the ventral path taken by migrating NCCs. Earlier, and during NCC migration, Tsp1 is expressed in the myotome of developing somites. TNCCs cultured with Tsp1 migrate significantly faster and more NCCs remain attached to the substrate than when NCCs are cultured on plastic. In supporting cell–substrate attachment, Tsp1 has a similar effect on fibronectin when both are assessed at physiological concentrations. Thus, Tsp1 and fibronectin are both permissive for and promote NCC migration, and the myotome (somite) plays an important role in TNCCs migration by synthesizing and secreting Tsp1 into the ECM.[27]

Tyrosine Kinases Receptors (Trk)

Trks are a family of *ty*rosine *k*inases receptors for neurotropins such as Nt3. TrkC is the family member most associated with neuronal cells, although some cells or tissues express TrkB and TrkC. Migrating neurogenic TNCCs, neural tube, and dermamyotome in chicken embryos express mRNAs for multiple isoforms of TrkC, expression occurring in waves, TrkC mRNA is then expressed in subpopulations of sensory ganglia and later in postmitotic cells of the DRG. Nt3 is required for the same subpopulation of neurogenic TNCCs to initiate neurogenesis—nonneurogenic NCCs do not express TrkC—indicating that expression of neurotrophin and receptor are both lineage specific. Dermomyotomal expression, like myotomal expression of thrombospondin, may play a role in directing Nt3 expression to those pathways along which neurogenic cells migrate, especially as Nt3 is mitogenic for both NC and somitic cells.[28]

Ephrins and Eph Receptors

Ephrins (*Eph*receptor *in*teracting *p*roteins) as transmembrane ligands and their Eph receptor tyrosine kinases are emerging as performing multiple roles in development, physiological regulation, and onset of disease states, not the least because they provide bidirectional signaling affecting both Eph-expressing and EphR-expressing cells.

Both ephrins and their receptors are differentially distributed along routes taken by migrating TNCCs. A major role is in mediating interactions between sclerotome and migrating TNCCs, confining the latter to specific rostrocaudal territories. EphB is expressed along the dorsolateral pathway taken by migrating CNCCs in chicken embryos, blocking early migrating cells but, along with several EphB receptors, promoting the dorsolateral migration of melanoblasts and their precursors, in part by increasing attachment of melanoblasts to fibronectin, by modifying the actin cytoskeleton.[29]

Ephrin receptors also guide CNCCs. As shown in Fig. 10.6 [Color Plate 9], complementary expression of EphA4/EphB1 receptors and the ligand prevents cells from the second and third arches from intermingling, and so targets third-arch CNCCS to their destination.

Inhibiting Migration

Components of ECMs

Ectoderm in the white axolotl mutant produces an inhibitor of NCC migration, differentiation, and survival, which along with intrinsic deficiencies in the NCCs and in collagen II and chondroitin-6-sulfate of the ECM through which NCCs migrate, renders the animals white; collagens I and IV are normal. In contrast, pigment cell migration is normal in albino mutant axolotls.[30]

Integrins, a family of transmembrane heterodimeric receptors with α- and β-subunits and present on many NCCs, bind fibronectin, laminin, type I collagen, and other components of the ECM. The α-subunit binds to laminin and type I collagen, the β-subunit to fibronectin and laminin. Integrin antisense oligonucleotides block integrin attachment to fibronectin or laminin and block NCC migration; $\alpha6$-integrin is required for later stages of neurulation in *Xenopus* but not for neural induction. $\alpha4\beta1$ Integrin, which binds fibronectin, is expanded in NCCs as migration is initiated. Blocking $\alpha4\beta1$ blocks migration and leads to increased NCC death. Integrins play other roles later in development; apoptosis of CNCCs is increased in $\alpha5$-integrin-null mice.[31]

Interactions between NCCs and matrix molecules are regulated in time and space. In rat embryos, for example, chondroitin sulfate proteoglycans, which retard migration, are at low levels at 9 days of gestation but at higher levels on day 10, and so are associated with migrating but not postmigratory cells. Treating the extracellular sheath that surrounds the avian notochord with chondroitinase removes chondroitin sulfate proteoglycan and permits NCCs to invade this matrix, into which they (unlike sclerotomal mesenchymal cells) normally cannot penetrate. Chondroitin sulfate proteoglycans inhibit the migration of NCCs through cell–surface interactions mediated by the hyaluronan-binding region.[32]

In summary, by inhibiting or promoting cell-to-cell and cell-to-matrix adhesion, extracellular glycosaminoglycans and proteoglycans, pericellular galactosyltransferases, and especially extracellular fibronectin, control NCC migration. An extracellular environment with adhesive molecules such as fibronectin on the migration pathway and nonadhesive molecules, such as hyaluronan and chondroitin sulfate on the nonmigratory pathway, seems to be sufficient to determine where NCCs will migrate.

Guiding Migrating NCCs

NCCs could merely 'run into' barriers and accumulate. This is known not to be the case, as beautifully demonstrated in a study in chicken embryos by Kulesa and colleagues (2005) in which physical barriers were implanted into the mesoderm adjacent to pathways of NCC migration and the response of migrating NCCs followed with time-lapse confocal imaging. NCCs are amazingly dynamic and 'inventive' in moving around the barrier and forming the lead cell in a new stream, even migrating into regions normally 'avoided' by migrating NCCs.

Barriers and Components of ECMs

Physical barriers—basal laminae, blood vessels, other cells (mesoderm-derived mesenchyme, somitomeres in the head, somites in the trunk, mesenchyme surrounding the notochord)—could direct NCC migration and/or influence cessation of migration and cell accumulation. The embryonic locations of NCCs, however, appear too precise to be explained by such a 'sloppy' mechanism, although physical barriers could carry specific molecular information.

NCCs grafted into the lumen of the neural tube cannot penetrate the basal lamina and thus accumulate against it. Organized extracellular matrices such as basal laminae could act as nonspecific barriers that direct cells passively, or they could provide specificity because of specific biochemical components in the cell membranes or within their pericellular matrices to which migrating populations or subpopulations of NCCs respond selectively. Blood vessels in developing Japanese quail are associated with a fibronectin-rich ECM that provides the substratum for migration. Other apparent 'physical' barriers may have similar molecular bases. Indeed, cessation of migration and localized accumulation of specific subpopulations of NCCs may be controlled in as complex a manner as are epithelial —> mesenchymal transformation and directionality of migration, sometimes by factors intrinsic to NCCs and sometimes by the extracellular environments they encounter.[33]

NCCs migrate as chains of attached cells. Hyaluronan-, chondroitin sulfate-, and aggregan-rich ECM encountered by deeply migrating NCCs inhibits the formation of cell-to-cell attachments. Consequently, chondroitin sulfate and other glycoconjugates constitute a barrier to crest cell migration. Microinjecting chondroitin sulfate or a xyloside to block chondroitin sulfate into rhombomeres of chicken hindbrains at H.H. stage 9 (immediately before migration; Fig. 3.11) inhibits NCC migration; and can result in cells moving into the neural epithelium instead of migrating laterally. Injecting vitamin A (retinoic acid) has similar effects (see Chapter 10).[34]

The Environment at the Final Destination

The microenvironment of the final site occupied by NCCs plays an important role in determining where and when they cease migrating, accumulating, and differentiating.

Versican (a large chondroitin sulfate proteoglycan with a hyaluronan-binding domain, a lectin and two Egf repeats) is expressed selectively in tissues that act as barriers to NCC migration or axonal outgrowth and is absent from tissues that are invaded by NCCs or axons. Thus, versican is found in caudal sclerotome, in the ECM surrounding the notochord, and beneath ectodermal cells. Versican inhibits migration because it inhibits molecules required for migration such as fibronectin, laminin, and type I collagen. (Tenascin and cytotactin, which facilitate migration, are found in the cranial sclerotome.) Expression of versican is reduced in mutants such as the white axolotl in which NCC migration along the dorsolateral pathway is reduced, but in which ventromedial migration is normal (Epperlein *et al.*, 2000a, 2007a).

The notochord produces a trypsin- and chondroitinase-labile molecule that inhibits NCC migration and thus prevents NCCs from colonizing the perinotochordal sheath. This is true even if the notochord is implanted ectopically *in situ* or if notochord and neural tube are extirpated or inverted dorsoventrally and

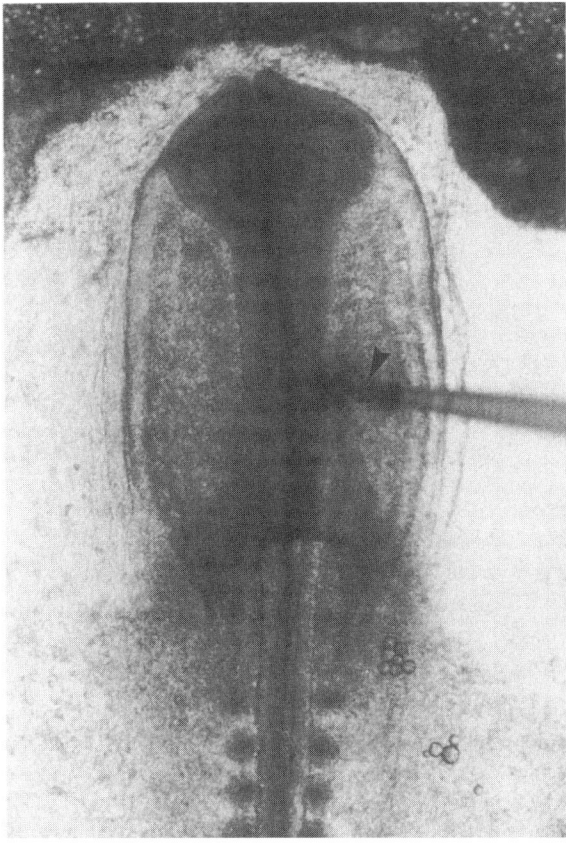

Fig. 3.11 This dorsal view of a chicken embryo shows the tip of a pipette (*arrowhead*) through which chondroitin sulfate was delivered to the rostral rhombencephalon before onset of neural crest cell migration. Reproduced from Moro-Balbás *et al.* (1998) from a figure kindly supplied by Jose Moro-Balbás. Reprinted by permission of the author and the Servicio Editorial of the International Journal of Developmental Biology

reimplanted, in which case NCCs delaminate from the new ventral (non-notochord-associated) surface of the neural tube. Aggrecan is one molecule whose removal with hyaluronidase or chondroitinase permits NCCs to invade the perinotochordal matrix.[35]

Factors responsible for the localization and accumulation of specific subpopulations of CNCCs were investigated by Peter Thorogood, who demonstrated that NCCs destined to form the cartilaginous craniofacial skeleton of embryonic chickens accumulate specifically along the neuroepithelium of the developing brain at sites rich in type II collagen. Subsequent epithelial–mesenchymal interactions at these sites elicit chondrogenesis from the accumulated mesenchymal cells.

Based on the pattern in chicken embryos, Thorogood (1993) developed a 'flypaper model' in which morphogenesis of the developing chondrocranium is the

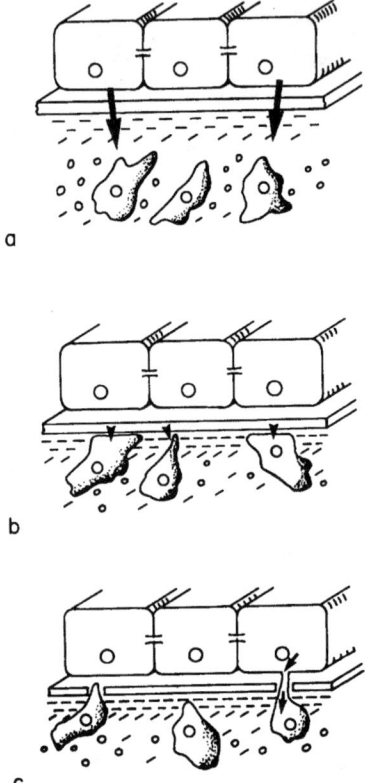

Fig. 3.12 Epithelia (shown as the connected row of three cells) may transmit signals to mesenchyme (shown as the three isolated cells) by one of three mechanisms: (**a**) long-range, diffusion-mediated interaction (*arrows*) affecting cells located some distance from the basal lamina; (**b**) short-range, matrix-mediated interaction affecting cells that approximate the basal lamina (*arrowheads*); or (c) short-range, contact mediated interaction affecting cells that have penetrated the basal lamina (*arrows*). See text for detail

result of a combination of NCC migration and neuroepithelium folding, the latter determining where type II collagen will be deposited, and therefore where NCCs will become trapped, although the 'trapping' is less random than the term applies; prechondrogenic cells express the receptor for type II collagen. The mechanisms cannot be as simple as type II-collagen binding, however. Type II collagen is more ubiquitously distributed in epithelial basement membranes, including locations such as the perinotochordal ECM where NCCs fail to (cannot?) settle down; nor can a requirement for type II collagen be present in all taxa; in *Xenopus* embryos, type II collagen is expressed after NCC migration (Seufert *et al.*, 1994).

Lethal spotted (*Ls/Ls*) mutant mice lack ganglia in the terminal 2 mm of the bowel. Mesenchyme of the presumptive aganglionic bowel prevents migrating NCCs from colonizing the bowel, resulting in the abnormal, aganglionic bowel

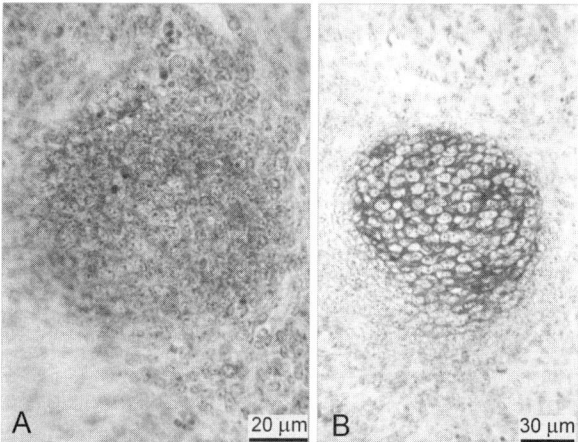

Fig. 3.13 Condensation (**A**) and chondrogenesis (**B**) as seen in a hyoid arch cartilage from mouse embryos of 12.5 days gestation (**A**) and 13.5 (**B**) days gestation. Synthesis and deposition of extracellular matrix is initiated with chondrogenesis (**B**). Modified from Hall and Miyake (1992)

segment. A close study of the basal lamina of the gut of *Ls/Ls* mutant embryos reveals abnormalities in laminin, collagen type IV, and proteoglycans from 11 days of gestation onward. Excess matrix components lead to the development of an abnormally thick basal lamina, which blocks colonization, although only in the future aganglionic or hypoganglionic regions of the bowel (Rothman *et al.*, 1993*).

NCCs from the developing quail bowel back-transplanted into a position between the somites and the neural tube reinitiate migration and colonize the neural tube, spinal ganglia, peripheral and sympathetic nervous systems, and adrenal glands, but not the bowel. Timing of entry to the gut may be critical; back-transplanting the embryonic gut wall adjacent to the neural tube enhances proliferation of neural tube but not ganglion cells, whose ability to respond is restricted temporally. Aganglionic bowel is further discussed as a human syndrome in Chapter 9.

Endothelins

The mitogenic peptide endothelin-3 (Edn3) was introduced when discussing pathways of migration of TNCCs. Edn3 promotes the differentiation of dorsolaterally migrating TNCCs as melanocytes. The endothelin family (Edn1–Edn3) is discussed in Box 3.2 in the context of their dual role in NCC migration and melanocyte differentiation, and in Chapter 9 in the context of the involvement of EDNRB in Hirschsprung disease.

As discussed in Chapter 10, the mouse mutants *Lethal Spotted* and *Piebald Lethal* display megacolon, not because VNCCs cannot migrate normally, for they can and

CONDENSATION

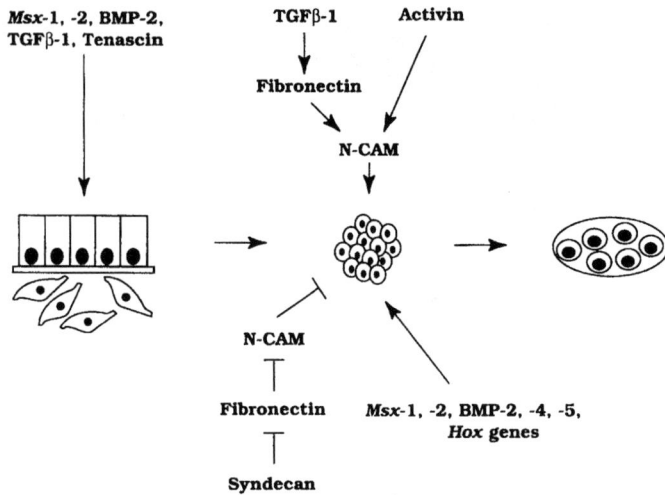

DIFFERENTIATION

Fig. 3.14 A summary of the major classes of signaling molecules associated with epithelial–mesenchymal interaction (*left*), condensation (*center*), and differentiation (*right*) of prechondrogenic cells. Signals involved in condensation are shown above the three phases of chondrogenesis; signals involved in differentiation are shown below. Epithelial–mesenchymal interactions, which are regulated by *Msx1*, *Msx2*, Bmp2, Tgfβ1, and tenascin, are responsible for formation of a condensation of prechondrogenic cells. Condensation is enhanced by upregulation of N-CAM, either directly through activin or indirectly through Tgfβ1 and fibronectin. The switch from condensation to overt differentiation of chondroblasts is controlled by two pathways. Upregulation of syndecan blocks fibronectin and N-CAM and so blocks condensation. Upregulation of *Msx1*, *Msx2*, Bmp-2, -4, and -5, and *Hox* genes, such as genes of the *Pax* family, provides positive signals for differentiation of chondroblasts. For further details, see Hall and Miyake (2000*) and Hall (2005b*)

do, but because a mutation in *Edn3* (*Lethal Spotted*) or in the gene for the receptor, *EdnRB* (*Piebald Lethal*) prevents NCCs from colonizing the bowel.[36]

Semaphorins, Delamination, and Migration

Semaphorins (Semas), a large family of secreted and membrane-bound proteins, function through a family of transmembrane receptors and guidance factors, the **neuropilins** (chiefly Nrp1 and Nrp2), which, along with **plexins** are co-receptors of semaphorins.

Distributed widely throughout the animal kingdom, semaphorins have been known for some time as important components in axon guidance and cell migration.

Fig. 3.15 This electron micrograph shows an intact basal lamina (*arrowheads*) underlying mandibular epithelial cells (*top*) that lie adjacent to mandibular mesenchyme, seen at the *bottom*

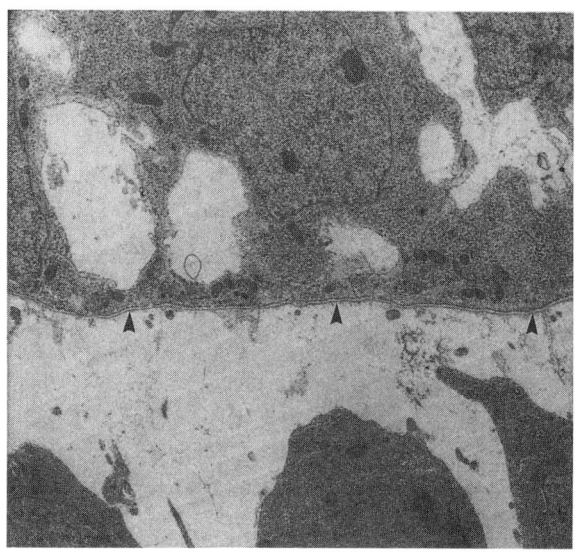

Fig. 3.16 The sheet-like basal lamina (*left*) overlying mesenchyme of the mandibular arch of a chicken embryo remains intact after the epithelium is removed using a chelating agent

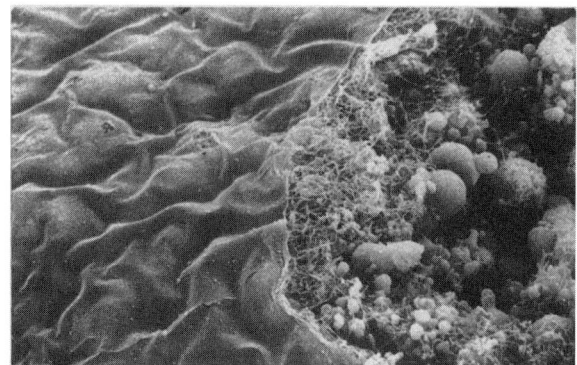

Class 3 semaphorins are secreted molecules that bind to neuropilins in the ECM, providing a substrate that guides the growth of axonal growth cones. From studies of chicken and zebrafish embryos, semaphorins and their interaction with neuropilins have emerged as guiding the delamination and migration of NCCs.

Chicken Embryos: *Sema3F* and *Sema3G* are expressed in odd-number rhombomeres in chicken embryos. In the absence of Nrp1, NCCs delaminate from these rhombomeres but fail to populate the pharyngeal arches. CNCCs expressing Nrp-1 fail to migrate on substrates containing Sema3A and fail to fully invade the pharyngeal arches *in ovo*; filopodia collapse and NCCs accumulate at the base of the arches (Fig. 3.4; McLennan and Kulesa, 2007).

Zebrafish Embryos: Two class 3 semaphorin genes, *Sema3F* and *Sema3G*, are expressed in r3 and r5 but their receptors, Nrp2a and Nrp2b, are expressed in NCCs that delaminate from r1, r2, r4, and r6.

Three lines of evidence demonstrate a role for semaphorin–neuropilin interaction in patterning CNCCs in zebrafish:

- In a mutant (*lzr/pbx4*) in which the three streams of CNCCs migrate as a single fused stream, both *Sema* genes show expanded expression throughout the region through which NCCs are migrating.
- Knockdown of *Sema* genes or of *Nrp2* restores normal patterns of migration in the mutant embryos.
- Knockdown of endogenous Sema3 ligands coupled with overexpressing Sema-3Gb produces a phenocopy of the mutant phenotype in wild-type embryos.[37]

Berndt and Halloran (2006) demonstrated a role for Sema3D as an inhibitor of the proliferation of neuroepithelial cells of the hindbrain in a gene cascade that places Sema3D downstream of Wnt/Tcf signaling, *Wnt* signaling having previously been shown to play a role in the transition from G1 to the S phase of mitosis in TNCCs. Sema3D inhibits *CyclinD1* (see Fig. 2.5), which is expressed during G, and so blocks hindbrain NCCs at G1. The results of this study are not inconsistent with the possibility that Sema3D may also regulate the cell cycle independently of Wnt signaling, although regulation of delamination by Wnt signaling is dependent on Bmp; *Wnt* functions downstream of *Bmp* signaling in regulating NCC migration at the G1-to-S transition of the cell cycle (Burstyn-Cohen *et al.*, 2004).

Mouse Embryos: CarNCC migration is severely disrupted in *Sema3C* mutant mice, which die soon after birth with cardiovascular defects equivalent to those seen after ablating the CarNC. Expression of Sema3C in the outflow tract is required for NCCs to enter the tract, which they fail to do in mutant embryos. *PlexinA2* is expressed in migratory and postmigratory CarNCCs in mouse embryos, while *PlexinA2*-positive cells develop abnormally in a number of mouse mutants with congenital heart disease.[38]

Taking advantage of *Sema3a-*, *Nrp1-* and *Nrp2*-null mice in combination with CRE recombinase, Schwarz and colleagues (2008) showed that loss of either neuropilin disrupts sensory neuronal patterning in the head, in part, because of misdirected CNN migration, and that ectopically positioned NCCs recruit sensory neurons from the otic placode and guide these neurons to lay down misdirected axonal projections, implicating Sema3a/Nrp signaling in these developmental events.

Subpopulations of NCCs

The preceding discussion of the precise localization of subpopulations of NCCs and the role of the environment in ensuring that cells arrive at precise locations and develop normally at those sites leads logically into a consideration of the control of the differentiation of subpopulations of NCCs. The last section of this chapter introduces four issues that are taken up in greater detail in Chapters 5, 6, 7, and 8 in the context of the differentiation of particular classes of NCCs. The four issues are: (i) subpopulations of NCCs; (ii) whether the fate of such cells is restricted before, during, or after migration; (iii) whether such cells are uni-, bi-, or multipotential; and (iv) the role of growth factors in eliciting a particular pathway of differentiation (or of de- and redifferentiation) from such cells.

Are NCCs:

- a homogenous multipotent population that is selected for or switched into particular pathways of differentiation? or
- a heterogeneous assembly of subpopulations, each behaving as lineage-restricted cells with perhaps only one, or at most two or three possible cell fates (uni-, bi-, or tripotential)?[39]

A third important question that has emerged over the past decade is whether NCCs are stem cells and, if so, whether stem cells persist in adult organs and can be activated in situations of repair or regeneration. As discussed at the end of Chapter 10, some recent studies regard the NC as comprised of multipotential, self-renewing populations of stem cells.

We know that the entire NC is not a single homogeneous population of cells. Although there is some overlap in their potential, cranial, cardiac, and trunk crest cells **each** form different cell types and contribute to different tissues and organs (see Chapters 1, 5, 6, 7, and 8). Heterogeneity for differentiative fate also exists within cranial, cardiac, and vagal and sacral crests, both in terms of the types of cells they can form (chondrocytes, smooth muscle cells, enteric ganglia, respectively) and in terms of the tissues or organs to which they contribute (skull, heart, peripheral nervous system, respectively). Schilling and Kimmel (1994) demonstrated this in zebrafish embryos by labeling individual CNCCs destined for the pharyngeal arches. Progeny of a single cell migrated into a single arch where they produced a single cell type. At the same time, Raible and Eisen (1994) provided evidence that most TNCCs in zebrafish are lineage-restricted before the onset of migration. These two studies provide among the best lines of direct evidence for restriction of premigratory NCCs *in vivo*.[40]

Further heterogeneity is seen with respect to morphogenesis as evidenced by the different structures produced by (what appear to be) equivalent cell types—for example, in the regionalized potential of the VNC to produce vagal or enteric ganglia (see Chapter 6), of CNCC-derived chondrocytes to produce different elements of the craniofacial skeleton (see Chapter 7), and of the odontogenic neural crest to produce incisor or molar teeth (see Chapter 8).

Heterogeneity at the level of subpopulations of NCCs is also seen in the segregation of cell lines (such as neuronal cells) within premigratory NC (as evidenced using quail/chicken chimeras and cell culture, discussed in the next section) or during and after NCC migration (as demonstrated using monoclonal antibodies against specific NCC types, as discussed in the subsequent section).[41]

Other subpopulations—for example, those forming the tissues of the mandibular arch of chicken embryos or the trabecular and pharyngeal arch cartilages in amphibians—possess regional specificity before delaminating from the neural tube. Thus, even if forced by an ingenious experimenter (Drew Noden) to migrate into the second or third pharyngeal arches, these cells still form first-arch structures, as evidenced by the development of an extra external auditory meatus and ectopic mandibular skeleton in the hyoid arch (Noden, 1983a, 1984b). The localization of such cells during normal development is controlled precisely, thereby avoiding the abnormal development that would follow from mismigration. As discussed in Chapters 9 and 10, major craniofacial defects can occur if NCC migration is disrupted.

When Noden (1983a) transplanted lower jaw (first arch)-destined NCCs into a more caudal region of the neural tube, the transplanted cells migrated into the second arch (from which the hyoid skeleton normally develops), where they formed a second set of lower jaw elements. Noden was well aware of the evolutionary implications of his results. In a letter penned to the late Peter Thorogood on April 15 1982 to accompany photographs of the results, Noden noted: 'It is important to state here that 2nd arch crest cells normally form the retroarticular process of Meckel's and the caudal part of the angular bone [in avian embryos]; these structures form in the mesenchymal "bridge" connecting 1st and 2nd arch crest mesenchyme. I'm sure you and Brian will recognize the evolutionary implications of this *vis-à-vis* the ear ossicles.' See Boxes 7.1 and 10.3 for further information on the development and evolution of middle ear ossicles, and how they demonstrate the prepatterning of NCCs.

To state that the NC consists of subpopulations does not imply that those subpopulations can only express one possible cell fate—that they are unipotential (see Table 1.2). **Importantly, identification of a cell lineage or a cell type is not necessarily a demonstration of a single fate for the cells in the lineage**. Although cells only ever do express one differentiative phenotype, many subpopulations are at least bipotential before or during migration (see Table 1.2). Some remain bipotential even after differentiating along one pathway, which is not the same as the two differentiated cells having shared a common pathway before diverging. Furthermore, and as discussed in Chapter 10, stem cells may reside within populations of otherwise restricted NCCs.[42]

Restricted Premigratory and Early Migrating Populations of TNCCs

A classic way to demonstrate the bipotentiality of a cell population is to determine whether the progeny of an individual cell can differentiate into more than one cell type. The least ambiguous way to obtain the progeny of a single cell is through clonal cell culture, although it is important to keep in mind that formation of more than one cell type *in vitro* need not mean that the cells necessarily form those cell types *in vivo*; potential exceeds actuality for many cells in early embryos. Because selection for or against particular cell types can occur *in vitro*, the full potential of a cell lineage may not be revealed *in vitro*. Nevertheless, cell culture, especially clonal cell culture, is a powerful technique with which to study cell differentiation.

In 1975, Cohen and Konigsberg devised a way to obtain clonal cultures of Japanese quail TNCCs.[43] A small proportion of the clones—each clone consisting of the progeny of a single cell—differentiated into two distinct cell types:

- adrenergic neurons, distinguished by the production of catecholamines, and
- pigmented cells, distinguished by the production of melanin.

Adrenergic neurons differentiate from TNCCs that migrate across the neural tube and through the somites, from which they receive a diffusible signal (probably Bmp4) permitting adrenergic neuronal differentiation. Adrenergic neuronal differentiation can also be elicited from cloned TNCCs not expressing a neuronal phenotype if the cells are exposed to extracellular products secreted by somitic cells or if they are grafted in association with somites. The differentiation of adrenergic neurons *in situ,* ectopically *in ovo,* and from cloned cells, demonstrates the bipotentiality of these TNCCs.[44]

A classic series of studies, performed by Bronner-Fraser and Fraser in the late 1980s, consisted of injecting single cells of the TNC of avian embryos with lysinated rhodamine dextran to monitor migration and cell fate. Single TNCCs gave rise to sensory neurons, pigment cells, ganglionic support cells (Schwann cells), adrenomedullary cells (sympathetic neurons), **and** neurons of the central nervous system (Fig. 3.17; and see Fig. 1.2). These results are fascinating for at least two reasons: they reveal the multipotentiality of single NCCs, and they reveal that NC and central nervous system neurons share a common lineage. The latter also speaks directly to the developmental origin of the NC and to its close connection to neural ectoderm, a topic discussed in Chapter 2.

This and further clonal analyses undertaken in the early 1990s established heterogeneity within a diversity of NCCs in mouse and quail embryos:

- *Premigratory or early migrating mouse* NCCs consist of heterogeneous populations of pluripotential and restricted cells.
- *Premigratory or early migrating quail CarNC* contains five clonal cell lines:

  (i)   unipotential clones that form only smooth muscle;
  (ii) unipotential clones that form only pigment cells;
  (iii) bipotential clones that form chondrocytes and sensory neurons;
  (iv) multipotential mixed clones that form pigment and other cell types;
  (v)   multipotential clones that form all these cell types except pigment cells.

- *Premigratory or early migrating quail CNCCs* (9- to 13-somite-stage) consist of:

  (i) precursors capable of producing neurons, glia, and chondrocytes, and
  (ii) small number of precursors from which neurons, glia, chondrocytes, and pigment cells arise (Fig. 3.17).[45]

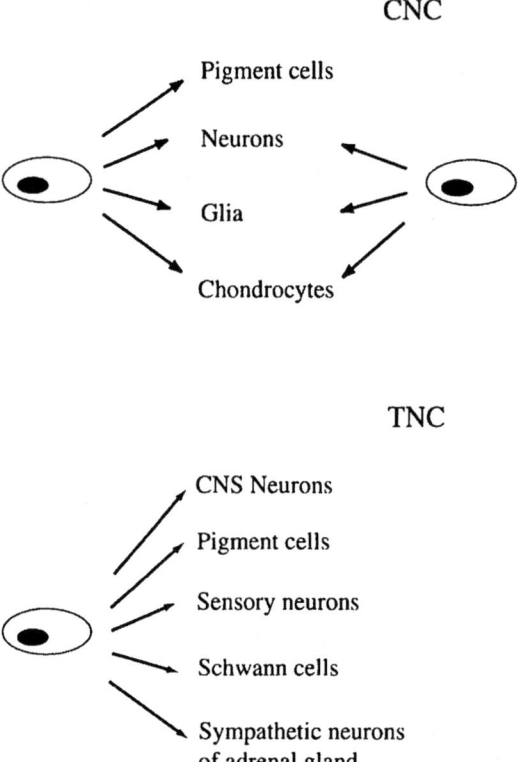

Fig. 3.17 Premigratory CNC of quail embryos consists of precursors that can produce neurons, glia, and chondrocytes and a small number of precursors that give rise to pigment cells, neurons, glia, and chondrocytes. Premigratory TNC of avian embryos contains individual precursor cells that give rise to neurons of the central nervous system (CNS), pigment cells, sensory neurons, Schwann cells, and sympathetic neurons. See text for details

The subpopulation of sympathoadrenal precursors in the NC gives rise chromaffin cells of the adrenal medulla, chromaffin cells outside the adrenal gland (extra-adrenal chromaffin cells) and to neurons of the secondary sympathetic ganglia. Injecting the proto-oncogene *V-myc* into the TNC of 10.5-day-old rat embryos immortalizes a clonal cell line that then gives rise to glial progenitors and sympathoadrenal cells.

Mouse achaete-scute homologue 1 (*Mash1*), expressed transiently in a subset of neural precursors, may be a useful marker for the sympathoadrenal lineage and may play a role in the determination of this lineage; expressing *Mash1* in *Xenopus* converts NC or epidermal ectoderm cells to a neuronal fate. *Insm1* (*insulinoma-associated 1*) encodes a zing-finger factor. *Insm1* mutant mice exhibit major underdevelopment of the sympathoadrenal lineage (sympathetic ganglia are small and adrenal chromaffin cells fail to differentiate) because of reduced proliferation and differentiation of NC precursor cells. *Insm1* has a sufficiently similar action to *Mash1* that Wildner and colleagues (2008) concluded that *Insm1* mediates some of the functions of *Mash1* on early specification of the sympathoadrenal lineage.[46]

Restriction During Migration

Clearly, premigratory NC or early migrating NCCs consists of subpopulations of cells.

How does such heterogeneity arise? Several possible mechanisms can be entertained:

- NCCs might be restricted by some intrinsic mechanism according to when they delaminate from the neural tube.
- Restriction might occur during migration (because, for example, only cells with particular potentials take particular migration routes).
- Restriction might occur during migration because cells encounter and respond to different gene products and environmental cues along different pathways.

This problem may be part and parcel of understanding how the NC itself arises. Early regionalization may be a component of the primary inductive events that segregate presumptive ectoderm into neural and epidermal ectoderm and neural crest, and that then regionalize neural ectoderm into fore-, mid-, and hindbrain and spinal cord. The potentiality of NCCs could be restricted (or further restricted) during their migration; monoclonal antibodies raised by Heath *et al.* (1992) against premigratory avian NCCs revealed heterogeneity but only after the cells had been cultured for some 15 h, which was when subpopulations began to diverge from a less heterogeneous premigratory population.

Restriction Along the Neural Axis

Perhaps not surprisingly, NCCs in different regions of the neural tube and/or from different species use a variety of mechanisms to limit their potentialities for differentiation. The sections below briefly consider such differences two major divisions of the NC: trunk and cranial NCCs.

TNCCs: Avian TNCCs following a ventral pathway of migration become neuronal and contribute to the dorsal root ganglia. Cells that take a more dorsolateral pathway differentiate as pigment cells. Late migrating cells preferentially take the dorsolateral pathway and have a more restricted set of developmental options than those cells that migrate early and can take either the ventral or the dorsolateral pathway. If early migrating cells are placed into the dorsolateral pathway, only those that would have become melanocytes survive; other cells undergo apoptosis (Erickson and Goins, 1995), indicating that selective survival is one way in which the fate of cells in particular locations is fixed.

In a seminal paper, Henion and Weston (1997) used clonal cell culture to demonstrate that most TNCCs are restricted when they emerge from the neural tube, early migrating cells forming neurons and glial cells, later migrating cells forming pigment cells. Restriction of early migrating cells as ganglionic varies along the rostrocaudal axis (see Chapter 6). Interestingly, lampreys, which lack sympathetic ganglia, lack a ventrally migrating population of NCCs, a conclusion based on the lack of HNK-1-positive cells in what would be the ventral pathway of migration for NCCs.[47]

A monoclonal antibody designated E/C 8 and directed against NC-derived dorsal root ganglia of chicken embryos identifies a subpopulation of postmigratory cells that form neurons but not pigment cells. Early in their migration, individual lineages of sensory and autonomic neurons are segregated from pluripotential cells that can form sensory neurons, autonomic neurons, or melanocytes (see Fig. 1.2). Clonal analysis of cells derived from dorsal root or sympathetic ganglia of quail embryos demonstrates:

- tripotential cells capable of forming pigment cells, sensory neurons, and adrenal medullary cells, and
- cells that are bipotential, either for sensory neuronal/adrenal medullary cells or for pigment cells/sensory neurons (Fig. 3.18).

At least some TNCCs in avian embryos retain their bipotentiality as they migrate. Fraser and Bronner-Fraser (1991) **labeled migrating cells** (in previous studies, only premigratory cells had been labeled) and showed that some cells formed sensory and sympathetic neurons, while others formed Schwann and nonneuronal cells. Furthermore, avian NCCs exposed to antisense probes against Fgf2 transdifferentiate from Schwann cell precursors into melanocytes.[48]

Early in development, sympathetic (adrenergic) neurons can be traced to a lineage that also forms melanocytes, neurons of the neural tube, and sensory neurons of dorsal root ganglia. Sympathetic ganglia contain preganglionic neurons that release acetylcholine that binds to the nicotinic acetylcholine receptors on postganglionic neurons, which then release noradrenaline (norepinephrine). Sympathetic neuronal differentiation is initiated by Bmp2 and blocked by the Bmp-antagonist Noggin. Later in development, Bmps also play roles in proliferation, apoptosis, and responsiveness of sympathetic neurons (Raible and Ragland, 2005*).

CNCCs: What applies to trunk does not apply to cranial or cardiac NCCs, alerting us to be cautious when extrapolating mechanisms from one region (subpopulation) of the NC to another region or subpopulation.

Early- and late-migrating CNCCs of chicken embryos appear to have equivalent developmental potential; exchanging early- and late-migrating cell populations does not produce the deficiencies that would be expected if cell fate were fixed early. Early-migrating cells are defined as those migrating from embryos with eight pairs of somites, late-migrating cells as those from embryos with 12 pairs of somites (Baker *et al.*, 1997). Most late-migrating cells migrate dorsally and therefore normally contribute fewer cells to skeletal tissues than do early-migrating cells.

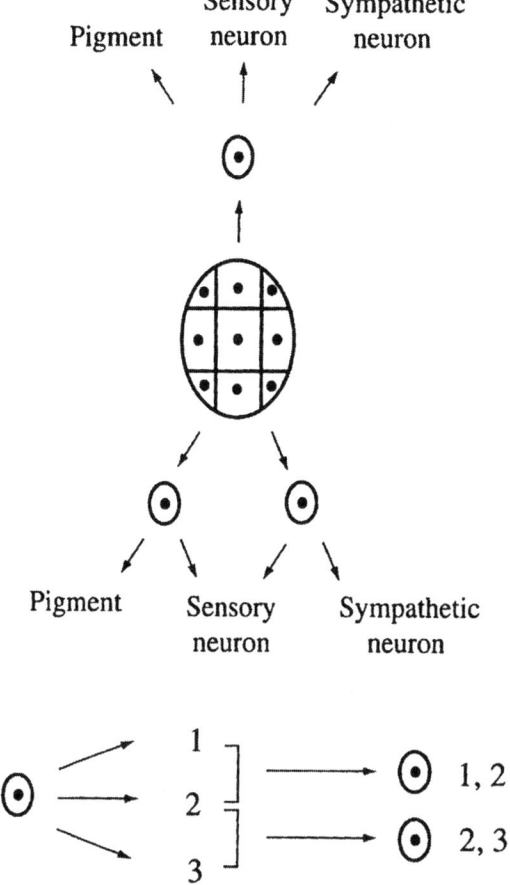

Fig. 3.18 Quail ganglia (*center*) contain cells capable of forming pigment cells, sensory and sympathetic neurons (*top*), and bipotential cells that can form either pigment cells or sensory neurons, or sensory or sympathetic neurons. These bi- and tripotential cell lineages are shown diagrammatically in the flow chart at the bottom. See text for further details

Early-migrating cells either migrate more medially and/or occupy spaces that late-migrating cells fail to fill. Chapters 5 and 7 contain further discussion of cardiac and skeletogenic CNCCs.

Differentiation

NCCs do not self-differentiate. Rather, they differentiate following interactions with transcription factors, growth factors, receptors, or other gene products that themselves are the products of other cells, often products secreted into an ECM, either a basement membrane or the matrix around mesenchymal cells. These signals may be encountered at any of the stages discussed above—within the neural tube, while NCCs are migrating, or at the site where they differentiate.

Until perhaps 10 or 15 years ago, the interactions required to initiate the differentiation of NCCs were known primarily on the basis of studies of cell-to-cell or tissue-to-tissue interactions, virtually all of which are epithelial —> mesenchymal. In the first such interaction identified and introduced in Chapter 2, amphibian pharyngeal endoderm interacts with CNCCs migrating over the developing foregut to initiate chondrogenesis. Although epithelial–mesenchymal interactions are conserved across taxa, it was discovered that NCCs at **different stages during migration interact with different epithelia in different taxa** to elicit the same differentiative fate, chondrogenesis, and to form an homologous element, Meckel's cartilage. Interactions occur between:

- migratory NCCs and pharyngeal endoderm in urodele and anuran amphibians;
- premigratory NCCs and cranial ectoderm in chicken embryos; and
- postmigratory NCCs and mandibular arch epithelium in mouse embryos.[49]

Is (are) the same gene product(s) involved in these interactions in different taxa and do similar NCC populations produce (in this case) Meckel's cartilage in different taxa? If the answer to this question is yes, then how do morphological differences evolve from ancestral conditions and persist in derived taxa as different as birds, frogs, and mice. These issues are addressed in Chapters 4 and 7.

The second of the interactions above, the matrix-mediated interactions that initiate chondrogenesis (and osteogenesis) in the craniofacial skeleton of avian embryos are outlined in Box 3.4, which contains the evidence that these are matrix-mediated, and the genes involved in signaling from epithelium to mesenchyme. From the latter, it can be seen that several classes of signaling molecules—growth factors (Bmps), homeobox genes (*Msx1*), and transcription factors (*Ap2*)—are involved in the initiation of chondro- or osteogenesis.

Box 3.3 The Mesencephalic Trigeminal Nucleus (MTN)

The first neurons to arise in the mesencephalon and the largest sensory neurons in the central nervous system are a population that forms the mesencephalic trigeminal nucleus (MTN). These, the only primary sensory afferent neurons within the amniote brain,[a] function as proprioceptive sensory neurons innervating the lower jaw, projecting to the trigeminal motor nucleus and transmitting mechanosensory input from the muscles that open and close the jaws, the extraocular muscles, and from the periodontal ligament of mammalian teeth. Their axons act as pioneering pathways for some major nerve tracts within the brain.

The MTN is found in all jawed vertebrates (gnathostomes) but not in hagfish or in lampreys (extant jawless vertebrates), raising the entrancing possibility that the evolution of the MTN was associated with the origin of sensory and integrative control required for the evolution of jaws.[b]

As is typical if not universal in the development of populations of neurons, neurons of the MTN undergo considerable programmed cell death. For example, in the Peking duck, *Anas platyrhynchos*, a population of almost 16,000 neurons at 9 days of incubation is reduced to just over 4,300 from 16 days of incubation onward, a loss of almost 70%. No further significant changes in neuron numbers occur after 16 days (Fig. 3.8).

As introduced in Chapter 3, an NC origin for the neurons of the MTN have been proposed. If they are, they are the only NCCs that do not migrate from the neural tube but remain in, or migrate into, the brainstem. On the face of it, this is unlikely behavior for NCCs, which we define as cells that arise from the NC, delaminate, and migrate away from the neural tube. As you might guess, evidence for an NC origin of the MTN is either equivocal, and/or these neurons are of NC origin in some taxa but not in others; extirpation with or without transplantation in different vertebrates (birds, frogs) having yielded conflicting results.

Experimental evidence of an NC origin of the MTN in birds comes from a single study in which quail MTN neurons developed after Japanese quail mesencephalic neural tube containing premigratory NCCs was grafted into the equivalent position in duck embryos. Grafting NCCs that had already migrated from the mesencephalon did not generate MTN neurons.[c]

More recently, Hunter and colleagues (2001) examined the MTN in chicken embryos with a battery of genes that mark NCCs—*Snail2*, *Hnk-1*, *Frzb*, *Ap2α*, *RXRγ* —none of which were expressed in the MTN. These neurons did express the anti-apoptotic transcription factor, *Brn3a*, a sensory neuron marker otherwise expressed in cells arising on either side of the dorsal midline of the neural tube. *Brn3a*-positive neurons move rostrally from the isthmus (the junction between mid- and hindbrain; see below) along the roof of the developing mesencephalon. With this in mind, and given that the quail mesencephalon grafted into duck embryos, discussed above, would have included dorsal tube, the MTN neurons that developed from the graft could have arisen from dorsal neural tube rather than from NCCs.

The **isthmus** functions as a developmental organizer and patterns the mesencephalon, a function that resides in the synthesis and secretion of *Fgf8*; *Fgf8* grafted into the caudal diencephalon functions as the isthmus and induces an ectopic midbrain. Hunter and colleagues (2001) demonstrated that the MTN develops in response to *Fgf8* secreted by cells in the isthmus; exogenous *Fgf8* enhances and inhibiting *Fgf8* reduces MTN formation. Significantly, NCCs do not develop in response to isthmus *Fgf8*. In chickens, therefore, and perhaps in all birds, the neurons of the MTN are not NC in origin but rather develop under the influence of the isthmus from neural ectoderm adjacent to the neural tube and as an integral part of the mesencephalon.

Fritzsch and Northcutt (1993*) proposed that R–B neurons in lampreys and amphioxus may be evolutionary precursors of MTNs, especially if R–B neurons are derivatives of lateral neural ectoderm and not NC (see Chapter 6). Equally parsimonious at this stage is the possibility that NCCs, R–B and MTN neurons had either a single origin in basal chordates or independent but contemporaneous or near-contemporaneous origins, although the presence of R–B neurons in amphioxus could be taken as indicating that R–B neurons preceded the origin of NCCs. The cells in the MTN, their relationship to the NC and placodal cells that form the trigeminal ganglion, and their pattern of migration, all deserve further study.[d]

---

[a] All others neurons in amniote brains are outside the medulla. Rohon–Béard neurons, which also transduce mechanosensory information, are also intramedullary

[b] Once thought to be primitive vertebrates, little changed since the appearance of the first jawless vertebrates in the Ordovician some 500 million years ago, lampreys and hagfish (Box 7.2) are as diverse, highly specialized and well adapted to their modes of life as we are to ours (Hardisty, 1979; Jørgensen et al., 1998). Both are speciose groups; 45 species of lampreys and 69 species of hagfish (hags).

[c] Narayanan and Narayanan (1978) transplanted the mesencephalic neural tube, Baker et al. (1997) grafted the migrating NCCs.

[d] See Rao and Jacobson (2005*) for a summary of earlier work on the MTN and Hunter et al. (2001*) and Petrosino et al. (2003*) for studies on chicken and duck embryos. Suggesting an even greater source of variation in origins, Sohal et al. (1996) and Stark et al. (1997) showed that the ventral neural tube provides an additional source of cells to the trigeminal ganglion in chicken embryos (see Box 6.3).

Box 3.4 Matrix-mediated interactions initiate chondrogenesis and osteogenesis

The discovery that chondrogenesis in amphibian pharyngeal arches is initiated only following interaction between pharyngeal endoderm and NC-derived mesenchyme (this chapter) has been augmented by a considerable body of knowledge demonstrating that all craniofacial cartilages *and bones* begin to differentiate (i.e., chondroblasts or osteoblasts form) as a consequence of one or more epithelial–mesenchymal interactions. We have perhaps the greatest knowledge for chicken embryos, for which we can identify the epithelia that evoke the differentiation of all the major cartilages and bones of the craniofacial skeleton (Table 3.2). As shown in Table 3.3, the timing of each of these interactions is tightly regulated, reflecting the stage during CNCC development at which the interaction occurs (Table 3.2).

Table 3.2 Epithelial–mesenchymal interactions that initiate the differentiation of cartilages and bones of the avian craniofacial skeleton[a]

Skeletal element	Epithelium	Timing of the interaction (days of incubation)
		Cartilages
Meckel's	Dorsal cranial ectoderm	1.5
Scleral	Pigmented retinal epithelium	2–3
Otic capsule	Otic vesicle	3–5.5
		Bones
Basisphenoid	Rhombencephalon, notochord	2–2.5
Parasphenoid	Notochord	2–2.5
Squamosal	mesencephalon	2–2.5
Occipital	Rhombencephalon	2–2.5
Parietal	Mesencephalon, rhombencephalon	2–2.5
Frontal	Prosencephalon, mesencephalon	2–2.5
	Cranial ectoderm	3.5–7
Maxilla	Maxillary	3–4
Mandible	Mandibular	3–4.5
Palate	Palatal	5–6
Scleral ossicles	Scleral	7–10

[a] The interactions are arranged in the temporal sequence in which they occur in the embryo. Note that two epithelia are involved in the differentiation of the basisphenoid and that two separate epithelial–mesenchymal interactions are involved in the differentiation of the frontal. See Hall (1987) for details and literature.

Modes of Interaction

Epithelia can transmit signals to adjacent mesenchymal cells using one of three modes of cellular transmission outlined below and in Fig. 3.12:

(1) **Diffusion-mediated interaction,** by producing a diffusible molecule to which cells up to 300 μm away respond (Fig. 3.12a).

(2) **Matrix-mediated interaction,** by depositing a molecule(s) into the basal lamina to which mesenchymal cells respond after establishing close contact with the basal lamina (Fig. 3.12a).

(3) **Cell contact-mediated interaction** after epithelial cells establish direct contacts with mesenchymal cells through perforations in the basal lamina and form gap junctional connections with mesenchymal cells (Fig. 3.12).

Matrix mediation is the most common basis for the epithelial —> mesenchymal interactions responsible for initiating chondro- or osteogenesis. Table 3.3 lists those interactions that are understood in the greatest depth.

Matrix-Mediated Interactions

Mesenchymal response to epithelial signaling is a proliferative one, resulting in the formation of a **cellular condensation**. The evidence that an interaction has occurred is thus the formation of a condensation (Fig. 3.13). Differentiation is initiated within, and a skeletal element arises from, the condensation.

The three essential processes between epithelial signal and overt differentiation therefore are:

(i) epithelial–mesenchymal interaction —> (ii) condensation —> (iii) differentiation,

each of which is under the control of separate (but overlapping) genes and gene networks.

The signaling molecules involved are most well understood for chondrogenesis. Figure 3.14 summarizes condensation of prechondrogenic cells and the switch from condensation to overt chondrogenesis.

What of the differentiation of osteoblasts and the deposition of bone in avian embryos? The signal provided by the mandibular epithelium to initiate osteogenesis (Table 3.3) is also a proliferative one (see below); enhancement of mesenchymal proliferation and condensation formation beneath and not between epithelial scleral papillae are the initial steps in the differentiation of the scleral bones that surround the eyes of embryonic chickens.[a]

Several independent lines of evidence from my laboratory demonstrated that the osteogenic interaction between mandibular epithelium and mesenchyme is mediated by a component(s) in the epithelial basal lamina, that is, is matrix mediated (Fig. 3.12a)[b]:

- A continuous and intact basal lamina (Fig. 3.15) rules out direct cell-to-cell contact.
- Mesenchymal cells and cell processes flatten out along some 60–70% of the basal lamina.
- Recombining mandibular mesenchyme and epithelium across Millipore or Nuclepore filters demonstrates that bone only forms if filters are less than 10 μm in thickness and have pores small enough to allow cell processes to penetrate, eliminating a diffusion-mediated interaction.
- Mandibular mesenchymal cells differentiate into osteoblasts and deposit bone if cultured on ECM previously deposited by mandibular epithelia.
- Mesenchyme separated from the epithelium using a chelating agent to retain the basal lamina on the mesenchyme (Fig. 3.16) differentiates as osteoblasts and deposits bone.

Genes-Mediating Interactions

Bmps

Growth factors bind to the basal lamina or to structural components of basal laminae. Some of these growth factors contain Egf repeats could function as mitogens and so are candidate signaling molecules (Hall and Coffin-Collins, 1990).[c] Bmp, a mitogen that acts as an inducer of cartilage and ultimately of bone from mesenchymal cells, is localized at the epithelial–mesenchymal interface when avian mandibular bone is elicited. Ekanayake and Hall (1997) and Barlow and Francis-West (1997) demonstrated that Bmp2 and Bmp4 are involved in the morphogenesis of skeletal tissues of the mandibular and other facial processes. Furthermore, we now have studies consistent with alteration in *Bmp* signaling during evolution during changes in beak shape in birds, presence or absence of teeth in teleosts, and correlating mechanical forces associated with feeding in teleosts (Box 3.5).

Msx1

Bmps exert their control over differentiation by upregulating *Hox* genes such as *Msx1,* which is expressed medioventrally in quail mandibular processes early in development and then in preosteogenic cells. Bmp-mediated *Msx1* signaling is required for the epithelial–mesenchymal interaction that initiates differentiation of mandibular membrane bone. Similar signaling operates during tooth development in mouse embryos in which *Msx1* is expressed in neuroepithelium, Rathke's pouch, limb bud mesenchyme, the NC, and in NC-derived craniofacial mesenchyme at sites of known epithelial–mesenchymal interactions for bone and tooth development; Bmp4 can substitute for dental epithelium and upregulate *Msx1* in dental mesenchyme (see Box 8.1).[d]

Ap2

Mice that are null mutants for the transcription factor *Ap2* have defective midline fusion, underdeveloped mandibular skeletons and abnormal cranial ganglia, and lack the malleus and incus of the middle ear.[e] Apoptosis is also increased in brain and proximal arch mesenchyme at 9–9.5 days postconception, often resulting in a single maxillary-mandibular element.

The neural folds fail to close in these mutants, although NCC migration is normal. *Ap2* is expressed in neural ectoderm, NC, and facial ectoderm, and may be a useful marker for lineage-related epidermal derivatives: Ap2α and Ap2γ are required to specify the NC in zebrafish; eliminating both eliminates all NCCs (Li and Cornell, 2007; and see Fig. 2.4). *Ap2* is also expressed in ganglia and facial, limb, and kidney mesenchyme, so its function is not unique to ectoderm or ectodermal derivatives. Nevertheless, with care, it could be used as a marker for early ectodermal derivatives.

Inhibition of mandibular skeletogenesis following normal NCC migration, and involvement of *Ap2* in the determination of basement membrane components, are consistent with the mutant gene acting on the epithelial–mesenchymal interactions responsible for the differentiation of murine mandibular bone and cartilage. Inca, a novel, p21-associated kinase-associated protein is expressed in premigratory and migratory NCCs. Induced in NCCs by *Ap2*, Inca is required for NCCs to condense as skeletal primordia, apparently regulating the actin cytoskeleton.[f]

Shh

The molecular basis of the role of pharyngeal endoderm is being revealed, especially in avian embryos from which pharyngeal endoderm has been removed or into which pharyngeal endoderm has been transplanted. The major findings are that:

- Transplanting future pharyngeal endoderm into the pathway of NCCs migrating into the first pharyngeal arch results in embryos with duplicated lower jaws,
- Removing pharyngeal endoderm before the seven-somite stage (but not between the eight and ten somite stages) results in failure of development of the lower jaw,[g]
- Lack of the lower jaw after endoderm is removed is not because of arrested NCC migration. Migration is normal but massive cell death removes the NCCs.
- Cell death is not initiated after removing the pharyngeal endoderm from chicken embryos with between eight and ten pairs of somites (see the second point above).
- In the latter embryos, expression of *Shh* spreads to include the excised area, implicating *Shh* from pharyngeal endoderm to prevent cell death in these CNCCs, that is, is required for their survival. This conclusion is reinforced by the finding that lower jaw development is rescued in embryos given exogenous *Shh* after pharyngeal endoderm is removed at the seven-somite stage, the development of specific skeletal elements being dependent on specific regions of the endoderm.[h]

PdgfRα

Expression of *PdgfRα* in pharyngeal endoderm and in pre- and postmigratory neural crest is consistent with Pdgf being involved in signaling cartilage differentiation in *Xenopus*; a recent analysis indicates that the *PdgfRα* gene in zebrafish is regulated by a microRNA (Mirn140), whose disruption results in cleft palate (Eberhart *et al.*, 2008).

In mice, PdgfRα is expressed in mesodermal and NC mesenchyme and in such ectodermal derivatives as the lens and choroid plexus; the mutation *Patch* (*Ph*), a deletion of the gene for PdgfRα, is associated with mesodermal and NC deficiencies—not because of lack of formation or proliferation of NCCs but because of lack of cell survival and inability to deposit ECM. PdgfRα and its receptor are expressed in separate but adjacent cell layers (Pdgf in epithelial cells, the receptor in mesenchymal cells), in regions that depend on epithelial–mesenchymal interactions for their initiation such as pharyngeal arches, the otic vesicle, sclerotome, hair, and mammary glands.[i]

---

[a] See Hall and Miyake (2000), Miyake *et al.* (1996), and Hall (2003a*, 2005b*, 2007) for the importance of condensation in development and Atchley and Hall (1991), MacDonald and Hall (2001), Schlosser (2002c), Hall (2003a-c*), and Gass and Hall (2007) for condensations as modules of evolutionary change. See Dunlop and Hall (1995) and Hall (2005b*) for condensations and osteogenesis.

[b] For overviews of the research on which this conclusion is based, see Hall (2005b*).

[c] NCCs from Japanese quail possess receptors for Egf (10^5 receptors/NCC) and Egf stimulates the release of proteoglycans and hyaluronan (and so ECM stability) and the incorporation of 3H-thymidine into NCCs (and so cell division).

[d] For patterns of expression and the role of *Msx1* in avian craniofacial development, see Y. Takahashi and Le Douarin (1990), Takahashi *et al.* (1991), and Mina *et al.* (1995). See Hall (1980) for the induction of murine mandibular bone, MacDonald and Hall (2001) for altered timing of osteogenic induction between inbred strains of mice, MacKenzie *et al.* (1991) for murine *Msx1*, Mahmood *et al.* (1996) for *Fgf* expression.

[e] See Box 7.1 for an evaluation of gene knockouts in the context of the evolutionary origin of middle ear ossicles.

[f] See Morriss-Kay (1996) and Zhang *et al.* (1996) for the transcription factor *Ap2α* and T. Luo *et al.* (2007) for Inca.

[g] Similarly, in the *Casanova* and *Van Gogh* mutants in zebrafish, which lack endoderm, the visceral skeleton fails to develop, although the neurocranium develops normally (Piotrowski and Nüsslein-Volhard, 2000; Holzschuh *et al.*, 2005*).

[h] See Couly *et al.* (2002) and Britto *et al.* (2000*, 2006*) for *Shh* and NCC death, and Le Douarin *et al.* (2007) for recent studies in chicken embryos showing a central role for *Shh* in facial and brain development.

[i] See Soriano (1997) for expression of PDGFα and its receptor in mesenchyme and/or in adjacent cell layers, and Tallqvist and Soriano (2003) for pharyngeal arch defects with loss of the receptor in chimeric mice.

Box 3.5 Bmp4, bird beaks, and fish teeth

Darwin's Finches

A canonical example of morphological change in evolution is the transformations of the 13 species of Darwin's finch on the Galapagos Islands. Perhaps the most obvious morphological change is the differences in the shapes and sizes of the beaks of different species on different islands. From direct observations of morphological change over a relatively small number of generations we know that variation in beak shape arise in response to environmental or dietary changes (Grant and Grant, 2002).

As lower jaws, beaks arise from CNCCs. As discussed in Box 3.4, *Bmp4* plays a critical role in directing the differentiation and morphogenesis of skeletogenic CNCC. Abzhanov and colleagues (2004) demonstrated that the size and shapes of the beaks of Darwin's finches correlate with the amount and pattern of expression of *Bmp4* in beak primordia. The obvious interpretation is that *Bmp4* has played a key role in the evolution of beak shape and size in Darwin's finches.

Fish

Fish use their jaws and their teeth during feeding. *Bmp4* is conserved in the tooth-forming regions of the jaws but absent from the toothless regions in three species of teleosts studied by Wise and Stock (2006): zebrafish, Japanese medaka, and Mexican tetra. The interpretation is that *Bmp4* signaling is responsible for tooth formation and integrates jaw and tooth development.

Cichlid fish in Lake Malawi in Africa are specialized to catch and consume an enormous range of prey items. Embryos of species that feed by biting and crushing algae have higher expression of *Bmp4* in their jaw primordia than do species that feed on plankton they suck from the water (Albertson *et al.*, 2005), consistent with robust jaws developing in the presence of higher levels of *Bmp4*.

The next and final section of the chapter introduces situations in which signaling molecules, often growth factors, are responsible for eliciting a particular pathway of differentiation from bipotential NCCs.

Table 3.3 Three epithelial–mesenchymal interactions in chicken embryos initiating mandibular and scleral chondrogenesis and osteogenesis[a]

Epithelium		CNCCs		Condensation		Skeletal element
Mandibular						
Epidermal ectoderm	—>	Premigratory CNCCs[b]	—>	Meckelian chondrogenic	—>	Meckel's cartilage
Mandibular arch	—>	Post-migratory mandibular[c]	—>	Mandibular osteogenic	—>	Mandibular membrane bones.
Periocular						
Pigmented retinal	—>	Post-migratory periocular	—>	Scleral chondrogenic	—>	Scleral cartilage
Scleral papillae	—>	Post-migratory periocular[d]	—>	Scleral osteogenic	—>	Scleral ossicles

[a] See Hall (2005b*) for epithelial–mesenchymal interactions, and Hall (2000c) for how to set up such tissue recombinations.
[b] Epithelial signaling occurs early in development when the interaction is with premigratory CNCCs (1.5 days of incubation).
[c] Interaction occurs progressively later in development when postmigratory CNCCs are involved: 2–4.5 days for mandibular membrane bone, 2–3 days for scleral cartilage, and 7–10 days for scleral ossicles.
[d] Groups of scleral ossicles are induced progressively between 7 and 10 days of incubation as scleral papillae arise sequentially around the circumference of the developing eye.

Differentiation of Bipotential Cells

Products localized within extracellular matrices modulate the differentiation of bipotential cells; TNCCs differentiate into pigment cells if grown on plastic but into catecholamine-containing adrenergic neurons if cultured in a media conditioned by somitic cells. During the normal development of Mexican axolotl embryos, components of the ECM along the dorsolateral migratory route elicit the differentiation of pigment cells, specifically melanophores and xanthophores, which in *Xenopus* — as determined by clonal cell culture—can arise from separate lineages or from a common cell lineage. One active factor is melanocyte-stimulating hormone (Msh), which, by acting through a cAMP-mediated pathway, regulates melanocyte differentiation in a dose-dependent manner.[50]

Even if specified as melanocytes before migration, labeled NCCs grafted back into the migration pathway still require specific environmental signals such as the mitogen endothelin (Edn3) to promote differentiation (see Box 3.2). Such restriction is reflected in the decline in the proportion of pluripotential NCCs in the skin as embryos age; the fate of pluripotential cells is determined late and in response to local environmental signals. In Japanese quail, for example, 20% of colonies established from the skin of H.H. stage 21 embryos form mixed populations of cells (melanocytes, sympathoadrenal cells, and sensory neurons), but only melanocyte colonies arise from cells isolated from older (H.H. stage 30) embryos.

From her studies on neuronal and melanocytic differentiation, Sieber-Blum (1990) proposed the model of two classes of signals: (i) positive signals for commitment of cell fate and (ii) negative signals to restrict cell fate. Numerous studies bear out this model:

- *Wnt* and *Bmp* pathways act antagonistically to specify melanocyte or glial cell fates: Wnt specifying melanocyte at the expense of glial or neuronal fates and Bmps repressing melanogenesis and promoting neurogenesis.
- A transient activation of *Notch* in mouse embryos irreversibly switches NCC precursors from neurogenesis to forming glial cells, overcoming *Bmp2*, which otherwise would retain the cells as neurogenic.
- *Notch* signaling determines whether NCCs within embryonic chicken DRG will become neural or glial.
- *Edn3* regulates the switch between glial and pigment cells.

As already mentioned, the signaling systems are even more complex, multilayered, interactive, and species specific than anticipated in 1990. For example, while *Edn3* can transform glial cells into melanocytes precursors in chicken embryos, its primary role *in ovo* is in NCC proliferation and migration.[51]

A Role for Growth Factors

Growth factors play important roles in regulating the pathways of differentiation expressed by NCCs (see Box 3.4). For example, in mice **Tgfβ** is distributed in NC and other mesenchymal derivatives in positions and at times when the epithelial–mesenchymal interactions required for the initiation of differentiation occur. NCCs themselves are sources of growth factors; CNCCs synthesize and secrete a latent Tgfβ that is activated by proteolysis. Hormones also enhance and stabilize chosen pathways of differentiation.[52]

The **nerve growth factor neurotrophin3** (Nt3), a product of the central nervous system, is present in the neural tube before NCC migration. Nt3 is mitogenic for NCCs and influences the survival and/or differentiation of postmitotic neuronal precursors. **Brain-derived neurotrophic factor** (Bdnf) and **nerve growth factor** (Ngf) promote the differentiation of sensory neurons from pluripotential NCCs maintained in a clonal culture. Ngf also switched medullary cells of the adrenal gland to sympathetic neurons. These pluripotential cells, which can give rise to sensory or autonomic neurons or to melanocytes, are also sensitive to other signals such as **neurotransmitters**; norepinephrine uptake inhibitors, such as lidocaine and chlorpromazine, inhibit expression of the adrenergic phenotype, implicating norepinephrine as a signaling molecule in adrenergic neurogenesis.[53]

NCCs can respond to multiple growth factors, each of which elicits a different cell fate: murine NCCs differentiate as glia, smooth muscle, or autonomic neurons under the influence of **glial growth factor** (Ggf), Tgfβ, and Bmp2/4, respectively;

Other gene products operating on other populations of NCCs confirm the bipotentiality of the cells and the role of environmental cues in switching cells from one pathway to another. Examples include the transformation:

- from adrenergic to cholinergic neuronal differentiation in response to a soluble factor released by heart cells or as a consequence of grafting NCCs to different locations along the neural axis, exposing cells that would normally form adrenergic neurons to signals that switch them into cholinergic neurogenesis;
- of avian periocular mesenchymal cells from chondrogenesis to neuronal differentiation if associated with embryonic hindgut, a tissue that they would normally not 'see' *in ovo* but to which they can respond; and
- of periosteal cells on avian membrane bones from osteogenesis to chondrogenesis at joints, where ligaments or muscles insert or attach, or in fracture repair, where mechanical conditions favor cartilage formation by up- or downregulating N-CAM.[54]

These examples—and there are many more in Chapters 5, 6, 7, and 8—show that NCCs can express more than one cell fate *in vivo* and can be 'made' to express cell fates *in vitro* cell fates that they normally do not express *in vivo*. In most NCCs, potential exceeds the actual pathway of differentiation taken.

Dedifferentiate and Redifferentiate

Expressing an alternate cell fate **after** initially differentiating along another pathway requires an individual differentiated cell such as a chondrocyte or neuron to **dedifferentiate** and then to **redifferentiate** into a different cell type; differentiated neuronal cells from the dorsal root ganglion can be transformed into pigmented cells in response to the phorbol ester 12-0-tetradecanoylphorbol-13-acetate.

Often the switch in pathway of differentiation occurs in response to a growth factor. For example, exposing dorsal root ganglia to growth factors changes cell fate; Fgf2 causes some 20% of dorsal root ganglia cells to transform into pigment cells, a transformation consistent with at least some NCCs dedifferentiating and redifferentiating, although we do not know how far along the initial differentiation pathway the dorsal root ganglia neurons were when Fgf2 was introduced. The demonstration that Tgfβ1a blocks this transformation reaffirms that individual NCCs are responsive to a variety of signals.[55]

An alternative means of achieving the same ends as the dedifferentiation of differentiated or partly differentiated cells and their redifferentiation along a new pathway would be to retain bipotential progenitor cells or **stem cells** within a given population of NCCs. Although, little has been said about stem cells thus far, whether NCCs are stem cells is discussed in the last section of Chapter 10.

Summary

At this stage in our discussion and before we move into discussing in Chapters 5, 6, 7 and 8, the various cell types arising from NCCs, it might be helpful to summarize what has been said so far about the existence and differentiation of bipotential NCCs.

Earlier, an evidence was outlined for the conclusion that many subpopulations of NCCs are at least bipotential before or during migration. As described above, some of these cells remain bipotential even after differentiating along or part way along one pathway. NCC can be cloned, and the clones subdivided and produce more than one cell type. Cells derived from the dorsal root or sympathetic ganglia of quail embryos are bi- or tripotential, forming pigment cells/sensory neurons, sensory neurons/adrenal medullary cells, or all three (Fig. 3.18). Growth factors, extracellular matrix molecules, or genes such as *Wnts* can and do modulate the differentiation of bipotential cells. The picture of NCC differentiation that emerges is of lineages of cells:

- some of which are present in the premigratory NC;
- others of which appear early during NCC migration;
- many if not all of which are at least bipotential; and
- in which the particular pathway of differentiation expressed depends on interactions with ECM and cell products such as growth factors, hormones, and neurotransmitters encountered before, during, or after cell migration.

When did cells with such astonishing properties of migration and differentiation along multiple lines arise? Can we identify their precursors in the ancestors of the vertebrates? Which organisms were the ancestors of the vertebrates? What might we expect the properties of a NCC precursor to be? What genes would it express? These questions are taken up in the next chapter.

Notes

1. Erickson (1993b), Newgreen (1995*), and Bronner-Fraser (1995) provide excellent and extensive reviews of early studies on crest migration.
2. As noted in Chapter 6, placodal ectodermal cells in chicken embryos can delaminate without undergoing an epithelial-to-mesenchymal transformation.
3. See O'Rahilly and Müller (2006*, 2007*) for migration of NCCs in human embryos and for detailed analysis of the major subdivisions of the human NC, and Peterson et al. (1996) for the long-tailed macaque. For timing of onset of NCC migration, see Nichols (1986), Tan and Morriss-Kay (1986*), and Smits-van Prooije et al. (1988*).
4. For scanning electron microscopic analysis of mammalian NC, and for timing of the onset of NCC delamination, see Erickson and Weston (1983*), Hall (1999a*), and Le Douarin and Kalcheim (1999*).
5. These studies were published in a series of papers in the 1980s, as an atlas of developmental stages in 1994, and as a revised atlas and detailed paper (O'Rahilly and Müller, 2006*, 2007*), currently the best treatments of human brain development available.
6. See Nichols (1987) for pioneering studies on the delamination of mammalian NCCs, and Blankenship et al. (1996) and Peterson et al. (1996) for studies on the long-tailed macaque. For breakdown of the basal lamina in association with epithelial —> mesenchymal transformation and delamination, see Erickson and Weston (1983) and Erickson (1993a).
7. See Akitaya and Bronner-Fraser (1992) and Shoval et al. (2007) for the distribution of N-CAM and Cad2, Bronner-Fraser et al. (1992) for the use of antibodies to explore the roles of cell adhesion molecules in the migration of CNCCs, Shoval et al. (2007) for interaction of Bmp4 and Cad2 in epithelial —> mesenchymal transformation, and Borchers et al. (2001) for *XCad11*.
8. See Savagner (2001*) for Rho GTPases and Zondag et al. (2000) for *Ras–Rho* interactions.
9. See Shankar et al. (1994) for retention of N-CAM by NCCs in response to retinoic acid, and Nakagawa and Takeichi (1998) and Shoval et al. (2007) for studies with *Cad2* and *Cad7*.
10. See Taneyhill et al. (2007) and Coles et al. (2007) for Cad6B as a *Snail2* target, and for overviews.
11. See Epperlein et al. (2000) for the hyaluronan study, Falck et al. (2002) for the patterns of delamination and migration of CNCCs in the Mexican axolotl, Spicer and Tien (2004*) for an overview of hyaluronan and its multiple roles, and Camenisch et al. (2000) for the hyaluronan synthase knockout.
12. See Menoud et al. (1989) for plasminogen activator, and Newgreen and Minichiello (1995) for protein kinase inhibitors and epithelial —> mesenchymal transformation.
13. See Holmdahl (1928) and Bartelmez (1960*, 1962*) for pioneering studies on the mammalian NC, Holmdahl (1928) and O'Rahilly and Müller (2006*, 2007*) for studies on human embryos, and Hall and Hörstadius (1988*) for discussions of the earlier studies.
14. For migration of TNCCs through the somites, see Stern and Keynes (1987). For the migration of determined melanocytes along dorsolateral pathways associated with ectoderm, see Erickson and Goins (1995), Reedy et al. (1998), and Harris and Erickson (2007).
15. See Weston (1970*), Noden (1987*), Le Douarin (1982), Graveson et al. (1995), and North-cutt (1996) for the control of directionality.
16. See Yip (1986) and Teillet et al. (1987) for the initial studies on regionalization of ganglionic primordia, and see Kasemeier-Kulesa et al. (2005) for studies based on time-lapse confocal microscopy. See Kulesa et al. (2005) for rerouting of hindbrain NCCs during regulation in chicken embryos, George et al. (2007) for colonization of dorsal root ganglia on the opposite

sides of chicken embryos, and Vaglia and Hall (1999) for regulation in zebrafish and for an overview of regulation.

17. See Collazo et al. (1993) for single cell labeling of Xenopus, Sobkowa et al. (2006) for the transgenic axolotl studies, and Epperlein et al. (1996*, 2000a, 2007a*,b*) for reviews of the axolotl studies.

18. See Sadaghiani et al. (1994*) for platyfish and Japanese medaka, and Hirata et al. (1997) for the studies on the swordtail.

19. See Thisse et al. (1995) and Jesuthasan (1996) for studies with the Spadetail mutant.

20. See M. M. Smith et al. (1994) for the DiI-labeling, and Abe et al. (2007) for Fgf.

21. For the studies on the macaque, see Blankenship et al. (1996) and Peterson et al. (1996). See Poelman et al. (1990) for attachment of fragments of basal lamina to migrating NCCs.

22. See Newgreen (1995*) and Tucker (2004*) for ECM components and migration.

23. For the composition of the ECM through which NCCs migrate, see the literature discussed inDuband et al. (1986), Erickson and Perris (1993), Newgreen (1995*), Hall (1999a*), and Le Douarin and Kalcheim (1999*). For modification of that ECM by migrating NCCs, see Brauer and Markwald (1987). For the binding of growth factors, such as Tgfβ1 and Bmp2, to collagen IV within basement membranes, see Vukicevic et al. (1994*).

24. See Duband et al. (1986*) and Loring and Erickson (1987) for the role of fibronectin, and Bronner-Fraser and Stern (1991) for migration through sclerotome.

25. For binding of NCCs to fibronectin, see Duband et al. (1986*) and Krotoski et al. (1986).

26. See Bronner-Fraser (1986) for coated latex beads and Pomeranz et al. (1993) for the synthetic decapeptide.

27. See Tucker et al. (1999) for the studies with chicken and Japanese quail embryos and Tucker (2004) for an overview.

28. See Kahane and Kalcheim (1994) and Henion et al. (1995) for trkC in avian embryos, and Kalcheim et al. (1992) for Nt3 as mitogenic.

29. See Pasquale (2008*) for an overview of Eph–Ephrin bidirectional signaling, Davy and Soriano (2007) for double knockout of EphRB2 receptors, and Santiago and Erickson (2002) the role of EphB and EphB receptors in migrating premelanocytes and melanocytes.

30. For data on ECM control over migration of amphibian NCCs and region-specific differentiation in wild type and albino, see Brauer and Markwald (1987), Olsson et al. (1996*), Parichy (1996) and Epperlein et al. (1996, 2007a*). For the older literature on pigment patterning in amphibians by Twitty, see Hörstadius (1950, especially pp. 75–78) and Hall (1999a*).

31. See Krotoski and Bronner-Fraser (1990), Goh et al. (1997), Lallier and De Simone (2000), and Testaz and Duband (2001) for the distribution of integrins, integrin ligands, and receptors, and for their role in Xenopus neurulation, TNCCs, migration, and apoptosis of NCCs.

32. See Kerr and Newgreen (1997) for the binding properties of fibronectin, tenascin, and chondroitin sulfate proteoglycan, and Perris et al. (1991*) for the effects of collagens on NC and sclerotomal migration.

33. See Erickson (1987), Erickson and Perris (1993), Newgreen (1995*), and Erickson and Goins (2000) for barriers to migration; Morris-Wiman and Brinkley (1990) for hyaluronan, early crest migration, and elevation of the neural folds; Forgacs and Newman (2005*) for physical mechanisms associated with morphogenesis and pattern formation; and Spence and Poole (1994) for fibronectin in the migration pathway and as the basis for vessel-mediated migration.

34. See Newgreen (1995*) and Tucker (2004*) for ECM components and migration, and Moro-Balbás et al. (1998) for the injection of chondroitin sulfate or retinoic acid into rhombomeres.

35. See Teillet et al. (1987) for migration of ganglionic precursors through the rostral halves of somites, Erickson et al. (1989) and Le Douarin and Kalcheim (1999*) for matrix products in the rostral sclerotome, and Pettway et al. (1990) and Stern et al. (1991) for the role of the notochord and perinotochordal matrix. PNA is also used as a marker of the condensation phase of mesenchymal cell differentiation (Miyake et al., 1996, Hall and Miyake 2000*).

36. See Rothman *et al.* (1993) and Puffenberger *et al.* (1994) for *Piebald Lethal* and *Lethal Spotted*.
37. See Pasterkamp and Kolodkin (2003*) for an overview of the functions of semaphorins, and Yu and Moens (2005), Berndt and Halloran (2006), and McLennan and Kulesa (2007) for the role of semaphorins and neuropilins in NCC migration in zebrafish and chick.
38. See Feiner *et al.* (2001) for *Sema3C*, Brown *et al.* (2001) for *PlexinA2*, and Chapter 8 for the CarNC.
39. See Le Douarin (1986), Anderson (1997), Sieber-Blum (1990), and Bronner-Fraser (1995) for the heterogeneity of premigratory and early migrating NCCs.
40. See Dorsky *et al.* (2000) and Raible and Ragland (2005*) for models for the specification of premigratory NCCs.
41. See Chapter 3 in Hörstadius (1950), Noden (1983a, 1984b), and Chapter 5 herein for transplantation of NC between amphibian and avian embryos, and see Thorogood (1993*) for a comparison patterning in amphibians and birds. For segregation of cell lines, see Ziller *et al.* (1983), Le Douarin (1986), and Atchley and Hall (1991).
42. See MacLean and Hall (1987) and Hall (1997, 1999a*, 2005b*) for the observation that lineage tracing does not necessarily equate with determination of cell fate.
43. Boisseau and Simonneau (1989) used Cohen and Kongsberg's method to establish murine neural tubes in a chemically defined medium so as to follow the migration of NCCs and neuronal differentiation.
44. See Cohen and Konigsberg (1975) and Ito and Sieber-Blum (1993a,b) for clonal cell culture, Kalcheim and Le Douarin (1986) for the differentiation of adrenergic neurons, and Bronner-Fraser *et al.* (1980) and Sieber-Blum and Cohen (1980) for adrenergic neuronal differentiation following exposure to somitic ECM.
45. See Baroffio *et al.* (1991), Ito and Sieber-Blum (1993), and Ito *et al.* (1993) for these clonal analyses.
46. See Baroffio *et al.* (1991) for the lineage studies, and Lo *et al.* (1991a,b) for the studies with *V-myc* and the *achaete-scute* orthologue. See Krotoski *et al.* (1988) and Bronner-Fraser and Fraser (1997*) for labeling studies, and Le Douarin (1986), Patterson (1990), Mujtaba *et al.* (1998), and Schubert *et al.* (2000) for neuronal lineages.
47. See Asamoto *et al.* (1995) and Bronner-Fraser and Fraser (1997) for studies on early- and late-migrating avian TNCCs, Serbedzija *et al.* (1997) for labeling of cells in the ventral pathway using DiI, Artinger and Bronner-Fraser (1992) for the influence of the notochord on the ventrally migrating cells, and Hirata *et al.* (1997) for the lamprey study.
48. See Sieber-Blum (1990*) for premigratory pluripotential cells, and see Sherman *et al.* (1993) for response to antisense probes against Fgf2.
49. For broader discussions of whether determination of cell fate occurs in one or several steps, for the temporal separation of commitment and differentiation, and for the existence of subpopulations within apparently homogeneous cell populations, see Maclean and Hall (1987), Anderson (1997, 2000), Sieber-Blum (1990*), Bronner-Fraser and Fraser (1997), Weston (1991), Stemple and Anderson (1993), Bronner-Fraser (1995), Sieber-Blum and Zhang (1997), and Moody (1999) and Morrison *et al.* (2000).
50. For differentiation of pigment cells in response to extracellular matricial components, see Epperlein *et al.* (1996, 2000a) and Olsson *et al.* (1996*).
51. See Erickson and Goins (1995) for back-transplantation, Lahav *et al.* (1996) and Dupin *et al.* (2000) for endothelin3 and melanocyte proliferation, differentiation, and reversion to glia, Morrison *et al.* (2000), and Endo *et al.* (2002) for Notch, Wnt-Bmp pathways, and Richardson and Sieber-Blum (1993) and Sieber-Blum and Zhang (1997) for determination of fate late in migration.
52. For the role of growth factors and hormones in generating and maintaining diversity among NCCs, see Le Douarin (1986), Hall and Ekanayake (1991), Sieber-Blum (1991), and Sieber-Blum and Zhang (1997). See Dudas and Kaartinen (2005) for a review of the Tgfβ superfamily and their receptors.

53. See Pinco *et al.* (1993) for the role of Nt3. See Sieber-Blum *et al.* (1988) for the differentiation of sensory neurons from avian NCCs maintained *in vitro*, Sieber-Blum (1991) and Sieber-Blum and Zhang (1997) for the studies with growth factors and neurotransmitter-uptake inhibitors, and Hall and Ekanayake (1991) for a discussion of the evidence for the role of Bdnf and Ngf in sensory neuronal differentiation.

54. See Landis and Patterson (1981) for an early study on Ngf and sympathetic neuronal/glial differentiation, Anderson (1997, 2000) for studies with the mammalian cells, and Tran and Hall (1989), Fang and Hall (1997, 1999*), Buxton *et al.* (2003), and Hall (2005b*) for chondrogenesis from periosteal cells, including the role of N-CAM. A monoclonal antibody against the Ngf receptor has been used to isolate multipotential precursors from mammalian NC that form neurons and Schwann cells as they transit from self-renewing stem cells to clonal progeny.

55. See Stocker *et al.* (1991) for the studies with growth factors and DRG, Smith-Thomas *et al.* (1986) for transformation of chondrogenic to neuronal cells, and Fang and Hall (1997) for an overview of de- and redifferentiation.

Chapter 4
Evolutionary Origins

Invertebrate nervous systems consist of either a chain of ganglia, a ventral nerve cord, or a diffuse nerve net. In contrast, chordate body plans are organized around a dorsal nerve cord and the dorsal notochord.

Despite these fundamental structural differences, and to our considerable surprise when they were discovered, common and conserved genetic signaling systems establish the dorsoventral axis in vertebrates *and* the ventrodorsal axis in invertebrate. Furthermore, these signaling systems pattern the dorsal nervous systems in *Xenopus* (and by inference and extension, all vertebrates) and pattern the ventral nervous system in *Drosophila* (and by inference and extension, all arthropods). Discovery of these conserved genetic elements (Box 4.1) makes it likely that a search for NC precursors in invertebrate ancestors of the vertebrates will be fruitful.

Box 4.1 Dorsoventral and ventrodorsal axis formation

All bilaterally symmetrical invertebrates are organized around a ventral body axis and have nervous systems consisting of a chain(s) of ganglia, a ventral nerve cord, or a diffuse nerve net. All chordates are organized around a dorsal body axis and have a dorsal nervous system and notochord. Long regarded as nonhomologous, invertebrate, and vertebrate nervous systems were assumed not to share genetic pathways initiating their respective nervous systems. Astoundingly, they do.

The major (ventral) body axis and initiation of the ventral nervous system in *Drosophila* is established by activation of the gene *Sog* (*short gastrulation*). Comparative genomic analyses revealed *Sog* to be the ortholog of the vertebrate gene *Chordin*, which specifies the dorsal surface in *Xenopus* and in other vertebrates.[a] Amazingly, *Drosophila Sog* mRNA injected into *Xenopus* eggs specifies dorsal, and *Xenopus Chordin* mRNA injected into *Drosophila* eggs specifies ventral.

What of the dorsal body surface in *Drosophila* and the ventral body surface in *Xenopus*? Dorsal in *Drosophila* is established by a gene product, decapentaplegic protein that antagonizes *Sog*. Ventral in *Xenopus* is established by Bmp4,

---

B.K. Hall, *The Neural Crest and Neural Crest Cells in Vertebrate Development and Evolution*, DOI 10.1007/978-0-387-09846-3_4,
© Springer Science+Business Media, LLC 2009

which antagonizes *Chordin*. And, yes, decapentaplegic protein is the inverte-brate ortholog of Bmp, further evidence for conservation of an ancient genetic network. The presence of these signaling molecules in the common inverte-brate/chordate ancestor is the most parsimonious explanation for such conser-vation. The even more recent demonstration that a marine polychaete worm expresses orthologs of NK homeobox genes used in segmentation and pattern-ing in *Drosophila* and vertebrates further reinforces the ancestry of such genes before the origin of invertebrates and vertebrates.[b]

---

[a] We saw in Chapter 2 that Chordin (along with Noggin) binds to Bmp4 and prevents Bmp4-receptor interactions, thereby specifying the most rostral neural ectoderm from which fore-brain and hindbrain arise. Targeted inactivation of *Chordin* in mice results in major craniofa-cial anomalies, equivalent to those seen in DiGeorge syndrome in humans. The mice die as embryos (Bachiller *et al.*, 2003).

[b] For *Sog* and *Chordin* and their roles in axial patterning, see Nübler-Jung and Arendt (1994), Holley *et al.* (1995), Arendt and Nübler-Jung (1996), Hall (1999a,b*), and Lacalli (2006*). For the polychaete study, see Saudemont *et al.* (2008)

What was that invertebrate ancestor?

All vertebrates are chordates, animals with a notochord and a dorsal nerve cord. The most recent common ancestor of vertebrates, therefore, was a chordate. The coupling of NC induction to the dorsal nerve cord (see Chapter 3) makes it likely that a search for NC precursors in vertebrate chordate ancestors will be fruitful.

What was the chordate ancestor?

As the chapter unfolds, we will see that the most likely chordate ancestors are organisms possessing properties seen in two extant chordate groups, cephalochor-dates and ascidians. Evidence, opinion, theory, debate, argument, and downright antagonism over which of the two is the more likely ancestor, or indeed whether the ancestor was an annelid or an echinoderm-like organism, have raged for at least 150 years. While this is a fascinating story, this chapter does not propose to deal with the many past views.[1] Today the weight of evidence indicates that the relevant extant organisms are cephalochordates and ascidians (Figs. 4.1 and 4.2). Having discussed the evidence for precursors of the NC in the vertebrate ancestor, the balance of the chapter is devoted to a second great transition, the evolution of jawed vertebrates (gnathostomes) from jawless vertebrates (agnathans, broadly defined.

Precursors of the Neural Crest

Possession of a neural crest (NC) derived from dorsal neural ectoderm is a quintessential vertebrate characteristic, according to some, *the* quintessential ver-tebrate characteristic (Figs. 4.1 and 4.2; and see Fig. 1.4). Maisey, for instance,

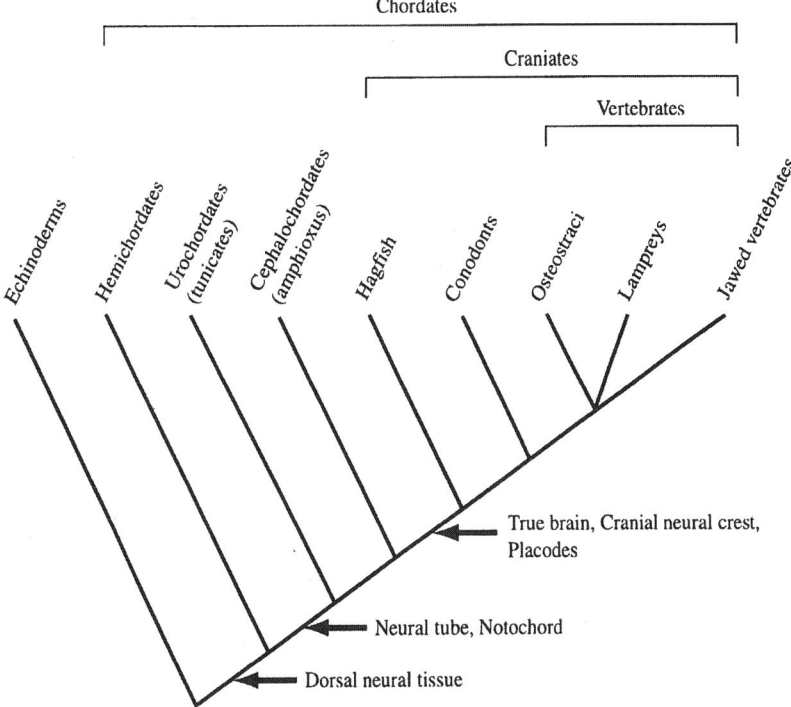

Fig. 4.1 Features associated with the evolution of the chordate/vertebrate dorsal neural tube are shown on this phylogenetic tree. Neural crest and placodes are vertebrate features that followed the evolution of a neural tube and notochord. Modified from P. W. H. Holland and Graham (1995)

identified many vertebrate features as consequences of the 'presence of neural crest, and all that that entails' (1986, p. 241).

According to Gans and Northcutt, what we now know as the NC could have existed initially as an epidermal nerve plexus or a nerve net controlling ciliary function during movement and filter feeding. This neural tube or peripheral nerve net would have produced neuronal and perhaps pigmented cells, as we now know is the case in extant ascidians. With increasing muscle-based locomotion, the dorsal nerve cord took over nervous control of locomotion, freeing the epidermal nerve cells (or their precursors) for other functions. Those new functions allowed modifications of existing cell products involved in mechanoreception to deposit an extracellular cartilaginous matrix in the pharynx.

Whether this scenario can accommodate an origin of the NC in direct association with the dorsal neural tube depends on the origin of the dorsal neural tube and its relationship to any epidermal nerve net. One of the two long-standing theories for the origin of the chordates—their origination from direct-developing hemichordates—posits that the chordate neural tube arose by the rolling up and

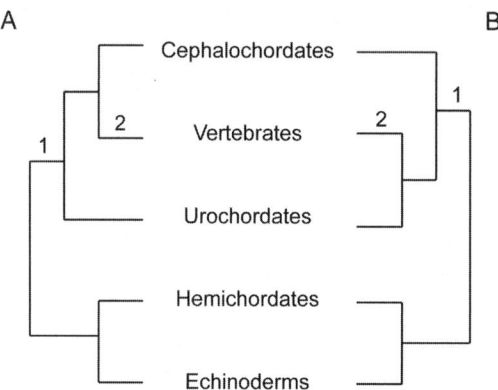

Fig. 4.2 Two phylogenetic relationships of extant chordates (urochordates and cephalochordates) and vertebrates with hemichordates (invertebrate chordates) and echinoderms as outgroups. (**A**) The long accepted view in which cephalochordates are the closest relatives of (sister group to) the vertebrates, with urochordates as a more basal group of chordates. (**B**) Recent molecular phylogenetic analyses place urochordates as the sister taxon to vertebrates and cephalochordates as more derived. Node 1 identifies the presence of a dorsal nerve cord, notochord, and postanal tail. Node 2 identifies the presence of the neural crest. Cephalochordates and urochordates lack a NC. Ascidians, a group within the urochordates, possess cells that Jeffery (2007) identifies as neural crest like

coalescence of a hemichordate epidermal nerve net. An alternate and equally long-standing theory sees the neural tube arising from the ciliary bands of an echinoderm dipleurula-like larva. As discussed in the following sections, current phylogenetic analyses make ascidians the closest group to vertebrates (Fig. 4.2), which would make the ciliary band origin more likely.[2]

Given that no cells with the range of potentials of NCCs occur in any inverte-brate, what might a NC precursor(s)—a proto-neural crest[⊕]—look like, and where in the vertebrate ancestor would we expect to find it? Six approaches are used in this chapter to answer these two fundamental questions.

(1) Given the intimate relationship between the NC and the dorsal nervous system during development, our analysis begins with the chordate dorsal nervous sys-tem, especially the nature and make-up of the brain. The issue is whether any

---

[⊕] In early drafts of this chapter, I referred to evolutionary precursors of the neural crest as a **proto-neural crest**. I still like the term, but do not use it, only because it has the connotation of pre-adaptation, as if cells or genes arose in vertebrate ancestors *so that they could become* a neural crest or NCCs. The term pre-adaptation has been replaced by *exaptation* in evolutionary biology for the same reason. Perhaps I am being too cautions for I do use the terms proto-vertebrate, protoHox, proto-oncogene, proto-tissues, and proto-chondrogenic (as do others) as terms that mean coming before and neither 'anticipating' nor 'developing in order to. . ..

cellular and/or molecular precursors (latent homologs) of the NC were present in ancestral cephalochordates or urochordates. Cells, genes, and gene regulatory networks associated with the dorsal neural tube in living (extant) members of the two groups are examined for signs of NC precursors.

(2) Having discussed the likely nature of NC precursors—the proto-NC—we turn to an examination of those fossils that have been interpreted as basal chordates or cephalochordates (Figs. 4.2 and 4.3) to see whether they shed any light on the origins of NCC derivatives. Do they have, for example, elements of a head skeleton or pharyngeal arches, two fundamental vertebrate characters (synapomorphies)?

(3) The third topic recognizes that the origin of the NC in the first vertebrates accompanied the evolution of a brain, a muscular pharynx, and paired sensory organs. In a paradigm-breaking hypothesis—often known as the **new head hypothesis** —Carl Gans and Glen Northcutt linked these evolutionary innovations to the evolution of the NC and ectodermal placodes (Gans and Northcutt, 1983; Northcutt and Gans, 1983). This chapter outlines the rationale behind the new head hypothesis; placodal ectoderm is discussed in Chapter 6.

(4) With the knowledge from (2) and (3) as background, we turn to an examination of the pivotal role played by NCCs in the evolution of pharyngeal arches, a topic also discussed in some detail in Chapter 7 in the context of the craniofacial skeleton. Why pharyngeal arches? Because integrations between the evolving vertebrate brain, muscular pharynx, and paired sensory organs may have necessitated that the pharyngeal arch skeletal system—and subsequently, the skeleton

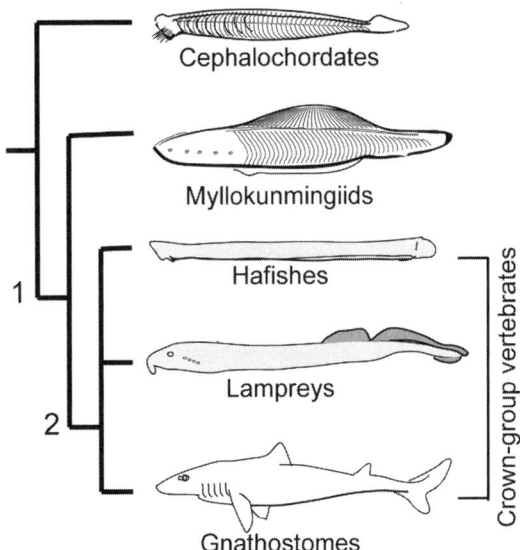

Fig. 4.3 A chordate phylogeny showing cephalochordates as basal chordates, *mykkokunmingiids* as more derived, and three crown group vertebrates with hagfish as the sister group of lampreys. Characters appearing at Node 1 are the neural crest, epidermal sensory placodes, cartilaginous capsule around the olfactory, optic and otic sense organs, gill filaments, and W-shaped myomeres. Cartilaginous radials in the paired fins appear at Node 2. Adapted from Janvier (2007)

of the jaws and much of the skull (the first vertebrates being jawless)—evolved from NCCs whose developmental connections were to neural ectoderm and neurons rather than to mesoderm and connective tissue; mesoderm produces much of the vertebrate skeleton, including virtually all the skeleton outside the head.

(5) The origination of the pharyngeal arch skeleton raises the issue of the group of organisms in which and how cartilage arose as a skeletal tissue. Did cartilage arise in the basal proto-vertebrate from a single germ layer, cell layer, or tissue, or were cells and/or genes co-opted from several layers or tissues?

(6) Two recent studies utilizing comparative genomics, bioinformatics, molecular fingerprinting, genetic labeling/cell selection, and GeneChip Microarray technologies are introduced as powerful ways to approach the questions that are central to this chapter.

We begin with the cephalochordates.

Cephalochordates

Cephalochordates (amphioxus, lancelets;[⊕] subphylum Cephalochordata), 25–50-mm-long sedentary, bottom-dwelling, marine chordates, have for almost 150 years been regarded as the least derived (most basal) extant chordates and therefore the closest chordate relatives (sister group) of the vertebrates (Figs. 4.1, 4.2, 4.3, and 4.4). Each has a notochord, a dorsal nerve cord, metamerically arranged muscles, a large perforated pharyngeal basket used to filter water and to direct food into the gut, and a thyroid gland—all features found in chordates. Further, the cephalochordate hepatic diverticulum is a structural homolog of the vertebrate liver, while the liver enzymes (plasminogens) of vertebrates have orthologs in amphioxus.

As is also true for ascidians—as seen in the following section—cephalochordate central nervous systems have been interpreted as sharing a number of homologous features with the vertebrate rostral nervous system/brain. These include:

- rostrocaudal organization of the brain;
- presence of an eye spot and regions of the brain that are homologs of vertebrate sensory organs and major divisions of the vertebrate brain, respectively;

[⊕] Amphioxus, like lancelet, is a common, not a generic name. Hence amphioxus should neither be capitalized nor italicized. The generic names are *Branchiostoma (12 species)* and *Asymmetron* (two species), although different authorities recognize as few as eight and as many as 30 species in the two genera. A detailed microscopical anatomy of all the developmental stages of *B. lanceolatum,* based on transmission and scanning electron microscopy is available (Stach, 2000).

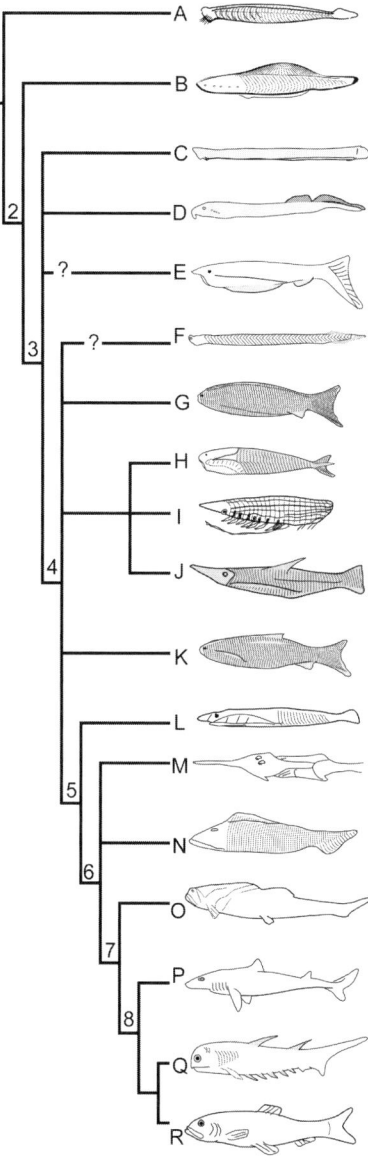

Fig. 4.4 Janvier's (2007) synthesis of chordate relationships using molecular and morphological data but not using a tree-building algorithm. Major groups are shown as **A–R**. Only groups referred to in the text are noted here: (**A**) cephalochordates; (**B**) *myllokunmingiids*; (**C**) hagfish; (**D**) lampreys; (**F**) conodonts; (**G**) anaspids; (**N**) osteostracans; (**P**) chondrichthyians; (**R**) osteichthyans; See Janvier (2007) for further details Numbers 1–8 identify nodes associated with the appearance of new characters and taxa. Only characters relevant to our discussion are listed here; see Janvier (2007) for further details. (1) Segmental spinal nerves. (2) Vertebrates—NC, epidermal placodes, sense organ capsules, gill filaments. (3) Cartilaginous radials in paired fins. (4) Gnathostomes—jaws, dermal skeleton, semicircular canals. (5) Perichondral ossification of endoskeleton. (6) Pectoral fins separate from shoulder girdle. Adapted from Janvier (2007)

- the presence and expression of orthologs of *Hox, Pax, engrailed* and *distalless* in homologous regions; and
- the expression and regulation of *Hox* genes by retinoic acid.

In the past, it was maintained that the most speciose cephalochordate genus, *Branchiostoma*— and the one most studied—does not possess anything resembling a NC. Cephalochordates lack even such basic NCC types as pigment cells, although they do possess a population of neurons, the dorsal cells of Rhode, that have been compared with vertebrate Rohon–Béard (R–B) neurons. R–B neurons are either NC in origin, share a developmental origin with NCCs, or arise from cells medial to the NC that can also become NCCs (see Chapter 6). The group of large cells associated with the dorsal region of the caudal neural tube in adult amphioxus, variously known as the dorsal cells of Rhode, the nucleus of Rhode, or the cells of Joseph, form part of the mechanosensory system providing input directly to the axial muscles and indirectly to the notochord. Any further relationships to R–B neurons await testing.

Amphioxus (and ascidians) possesses two classes of ectodermal sensory cells; primary neurons with axons and secondary neurons lacking axons. Amphioxus lacks neuromasts or lateral line nerves but does possess one placode homolog: the rostral ectoderm of amphioxus is regarded as homologous with the olfactory placode, which is the only vertebrate placode that produces only primary and no secondary neurons. Ascidians have a far greater range of placodes and placodal derivatives, as discussed in Box 4.2 and later in this chapter; see Chapter 6 for placodes and their origin.

Box 4.2 Ascidian placodes

Placodes and placodal ectoderm were introduced and discussed in Chapter 2.

The presence, nature, and homology of placodes in two ascidians, the sea vase, *Ciona intestinalis*, and the star ascidian, *Botryllus schlosseri*, have been investigated in some depth. Manni *et al.* (2004) came to the following conclusions regarding the homologies of ascidian and vertebrate placodes at the cellular level:

- the ascidian stomodeal placode contains elements that are homologous to elements of vertebrate adenohypophyseal and lateral line placodes;
- the ascidian neurohypophyseal placode contains elements that are homologous to elements within the vertebrate olfactory placode and within the hypothalamus;
- the atrial placode[a] is homologous to the otic placode;[b] while
- the rostral placode could be homologous to the epidermal cell populations that produce the adhesive/hatching glands of vertebrates.[c]

On this basis, placodes unique to (synapomorphies of) vertebrates are the lens, profundal, trigeminal, and epibranchial placodes.

---

[a] Kourakis and Smith (2007) demonstrated a conserved signaling role for *Fgf8/17/18* in the atrial siphon placode of the Pacific sea squirt, *Ciona savignyi*, and demonstrated *Fgf* signaling during development of the otic placode, findings that further reinforce the homology of the siphon and the placode (and providing evidence for an inductive interaction between invaginating atrial ectoderm and pharyngeal endoderm in gill slit formation).
[b] In the Atlantic cod, *Gadus morhua*, the ectodermal lining of the opercular cavity arises by invagination from the otic placode (Miyake *et al.*, 1997).
[c] In *Xenopus*, the hatching gland, NC, and preplacodal ectoderm all arise from the neural-plate border, the choice between the three fates being dependent on *Pax3* (hatching gland), *Zic1* (placodal), or a combination of the two (NC); manipulating levels of the two gene products shifts cell fate among the three alternatives (Hong and Saint-Jeannet, 2007).

On the basis of studies initiated some 15 years ago, Nicholas Holland, Linda Holland, and their colleagues maintained that, despite the lack of any HNK1-positive cells, two cellular features of amphioxus dorsolateral ectoderm are indicative of NCCs. One is topographical: epidermal cells lie at the lateral border of the neural plate. The second is the migration of these epidermal cells over the neural tube and toward the midline (Fig. 4.5).[3] Although the direction of migration is unlike that taken by NCCs, this association with and the ability to migrate away from the neural tube are features we might expect of neural crest precursor cells. Given that many invertebrate nerve cells are migratory, and that nerve cells in some taxa transform from epithelial —> mesenchymal to delaminate from the nervous system and migrate, these amphioxus cells could be convergent with NCCs. The shared chordate ancestry of cephalochordates and vertebrates makes this a less likely interpretation than common ancestry of both cell lineages.[⊕]

Genes and Gene Networks in Cephalochordates

Amphioxus contains a number of genes and regulatory pathways that foreshadow genes and pathways associated with NCCs. As you might expect, many of these genes are shared by amphioxus and vertebrates, because, as chordates, they share the presence of a dorsal neural tube. Consequently, amphioxus and vertebrates share the presence of a neural tube–epidermal ectodermal border. We might therefore expect to find shared genes in the dorsal neural tube and at the border that are not associated with NCC induction. Boundaries and induction are addressed below and in an insightful analysis by Meulemans and Bronner-Fraser (2004). Genes in amphioxus

---

[⊕] Convergence and the related issue of latent homology, are discussed by Stone and Hall (2004) and Gass and Hall (2007) with special reference to the evolution and organization (modularity) of the NC.

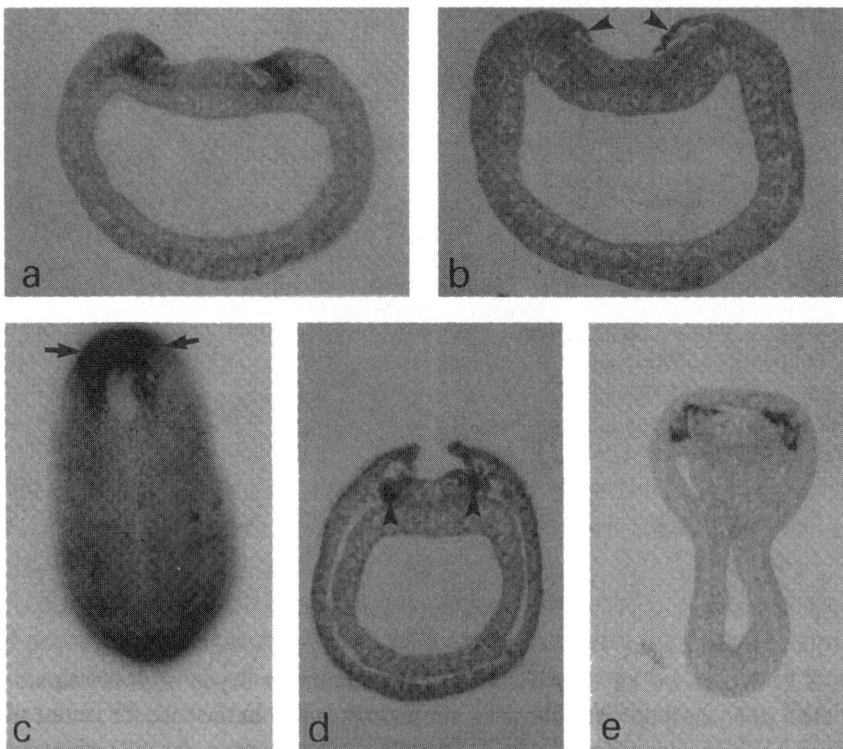

Fig. 4.5 Expression of *AmphiDll* in neurula-stage embryos and in larvae of the cephalochordate *Branchiostoma*. (**a**) Cross-section through the rostral region of an early neurula, showing expression (*black*) in epidermal ectoderm adjacent to the neural plate, which also contains positive *AmphiDll*-positive cells. (**b**) Cross-section through the caudal region of the same embryo as in (**a**), showing *AmphiDll*-positive epidermal ectoderm growing over the neural plate (*arrowheads*). (**c**) Expression in the rostral epidermal ectoderm (*arrows*) covering the neural plate as seen in a dorsal view of a whole mount at the end of the early-neurula stage; anterior to the *top*. (**d**) Cross-section through the rostral region of the embryo shown in (**c**), showing strong expression in lateral cells of the neural plate (*arrowheads*). (**e**) Cross-section through a hatching neurula-stage embryo, showing the strongly *AmphiDll*-positive cells on either side of the central neural plate. Modified from N. D. Holland *et al.* (1996), with the permission of Company of Biologists Ltd., from figures kindly provided by Nicholas Holland

that are orthologs of genes involved in vertebrate chondrogenesis are discussed separately in the section on the origin of cartilage.

AmphiSnail1

As we saw in Chapter 2, members of the *Snail* family of zinc-finger transcription factors are markers of the NC, although it is important to recall from Chapter 2 that *Snail* cannot induce NCCs in the absence of other signaling molecules.

The isolation and expression of an amphioxus *Snail1* gene, *AmphiSnail1* (and of an ascidian gene, *CiSnail1*; see below), at the lateral borders of the neural plate and in the dorsal neural tube, precisely where we would expect markers for neural crest precursors to lie, demonstrates that *Snail1* evolved before the NC and so was available to be co-opted into the NC. In this sense, we can consider *AmphiSnail1* as a precursor gene of the NC. The transcription factor Ap2α regulates the expression of *Snail2* after NCCs arise in vertebrates, that is, Ap2α acts upstream of those genes that specify the NC; Ap2α is expressed in epidermal but not neural ectoderm of amphioxus larvae (Meulemans and Bronner-Fraser, 2002), but as an *AmphiSnail2* has not been reported, this pathway is presumed not to be present in amphioxus.

Depending on which group of vertebrates is examined, either *Snail1* or *Snail2* is expressed in the paraxial mesoderm. Consistent with *Snail1* having evolved in a context other than NCCs, *AmphiSnail1* is expressed in presomitic mesoderm; that is, dorsal neural tube expression of *Snail1* was enhanced and co-opted and the phylogenetically older mesodermal expression lost with the origin of the NC. In the sea vase, *Ciona intestinalis* (an ascidian), *CiSnail1*-positive cells form the ependyma, a layer of cells lining the spinal cord. In ascidians, as in amphioxus, therefore, *CiSnail1*-positive cells have diverged from a strictly neuronal lineage, as, of course, have NCCs. In an insightful recent analysis of these ascidian cells, which he named 'neural crest-like cells,' Jeffery (2007) emphasized that their primary role is to differentiate as pigment cells. If as recent molecular phylogenies suggest, ascidians are the sister group to vertebrates (Fig. 4.2 and see the section on ascidians below), and if the NC-like cells reported in ascidians presage NCCs, then the original role for these NC precursors was to form pigment cells.[4]

Hox genes: AmphiHox1–AmphiHox12

Expression patterns of amphioxus *Hox* (*AmphiHox*) genes provide a further source of data relevant to the origin of the NC.

As discussed in Box 1.2, gene duplication allows one copy of a gene to retain its original function and the second to diverge in structure/regulation, potentially to serve a new function. The chordate ancestor is postulated to have had a single cluster of *Hox* genes (Box 1.2). The presence of single *Hox* genes in amphioxus and multiple *Hox* clusters in vertebrates (including lampreys) is consistent with new genes or expanded gene functions at the outset of vertebrate evolution having been instrumental in the origin and diversification of NCCs and other vertebrate synapomorphies (see Box 1.2).

Gene duplication in vertebrates includes other regulatory genes such as *engrailed*, *Msx*, *Otx*, *MyoD*, the globin gene family, and insulin-like growth factors. Genes that exist as separate genes in vertebrates but as a single gene in amphioxus include:

- *AmphiBmp2/4* —the amphioxus ortholog of *Bmp2* and *Bmp4*,
- *AmphiPax3/7* —the amphioxus ortholog of *Pax3* and *Pax7*, and
- *AmphiPax2/5/8* —the amphioxus ortholog of *Pax2, Pax5*, and *Pax7*.

Such a pattern of gene duplication is most readily interpreted as early duplication of the entire genome with subsequent selective losses in particular lineages. Duplication does not necessarily mean later divergence in gene regulation; *cis-* acting elements of *Otx2* that regulate gene expression in mesencephalic NC (but which differ between premandibular and mandibular segments) are conserved between puffer fish and mice (Kimura *et al.*, 1997). Furthermore, gene duplication continued in cephalochordates after their separation from other chordates. The gene *Brachyury* is duplicated in amphioxus but not in vertebrates.[5] As discussed in the following section, a similar situation exists in ascidians.

Amphioxus has at least 12 *Hox* genes (*AmphiHox1–AmphiHox12*) with rostro-caudal boundaries of expression that can be equated with those seen in vertebrates (see Chapter 2). In vertebrates, *Hoxa1*, *Hoxb1*, and *Hoxd1* have their most rostral boundaries of expression between r3 and r4, while the most rostral expression boundaries of *Hoxa3*, *Hoxb3*, and *Hoxd3* lie between r4 and r5. In contrast, *Amphi-Hox1* and *AmphiHox3* are expressed only in a circumscribed region of the nerve cord. *Hox3* genes are expressed in all but the most rostral region of the vertebrate neural tube; *AmphiHox3* has a similar pattern of expression in the nerve cord.

These shared expression boundaries have been used to support the homology of regions of the amphioxus rostral nerve cord with the vertebrate hindbrain. It has especially been concluded, in part from the expression of *AmphiHox3* in caudal mesoderm *and* in the rostral dorsal nerve tube, that amphioxus has homologs of the vertebrate pre- and postotic hindbrain, an important boundary in vertebrates; the postotic neural crest defines the boundary between head and trunk (see Chapter 2).

A second way of changing gene function is by evolutionary changes in the regulatory regions of a single gene or class of genes such as the *Hox* genes. Shared regulatory regions, which provide *prima facie* evidence for shared function, can be detected by incorporating the regulatory region of the gene from one species into the orthologous genes of another. In this way it has been shown that reporter constructs of the regulatory regions of *AmphiHox 1, 2*, and *3* can direct gene expression in vertebrate embryos. Perhaps the most revealing result in the present context is a construct that directs gene expression in NC and placodal cells. Other constructs reveal regulatory regions for the retinoic acid receptor (see below). Therefore, we can conclude that 'some of the regulatory elements that direct Hox gene expression in NCCs evolved before the neural crest (or vertebrates) evolved', making a search for precursors of NC pathways a realistic occupation.[6]

AmphiDll

Distal-less (*Dlx*) genes are expressed in NCCs and in pharyngeal endoderm. As discussed in Chapters 2 and 7, *Dlx* genes are markers of and pattern the forebrain. They also play a crucial role in establishing the dorsoventral patterning of the pharyngeal endoderm, which, in turn, provide the dorsoventral patterning of the cartilages of the first and second pharyngeal arches.

AmphiDll, the amphioxus ortholog of *Dlx*, is expressed in the rostral neural plate, in two dorsal clusters beside the neural tube near the cerebral vesicle, and in premigratory and migratory ciliated epidermal cells (Fig. 4.5; N. D. Holland *et al.*, 1996). While the gene is orthologous to *Dlx*, the patterns of expression of Dlx and *AmphiDll* are more difficult to homologize. In one hemichordate studied, the Hawaiian acorn worm, *Ptychodera flava, distal-less* (*PfDll*) is expressed in the aboral embryonic ectoderm and then in cells of the larval ciliary band.[7]

AmphiOtx

Otx genes are members of a family of genes that are the vertebrate ortholog of the *Drosophila* gap gene *Orthodenticle*.

   Otx genes are essential for the development of the rostral portion of the head. *Otx1* and *Otx2* have nested expression domains in the murine rostral hindbrain. *AmphiOtx*, a single *Otx* gene in amphioxus, has a pattern of expression similar to *Otx1* and *Otx2*, initially in the rostral neurectoderm and in mesendoderm (as in mice), subsequently in the rostral tip of the cerebral vesicle and in the frontal eye (Fig. 4.6).

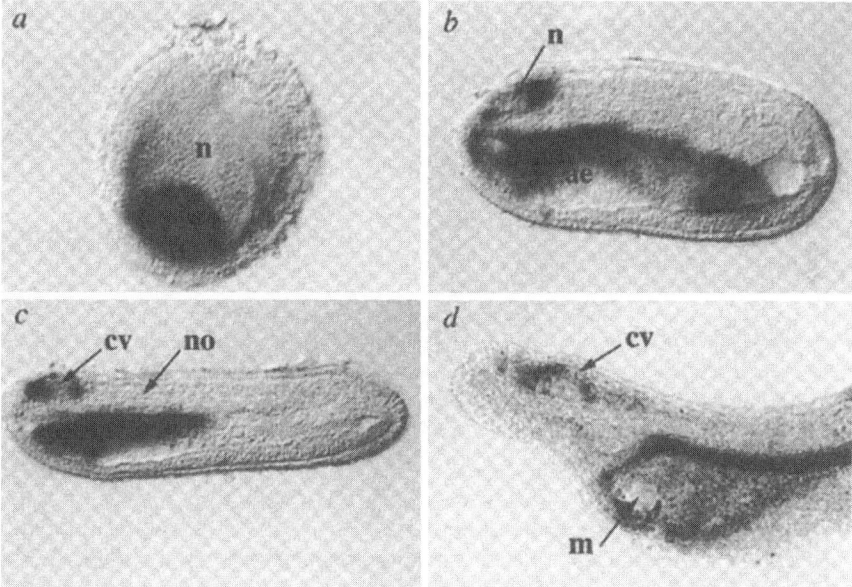

Fig. 4.6 Expression of *AmphiOtx* in whole-mounted *Branchiostoma* embryos. (**a**) Dorsal view of an early neurula showing strong expression in the rostral neural plate. (**b** and **c**) Mid- and late-neurula stages, seen in lateral view (rostral to the *left*), show localized expression in the neural plate (n), cerebral vesicle (cv) and anterior endoderm (ae). (**d**) An early larval stage with neural expression confined to clusters of cells in the anterior cerebral vesicle (cv) and a few epidermal ectodermal cells rostral to the vesicles. Expression continues to be strong in the endoderm of the anterior pharynx. m, mouth region. Reprinted with permission from Williams and Holland, Copyright © (1992), Macmillan Magazines Limited, from a figure kindly provided by Nick Williams

Otx2 is expressed initially throughout the mouse epiblast, then in first arch mesenchyme. *Otx2* mediates fore- and midbrain regionalization through expression in pharyngeal endoderm. Mice with homozygous mutations of *Otx2* ($Otx^{-/-}$) fail to form any structures rostral to r3. Heterozygotes ($Otx^{+/-}$) display otocephaly and lack eyes and mandibles.[8] As discussed in Box 2.3, *XOtx2* is expressed in rostral neurectoderm during gastrulation, ectopic expression inducing an ectopic cement gland.

Recent analysis demonstrates an evolutionarily deep conservation of two *Otx2* enhancers, whose expression and function is tissue specific. One, designated AN, regulates *Otx2* in rostral neurectoderm. The other, designated FM (fore- and midbrain), regulates *Otx2* in the caudal fore- and midbrain. Tetrapods contain both enhancers, but teleosts have only the forebrain–midbrain enhancer FM. The presence of both enhancers in the Indonesian coelacanth and in the clearnose skate was used by Kurokawa and colleagues (2006) to conclude that presence of both enhancers is the basal vertebrate condition, the loss of AN in teleosts being secondary.

AmphiBmp

As indicated above when discussing *AmphiHox* genes, amphioxus has a single *AmphiBmp2/4* gene, the ortholog of *Bmp2* and *Bmp4*. *AmphiBmp2/4* may inhibit specification of the cephalochordate dorsal neural plate as the vertebrate orthologs do; *AmphiBmp2/4* is expressed in many early tissues, consistent with a role in establishing the dorsoventral polarity of the ectoderm.[9]

AmphiPax

As indicated above when discussing *AmphiHox* genes, a single gene, *AmphiPax3/7* is the amphioxus ortholog of *Pax3* and *Pax7*. *AmphiPax3/7* and the single ascidian ortholog (see below) form a separate cluster outside the vertebrate genes (see Figs. 4.7 [Color Plate 4] and 4.8A), indicating either convergence or deep divergence of the cephalochordate/ascidian and vertebrate genes.

Amphioxus also has a single *AmphiPax2/5/8* paired box gene, which is expressed in the rostral hindbrain (compare with expression in ascidians discussed below), pharyngeal gill slits, and several other structures, for which see Kozmik *et al.* (1999).

In vertebrates and in amphioxus, as summarized in the following section, pharyngeal cleft formation is controlled by retinoic acid. *Pax1* and *Pax9* represent a family of vertebrate *Pax* genes, a family represented in amphioxus by a single gene, *AmphiPax1/9* expressed in pharyngeal endoderm. Expression of *AmphiPax1/9* and *Pax1* in vertebrates in the anterior gut endodermal is required as an upstream regulator of retinoic acid. By extension, anterior gut (pharyngeal) endodermal–retinoic acid interactions under *Pax* gene control were features of the common ancestor of

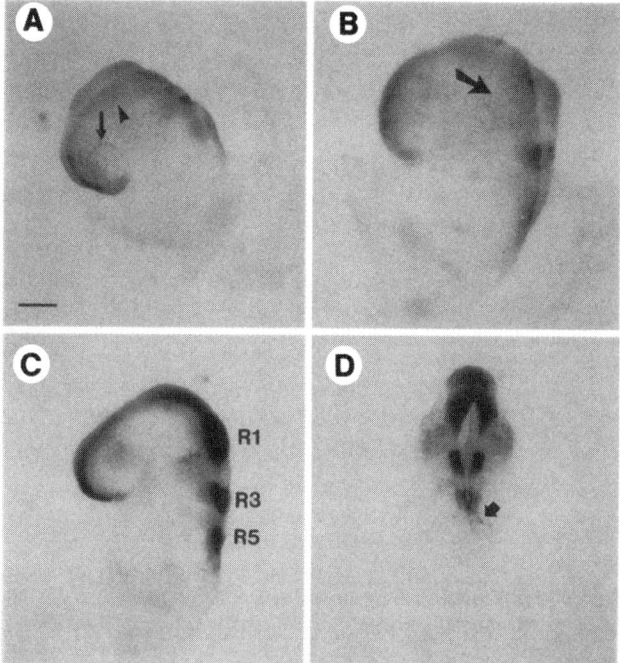

Fig. 4.7 Expression of *Pax7* in CNC as seen in lateral (**A**, **B**, and **C**) or dorsal (**D**) views of whole mounts of 8.5- (**A** and **B**) and 9.5- (**C** and **D**) day-old mouse embryos. (**A**) Neural crest cells surrounding the optic vesicle (*arrow*) and at the level of the midbrain (*arrowhead*) are highlighted. (**B**) The arrow indicates neural crest cells migrating from the hindbrain. (**C**) Rhombomeres 1, 3, and 5 are identified and strongly express *Pax7*. (**D**) This dorsal view highlights expression in alternating rhombomeres. R5 is marked by the *arrow*. Bar = 200 μm (**A** and **C**); 240 μm (**B**); and 160 μm (**D**). Reproduced from the color original in Mansouri *et al.* (1996), with the permission of Company of Biologists Ltd., from a figure kindly provided by A. Mansouri (see Color Plate 4)

amphioxus and vertebrates. Linda and Nicholas Holland, who published a seminal study on amphioxus in 1996, conclude that genes such as *AmphiPax1* originated in pharyngeal endoderm and were co-opted by neural ectoderm and NC at the origin of the vertebrates (L. Z. Holland and N. D. Holland, 2001*).

Retinoic Acid and Retinoic Acid Receptors

A further body of evidence for an early origin in cephalochordates of what became NC activity has come from major affects of **retinoic acid** on NCCs and their derivatives in vertebrates and the presence and functions of retinoic acid and retinoic acid receptors in amphioxus.

As just outlined, vertebrates and amphioxus express *Pax* genes in the pharyngeal endoderm, and retinoic acid is regulated by members of the *Pax* gene families.

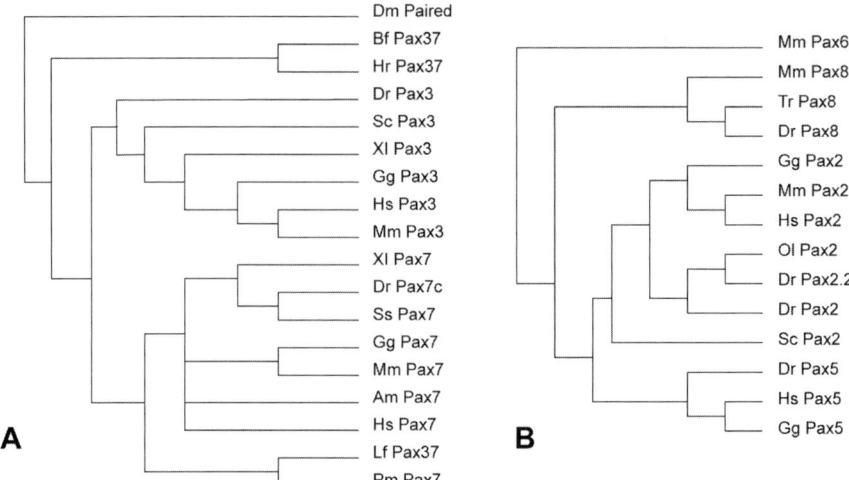

Fig. 4.8 Relationships between *Pax3* (**A**) in ascidians, amphioxus, lampreys, and jawed vertebrates and of *Pax2, 5, 6,* and *8* in jawed vertebrates (**B**) based on predicted protein sequences. With respect to *Pax3*, amphioxus and the ascidian (*Branchiostoma floridae* [Bf Pax37] and *Halocynthia roretzi* [Hr Pax37], respectively) form one cluster (*top* in **A**), lampreys (*Lampetra fluviatilis* [Lf Pax37] and *Petromyzon marinus* [*PmPax7*], respectively) a separate and distantly related cluster (*bottom* in **A**). The outgroup is represented by the gene *Paired* in *Drosophila melanogaster* (Dm paired). Species of jawed vertebrates represented (from *top* to *bottom* in **A**) are: the zebrafish (Dr), the small spotted catfish (Sc), *Xenopus laevis* (Xl), the common chicken (Gg), humans (Hs), mouse (Mm), salmon (Ss), Mexican axolotl (Am). The close phylogenetic relationships between *Pax2, 5, 6,* and *8* in jawed vertebrates are shown in (**B**). Species not listed above are (from *top* to *bottom* in **B**) are fugu (Tr) and Japanese medaka (Ol). Adapted from O'Neill *et al.* (2007)

Retinoic acid is synthesized in trunk mesoderm, pharyngeal arches, and fin buds of zebrafish embryos. Mutations of either of the two genes for retinaldehyde dehydrogenase—which is required for the synthesis of retinoic acid—disrupt gill and fin development. *No-fin* (*nof*) mutants, a phenotypic mutation that maps to the *Raldh2* gene, form neither pharyngeal arch nor pectoral fin cartilages (Grandel *et al.*, 2002). *Raldh2*$^{-/-}$ mice die as embryos because of heart defects. Providing retinoic acid to the females carrying these mutant embryos results in normal first and second arches, but the more posterior arches fail to form so that all their derivatives all affected; post-otic NCCs, aortic arches, thymus, parathyroid glands, and enteric ganglia.[10]

Until this millennium, defects in the vertebrate craniofacial or pharyngeal skeletons seen after administering retinoic acid were interpreted entirely as a consequence of the *direct* action of retinoic acid on migrating, proliferating, and differentiating NCCs. Now we know that retinoic-acid-induced defects in NC-derived craniofacial tissues in mice are mediated via activation of receptors (RaRβ/RxR heterodimers) in the pharyngeal endoderm (Matt *et al.*, 2003). Further recent studies on the patterning role of anterior gut endoderm in amphioxus (introduced above) and pharyngeal endoderm in vertebrates, in the presence of retinoic acid or with excess retinoic acid, have revealed that, while NCCs pattern the pharynx, their role is developmentally and evolutionarily secondary to the patterning role exercised by pharyngeal endoderm; pharyngeal arches form and are normally patterned in embryos from which NCCs have been removed. Consequently, we must take into account the patterning role of pharyngeal endoderm when we consider the role(s) of retinoic acid on NCCs or their derivatives (see Chapter 10). Given the presence of pharyngeal endoderm, retinoic acid, and retinoic acid receptors in amphioxus, this signaling pathway is relevant to our search for NC precursors.[11]

Escriva *et al.* (2002) cloned the genes for a single retinoic acid receptor, *Amphi-RAR*, a retinoid X receptor, *AmphiRXR*, and an orphan receptor, *AmphiTR2/4*. Administering excess retinoic acid to Florida lancelet embryos results in changes comparable to those elicited by excess retinoic acid in vertebrates. In vertebrates, however, the changes are mediated through effects on NCCs (see Chapter 10 for the details). What is the pathway in amphioxus?

In retinoic-acid-treated amphioxus embryos, expression of *AmphiHox1* is extended more rostrally than in untreated embryos. Mediated via the action of retinoic acid and its receptors on amphioxus pharyngeal endoderm, the mouth and the gill slits fail to develop. Similar defects elicited in vertebrate embryos are mediated by defective induction of the NC or by defective NCCs under the influence of pharyngeal endoderm. To reiterate:

- pharyngeal arches form with the correct patterning in vertebrate embryos from which NCCs have been removed;
- the patterning, which is primarily based in pharyngeal endoderm and only secondarily in NCCs in vertebrates, is solely under pharyngeal endodermal control in amphioxus.

The obvious conclusion is that the more ancient mechanism, based in pharyngeal endoderm in cephalochordates, has been expanded to include (or be co-opted by) NCCs in vertebrates; *cis*-regulation of *Hox* genes in the neural tube and dependence on retinoic acid are shared by amphioxus, chickens, and mice, indicative of a very ancient mechanism.[12] As discussed in Box 3.4 and as we look ahead to Chapter 7, the NC cartilages of many extant vertebrates develop in response to inductive interactions from pharyngeal endoderm or other ectodermal or endodermal epithelia. In addition, as discussed in the section on the origin of cartilage, the ability of pharyngeal endodermal to signal to other cell types also evolved *before* the NC.[13]

In summary, in extant cephalochordates, the expression patterns (and where known, the functions) of *AmphiHox,AmphiOtx*, *AmphiBmp*, *AmphiPax*, retinoic acid, and its receptors are consistent with the presence and activity of these genes and pathways in basal cephalochordates.

Did the basal proto-vertebrate in which the NC originated have these cephalochordate features or, alternately, do the (different) features found in extant ascidians correspond to the proto-vertebrate NC precursor? Extant ascidians are not cephalochordates but a group with the urochordates.

Urochordates—Ascidians

In the traditional Linnean classification, Urochordata (urochordates) are a subphylum of the chordates comprised of three classes:

- ascidians (tunicates or sea squirts),
- larvaceans (appendicularians), and
- thaliaceans (salps).

Almost all the research on urochordates relevant to the NC has been conducted on two species of ascidians, the sea vase, *Ciona intestinalis*$^{\oplus}$ and the sea squirt, *Halocynthia roretzi.*

Ascidians (from the Greek *askidion*, wineskins) are small, sessile marine filter feeders, some living at depths as great as 8,000 m. Also known as tunicates (from the Latin *tunica*, a tunic), sea squirts or sea vases, these animals have been given a prominent place in vertebrate origins since the 1860s, when Alexandr Kovalevsky discovered a notochord in ascidian larvae (Fig. 4.9). The relevance of ascidians to vertebrate origins has increased considerably with the recent publications of molecular phylogenies that place urochordates rather than cephalochordates as the sister group of vertebrates (Fig. 4.2).[14] A larval CNS consisting only of 330 cells, the

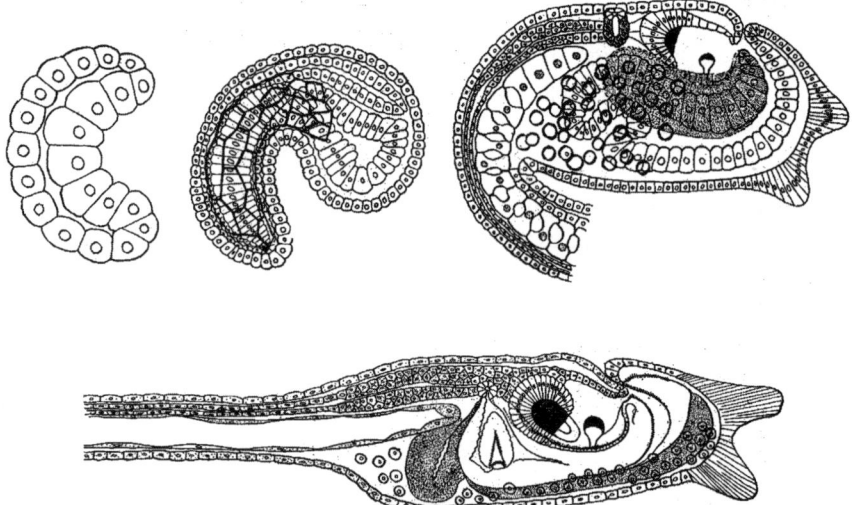

Fig. 4.9 Ascidian larvae for notochord and neural tube and metamorphosis

---

$^{\oplus}$ Although only one species of *Ciona*, *C. intestinalis*, has been described, four subspecies are known. Consequently, there may be undocumented variation in data from *Ciona* reflecting evolution among the subspecies. Genes in *C. intestinalis* are prefixed *Ci*, genes in *H. roretzi*, *Hr*.

ability to trace the lineage of those cells (13 generations), the phylogenetic position occupied by ascidians, and the sequencing of the *Ciona* genome, all make ascidians an ideal group in which to search for origins of the NC.

What about the other urochordates?

As discussed in the last section on cephalochordates, the chordate body plan is established by retinoic acid activation of *Hox* genes in vertebrates and cephalochordates, a signaling pathway that was a major innovation at the origin of the chordates. However, in the larvacean *Oikopleura dioica*, as in ascidians, retinoic acid does not influence the anterior–posterior axis. Cañestroa and Postlethwait (2007) interpret the difference as a consequence of the lack of clustering or co-linear expression of *Hox* genes in *Oikopleura*. These fundamental differences demonstrate the need to investigate more species of ascidians, as well as larvaceans and salps, to appreciate urochordate diversity and the insights urochordates can provide in understanding chordate origins and evolution.[15] That said, the discussion below is limited to ascidians.

Nervous System and Notochord

A dorsal nerve cord and notochord are present in free-swimming ascidian larvae (Fig. 4.9), but not in the sessile adults; larval swimming is by segmental muscles attached to the undulating and flexible notochord. The larval notochord and a pharyngeal basket in adults affirm that ascidians are chordates.

Structural features in vertebrate and ascidian brains along with shared processes by which their neural plates and dorsal nerve cords are induced are consistent with **a prevertebrate origin of neural induction**. Shared processes at the molecular level include:

- duplicated *Hox* genes regulated by retinoic acid,
- expression of the gene *Brachyury* in the notochord under the control of a regulatory element related to the vertebrate Notch signaling pathway, along with *Brachyury* regulating downstream genes in the *Wnt* and *Tgfβ* pathways,[16]
- expression of *Pax2/5/8* (see Fig. 4.8) in a region of the brain of the sea vase, *Ciona intestinalis*, that is homologous to the midbrain–hindbrain organizer in vertebrates, and
- expression of the transcription factors *CiPhox* 2 and *CiTbx20* in motoneurons that originate from a region of the brain equivalent to the vertebrate hindbrain.

The latter point was used by Dufour and colleagues (2006) as evidence for the precraniate origin of cranial nerves against the background that cranial sensory neurons arose with the new vertebrate head and that vertebrate cranial motor neurons are homologous to the motoneurons found in adult *Ciona*.

Ascidians also possess a population of neuronal and neuroendocrine cells associated with the dorsal strand of the nervous system that may arise from (and therefore migrate away from) the dorsal strand.

Pigment Cells

The presence in ascidians of sensory pigment cells induced by the dorsal neural tube, and of ectodermal sensory cells with primary neurons that send axons to the central nervous system, provides evidence of the ancient evolutionary origins of vertebrate sensory systems.[17]

Recall that cephalochordates lack pigment cells. Ascidians, however, possess a band of pigmented cells along the dorsal aspect of the neural tube that has been regarded as presaging the NC origin of vertebrate pigment cells. Lineage labeling of early embryonic cells of the mangrove tunicate with the lipophilic dye DiI—which is incorporated into cell membranes and so passed along at each cell division—demonstrates that these cells arise near the neural tube, and then migrate into the body wall and the primordia of the siphon, where they differentiate into pigment cells. The earliest labeled cells are mesenchymal in appearance, and, while they may delaminate from the neural tube, delamination has not been conclusively demonstrated.

Even more tantalizing, and in contrast to any cells within amphioxus, these cells express two gene markers of NCCs. One is the HNK-1 antigen, and the second is a *Zic* gene; recall from Chapter 2 that *Zic3* is one of the earliest genes involved in neural ectoderm and NC induction. Jeffery and colleagues (2004) conclude that migratory cells with pigment-forming capability arose in the neural tubes of the common ancestor of ascidians and vertebrates. The alternative interpretation—that these HNK-1-positive cells evolved independently in ascidians and vertebrates—is rendered much less likely by the finding that four other species of ascidians, representing evolutionarily divergent families, also contain HNK-1-positive cells (Jeffery, 2007).

In a scenario with ascidians as the sister group to vertebrates, and NC-like cells in ascidians before the origin of NCCs, we should reevaluate the original function of the first NC cells in vertebrates, ask whether their precursors had a pigment cell function, and consider whether the NC-like cells in ascidians functioned as a 'proto-neural crest' that co-opted genes and gene pathways from mesoderm, nonneural ectoderm, or pharyngeal endoderm, just as basal cephalochordates are proposed to have done. The absence of HNK-1-positive cells (and of pigment) from the more derived cephalochordates is most parsimoniously interpreted as a secondary loss, possibly associated with the amphioxus way of life, buried in marine sediments, an environment in which selection for pigmentation is unlikely to have been strong. That said, before too many conclusions concerning evolutionary patterns can be drawn, we need data from many more ascidians, especially to demonstrate links to the neural tube and to examine larvaceans and thaliaceans.[18]

Calcitonin

A further sign of a NC precursors in ascidians comes from cells in the floor of the pharynx that are reactive for calcitonin, a vertebrate calcium-regulating hormone synthesized and secreted by cells of NC origin. While not an evidence for an ascidian NC, these cells are certainly evidence of cellular activity comparable to that found in derivatives of vertebrate NCCs, activity that could have been transferred from the pharynx to the NC, as proposed for genes expressed in pharyngeal endoderm in amphioxus.

Research conducted into the role of β-catenin in ascidian endoderm may provide some hints of the basis from which endodermal gene transfer could have occurred. β-Catenin has been shown to be involved in the specification of endodermal cell fate in two ascidians, the sea vase, *Ciona intestinalis* and its congener, the Pacific sea squirt, *Ciona savignyi*. Misexpression or overexpression of β-catenin is followed by the upregulation of an endodermal-specific alkaline phosphatase in notochordal and epidermal cells. Downregulation of β-catenin is followed by inhibition of the endodermal cell lineage and the deployment of additional cells into the epidermal lineage (Imai *et al.*, 2000). Such lability of lineage specification coupled to a gene, β-catenin, that plays a central role in cell specification in vertebrates, encourages us in the search for ancestral genetic pathways.

Bipotentiality and Conditional Specification

Two basic (although not exclusive) properties of NCCs are bi/multipotentiality, and conditional specification of cell fate (see Chapters 3, 6, and 7).

Ascidians exhibit determinate development and specify cell fates autonomously. The fate of the majority of their cells is fixed through the inheritance of specific cytoplasmic constituents, rather than by cell–cell interactions, although the fate of a minority of muscle cells is set by inductive interactions and not autonomously. One would not expect such determinate cells to be able to change their fate. Whittaker (1987), however, showed that presumptive pigment cells could be induced to express a different cell fate if allowed to interact with endoderm. Bipotentiality (see Table 1.2) and the ability to respond to inductive interaction therefore preceded the origin of the vertebrates.[19]

Genetic Control of Ascidian Neural Development

We saw in the previous sections that the dorsal nervous system and pharyngeal endoderm of extant cephalochordates express numerous genes and pathways associated with NCC in vertebrates. Genes associated with neural development in ascidians are therefore of considerable interest, especially if ascidians rather than cephalochordates are the nearest vertebrate relatives.

The *Ciona intestinalis* genome, some 160 Mb in size, contains around 16,000 protein-coding genes, about 80% of which are found in vertebrates (Dehal *et al.*, 2002). Some of these shared genes specify and/or regulate the formation of vertebrate eyes, heart, thyroid gland, and the immune system—organs not present in ascidians—pointing to the ancient origins of these genes and to roles that predate the origin of the vertebrate organs. Given that duplication of the vertebrate genome occurred after the origin of ascidians, ascidians have the potential to provide us with a glimpse of the ancestral, nonduplicated genome. In the course of examining ascidian genes, many genes familiar from vertebrates have been found.

Bmps

Bmp7 plays an important role in the induction of the NC (see Figs. 2.4 and 2.5). An ortholog of *Bmp7*, *HrBmp7* (*HrBmpa*) is expressed in neurectoderm at the boundary of neural and epidermal ectoderm in the sea squirt, *Halocynthia roretzi*. Should ascidians be shown to use this ortholog of *Bmp7* to establish neural ectoderm, this fundamental patterning mechanism would predate the ascidian/chordate split.[20]

Snail and *Hnf3β*

Ascidians and vertebrates share other important molecular regulators of neural tube patterning. *Ciona* contains an ortholog of the vertebrate forkhead gene *Hepatocyte nuclear factor3β* (*Hnf3β*) that, like the vertebrate gene, is expressed in ventral midline cells of the developing neural tube. *CiSnail2,* the ascidian ortholog of vertebrate *Snail2,* a marker for dorsal neural cells and NC, is expressed in *Ciona* at the border of neural and dorsolateral ectoderm, an equivalent position to the expression of *Snail2* in vertebrates.

As noted when discussing *AmphiSnail1*, *CiSnail1*- positive cells form the ependymal lining of the spinal cord. In ascidians, as in amphioxus, *Snail1*-positive cells have therefore diverged from a strictly neuronal lineage.[21]

Pax Genes

In vertebrates, the homeobox-containing gene, *Pax7* is expressed in midbrain NC and in r1, r3, and r5 (Fig. 4.7); in zebrafish, *Pax7* is expressed even more broadly within NCCs. Its importance is illustrated by the finding that malformations of facial structures with a NC origin occur in mice after *Pax7* is disrupted.

An earlier and even more major role for *Pax3* has been identified during gastrulation in chicken embryos when *Pax3* is involved in the specification of **a NC field independent of the specification of neural ectoderm or mesoderm**. Further study is required to determine whether a NC field is specified in other vertebrates, and if it is, whether specification occurs before neurulation. Such an early role for *Pax7* would be consistent with its regulation of other genes expressed in NCCs: *Snail1*, *Sox9*, *Sox10*, and *HNK-1*.[22]

Pax7 and *Pax3* are functionally redundant in vertebrates. *Ciona* and *Halocynthia* both express the genes *Pax3/7* and *Pax 2/5/8* in neural tube cells. *HrPax3/7* is expressed in future dorsal neural tube, dorsal epidermis, sensory neurons, and in 15 segmental spots within the neural tube, perhaps reflecting an ancestral segmental pattern.[23]

Wada *et al.* (1998) used expression domains of *HrPax2/5/8* to divide the brain of *H. roretzi* into regions they consider homologs of vertebrate fore-, mid-, and hindbrain and spinal cord. Expression in the primordia of the peribranchial space (atrium) in ascidians was taken to indicate that this sensory organs, with its origin in an ectodermal thickening, is a homolog of a placode or placode precursor in ascidians (see Box 4.2 for placodes in ascidians), although recall that not all agree that structural homology can be established solely on the basis of expression boundaries of single genes, even important single genes.[24]

Although *HrPax3/7, AmphiPax3/7,* and *Pax3* and *Pax7* are orthologous genes shared between ascidians, cephalochordates, and vertebrates, respectively, comparative molecular and phylogenetic analysis shows that the single *Pax3/7* genes found in ascidians and in amphioxus form a separate cluster outside the vertebrate genes (Fig. 4.8A). Two species of lampreys investigated—the sea lamprey and the European river lamprey—each have a single *Pax3/7* gene, the ortholog of vertebrate *Pax3* and *Pax7* that also lies outside the gnathostome cluster (Fig 4.8A). As elaborated below when discussing the evolutionary origin of cartilage, far more is involved in gene evolution than whether orthologous are expressed or not.

Fossil Chordates

Does the fossil record, which offers an entirely different type of evidence, provide any information about the origin of NC derivatives? We do not expect to see embryos with migrating NCCs, but we might expect to see pharyngeal cartilages, cartilages around the sense organs of the head, and even pigment, features known to be of NC origin in vertebrates.

A number of putative[⊕] cephalochordates have been identified in Cambrian fauna. The fossils from two faunas are examined here, the Burgess Shale Formation located in what is now Canada and the Chengjiang Formation from what is now China.

Burgess Shale

All the fossils in the Burgess Shale are invertebrates with one possible exception. Because of a putative notochord, the presence of segmental muscle blocks, and its amphioxus-like appearance, [†]*Pikaia gracilens,* from the Burgess Shale inshore marine fauna of British Columbia is thought to be a cephalochordate. Although the many specimens discovered are highly suggestive of cephalochordate characters, the chordate affinities of [†]*Pikaia* have yet to be subjected to a rigorous analysis.

⊕ The term putative in this and similar contexts is used to mean that the organism or feature looks as if should belong to the group but the evidence is inconclusive, not universally accepted, incomplete, or not sufficiently analyzed.

Chengjiang Formation

Discoveries in the 530-million-year-old, Early Cambrian Chengjiang Formation in Yunnan Province, China—a formation that is 10-million years older than the canonical Early Cambrian fauna from the Burgess Shale—allow us to see how animals identified as cephalochordates were constructed, and how interpreting those structures creates such problems in assigning organisms to major groups.

One species, [†]*Yunnanozoon lividum,* is classified by some paleontologists as a cephalochordate because of the interpreted presence of a notochord, filter-feeding pharynx, endostyle, and segmented muscles, pharyngeal arches, and gonads. Others interpret [†]*Yunnanozoon* as a hemichordate (acorn worm), the pharyngeal arches being branchial tubes and the notochord the upper portion of the pharynx.

Further discoveries of 14 specimens of animals up to 40 mm in length have been assigned to one or two species in the genus [†]*Haikouella.* More derived than [†]*Yunnanozoon* and interpreted as hemichordates because of their more derived features, [†]*H. lanceolata* and [†]*H. jianshanensis* have a notochord, six pharyngeal arches, a heart, pharyngeal teeth, and what has been interpreted as a tail with a fin.[25]

[†]*Cathaymyrus diadexus,* a cephalochordate from the Chengjiang formation, initially described from a single 22-mm-long specimen in 1996 by Shu and colleagues, has a pharynx, gill slits, segmented 'myomeres', and a putative notochord. A second species, [†]*C. haikoensis* has since been discovered.

Two further genera from the Chengjiang Formation are [†]*Myllokunmingia fengjiaoa* and [†]*Haikouichthys ercaicunensis.* Both have pharyngeal clefts, myomeres, and a well-developed dorsal fin containing skeletal elements. Both share many similarities with extant hagfish. Despite their lack of fossilized mineralized tissues, Janvier (2007) considers the [†]Myllokunmingiida (the group name for these animals) the most convincing Cambrian vertebrates known and the closest vertebrate group to the cephalochordates (Figs. 4.2 and 4.3). Why? Because of the evidence of skeletons associated with the gill slits, and because of cartilaginous olfactory, optic (and possible otic) capsules (Fig. 4.3). Parsimony allows us to conclude that these would be NC-derived skeletal elements.

Does the presence of a pharynx and gill slits in these organisms mean that the mesenchyme in the pharyngeal arches was NC in origin? Not necessarily. Mesoderm-derived mesenchyme may have filled the arches in these ancient chordates as it does in extant cephalochordates (see below).

Does the presence of a pharynx and gill slits mean that a cartilaginous skeleton was present but not fossilized. Not necessarily. As discussed below, a collagenous skeleton could have functioned equally well in the early chordates, and such a skeleton could have been NC or mesodermal in origin.

Despite the problems of interpretation, the Early Cambrian fossil record demonstrates that the evolutionary history of the NC and the nature of the first vertebrates are tied closely to the evolution of the muscular pharynx, brain, sensory apparatus, and associated skeletal capsules of a newly evolving head. As discussed below and in Chapter 7, skeletal elements within pharyngeal arches, jaws, and a skull were later evolutionary events.

The First Vertebrates

Given our understanding of the genes and cell types found in association with the nervous system in extant cephalochordates and ascidians, and given the evidence of the presence of a brain, muscular and skeletonized pharynx, and paired sensory organs with cartilaginous tissues in Cambrian vertebrates, how do we imagine that the NC arose as a germ layer producing neuronal *and* skeletal tissues?

An innovative synthesis and scenario that ushered in the modern era of investigation and prominence of the NC and NCCs was presented in the 1980s by Carl Gans and Glen Northcutt, building on foundations established by E. S. Goodrich, N. J. Berrill, A. S. Romer, B. Schaeffer, and others.[26] Their new-head hypothesis drew from and integrated three fields within biology:

- *Comparative anatomy and neurobiology*, especially the homology of NC-derived neurons with the epidermal nerve net of hemichordates—a dorsal nervous system could have arisen from the coalescence of an epidermal nerve net around the dorsal midline—and, as discussed in Chapter 2, the importance of ectodermal placodes in sense organ formation.
- *Developmental biology* especially detailed knowledge of the developmental origins and fates of NCCs, and findings on the age-old problem of segmentation of the vertebrate head, some of which were discussed earlier.[27]

As outlined in the first part of this chapter, in the intervening quarter of a century, molecular biology and paleontology have contributed enormously to our understanding of vertebrate origins. The following sections concentrate on the origin of the pharyngeal skeleton, cartilage, and jaws.

The Pharyngeal Skeleton

In a brief outline, the scenario for the origin of a skeletal system associated with the pharyngeal arches is as follows:

(1) Early chordates were soft bodied, lacking a mineralized skeleton and perhaps lacked any skeleton at all. The fossils discussed in the previous section bear this out.

(2) Early chordates probably lived in a near-shore or estuarine marine environment, obtaining food by filter feeding through a ciliated branchial basket. The fossils discussed in the previous section bear this out. One can imagine animals like †*Pikaia gracilens*, †*Myllokunmingia fengjiaoa*, and †*Haikouichthys ercaicunensis* fitting this description.

(3) The evolution and elaboration of a notochord, central nerve tube, and axial muscular system allowed these animals to become more mobile, use muscles rather than cilia for locomotion, use the notochord as a stiff rod to avoid

antero-posterior telescoping during bending of the body, and use the dorsal nerve cord to integrate sensory input and coordinate movement.

(4) With increasing muscularization of the body came the development of muscles in the wall of the pharynx facilitating the pumping of water and allowing the pharynx to function during gas exchange and for filter feeding.

(5) The advantage of storing energy by deforming the muscularized pharynx—allowing pumping by an elastic recoil mechanism—is hypothesized to have led to the differentiation of cartilage from cells that were previously exclusively neural. A number of molecular similarities of cartilage and nerve cells are consistent with such a switch (Table 4.1).

Given these shared features and that hemichordates and extinct cephalochordates possess skeletal pharyngeal bars but ascidians do not, what was the origin of vertebrate cranial and pharyngeal cartilages? Was it neural, as proposed in the new-head hypothesis (point 5 above), mesodermal or endodermal?

The Origin of Cartilage

Cartilage, a primary skeletal tissue in vertebrates, did not evolve with the vertebrates, at least not if we recognize cartilage as a cellular supporting tissue with an extracellular matrix composed of fibrous proteins and proteoglycans. Using this definition, cartilage is a surprisingly common tissue in a variety of invertebrates—coelenterates (medusae and polyps), annelids (sabellid or feather-duster worms), mollusks (snails, cephalopods), arthropods (the horseshoe crab, for example)—and

Table 4.1 A number of molecules shared by chondrocytes and neurons

Molecule	Shared distribution
Chondromucoid	A basic ECM component of primitive cartilage, orthologous to molecules expressed in mechano- and electroreceptors.
Type II collagen	Expressed in embryonic central nervous systems, epithelial basement membranes, and in the perinotochordal sheath.
Aggregan	A proteoglycan expressed in embryonic central nervous systems.
Link protein	The protein linking proteoglycans to other chondrocyte matrix components is expressed in embryonic central nervous systems.
The S100 acidic protein	Putative regulator of Ca^{++}-mediated cell functions, initially thought to be unique to nerve cells, is present in chondrocytes and chromatophores.
CSPG	'Cartilage-specific' chondroitin sulfate proteoglycan (CSPG) is also expressed within rat cerebellar astrocytes (unless they differentiate into oligodendrocytes, in which case they stop synthesizing CSPG), providing a nice example of a component shared by chondroblasts and neurons and of the modulation of an ECM product with altered differentiation.

is the only skeletal tissue in lampreys and hagfish.[28] (Cartilaginous fish possess bone but only in very small amounts and few locations.)

Invertebrate cartilages can be either epithelial or mesodermal in origin (see Cole and Hall, 2004a,b, for reviews). Basal chordates could have generated cartilage from ectodermal or mesodermal cells; pharyngeal-arch cartilages are mesodermal in origin in extant cephalochordates (amphioxus) but are NC in origin in all vertebrates. As we will see below from recent studies, endodermal contributions to the origin of cartilage cannot be excluded.

Lack of evidence for type II collagen in any of the invertebrate cartilages investigated is one reason why invertebrate cartilage has not been regarded as homologous to vertebrate cartilage. However, as discussed in Chapter 7, neither lamprey nor hagfish cartilages use type II collagen as their structural fibrous protein, although *Col2a1*, the type II collagen gene,[⊗] is expressed in perichondrial cells of some cartilages. Indeed both lampreys and hagfish display a diversity of cartilage types within a single species (see Fig. 4.12). Therefore, cartilage can be homologous as a tissue or organ, whether or not type II collagen is present.

In their scenario for the basal prevertebrate chordates, Gans and Northcutt saw:

* collagen as the primitive structural molecule of connective tissue,
* cartilage as the first skeletal tissue, and
* bone—whether NC or mesodermal—as a later evolutionary development(s).

A dermal osseous skeleton is seen in many radiations of the earliest jawless vertebrates but in no fossil or extant cephalochordates. Cartilage enhances pharyngeal gas exchange, while a bony exoskeleton limits diffusion. The early vertebrates could not give up gaseous diffusion across the skin until they possessed an alternative gas-exchange mechanism utilizing the muscular, cartilage-supported pharynx. Therefore, Gans and Northcutt concluded that cartilage must have existed before early vertebrates could afford to ossify the skin.

As discussed in the previous section, cartilaginous sense capsules and putative evidence for cartilaginous pharyngeal arches are found in Early Cambrian chordates, some of which are cephalochordates or putative cephalochordates (Fig. 4.2). Unmineralized cartilage would have fossilized only under rare and unusual circumstances, making confirmation of the presence of cartilage difficult. Extant cephalochordates are not helpful. They have skeletonized pharyngeal arches, but the tissue is acellular.

So, if we take the fossil record and at face value, the first skeletal tissue was cartilage associated with pharyngeal arches and sense organs. If we take the data on cell types and gene expression in amphioxus at face value, the skeleton of basal

---

[⊗] For many years, *Col2a1* has been regarded as the 'cartilage collagen gene'. However, we will see that *Col2a1* is expressed in the notochord and in endodermal cells. Furthermore, many NCCs express *Col2a1*, often in combination with genes such as *Sox9*, *Sox10*, and *LSox5* activate the collagen gene (see Fig. 4.10). So general is this combination of gene expressions that T. Suzuki and colleagues (2006) concluded that *Col2a1* is a 'general mesenchyme gene'.

Fig. 4.10 Downstream targets of *Sox9* as identified in chicken embryos. *Sox9* is an important regulator of *Sox10* in NCCs and of *Col2a1* in mesenchyme and chondrogenic cells (both directly and via *Sox10*). *Sox9* also regulates *LSox5* in the peripheral nervous system (PNS). Adapted from Suzuki *et al.* (2006), with permission of the author and the publisher, Blackwell Publishing Ltd.

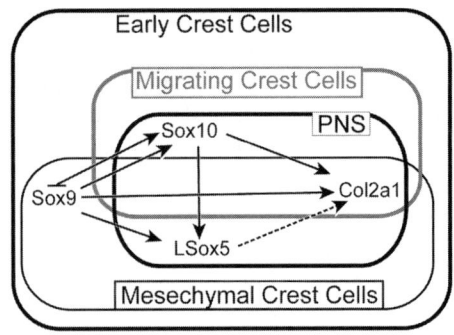

cephalochordates was not NC in origin. Although the jury is still out on whether that cartilage was neural or mesodermal, the weight of evidence supports a mesodermal skeleton in Cambrian cephalochordates (see below).

Given that the homologous vertebrate skeletal elements are NC in origin, how did the switch from mesoderm to NC occur? The answer is by gene co-option.

Meulemans and Bronner-Fraser (2007) addressed this issue head on by examining the expression patterns of 11 genes in amphioxus that are orthologs of genes known to be involved in NC-based chondrogenesis in vertebrates. Their search for proto-chondrogenesis genes—which parallels the search for a 'protoneural crest' in amphioxus discussed above—was very revealing, as can be seen from the data summarized in Table 4.2. Expression of the genes in cartilage, shown on the left in the table, can be compared with expression of the amphioxus orthologs in five developing tissues.

The search was successful: Meulemans and Bronner-Fraser found orthologs of four genes distributed among the five tissues. No single tissue expressed all vertebrate orthologs, meaning that no single tissue could be unambiguously identified as 'the' NC precursor. However, pharyngeal mesoderm expressed three genes that regulate or are associated with chondrogenesis in vertebrates (Table 4.2). Each of these is discussed in a little more detail below.

AmphiColA: Amphioxus has a single collagen gene (*AmphiColA*), which is the ortholog of vertebrate *Col2 a1*, the collagen gene specific to cartilage and also found in hagfish and lampreys. Interestingly, perhaps significantly, *Col2A* is expressed in cartilages of hagfish and lampreys that do not contain collagen type II.

ColA is expressed in amphioxus notochord, neural tube, and pharyngeal mesoderm. While *AmphiSoxE* is co-expressed with *AmphiColA* in the neural tube, and *AmphiSoxD* is co-expressed with *AmphiColA* in the notochord, neither gene is expressed in the pharynx of amphioxus. *SoxD* and *SoxE* regulate collagen gene expression and collagen deposition in vertebrate chondrogenesis and are markers for CNCCs (Fig. 4.10). The lack of co-expression of all three genes in a single amphioxus tissue is consistent with a nonchondrogenic role for *AmphiColA*, and interpretation that does not support wholesale co-option of a gene network from a

Table 4.2 Orthologs in amphioxus of genes expressed in the neural crest and in chondrogenic cells in vertebrates[a]

Vertebrate	Amphioxus			
	Notochord	Neural tube	Pharyngeal mesoderm[b]	Pharyngeal endoderm[c] medial somites
		CNCC markers		
CNCC				
SoxE → Twist1/2	SoxE,[d] Twist1/2	Twist 1/2		Twst1/2
↓				
SoxD	SoxD		SoxD	SoxD
Ets1/2		Ets1/2[e]	Ets1/2	
Id2/3		Id		
		Chondrogenesis markers		
Cartilage				
Col2a1	ColA	ColA		
Alx3/4		Alx3/Alx4/Cart1		
Barx1/2				
Cart1		Alx3/Alx4/Cart1		
Bapx1 → Gdf5				Bapx1
Runx1/2/3			Runx1/2/3	
Aggrecan			Fgf8/17/18	

[a] Based primarily on data in Meulemans and Bronner-Fraser (2007).
[b] No SoxD or SoxE in pharyngeal mesoderm of amphioxus.
[c] No Runx1/2/3 or Gdf5 in pharyngeal endoderm.
[d,c] Notochordal expression of SoxE is transient and confined to a narrow strip of ventral cells.
[e] Expression of Ets1/2 in pharyngeal mesoderm of first pharyngeal arch at late larval stage.

single cephalochordate tissue as the origination of NC-based chondrogenesis; also see Zhang and Cohn (2006) on this point.

AmphiSoxD: *AmphiSoxD*, however, is expressed in notochord, medial somitic mesoderm, and in pharyngeal endoderm (Table 4.2), while *AmphiSoxE* (which regulates *SoxD* and *Twist1/2*) and/or *Twist1/2* are expressed in all tissues other than pharyngeal endoderm. These patterns of expression do not support any amphioxus tissue as the proto-chondrogenic signaling tissue. Although not proof, the expression of the cartilage-regulatory genes in tissues adjacent to pharyngeal mesoderm and to notochord in amphioxus is consistent with the possible regulation of *AmphiColA* by tissue interaction. the alternate interpretation is that *AmphiColA* was not regulated by the *Sox* genes.

Alx3/4, Cart1: The vertebrate cartilage homeobox transcription factors, *Alx3/4* and *Cart1*, are present as a single gene in amphioxus, expressed in pharyngeal mesoderm (Table 4.2). This analysis shows the presence in amphioxus of elements of genes involved in vertebrate chondrogenesis, albeit distributed among several tissues, a distribution that is not consistent with a single proto-vertebrate tissue as a proto-chondrogenic signaling tissue. Meulemans and Bronner-Fraser (2007) concluded that the prevertebrate chordate would have co-opted genes from several basal cephalochordate tissues. If any tissue can be identified as the precursor of cartilage they believe it to be mesoderm, in part because of the patterns of gene expression shown in Table 4.2, and because of the presence of skeletal tissues in the pharyngeal arches of Cambrian cephalochordates discussed above. Their conclusion that the proto-chondrogenesis genes in the basal prevertebrate functioned primarily in the mesoderm, would mean that NCCs co-opted most of those genes from mesoderm rather than from neural or endodermal ectoderm.

As discussed earlier, given that the patterning role of pharyngeal endoderm in amphioxus predated the origination of NCCs, epithelial —> mesenchymal signaling to activate chondrogenesis (see Box 3.4) could have operated in the proto-vertebrate chordate. Of course, extant cephalochordates have undergone half a billion years of evolution since they split from vertebrates. Proto-chondrogenesis genes would be expected to have changed, been lost, or altered their function. Nevertheless, and despite these caveats, the data from extant amphioxus are consistent with the NC having co-opted genes from several tissues and having done so piecemeal rather than *en bloc*; proto-chondrogenic genes from pharyngeal arch mesoderm, and signaling molecules from the neural tube, mesoderm, and pharyngeal endoderm (Table 4.2).

Comparative Genomics and Bioinformatics

An indication of the power of genomic analysis when coupled with a bioinformatics approach may be seen in the recent study by Martinez-Morales *et al.* (2007), in which 615 NC genes were identified and compared across the animal kingdom. The approach, results, and significance of the study are outlined below.

All 23,658 known mouse protein sequences were sequentially blasted[⊗] against genomes from seven groups of organisms organized into prokaryotes, eukaryotes, metazoans, deuterostomes, chordates, vertebrates, and mammals. Orthology with a prokaryote protein was taken as evidence of a protein with an ancient evolutionary history. Successive comparisons with genes from the seven groups of organisms detected more and more recent proteins (and so genes); 'recent genes' are defined as new genes on the basis of their absence from the previous group of organisms to which they were compared. Frequency of gene emergence was then calculated against divergence times for the groups (Fig. 4.11). Exciting stuff, but that was only step one.

The second major step was to build a tissue-by-tissue list of gene functions by probing 14,000 genotype records associated with 6,442 genes in the Mammalian Phenotype Browser. These records of mutant mice associate genes with features (genotype with phenotype), using such categories (ontologies, in the new terminology) as pigmentation phenotype, bone derivatives phenotype, abnormal autonomic nervous system morphology, abnormal peripheral nervous system glia, and so forth. By combining the two vast data sets, Martinez-Morales and colleagues categorized the 615 NC gene and their functions. In support of the concept of co-option of genes into the NC from ancestral organisms, 91% of the NC genes were present in basal metazoans or even earlier in organismal evolution. These include genes that in vertebrates specify:

- the neural plate–epidermal ectodermal border (*Pax3*, *Zic3*, *Msx1*, *Msx2*),
- gene families involved in the induction of the neural plate (Fgfs, Wnts, Bmps), and
- genes that specify the NC itself (*Snail1*, *FoxD*, *Sox9*, *Sox10*, *Twist*).

This is a remarkable demonstration of the antiquity of genes we associate with the NC as a vertebrate innovation, and of the concept of co-option or gene recruitment in the evolution of novelties such as the NC. A similar analysis for NC lineages showed heightened emergence rates in the transition from chordates to vertebrates for NC derivatives as a whole and for bone, but not for neural or pigment derivatives (Fig. 4.11).

Further analysis established emergence rates for categories of genes: the higher the emergence rate, the more associated the genes are with the origin of a new feature. The genes could be divided into three groups: (i) When plotted against the groups of organisms, genes associated with ventral mesoderm and NC (and renal–urinary, hematopoietic, and immune systems) show a major shift in emergence rates around the time of the transition from deuterostomes to chordates to vertebrates. (ii) Genes associated with neural, muscle, and endocrine/exocrine systems show less rapid emergence rates. (iii) Genes associated with the intestinal, skin,

---

[⊗] BLAST (**B**asic **L**ocal **A**lignment **S**earch **T**ool) refers to a number of computer programs developed to compare genomes on the basis of alignment of short sequences of genes of interest.

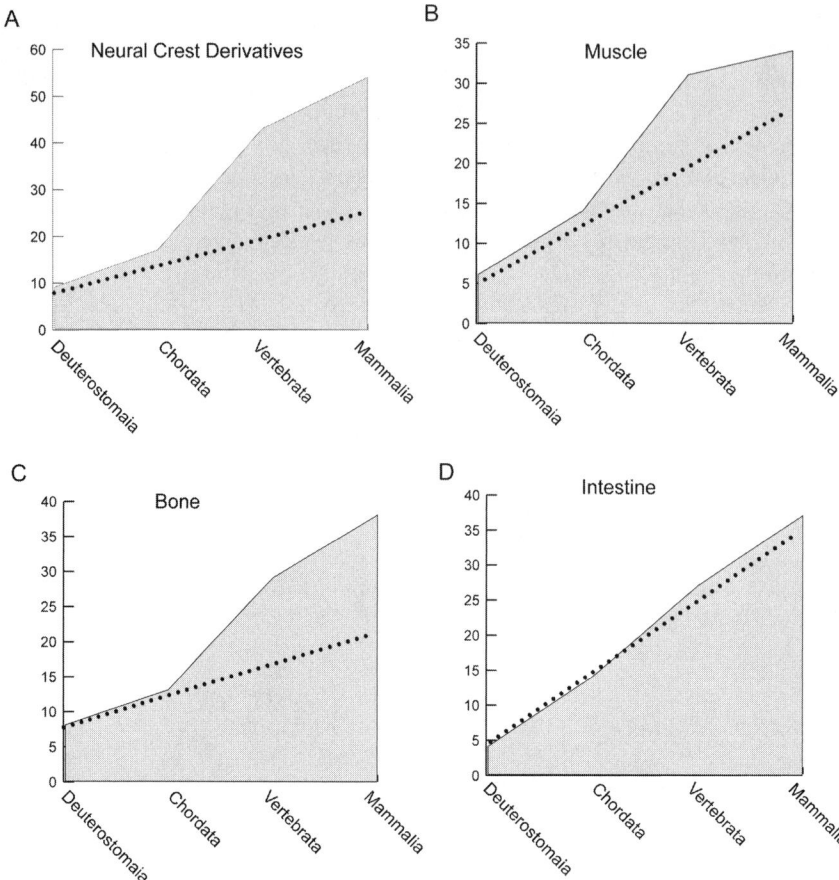

Fig. 4.11 Gene emergence rates plotted against four higher taxa to show the increase in the rates of origin of new genes with functions in NC, muscle, and bone (**A**, **B**, and **C**) in comparison the much lower rate of emergence of genes with functions in the intestine (**D**). Modified from Martinez-Morales *et al.* (2007)

and liver systems show no change in emergence rate across the same organismal transitions (Fig. 4.11).

The final analysis was of the classes of 'NC genes,' associated with the organismal transitions, a NC gene being defined as first appearing in vertebrates and known to be associated with the NC or NCC. A major finding was that 50% of the NC genes encode extracellular ligands that function in the commitment of NCCs to individual lineages—*Bmp2/4*, *Wnt* signaling, kit ligand, *endothelin1* and *3*, neuroregulins, and *neurotropin3* and *4*. Their interpretation of these findings is that origination of receptor ligands was an important event in the origin of the NC, as it is likely to have been in the origin of other novelties.

Molecular Fingerprinting: Genetic Labeling/Selection and GeneChip Microarray Technology

A second approach, utilizing a combination of technologies unavailable a few years ago, is an analysis of patterns of gene expression in mesodermal and NC-derived mesenchyme in the first (mandibular) pharyngeal arches of mouse embryos, undertaken by Bhattacherjee *et al.* (2007).

NCCs and NC derivatives in mouse embryos were labeled using a Wnt1Cre/ZEG transgenic marker that enabled GFP-labeled NCCs to be discriminated from unlabeled mesodermal mesenchyme in frozen sections of mandibular arches from 9.5-day-old mouse embryos (see Table 2.1). The two mesenchymal populations were then isolated from frozen sections under epifluorescence illumination, using the technique of laser-capture microdissection, essentially using the fluorescence as the marker to identify cells for isolation (see Figure 1 in their paper). RNA was then extracted, amplified, and transcribed, gene expression profiles were generated by hybridization to oligo-based GeneChip microarrays, and the gene profiles were compared using real-time PCR.

Over 140 genes with statistically different levels of expression between the two populations of mesenchymal cells were identified. Some were genes previously known to be involved in NC or mesodermal differentiation; others were uncharacterized coding sequences. With respect to genes discussed in the text to now, levels of *Msx1* and *BmpR1* were higher in NC and mesodermal cells, respectively.

As in the comparative genomics/bioinformatics approach described above, patterns of expression of classes of genes emerged; mesenchymal cells of mesodermal origin—which, in the mandibular arch are primarily muscle precursors—expressed more than twice the number of genes associated with growth and differentiation and 3.4 times the number of transcription factors as did NC-derived mesenchymal cells. NC-derived cells, on the other hand, expressed significantly more kinases and phosphatases. The potential of this approach for analysis of development stage by stage or tissue by tissue is evident, as is its potential for application to comparisons of cephalochordates/ascidians with vertebrates.

Jawless Vertebrates and the Origin of Jaws

How can we speak of the precursors of jaws in animals (jawless vertebrates) that have not evolved jaws? Developmentally, the question is how do we identify a mandibular arch in agnathans, such as hagfish and lampreys, that have not evolved mandibles? The answer to the question—which applies equally to the evolutionary origin(s) of the NC itself—depends upon morphological, developmental, and molecular evidence.

The ability to recognize and identify such features as mandibular arches in jawless vertebrates (this chapter), precursors of middle ear ossicles in 'mammal-like reptiles' (see Box 7.1), 'prototissue' in sponges, or a 'protoneural crest' in ascidians and amphioxus (this chapter), goes to the heart of the concept of **homology**. Fortunately, homologies can be determined even when features have diverged substantially during evolution, allowing us to equate features even though they are not morphologically similar. This is because homology is a statement about shared inheritance not about structural identity, as illustrated by the evolution of the mammalian middle ear ossicles (Box 7.1).[29]

The origination of vertebrate jaws from the most rostral gill arches of jawless vertebrates—jaws evolved from gill arches—is the traditional view of the transition from jawless (agnathan) → jawed (gnathostome) vertebrates (see Janvier, 1996*, 2007* for reviews). However, although traditional, it may not stand up in the light of current knowledge.

Jaws from Gill Arches?

Whether jaws evolved from the most rostral gill arches of jawless ancestor is being reevaluated on the basis of studies of lampreys, one of the two groups of extant agnathan whose skeletons are discussed in Chapter 7. To understand the evolutionary origin of the jaws, craniofacial, and visceral skeletons of jawed vertebrates, we need to discuss what is known of the developmental origin of the head and pharyngeal arch skeletons in hagfish and lampreys.

Lampreys possess a **rostral velum** that functions as a pharyngeal pump (Fig. 4.12). The velum is derived from the mandibular arch but *not from a gill arch*. The rostral arches of hagfish are also associated with a velum rather than with the gills, although some question the homology of hagfish and lamprey arches, and therefore the homology of the velum in hagfish and lampreys (Fig. 4.12).

Fig. 4.12 The craniofacial skeletons of hagfish, lampreys, and gnathostomes (represented by a shark) as seen from the left-hand side, anterior to the *left*. Cartilaginous olfactory and otic capsules (*pale gray*) are homologous between the three groups. The lingual (*dotted*) and velar (*black*) skeletons are found in hagfish and lampreys. Dorsal elements of vertebrae (*dark gray*) fully developed in gnathostomes, are shown in the lamprey as arcualia but their homology to vertebrae is inconclusive. Jaws (shown with teeth) define gnathostomes. Adapted from Janvier (2007)

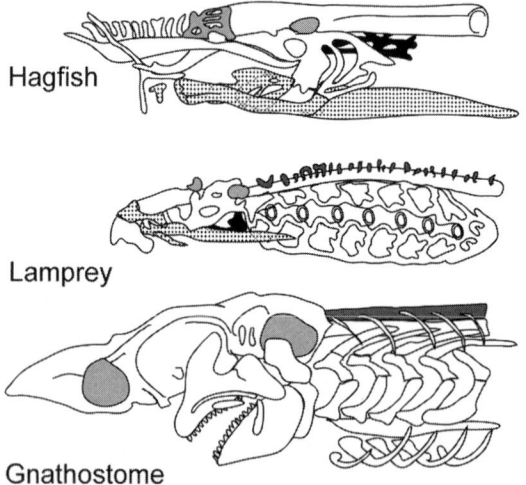

Hagfish

Lamprey

Gnathostome

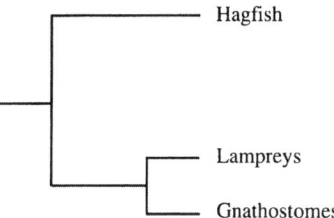

Fig. 4.13 Lampreys are more closely related to gnathostomes (jawed vertebrates) than either is to hagfish

An important issue is the phylogenetic relationships between extant agnathans, on the one hand, and jawed (gnathostome) vertebrates, on the other: are lampreys more closely related to jawed vertebrates than are hagfish? Although debate continues, the weight of evidence supports the view that lampreys and hagfish had quite different evolutionary histories, only lampreys sharing a close evolutionary relationship with jawed vertebrates (Figs. 4.2, 4.3 and 4.13). Indeed, while lampreys and hagfish are **craniates**; that is, chordates with a head, hagfish are often not included with the vertebrates because of their lack of a vertebral column (Fig. 4.1), although traces of elements dorsal to the nervous system may represent degenerate hemal arches, that is, remnants of ancestral vertebrae (Fig. 4.12). As Ota and Kuratani (2007) commented in their evaluation of hagfish embryology, if the hagfish condition did not arise secondarily from degeneration, then 'In the strict sense, the hagfish is an invertebrate' (p. 329). Lampreys are vertebrates (Fig. 4.12).

Evidence from lampreys can be interpreted as supporting one of two opposing theories:

- jaws arose via the transformation of a preexisting pharyngeal (mandibular) arch, with the consequence that jaws and agnathan pharyngeal arches are homologs (Figs. 4.14 and 4.15); or
- the velum rather than a pharyngeal arch is the homolog of the jaws, with the consequence that the mandibular skeleton evolved by modification of the velar skeleton (Fig. 4.12), with the consequent addition of a new anterior arch.[30]

If the mandibular arch arose from the first pharyngeal arch of jawless vertebrates, *then* the *first pharyngeal arch of jawless vertebrates is a homolog of the mandibular arch of jawed vertebrates*, and the skeleton of the first pharyngeal arch is a homolog of the skeleton of the mandible, a theory that has prevailed since the 19th century.

If, however, the mandible did not arise from a pharyngeal arch—whether the most rostral (first) or not—*then*, despite their strong similarity, *mandibular and pharyngeal arches are not homologous*, and even though they share many genetic, developmental, and structural features, vertebrate jaws are an evolutionary novelty.

Knowledge of the embryological origin of cell populations (Fig. 4.15), elements of the craniofacial skeleton (Fig. 4.12), and patterns of gene expression all are critical when determining homologies between hagfish, lampreys, and gnathostomes.

A

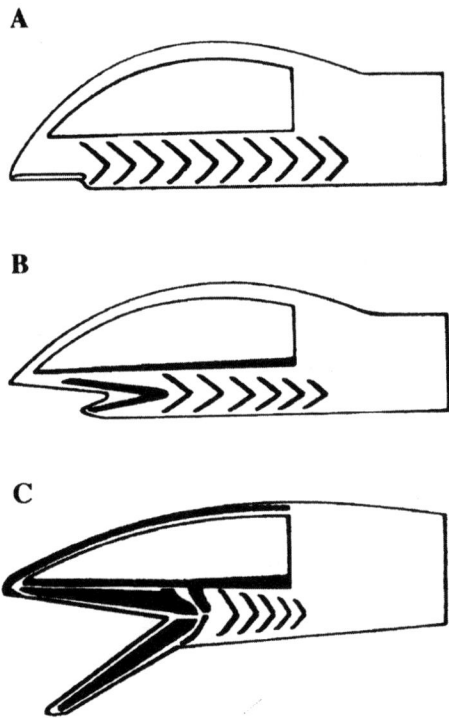

B

C

Fig. 4.14 (**A**) An agnathan, showing the serially arranged gill arches. (**B**) Transformations of the first of the serial gill arches to produce the jaws in gnathostomes. (**C**) The jaws were further strengthened, elongated, and supported by the skeleton of the second pair of gill arches. Modified from du Brul (1964)

Genes, cell populations, and skeletal elements of hagfish and lamprey are discussed more fully in Chapter 7. As one example, the *engrailed* gene in the Japanese lamprey is expressed in a mandibular arch muscle, the velothyroideus, which activate the velum. This muscle has been homologized with the *levator arcus palatini* and *dilator operculi* muscles of vertebrate mandibular arches, and the shared gene expression patterns used to conclude that lamprey velar and gnathostome mandibular arches are homologous.[31]

In summary, despite the apparent difficulties of knowing what really happened half a billion years ago, we have made surprising and exciting advances in our understanding of the origin of the NC, of NC skeletal derivatives, and of vertebrates themselves. These advances demonstrate the power of the synthesis of comparative anatomy and paleontology, developmental and molecular biology, systematics and evolution, and of genomics and bioinformatics (Hall, 1999a,b).

The next five chapters are devoted to discussions of the origins and differentiation of major classes of NCCs—pigment cells (chromatophores), neurons, mesenchyme, and the craniofacial skeleton, tooth-forming and cardiac NCCs.

Fig. 4.15 Premandibular (pm) and mandibular (mm) NC-derived mesenchyme (*pale gray*) in the developing head of a lamprey (**A**1) and basal gnathostome (**A**2) as seen in *side view*, with anterior to the *left* and the direction of growth of the *upper* and *lower* lips indicated by the *arrows*. Oro marks the position of the opening of the oral cavity. (**B**1) and (**B**2) are *en face* views of the same embryos with premandibular mesenchyme in *black* and mandibular mesenchyme cross-hatched. The *solid arrows* indicate growth and rostral extension of premandibular mesenchyme, the *dashed arrow* in (**B**2) the rostral extension of the mandibular mesenchyme (cross-hatched) in the gnathostome. Olfplac, olfactory placode; Rp, Rathke's pouch. Modified from Janvier (2007)

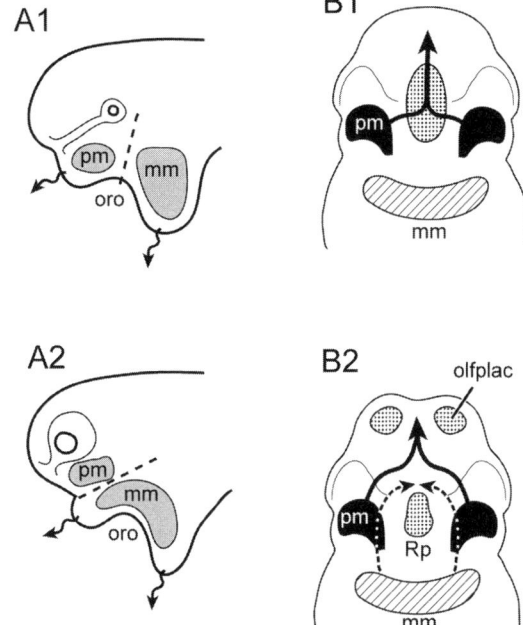

Notes

1. No extensive discussion is provided of the various theories for the origin of the vertebrates, which have been the topic of several major treatments, especially Gans and Northcutt (1983), Mallatt (1984, 1996), Maderson (1987), Mallatt and Chen (2003) and Janvier (2007).
2. See Northcutt (1996), Baker and Bronner-Fraser (1997b) and the papers in Hall and Wake (1999*) for evaluations of the various theories of chordate origins.
3. See L. Z. Holland and N. D. Holland (1996, 2001*) and N. D. Holland *et al.* (1996) for their studies.
4. See Corbo *et al.* (1997) and Langeland *et al.* (1998) for *Snail1* in amphioxus and in ascidians, and see Martinez-Morales *et al.* (2007) for an analysis of 'gene-emergence plots' using a bioinformatics approach.
5. See P. W. H. Holland and Graham (1995), P. W. H. Holland (1996), N. D. Holland (1996), Sharman and Holland (1998), Hall (1999a*,b*) and L. Z. Holland and N. D. Holland (2001*) for the significance of *Hox* gene clusters for vertebrate evolution, and P. W. H. Holland (1996) for *Hox* and other genes cloned from *Branchiostoma*.
6. See Stone and Hall (2004) for potential difficulties associated with using the expression of single genes to identify homologs or latent homologs.
7. See Holland *et al.* (1996) for *AmphiDll*, and see Harada *et al.* (2001) for *PfDll* (which they call *PfDlx*). The Hawaiian acorn worm also has an Otx gene (*PfOtx*), expressed in phases—in the blastula and the archenteron of the gastrula, expression then being lost and a new pattern of expression appearing in the ciliary bands of the larva (Harada *et al.*, 2000).
8. See Simeone *et al.* (1992), Matsuo *et al.* (1995) and Rhinn *et al.* (1998) for *Otx2* in murine embryos, and Williams and Holland (1998) for the expression and organization of *AmphiOtx*.
9. See Shimeld (1999) and L. Z. Holland and N. D. Holland (2001*) for *AmphiBmp2/4*.

10. See Grandel *et al.* (2002) for the *No-fin* mutant, and Niederreither *et al.* (2003) for the *Raldh2*^{-/-} mice.

11. See Kopinke *et al.* (2006) for patterning by pharyngeal endoderm in zebrafish, and Gavalas *et al.* (2001) and Graham and Smith (2001*) for normal arch formation after NCC removal.

12. See Gavalas *et al.* (2001), Graham and Smith (2001*), and Kopinke *et al.* (2006) for the patterning studies, and Manzanares *et al.* (2001) for the retinoic acid receptor.

13. For the evolution of tissue interactions at the outset of vertebrate evolution, see Hall (2000a, 2005a,b*), Maisey (1986), M. M. Smith and Hall (1990, 1993), and Donoghue (2002).

14. A recent issue of the journal *Developmental Dynamics* (2007, 236: 1695–1747) contains seven reviews on the state of knowledge of ascidian biology. Also see Hall and Hallgrímsson (2008) for current views on ascidian relationships, and Bourlat *et al.* (2006) and Delsuc *et al.* (2006) for molecular phylogenies.

15. See Sardet *et al.* (2008) for a perspective on current activity and future prospects of urochordate research from a recent meeting of 150 researchers.

16. A Notch gene, *AmphiNotch*, has been cloned from Amphioxus, shown to be expressed in the notochord and pharyngeal endoderm, and to regulate Wnt genes and *Brachyury*, consistent with genetic control of specification of the A–P axis neural ectoderm (induction?) predating the amphioxus/deuterostome split (L. Z. Holland *et al.*, 2001).

17. See Satoh (1994), Shimeld (1999), Meinertzhagen and Okamura (2001), Northcutt (2004), Mazet and Shimeld (2005), Dufour *et al.* (2006), Lacalli (2006*), Pasini *et al.* (2006*), and Kumano and Nishida (2007*) for studies on which this summary is based. See Stach (2000) for the microscopic anatomy of amphioxus development.

18. See Jeffery (2007) for NC-like cells; Meulemans and Bronner-Fraser (2004) and Rychel and Swalla (2007) for gene expression in amphioxus and in hemichordates; Martinez-Morales *et al.* (2007) for the recruitment of ancestral regulatory genes and signaling peptide; and Hall (2008) for a recent commentary.

19. See Maclean and Hall (1987), Whittaker (1987), and Jeffery *et al.* (2004) for determination of cell fate in ascidians, and Satoh (1994), Jeffery (1997), Meinertzhagen and Okamura (2001), and Kumano and Nishida (2007) for discussions of ascidian development.

20. For suppression of *Bmp* dorsally and for *HrBmp2*, *4*, and *7*, see Hammerschmidt *et al.* (1996) and Miya *et al.* (1997).

21. See Corbo *et al.* (1997), Langeland *et al.* (1998), and Meulemans and Bronner-Fraser (2004) for *Snail1* in amphioxus and ascidians.

22. See Lacosta *et al.* (2007) for *Pax7* expression, Mansouri *et al.* (1996) for *Pax7*– mutant mice, and Basch *et al.* (2006) for *Pax7* and the NC field.

23. For orthologs of vertebrate *Pax* genes in the nervous systems of *Ciona* and *Halocynthia*, see Wada *et al.* (1998) and Corbo *et al.* (1997).

24. See Stone and Hall (2004) for using expression of single genes to identify homologs or latent homologs. See Kozmik *et al.* (1999) for expression of *AmphiPax2/5/8* in the rostral brain amphioxus, a location not equivalent to the midbrain–hindbrain boundary in vertebrates.

25. See J.-Y Chen *et al.* (1995) and Shu *et al.* (1996a,b), respectively, for the original description and reinterpretation of [†]*Yunnanozoon*, Chen *et al.* (1999) for [†]*Haikouella*, and Mallatt and Chen (2003) for a detailed review and synthesis of both genera. The pharyngeal arches of extant hemichordates contain an acellular matrix, the characteristics of which are under investigation and which are secreted by pharyngeal endoderm (Rychel *et al.*, 2006; Rychel and Swalla, 2007).

26. See Gans and Northcutt (1983, 1985) and Northcutt and Gans (1983) for their pivotal scenario of the evolution of the NC, and Goodrich (1930*), Romer (1972), and Schaeffer (1977*) for earlier studies. The symposium proceedings edited by Maderson (1987), the three-volume series on the vertebrate skull edited by Hanken and Hall (1993), and Donoghue (2002) and Hall (2005b) review and synthesize much of this evidence.

27. See Mallatt (1984), Maisey (1986), Hanken and Hall (1993), Mallatt and Chen (2003), Kuratani (1997, 2004, 2005), Kuratani *et al.* (2001), Erickson *et al.* (2004), and Olsson *et al.*

(2005) for studies on segmentation of the vertebrate head, and Stern (1990) for the different bases for segmentation of the hindbrain (which is based in the neural ectoderm) and the trunk (which is based in mesoderm).

28. See Cole and Hall (2004a,b) for overviews of invertebrate cartilages, and Cole and Hall (2008) for the development of cartilages in cephalopods.

29. See a text on vertebrate anatomy such as Liem *et al.* (2001), or Hall (2005b) and Kardong (2006) for the morphological evidence, which has been available for 140 years, and see Kuratani (1997, 2004, 2005), Kuratani *et al.* (2001), and Kimmel *et al.* (2001) for developmental and molecular evidence for the origin of jaws. See Hall (1994, 1995, 2003a, 2007b), Kuratani (2004, 2005), and Cracraft (2005) for homology.

30. For discussions of the possibility that the velum of lampreys may be a homolog of the mandibular arches of gnathostomes, see Hardisty (1979), Forey (1995), Mallatt (1996), and Janvier (2007). For mandibular and velar skeletons, see Cohn (2002).

31. See L. Z. Holland and N. D. Holland (2001*) for *engrailed* expression in lamprey and gnathostome muscles, and P. W. H. Holland (1996), Miyake *et al.* (1992), Hall (1999b*), and Stone and Hall (2004) for more general discussions of the issues raised by such shared specific patterns of gene expression.

Part II
Neural-Crest Derivatives

Part II summarizes and analyzes our knowledge and understanding of the NC, NCCs, and their derivatives. Unlike the approach used in the first edition, where vertebrates were presented group by group, in this edition I have organized each of the four chapters around a class or classes of NC-derived cells and the tissues and organs they form or to which they contribute. The four groupings are:

- pigment cells and color patterns (Chapter 5);
- neurons and the nervous system (Chapter 6);
- cartilage and skeletal systems (Chapter 7); and
- dentine-forming cells (teeth) and smooth muscle, septa- and valve-forming cells of the heart (Chapter 8).

These chapters cover the four major subpopulations of NCCs and their derivatives:

- *TNCCs* (Chapter 5);
- the *vagal and sacral* enteric ganglionic neural crest, spinal and cranial ganglia, the autonomic nervous system, Schwann and glial cells, and Rohon–Béard neurons (Chapter 6);
- *CNCC*, chondroblasts and osteoblasts, mesenchyme, the skeletogenic (chondrogenic) neural crest, and epithelial–mesenchymal interactions (Chapter 7); and
- the *odontogenic NC*(tooth formation) and the *cardiac NC* (heart, development; Chapter 8).

Chapter 5
Pigment Cells (Chromatophores)

The following major points concerning the embryological origins of pigment cells (chromatophores) were made in earlier chapters:

- Using amphibian embryos, Kölliker (1884) demonstrated that pigment, epithelial, and neuronal cells arise from ectoderm.
- Using mouse embryos, Griffith *et al.* (1992) demonstrated that pigment cells (and neurons, myoblasts, and chondroblasts) all arise from the tail buds without segregation of the caudal region into germ layers.[⊕]
- Pigment cells can arise without neural derivatives when a neural crest (NC) is induced ectopically in the lateral epiblast of chicken embryos.
- Because all pigment cells other than those from the pigmented layer of the retina[1] and substantia nigra of the midbrain arise from neural crest cells (NCCs), the differentiation of pigment cells can be used as a marker of NC origin, even when NC arises ectopically.

The following major points concerning pigment cells in nonvertebrate chordates (cephalochordates and ascidians) were made in earlier chapters:

- Cephalochordates lack pigment cells.
- Sensory pigment cells arise along the dorsal aspect of the neural tube in association with neural tube induction in ascidians such as *Ciona* spp.
- Cells arising in the dorsal neural tube of ascidians and expressing the HNK-1 antigen and a *Zic* gene (both of which are markers of NCCs) migrate into the body wall and then into the primordia of the siphon where they differentiate into pigment cells.

[⊕] As the pigment cells that arise from tail buds do so without the caudal cells segregating into germ layers, it could be claimed that these pigment cells do not arise from NC, because an NC does not form. On this point, see the discussion of secondary neurulation/induction in Chapter 2.

B.K. Hall, *The Neural Crest and Neural Crest Cells in Vertebrate Development and Evolution*, DOI 10.1007/978-0-387-09846-3_5,
© Springer Science+Business Media, LLC 2009

If ascidians rather than cephalochordates are the sister groups of the vertebrates, as indicated by the recent molecular phylogenies (see Fig. 4.2), then pigment cells were the first derivatives on the lineage that became NCCs. This chapter provides an overview of basic knowledge on the origin and function of pigment cells (chromatophores).

Types of Chromatophores

Three basic types of chromatophores are distinguished on the basis of the color of the pigment they produce or because they reflect light:

- **Melanophores/melanocytes**[⊕] produce or contain the black pigment, **melanin** (Figs. 5.1, 5.2 and 5.3). Because of a relationship between ultraviolet light and folic acid, melanin may protect against neural tube defects (see Chapter 9).

  Dermal melanophores are responsible for the rapid color changes seen in many fish, some amphibians, and some reptiles, in which melanophores respond to light by redistributing pigment granules within the cytoplasm.

  Epidermal melanocytes (Fig. 5.2) deposit granules of melanin into keratinocytes, the most prominent epidermal cell type. Mutations such as *albino* either disrupt the ability of melanocytes to synthesize or deposit melanin (mammals)[2] or prevent melanophore differentiation by blocking signals from the surrounding ECM (amphibians).

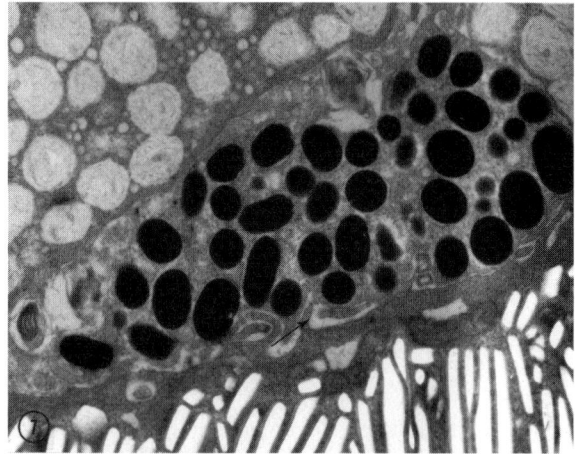

Fig. 5.1 Histological section through the skin of a green tree frog, *Hyla cinerea*, showing a xanthophore (*top left*), a melanophore filled with melanin granules (*middle*), and an iridophore (*bottom*) with reflecting platelets and the finger-like projections interdigitating with the melanophore (*arrow*). Figure courtesy of Joseph T. Bagnara

---

⊕ Pigment cells in fish, amphibians, and reptiles (poikilothermic vertebrates) are known as **melanophores**. The homologous cells in birds and mammals (homeothermic vertebrates) are known as **melanocytes**.

Fig. 5.2 Epidermal melanophores from the dorsal surface of a Northern leopard frog, *Rana pipiens*, in the dispersed state typical of these cells. Figure courtesy of Joseph T. Bagnara

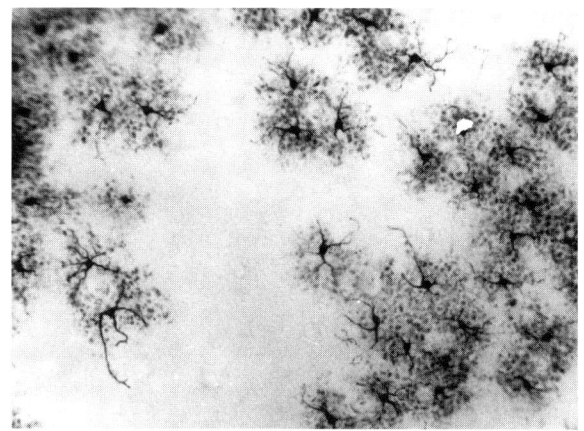

Fig. 5.3 The relationship between an epidermal melanophore (EM) in the epidermis, a xanthophore (X), iridophore (I), and melanophores as seen in cross-section through the skin of a frog. Figure courtesy of Joseph T. Bagnara

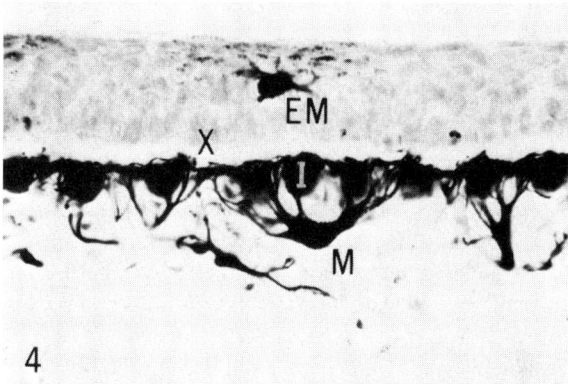

- **Iridophores**, which are found as stacks of cells in the dermis, reflect light and so are iridescent (Fig. 5.1), usually located atop a layer of melanin.
- A family of **dermal chromatophores** containing **carotenoids** produce different colors. Four types are usually recognized: *xanthophores* (yellow/orange), *erythrophores* (red), *cyanophores* (blue), and *leucophores* (white).

All three families of chromatophores are found in amphibians and fish. Birds retain the ability to form all types of chromatophores in the iris. Mammals have lost the ability to produce iridophores, xanthophores, and erythrophores. Interestingly, some chromatophores in most if not all vertebrates groups contain more than one pigment type, a phenomenon known as **chromatophore mosaicism**. Mosaic cells are given compound names such as irido-melanophores for chromatophores that reflect light and produce melanin (Bagnara *et al.*, 1979; Bagnara, 1999). Melanophores, xanthophores, and iridophores interact both morphologically and physiologically in what Bagnara *et al.* (1968) defined on the basis of his studies on frogs as the **dermal chromatophore unit**. The concept of the origin of these classes of pigment cells from a common stem cell also goes back to studies by Bagnara and his colleagues (1979).

Melanosomes

Melanocytes synthesize pigment granules (**Melanosomes**; Fig. 5.1), which bud from the endoplasmic reticulum. Melanosomes do not remain within the melanocytes that produce them but are transferred along melanocyte dendrites to nearby keratinocytes. A melanocyte and its neighboring keratinocytes form an epidermal melanin unit.

Given this transfer of melanosomes to keratinocytes, several mechanisms can affect color patterns:

- regulation of melanocyte numbers,
- regulation of pigment synthesis within melanocytes, and/or
- alterations in the transfer of melanosomes to keratinocytes.

All three processes have been studied most intensely in feathers and in hair, especially through the use of mutations.[3]

A mutation that exerts its affect by regulating pigment synthesis is the *dilute* (*dil*) allele in the budgerigar. *Dilute* almost completely obliterates melanocyte dendrite development or reduces dendrite size to such an extent that melanocytes cannot interface with their neighboring keratinocytes, with the consequence that melanosome numbers in keratinocytes are reduced by 80%. Other mutations in budgerigars influence either pigment production or the number of melanosomes in each melanocyte. At the other extreme, the sex-linked *pastel factor* in canaries results in uncontrolled production of pigment granules in melanocytes that cannot transfer melanosomes to keratinocytes.

In humans, the degree of pigmentation of the skin correlates with the numbers, size, and density of melanosomes. Sun tanning works because UV irradiation increases the rate of transfer of melanosomes to keratinocytes and, to a much lower extent, enhances melanocyte production.

As discussed below, progenitors of pigment cells originate from NCCs, mostly from TNCCs, although CNCCs do produce pigment cells in some taxa. Pigment cell progenitors are migratory and can home to target sites such as the body wall, feathers, or hairs, providing color and/or pattern to larvae and adults alike. Studies establishing the NC origin of pigment cells are discussed below, as are studies concerning the origins of larval and adult pigmentation patterns and how chromatophore differentiate.

Lampreys

Almost 100 years ago Koltzoff (1901) described a NC from histologically sectioned embryos of the European brook lamprey. In the 1930s, Bytinski-Salz concluded that any pigment cells differentiating in grafts of lamprey cranial neural folds must have originated from NCCs included in the grafts, a conclusion that he confirmed when extirpating the NC resulted in pigment cell deficiencies.

After removing the CNC from three-somite-stage embryos of the European brook lamprey, pigment cells (along with dorsal root ganglia and some head mesenchyme) failed to form, results interpreted by Newth (1950) as showing that these three cell types arise from the NC. Removing one of seven regions of CNC (each 250 μm in cranio-caudal extent as shown as I–VII in Fig. 7.2a) from early-neurula-stage North American sea lamprey embryos resulted in a substantial reduction in pigmentation of the heads of the resulting larvae.[4]

Urodele and Anuran Amphibians

From *in vivo* analyses and the emergence and development of pigment cells from cultured frog spinal cord, numerous researchers in the first quarter of the 20th century posited an NC origin for anuran pigment cells, an origin confirmed in the 1930s and 1940s by extirpation and transplantation of the N. C. DuShane's (1943, 1944) studies are representative of evidence based on transplantation *between individuals of a single species*:

- pigment cells appear at the graft site after transplanting neural folds into the ventral body wall;
- transplanted limb buds acquire pigmentation through the migration of NCCs into the transplant; and
- bilateral extirpations of small regions of the TNC result in local loss of pigment cells (and of spinal ganglia, dorsal fin, and Rohon–Béard neurons) in regions corresponding to the ablated NC.

Using NCC markers in combination with transplantation, Epperlein *et al.* (2000a, 2007a) confirmed these results in the Mexican axolotl.

Studies by Twitty and Raven are typical of those in which NC was transplanted *between embryos of different species*, usually between embryos of two species of urodeles, or from anuran to urodele embryos. The NC origin of pigment cells in the host embryos can be determined because donor and host tissues can be distinguished from one another, an approach that allowed Niu (1947) to demonstrate that the TNC is a more potent source of pigment cells than the CNC: as noted in Chapter 2, pigment cells provide excellent markers for NCC migration and are still used with profit to detect subpopulations of amphibian NCCs, especially after wild-type (pigmented) tissues are grafted into albino or into white mutant embryos.[5]

Labeling the NC of larvae of the Japanese Northeast salamander with vital dyes or India ink, coupled with transmission and scanning electron microscopy allowed Hirano and Shirai (1984) to track migrating NCCs. In a subsequent study, Hirano (1986) demonstrated that separate cell populations are present in the NC and that they follow *one of two paths*; future pigment cells migrate dorsally between the epidermis and the neural tube, prospective ganglia migrate lateral to the neural tube.

Patterns of Pigmentation

The extraordinary range of color patterns displayed by vertebrates can be reduced to two basic types:

- color difference seen when the dorsal surface is more pigmented than the ventral surface, as seen, for example, in frogs or in bottom-dwelling fish;
- a pattern of stripes, spots, or splotches of color against a background that is either unpigmented (and so white) or of a different color than the stripes, spots, or splotches (Fig. 5.4 [Color Plate 5]).

Major aspects of pigment patterns are species specific, as seen, for example, in the patterns of migration and dispersal of pigment cells in red-bellied and California newts. As determined *in vitro* and following heterospecific grafts of NCCs, red-bellied newt cells disperse, producing even pigmentation patterns; California newt cells aggregate, producing pigmented bands.

Within individuals, patterns of pigmentation respond to and are directed by extrinsic cues. During axolotl embryonic development, components of the ECM along the dorsolateral migratory route or other signals elicit the differentiation of pigment cells (see Fig. 3.9). As determined by clonal cell culture, *Xenopus* melanophores and xanthophores can arise from separate lineages or from a common

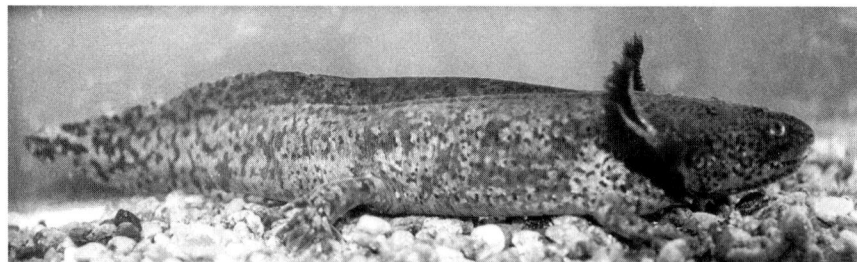

Fig. 5.4 An adult Mexican salamander to show the typical pattern of pigmentation. As metamorphosis does not normally occur adults retain the external gills and the dorsal median fin. Compare with the metamorphosed adult in Fig. 5.6. Figure kindly supplied by Lennart Olsson (see Color Plate 5)

cell lineage.[6] Collazo *et al.* (1993) injected vital dyes into groups or single TNCCs in *Xenopus* embryos and showed that pigment cells, spinal ganglia, mesenchyme, and adrenal medullary cells arise from the progeny of single TNCCs. TNCCs differentiate into pigment cells if grown on plastic but into catecholamine-containing adrenergic neurons if cultured in media conditioned by somitic cells.

Developmentally regulated lectins also influence pigment patterns in axolotls; an ectodermal defect and resulting low lectin concentrations in albino axolotls prevent melanophores from colonizing the skin. Migration of NCC along the dorsolateral pathway normally taken by pigment cell precursors is disrupted, but migration along the ventromedial pathway is unaffected. A different mechanism operates in albino mutant mammals, in which enzymes in the melanin synthesis pathway are blocked and melanin cannot be produced. Yet another mechanism is responsible for albinism in Mexican tetras. Blind, cave-dwelling populations possess melanophores, but population-specific deletions in the gene *oculocutaneous albinism2* (*Oca2*) prevent melanization of the melanophores.[7]

Larval-to-Adult Patterns

During metamorphosis, most if not all amphibians display characteristic color changes as larval patterns are replaced by adult patterns; Fig. 5.5 shows the larval patterns in three species. As indicated above, basic patterns are species specific. Indeed, adult patterns are sufficiently invariant to serve as diagnostic characters for species and subspecies.

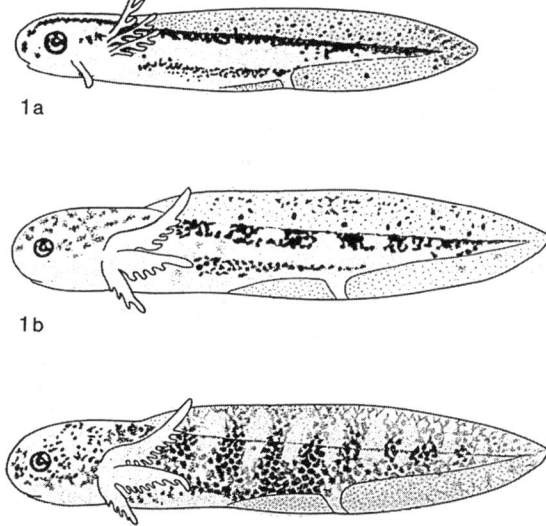

Fig. 5.5 Drawings (1**a**, 1**b**, and 1**c**) of newly hatched larvae of *Triturus alpestris* (1**a**), *Ambystoma tigrinum tigrinum* (1**b**), and *Ambystoma mexicanum* (1**c**), showing the differences in larval color pattern. The drawings are slightly idealized for clarity. Note the intermediate pattern (1**b**) between two major horizontal stripes (1**a**) and vertical stripes (1**c**). Horizontal stripes developing dorsal to the *midline* are shown in more detail in the image below. Figures kindly supplied by Lennart Olsson

1 a

1 b

1 c

Fig. 5.6 An adult Mexican axolotl that was induced to metamorphose retains the larval pigment pattern, loses the external gills and median dorsal fin, and has a differently shaped head than the normal unmetamorphosed adult shown in Fig. 5.4. Figure kindly supplied by Lennart Olsson (see Color Plate 5)

The melanophores that produce larval *and* adult pigmentation are NC in origin. On the basis of studies from the late 1930s, hormonal changes associated with metamorphosis are assumed to trigger the onset of adult pigmentation patterns.[8] An interesting exception that proves the rule, shown in Fig. 5.6 [Color Plate 5], is an axolotl in which metamorphosis took place. This 'adult' has no external gills (compare with the postmetamorphic adult in Fig. 5.4) and has a pigmentation pattern that in no way compares to the pattern found in 'normal', neotenic axolotls.

Twitty and Bodenstein (1939) transplanted NC from the North American spotted salamander, *Ambystoma maculatum*⊕ into the ventral regions of California newt embryos, which have pigmented tadpoles. Pigmentation did not appear in the grafts during tadpole development but pigment spots characteristic of the donor (spotted salamander) appeared with metamorphosis and persisted into the adult. The presumption that rising thyroxine levels associated with metamorphosis regulate the differentiation and patterning of chromatophores surprisingly remains just that, a presumption that has not been tested experimentally.

Comparative studies of the origin of larval pigmentation patterns in other larval urodeles took advantage of the difference between the horizontal melanophore stripes of larval alpine newts and the vertical melanophore bars of larval Mexican axolotls. Melanophores in alpine newts use environmental cues in organizing the horizontal stripes, while melanophores in axolotls depend upon cell-to-cell interactions. Comparative analyses in a phylogenetic context enabled the same group to show that the formation of horizontal stripes is the basal condition and the development of vertical bars, the result of later evolutionary changes in pigment cell aggregation (see Epperlein *et al.*, 1996* for a review).

Perhaps the most extensive study since 1939 (using a similar transplantation approach) is that by Parichy (1998), who investigated whether larval and adult patterns of pigmentation are coupled. Parichy showed that patterns of pigmentation in

---

⊕ *Ambystoma maculatum* is often referred to in the older literature as *Ambystoma punctatum*. However, *Ambystoma maculatum* is the correct name (Wake, 1976).

adult tiger salamanders are largely independent of larval patterns and of the lateral line cues involved in generating the larval patterns. As is typical for many salamanders, larval pigmentation in tiger salamanders consists of yellow horizontal stripes coupled with black dorsal and ventral stripes. This larval pattern is established during migration of the lateral line primordia, which deposit neuromasts along the position of the future lateral line (the distribution and development of lateral lines and their associated neuromasts are discussed in Box 6.1). Melanophores interact with melanophores, which retreat away from the neuromasts, creating a melanophore-free zone that is then filled by xanthophores to create the yellow stripes. Migrating NCCs (premelanophores) thus influence the patterning of the neuromasts deposited by lateral line primordia as they migrate along the body. Removing segments of TNC from Mexican axolotl embryos demonstrates that while neuromast precursors migrate through NC-free zones, they fail to deposit neuromast precursors in these regions, but resume depositing precursors once they emerge from the NCC-free zone.[9]

In addition to melanophores and xanthophores, iridophores contribute to the adult pattern, which is established during metamorphosis, and which consists of a dark background of melanophores against which spots produced by xanthophores and iridophores stand out. Although iridophores do not contribute to larval pigmentation, they do differentiate in association with neuromasts in tadpoles. Although it is therefore possible that lateral lines/neuromasts play a role in establishing adult patterns of pigmentation, Parichy's experiments show that they do not. Whether this decoupling is because different populations of chromatophores produce the larval and adult patterns or because a single population is differentially regulated over the life cycle remains as elusive as it did 50 years ago when Twitty and Bodenstein published their study.

Teleost Fish

Fish possess and display an enormous range of colors and patterns, which they use in many aspects of their life history (Fig. 5.7 [Color Plate 5]).

In his studies on the needlefish, *Belone acus* Borcea (1909) was one of the first to report that pigment cells in bony fish are derived from the NC. Having described pigment cells migrating between spinal cord and trunk epidermis, Borcea concluded that they must have arisen from the ectoderm, presumably from TNCCs. Lopashov (1944) extirpated and grafted neural tubes from several species of teleosts to determine whether pigment cells were NC in origin. Head pigmentation was reduced or eliminated after regions of the cranial neural tube were removed. Meanwhile, the yolk sacs onto which cranial neural tubes were grafted became pigmented. Cartilage also differentiated in these grafts, an early indication of the NC origin of cranial cartilages in teleosts. In a later study, pigment and neuronal cells were shown to differentiate from NCCs that had migrated from the neural tubes of swordtail and Japanese medaka embryos maintained *in vitro*.[10]

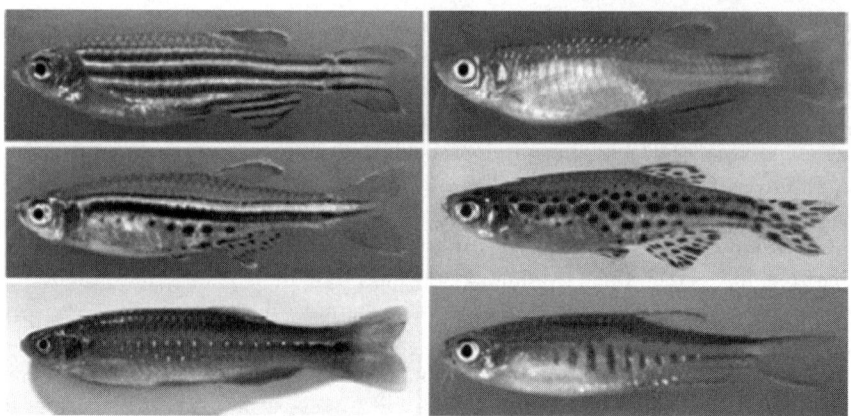

Fig. 5.7 The variation in patterning (horizontal stripes, vertical stripes, spots, and albino shown) and in color found within a single genus (*Danio*) is shown in these images of six species of zebrafish, which are, clockwise from the *top left*, the zebrafish, *Danio rerio*; pearl danio, *D. albolineatus*; *D. sp. Nov.*; *D. choprae*, *D. feegradei*; and the spotted danio, *D. nigrofasciatus*. Reprinted with permission from the Annual Review of Ecology, Evolution, and Systematics, Volume 38 © 2007 by Annual Reviews www.annualreviews.org. Images kindly provided by David Parichy (see Color Plate 5)

Migration and Cell Fate

An early attempt to label fish NC involved isolating ^{3}H-thymidine-labeled embryonic trunk neural tubes from the level of somites 5 and 6 in rosy barb embryos and grafting them either into the hindbrain or into the trunk neural tube of unlabeled embryos. ^{3}H-thymidine-labeled cells migrated away from the grafts in two streams: between the developing somites and the trunk epithelium as a superficial population, and as a deeper population medial to the somites and between somites and neural tube/notochord (Lamers *et al.*, 1981). Similar pathways of migration of trunk and sacral NCCs have been described in zebrafish and other vertebrate groups (see Chapter 6) and are an evolutionarily conserved feature of NCCs; labeled cells are seen as differentiated pigment cells, adrenal medullary cells, and neurons of spinal, sympathetic, and enteric ganglia. These three cell types can therefore be assigned to the list of NC derivatives in teleosts.

TNCCs in zebrafish are larger and fewer in number than are avian NCCs (see upcoming section) and migrate along two ventral paths (see Raible and Eisen, 1996* for a review). Labeling premigratory TNCCs by intracellular injection with lysinated rhodamine dextran (LRD) demonstrates cell populations with restricted cell fate. Indeed, most zebrafish TNCCs are lineage restricted, time spent in the neural tube before delamination affecting lineage potential. Early migrating cells contribute to pigment cells, dorsal root ganglia, and glia; sensory and sympathetic cells arise only from early migrating cells. Late migrating cells are nonneuronal.

Migration and survival of melanophores in zebrafish and of melanocytes in mice is dependent upon *c-kit* signaling regulated by *Sox10*, which directs NCCs into melanophore/melanocyte differentiation; zebrafish and mouse *Sox10* mutants lack

migrating pigment cell precursors. Inhibiting NCC migration in zebrafish by injecting a *Sox8* morpholino at the eight-cell stage reduces pigmentation in two-thirds of
treated embryos.

   c-Kit regulates the *microphthalmia-associated transcription factor* (*Mitf*) in mice
and its ortholog in zebrafish, *Mitfa*, which binds to the *Mitf* (*Mitfa*) promoter. The
cascade of transcriptional factors and signal transduction pathways regulated by *Mitf*
are important regulators of the differentiation of normal melanocyte. Consequently,
defects in these pathways are associated with (perhaps causally) the formation and
spread of melanomas (Goding, 2000*). *c-Kit* is regulated by the transcriptional activator protein Ap2α, which is regulated by and can substitute for Bmp in pigment
cell precursors. *c-Kit* is further discussed below and in the context of animal models
for human pigmentation defects in Chapter 9.

   Formation of pigment cells in zebrafish and in mice is at the expense of neuronal
and glial cells, a lineage specification that is promoted by *Wnt* signaling that stabi-
lizes β-catenin in the canonical Wnt pathway (see Fig. 2.5). Similarly, stabilization
of β-catenin by the product of the gene *Cullin1* (*Cul1*) in *Xenopus*, directs more
cells into the NCC lineage, and more NCCs into melanophores than into cranial
ganglion neurons.[11]

Larval Patterns

Building on the knowledge that melanophores arising early in metamorphosis are
initially dispersed widely and only later form stripes, and that melanophores arising
late in metamorphosis initially appear in a striped pattern, David Parichy's laboratory showed that the development of early melanophore populations in zebrafish is
dependent on the *kit* receptor tyrosine kinase; *kit* mutants lack early melanophores
but still develop late melanophores that form regular stripes.

   The same dependence of early melanophores on tyrosine kinase at the same late
metamorphic stage was shown in a second species, the pearl danio. Wild-type pearl
danio (Fig. 5.7) are uniform in color. *kit* mutant pearl danio have stripes similar to
those in wild-type zebrafish, a finding interpreted by Mills and colleagues (2007) as
indicating a latent stripe-forming potential in pearl danio.[12] As noted in discussing
animals models and mutations for neurocristopathies in Chapter 9, Miller *et al.*
(2007) provided important insights into the evolution of *kit* regulation across the
vertebrates when they showed that parallel evolution of *cis*-regulation of *Kit ligand*
(*Kitlg*), the gene for c-kit ligand, elicits light coloration in marine and freshwater species of threespine sticklebacks and in humans. In an even more recent study,
Budi *et al.* (2008) demonstrated that the zebrafish phenotypic mutant *picasso* results
from mutations in the gene encoding an egf receptor-like receptor tyrosine kinase
that is required during NCC migration for adult pigmentation patterns to develop
normally.

   Twenty connexin genes are known from humans and 40 in zebrafish. The
leopard (*leo*) gene in zebrafish, which results in a pigment pattern of lines of pigmented spots rather than stripes (Fig. 5.8 [Color Plate 6]; cf. Fig. 5.7), is an ortholog

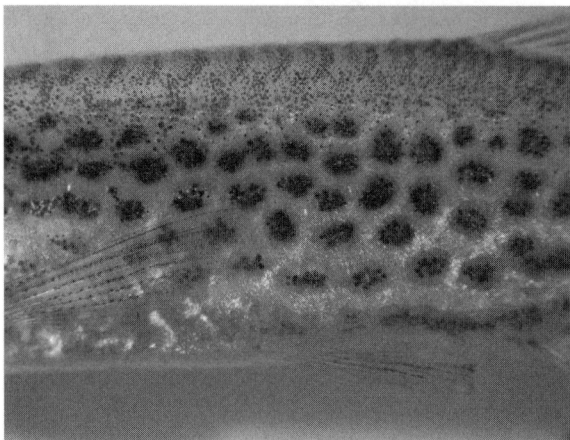

Fig. 5.8 A close-up view of the body of a *leopard* (*leo*) mutant zebrafish showing the lines of pigmented spots made up of melanophores (*blue–black*) that replace the horizontal stripes seen in wild-type zebrafish (Fig. 5.7). Individual melanophores and the pigment-free zone around each spot are clearly visible, as are the *yellow* xanthophores and the iridescent iridophores. Image kindly provided by Michelle Connolly (see Color Plate 6)

of the mammalian *connexin 40* gene. Of the many *leo* alleles, *leo*^{tq270} disrupts the cell-to-cell communication function of *connexin 41.8* (Watanabe *et al.*, 2006), illustrating the degree of genetic control that exists over the patterning of NC derivatives.

Genes and Cell Lines

In an early molecular approach to the analysis of NC derivatives in fish, Vielkind and colleagues (1993*) took advantage of their knowledge of the pathways of NCC migration, DNA technology, and genetic transformation to inject purified DNA of the *Tu* gene (which plays a role in the differentiation of T-melanophores) into the NC of embryos of a strain of swordtail lacking the *Tu* gene, and which, therefore cannot produce melanophores. The appearance and differentiation of T-melanophores in the genetically transformed strain demonstrated incorporation of the *Tu* gene into host cells and the NC origin of the T-melanophores. A monoclonal antibody directed against the melanoma gene from swordtail cross-reacts with human melanomas but not with other tumors derived from NCCs, demonstrating the utility of fish as model system for human tumors of NC origin, a topic revisited in Chapter 9.

A second ingenious experimental approach for delineating NC origins was taken by Matsumoto and colleagues (1989) when they established clonal cell lines from goldfish and Nibe croaker melanophore tumors—an erythrophoroma (a tumor of erythrophores) and an irido-melanophoroma. The latter is a chromatophore mosaic, a mixed tumor of iridophore/melanophore origin. Individual clonal cell lines differentiated into melanophores and cell types of the dermal skeleton, including osteoblasts (found in dermal bone), scleroblasts (found in scales and fin-rays; see

Box 5.1 for NCC contributions to the paired fins) and odontoblasts (dentine-forming cells, found in scales or teeth). Osteoblasts and scleroblasts differentiated spontaneously without supplementation of the culture medium; that is, without any exogenous factors acting as inducers. Odontoblasts formed only if serum or DMSO was added to the culture medium.

Box 5.1 Neural crest and paired fins

Two fundamental differences between paired fins and tetrapod limbs are: (1) the presence of putative NC-derived distal bony rays in fins and their absence from limbs and (2) mesoderm-derived distal elements (digits) in limbs and their absence from fins. Simplistically, you can think of **limbs as fins, minus fin rays, plus digits**.

The proximal fin endoskeleton of teleosts, elasmobranchs, and chondrichthyans is composed of cartilages that are mesodermal in origin.[a] Bony fin rays, however, are the major component of all paired and many median teleost fins. It is presumed but not proven that the lepidotrichia and actinotrichia from which fin rays are constructed are NC in origin; labeled cells have not been traced sufficiently far into development to demonstrate labeled fin rays but far enough to demonstrate labeled mesenchyme in the tail fin where fin rays later develop.

Limbs lack any NC (exoskeletal, dermal) elements and have expanded and elaborated the endodermal skeleton distally to form the digits. At the cellular level, the transformation from fins to limbs involved:

- eliminating the NC component (the dermal fin rays),
- distal growth of the proximal mesodermal endoskeleton,
- elaborating the new distal endoskeleton as wrist/ankle elements and proximal phalanges, and
- developing new distal endoskeletal elements as digits.

Developmentally, these changes are essentially loss of the NC component (dermal skeleton) and elaboration of the mesoderm-derived endoskeleton. In terms of gene regulation, the transition has been regarded as involving a second wave of *Hox* gene expression in limb buds not seen in fin buds. However, recent findings of this second wave in the fin buds of paddlefish and sharks requires that we reevaluate our ideas on the genetic changes involved in the transformation of fin —> limbs.

---

[a] See M. M. Smith and Hall (1993) and M. M. Smith *et al.* (1994) for the NC origin of fin mesenchyme and the dermal skeleton, and Davis *et al.* (2007) for *Hox* gene expression. For discussions of the cellular and molecular changes required in the transformation of fin —> limb, see Hall (2005b∗) and the chapters in Hall (2007a∗).

A follow-up study confirmed that goldfish erythrophoroma cells could initiate multiple pathways of differentiation, including melanin-synthesizing cells, platelets, neurons, osteoblasts, (all NCC types) and lens cells, with individual cell lines functioning as clonal lineages. Significantly, there was no evidence of the erythrophoroma tumor phenotype in any cell; the reprogramming associated with the initiation of new pathways of differentiation 'switched off' the genes associated with the tumor phenotype.

On the presumption that these melanophore tumors are homogeneous, several interesting conclusions may be drawn from this study:

- melanophores and mesenchymal cells of dermal bone, scales, and fin rays (Box 5.1) are NC derivatives;
- the cells of melanophore tumors (and of the NC itself?) are multipotent;
- some NC derivatives—melanophores, dermal, bone and scales—will differentiate from tumors without induction, while others—fin-rays and teeth—require environmental signals to trigger their differentiation; although
- in no instances do the cells remain tumorogenic.

Birds

The most extensive studies on avian embryos undertaken in the 1930s and 1940s were on the NC origin of chromatophores, with feathers as the preferred organ system of study.

In studies published in the late 1930s, Dorris found that pigment cells only formed *in vitro* if the neural tube was included in cultures, and that host limb buds became pigmented only if NCCs were grafted into them (a parallel to the studies with amphibian embryos discussed above). Such studies laid the basis for grafting between embryos of differently pigmented breeds of domestic fowl, or between embryos of different avian species (NC of robin, pheasant, or Japanese quail grafted into white leghorn chicken embryos, for example), to follow the fate of grafted NCCs. Rawles (1948) used such a technique to map the location of the presumptive NC in chicken embryos at the primitive-streak stage.[13]

Making use of the availability of radioisotopes as markers, Weston (1963) labeled a set of chicken embryos with ^{3}H-thymidine and then grafted TNCCs from embryos of 3 and 4 days of incubation into unlabeled embryos. He documented two major streams of migrating TNCCs: one migrated into the superficial ectoderm and differentiated into pigment cells; the other migrated ventrally and medially, moving between the developing spinal cord and somites to become the spinal ganglia and sympathetic neurons. We now know that avian TNCCs following a ventral pathway of migration become neuronal and contribute to the dorsal root ganglia. Cells that take a more dorsolateral pathway become pigment cells.[14]

In a classic study, Cohen and Konigsberg (1975) established clonal cultures from TNCCs of Japanese quail. A small proportion of the clones differentiated into two distinct cell types: melanin-producing pigmented cells and catecholamine-producing adrenergic neurons. Substantial evidence now supports the existence of restricted lineages capable of forming disparate cell types within premigratory and early migrating avian NCCs; in 9- to 13-somite-stage quail embryos, the CNC consists of precursors capable of producing neurons, glia, and chondrocytes and a small number of precursors from which neurons, glia, chondrocytes, and pigment cells arise (see Fig. 3.17).

A further classic series of studies performed by Bronner-Fraser and Fraser in the late 1980s, consisted of injecting single cells of the TNC from avian embryos with lysinated rhodamine dextran to monitor migration and cell fate. Single TNCCs gave rise to sensory neurons, pigment cells, ganglionic support cells (Schwann cells), adrenomedullary cells (sympathetic neurons), and neurons of the central nervous system (see Fig. 3.17).

Overexpressing *Cad2* or *Cad7* in the neural tubes of chicken embryos prevents the migration of many NCCs, completely prevents melanocytes from migrating along the normal dorsolateral migration pathway, and leads to an accumulation of melanocytes and melanocyte precursors within the neural tube (Nakagawa and Takeichi, 1998).

Patterning Feather Tracts

Pigment-forming NCCs of Japanese quail migrate along the epidermal–dermal junction after they are grafted into embryos of the white leghorn breed of domestic fowl but migrate below the epidermis if grafted into embryos of the silkie breed.

Beyond this indication that patterns of migration are not fixed, the mechanism patterning feather tracts may vary from region to region: quail NCCs do not respond to positional signals within the dorsal trunk feathers of chicken embryos, a finding consistent with prepatterns in the dorsal body wall that precede NCC invasion. However, quail NCCs do respond to local signals within the wings of chicken embryos, consistent with a prepattern within wing ectoderm and the presence of local cues to which NCCs respond. Within the wing plumage, therefore, local cues are specific: melanoblasts from one individual can 'read' local patterning signals in feather papillae of another. In experiments in which quail-duck feather chimeras were generated, craniofacial dermis was found able to induce host feathers with spatial patterns and timing of development that followed the source of the dermis, signaling operating through *Bmp2* and *4*, *Shh* and *Delta/Notch* signaling.[15]

Retinoic acid promotes the differentiation of melanocytes and may play a role in patterning pigmentation.

Non-Avian Reptiles

Investigation into the behavior of reptilian NC *in vitro* is limited. Hou and Takeuchi (1994) cultured NC from stage 9 and 10 Japanese turtle embryos. NCCs were

HNK-1 positive. TNC established *in vitro* underwent limited migration but did differentiate into melanophores and neurons.

Mammals

Rawles (1948) used the technique developed by Dorris (discussed under birds above) of grafting NC between species of birds to graft mouse NCCs into chicken embryos and demonstrate the NC origin of mouse melanocytes. Subsequently, organ culture of fragments of embryonic mouse neural folds was used to demonstrate that mouse midbrain NCCs do not form pigment cells.

Trunk neural tubes from 9-day-old mouse embryos established in organ culture delaminate TNCCs, which migrate and differentiate into melanocytes and adrenergic neurons. Melanocyte formation from NC has also been demonstrated after cultured murine NCCs are microinjected into 9-day-old embryos; the injected cells migrate extensively, contributing pigment-forming cells to the hair follicles of the chimeric embryos. Culturing fragments of murine midbrain neural folds, on the other hand, confirms that midbrain NCCs are not a source of pigment cells but do produce cranial mesenchyme.[16]

As with avian embryos (above), lysinated rhodamine dextran and DiI have been used with effect to label individual murine NCCs, follow their migration, and demonstrate that single cells from TNC are multipotential and can form pigment cells, neurons of the neural tube, dorsal root ganglia, and the sympathoadrenergic system, along with Schwann cells (Serbedzija *et al.*, 1991, 1997*). All sensory neurons of dorsal root ganglia require neurogenins, either *Ngn2* or *Ngn1* to mediate the first and second waves of neurogenesis, respectively, neurogenins being regulated by the canonical Wnt pathway (see Fig. 2.5).[17]

Pattern Formation

In murine embryos (and perhaps in other vertebrates as well), each pigmented stripe is the product of a clone of cells derived from a single melanoblast; retroviral labeling of single cells in neurulating albino embryos produce pigmented stripes in otherwise albino animals. Culturing NCCs from mutant embryos and then injecting the cultured cells into host embryos demonstrates that the host genotype modifies coat pattern, presumably by the type of extracellular-matrix-mediated interactions demonstrated to operate in amphibians. Genetic regulation also is being uncovered; *Wnt1* and *Wnt3a* enhance melanocyte numbers and melanocyte differentiation in mice although I do not propose to consider regulation of pigmentation in any depth.[18]

Some mouse mutants exhibit pleiotropic effects on several NC derivatives, such as pigment cells, nerve cells, and cells of the craniofacial skeleton. One mutant, *Patch*, results from a deletion in the gene encoding PdgfRα. *Patch* embryos show abnormalities of the glycosaminoglycans of the ECM, resulting in abnormal pigmentation, as well as defects of the craniofacial skeleton, thymus, heart, and teeth.[19]

Patch is discussed in more detail in the section on animal models and mutations for neurocristopathies in Chapter 9. A brief introduction to neurocristopathies in the context of pigment cells is provided below.

Neurocristopathies

As a look ahead to the more expanded discussion in Chapter 9, a sample of syndromes in human embryos, recognizable and classifiable because of shared NCC origins of the cells involved, is listed in Table 5.1. Such syndromes are known as **neurocristopathies** (see Chapter 9). A typical neurocristopathy, such as a malignant embryoma, may contain pigment cells, ganglia, connective tissue, cartilage, and Schwann cells, all NC derivatives.

Direct experimental evidence is available for the responsiveness to tumor-promoting agents of NCCs and of tumors derived from NCCs; the phorbol ester 12-0-tetradecanoylphorbol-13-acetate transforms neuronal cells of dorsal root ganglion into pigmented cells. Cultures of TNCCs or cell lines established from neuroblastomas, melanomas, or pheochromocytomas (the latter are tumors of the chromaffin cells of the adrenal medulla) exposed to the drug divide more rapidly, delay pigmentation, and block the adrenergic phenotype, showing that tumor promoters can redirect the differentiation of NCCs.[20]

Hirschsprung disease (see Chapter 10) and Waardenburg–Shah syndrome (Table 5.1) are two of a large number of syndromes traceable to defective populations of NCCs. Waardenburg–Shah syndrome, which combines the features of Waardenburg syndrome and Hirschsprung disease (see Chapter 9), is a dominantly inherited condition in humans with major effects on two populations of NCCs in which *SOX10* acts cell autonomously: pigment cells (hypopigmentation)

Table 5.1 Syndromes in human embryos recognized because of shared NCC origins of the cells involved

Syndrome	Incidence/10,000 live births
Neuroblastomas (sympathetic neurons of spinal ganglia)	2.8–1.1
Hirschsprung disease (aganglionic megacolon)	2
Goldenhar syndrome (a first arch syndrome)	1.78
Albinism (lack of pigmentation)	0.59
Waardenburg syndrome (aganglionic megacolon)	0.3
Treacher Collins syndrome (a first arch syndrome)	0.25–0.14
CHARGE association[a]	0.2–0.1
Waardenburg–Shah syndrome (Waardenburg syndrome IV)	0.1

[a] CHARGE is an acronym from **C**oloboma of the iris, **H**eart defects, **A**tresia of choanae, **R**etardation of physical and mental development, **G**enital anomalies, and **E**ar anomalies and/or deafness (see Chapter 9).

and enteric ganglia (aganglionic megacolon or Hirschsprung disease). Associated defects include deafness and eye involvement, although neither cardiac nor skeletogenic CNCCs are affected; a mouse mutant, *dominant megacolon* (*Dom*), terminates expression of *Sox10* prematurely and lacks NCCs able to form pigment (Potterf *et al.*, 2001).

These syndromes show the extent to which genetic regulation is specific to cell lineages, a conclusion borne out by studies in other vertebrates. Because of the link to Waardenburg–Shah syndrome, studies on *Sox 10* in zebrafish are used as an example. Mutations of *Sox10* in zebrafish such as *colorless* (*cls*) specifically affect pigment cell lineages by preventing the generation of nonmesenchymal NCCs (cartilages and fin mesenchyme develop normally), an effect that can be reversed by *Sox10*, which acts as a switching gene for mesenchymal or nonmesenchymal cell fates[21] (Dutton *et al.*, 2001). Loss-of-function studies of *Sox10* have been undertaken in zebrafish and mice, the latter prompted by the association of *SOX10* with Waardenburg–Shah syndrome. In both taxa, oligodendrocytes fail to differentiate and many NCCs—including melanocytes, enteric ganglia, and neurons and glia of both sympathetic and dorsal root ganglia—are reduced or absent, reflecting, in part, *abnormal migration* of NCCs along the lateral pathway. Mesenchymal derivatives of the NC (such as the craniofacial skeleton) and cardiac derivatives are unaffected, reflecting, in part, *normal migration* into the pharyngeal arches and heart. *Sox10* is either not expressed in mesenchymal, skeletogenic, and cardiac lineages and their derivatives or is only expressed transiently, depending on the taxon examined.

The cell-autonomous action of *SOX10* in enteric ganglia leading to Hirschsprung disease and of *Sox10* in pigment and neuronal cell lineages, provide a segue to Chapter 6, in which the origin and regulation of neuronal cells are examined.

Notes

1. Pigment cells in the eye have a dual origin: pigment of the choroid coat and outer iris is neural ectodermal in origin, while the pigmented retinal epithelium arises from the neuroepithelium of the optic cup. Periocular mesenchyme functions differently from other mesenchyme, interacting with melanocytic NCCs to modify the type of pigment cell produced.
2. Because of a relationship between ultraviolet light and folic acid, melanin may protect against neural tube defects (see Chapter 9).
3. See Yu *et al.* (2004) for feather, and Neste and Tobin (2004) and Schmidt-Ullrich and Paus (2005) for hair development and pigmentation.
4. See Langille and Hall (1988b, 1989) for the ablation studies, and McCauley and Bronner-Fraser (2003) and Sauka-Spengler *et al.* (2007) for more recent approaches.
5. See Hall and Hörstadius (1988*) for the older literature on NC extirpation and transplantation. For pigment cells as markers for transplanted NC, see Figures 37–41 in Hörstadius (1950), Graveson *et al.* (1995), Northcutt (1996), and Epperlein *et al.* (2000a, 2007a*). See Figures 3–5 in Graveson *et al.* (1995) and Figure 6 in Northcutt (1996) for illustrations of the transplantation of NC from pigmented into albino axolotls. Readers interested in further details of the origin of pigment cells from the amphibian NC will find that Chapters 2 and 4 of Hörstadius (1950*), comprising some 40% of his monograph, provide an excellent evaluation and synthesis of the earlier work.

6. See the studies in Hörstadius (1950*, pp. 73–78, especially Figures 38 and 39) for intrinsic patterns of specification of pigment cells, and Epperlein *et al.* (1996) and Olsson *et al.* (1996*) for the role of matrix components.

7. See Tuckett and Morriss-Kay (1986) and Smits-van Prooije *et al.* (1988), respectively, for the antibody and lectin studies of NCC migration, and Epperlein *et al.* (2007a*) for a recent overview. Galactose- and sialic-acid-containing cell surface carbohydrates also regulate adhesion of migrating NCCs and melanophores (Milos *et al.*, 1998). See Protas *et al.* (2006) for *Oca2* in Mexican tetras.

8. The differentiation of melanophores is under the control of the peptide melanophore-stimulating hormone (MSH) produced by the pituitary gland. MSH interacts with the melanocortin type 1 receptor to activate cAMP-mediated pathways.

9. For the influence of NCCs on neuromast patterning and of lateral lines on melanocytes, see S. C. Smith *et al.* (1990, 1994), Northcutt (1996), and Parichy (1996). See S. C. Smith *et al.* (1990) and Northcutt *et al.* (1994, 1995) for lateral line and neuromast development in the Mexican axolotl and the northern leopard frog.

10. Lopashov conducted his grafting experiments using groundling (weather loach), *Misgurnus fossilis*; the stone loach, *Nemacheilus barbatulus*; and European perch, *Perca fluviatilis*. See Sadaghiani *et al.* (1994*) and Hirata *et al.* (1997) for the *in vitro* studies.

11. See Raible and Ragland (2005) and Voigt and Papalopulu (2005) for these studies on β-catenin.

12. See Parichy, 2007* for an overview of extension of their studies on postembryonic late-onset precursors of pigment patterns using mutant screens and comparative analyses of a wider range of danio species. See Mellgren and Johnson (2002*) for mechanisms of zebrafish stripe formation.

13. See Hall (1999a*) for the demonstration by Dorris of NC formation of pigment cells in birds and Hörstadius (1950, pp. 81–88) for a discussion of pioneering studies grafting between and among breeds and species of avian embryos.

14. See Erickson and Goins (1995) and Harris and Erickson (2007) for models of NCC specification during migration.

15. See Richardson *et al.* (1989), Richardson and Hornbruch (1991), Richardson and Sieber-Blum (1993*), and Eames and Schneider (2005) for prepatterns versus local signaling in feather development, and Harris and Erickson (2007) for a model of the progressive acquisition by restricted melanoblast precursors of properties that guide their migration and differentiation, a model the authors tested against sensory neuronal, sympathoadrenal, and enteric precursors.

16. See Jaenisch (1985) and Osumi-Yamashita *et al.* (1994) for organ culture studies demonstrating the NC origin of melanocytes and adrenergic neurons in mice.

17. See Serbedzija *et al.* (1991, 1997*) for multipotential TNCCs, and Begbie *et al.* (2002) for *Ngn1* and *Ngn2*.

18. See Huszar *et al.* (1991) for studies on coat patterns in mice, and see Dunn *et al.* (2000) for the role played by Wnts. For mechanisms that may account for pigmentation patterns in vertebrates, see Epperlein *et al.* (1996, 2007a*), Olsson *et al.* (1996*), Parichy (1996, 1998), Mellgren and Johnson (2002*), and Mills *et al.* (2007).

19. See Morrison-Graham *et al.* (1992) and Soriano (1997) for *Patch*.

20. See Nishihira *et al.* (1981) for the response of NCCs to tumor promoters.

21. See Kelsh and Eisen (2000) and Dutton *et al.* (2001) for the *colorless* mutant.

Chapter 6
Neuronal Cells and Nervous Systems

The following major points concerning the embryological origins of neurons have been made so far:

- Neurons of spinal ganglia arise from the neural crest (NC).
- Neurons arise from the tail buds of embryos and not from a delaminated germ layer.
- All neuronal cells express the cell-adhesion molecule, N-CAM.
- Neural crest cells (NCCs) and cells of the central nervous system are so closely related that they share a common lineage; central neurons and NC derivatives can arise from the same cloned cells.
- Single cells from mouse TNC are multipotential and capable of forming neurons of the neural tube, dorsal root ganglia, and sympathoadrenergic system, in addition to Schwann and pigment cells.
- Early in NCC migration, lineages of sensory and autonomic neurons segregate from pluripotential cells that can form sensory and autonomic neurons or melanocytes.
- Single quail CNC are multipotential precursors capable of producing neurons, glia, and chondrocytes and a small number of precursors from which neurons, glia, chondrocytes, and pigment cells arise.

The following points concerning the evolutionary origins of neurons has been made:

- Ascidians possess pigment cells and are now regarded as the sister group to vertebrates.
- The new-head hypothesis proposed by Gans and Northcutt was introduced.
- Lampreys do not possess sympathetic ganglia and lack a ventrally migrating population of NCCs.

Neuronal differentiation from uni-, bi-, or multipotential cells was discussed in some detail in Chapter 3 in the context of the segregation of subpopulations of NCCs. The emphasis in the current chapter is on evidence for the NC origin of these neurons and of the nervous systems to which they belong.

B.K. Hall, *The Neural Crest and Neural Crest Cells in Vertebrate Development and Evolution*, DOI 10.1007/978-0-387-09846-3_6,
© Springer Science+Business Media, LLC 2009

The Neural Crest, Neurons, and Nervous Systems

Pioneering studies using amphibian embryos providing evidence for the NC origin of a variety of types of neurons, parts of the nervous system—the peripheral and autonomic nervous systems—and neuron-supporting cells—Schwann cells and glia. Indeed, many of the basic elements of modern-day neurobiology were established in studies, in which amphibian NCCs were labeled, excised, or transplanted. Subsequent studies on the neuronal derivatives of the NC concentrated on avian embryos, especially once the quail-chicken cell-marker system was discovered.[1]

Neuronal- and nervous-system derivatives discovered to be NC in origin include:

- spinal and cranial ganglia of the peripheral nervous system,
- adrenergic and cholinergic sensory neurons of the autonomic (sympathetic and parasympathetic) nervous system,
- Schwann cells,
- Glial cells, and
- Rohon–Béard neurons (see Table 1.1).

The **peripheral nervous system** consists of sensory neurons organized into sensory ganglia that transmit information from peripheral tissues and organs to the brain or spinal cord of the central nervous system. Sensory neurons extend one axon peripherally to the target organ and one centrally to the CNS. Two major classes of neurons comprise the peripheral nervous system: (i) spinal ganglia, also known as dorsal root ganglia, organized in pairs along the spinal cord and that transmit information to the spinal cord and (ii) cranial ganglia organized along the cranial nerves and that transmit information to the brain.

The **autonomic nervous system** (sometimes known as the visceral nervous system), which controls involuntary functions such as breathing, digestion, and heart rate, consists of sensory and motoneurons in three subsystems: the sympathetic, parasympathetic, and enteric nervous systems. Studies initiated in the 1970s greatly expanded our knowledge of the NC origin of the autonomic nervous system, primarily using chicken embryos, with confirmatory studies with Japanese quail. Transplanting quail NC into chicken embryos identified the regions of the NC from which particular portions of the autonomic nervous system and hormone-synthesizing cells arise:

- **Parasympathetic (cholinergic) enteric ganglia** of the gut are derivatives of the vagal neural crest (VNC), corresponding to NC at the levels of somites 1–7.
- NC adjacent to somites 8–27 does not produce enteric ganglia but gives rise to the **sympathetic (adrenergic) ganglia** of the adrenal gland.[2]
- Some regions of the NC—those adjacent to somites 6 and 7 and those caudal to somite 18—produce cholinergic and adrenergic neurons.
- **Adrenomedullary cells** develop from NC that originates at the level of somites 18–24.

- NCCs caudal to somite 28 (**the sacral neural crest**) produce enteric ganglia for the caudal (postumbilical) portion of the gut, including the ganglion of Remak of the ileum.[⊕]

Such precise cellular localizations, a necessary prelude to investigating the mechanisms of differentiation, neoplasia, and dysmorphogenesis (see Chapters 9 and 10), could not have been accomplished without the benefit of a 'label', such as the quail nuclear marker. Each of these cell types, NC regions, and nervous systems is now described in the context of the evidence for their NC origin and information on how their differentiation is controlled.

The Peripheral Nervous System—Spinal and Cranial Ganglia

The identification of **spinal ganglia** as NC derivatives goes back to Wilhelm His' discovery of the NC (see Chapter 1). Indeed, so readily was the claim of a crest origin for spinal ganglia accepted that this embryonic region was known for a time as the **ganglionic crest**.

The initial observations on the existence of the NC in elasmobranchs made in the 1870s and 1880 included descriptions of neuronal NCC derivatives.[3] Balfour (1876) identified a neural ridge (see Box 3.1) within elasmobranch neural tubes from which spinal ganglia arose. He thought that ganglia of the sympathetic nervous system arose as branches of the spinal nerves and therefore were products of the neural ectoderm. Balfour was adamant that His was wrong in his interpretation of the existence of an 'intermediate cord' (NC). By 1881, however, Balfour had changed his view and was illustrating and discussing ganglia as NC in origin (see Fig. 3.2 in Box 3.1).

Also in the 1880s, the dorsal roots of the spinal ganglia were described as arising from the spinal cord in sharks, although no distinction was made between cells of the neural tube or NCCs as the source of the neurogenic cells. In what was perhaps the next detailed study, Conel (1942) described the NC origin of the dorsal root ganglia of cranial and spinal nerves in the spiny dogfish, *Squalus acanthias*, and in the electric ray (Fig. 6.1).

Wilhelm His made his observations using fixed embryos. Techniques to visualize cells in living embryos were not developed the 1930s. Detwiler stained local regions of NC with pieces of agar impregnated with vital dyes to show that spinal ganglia arise from TNCCs that migrate between the somites and the spinal cord/notochord. Experimental proof was provided by Harrison and by DuShane, both of whom observed that spinal ganglia and sensory nerves fail to form in tadpoles if the

---

[⊕]The ganglion of Remak or Remak ganglion, which is only found in birds, originates in the lumbosacral NC and is the major parasympathetic nervous system element in the hindgut. It is named after Robert Remak (1815–1865) who first described histological characters for each germ layer and provided the first detailed description of mesoderm, Pander's 'middle vessel' layer (see Chapter 2).

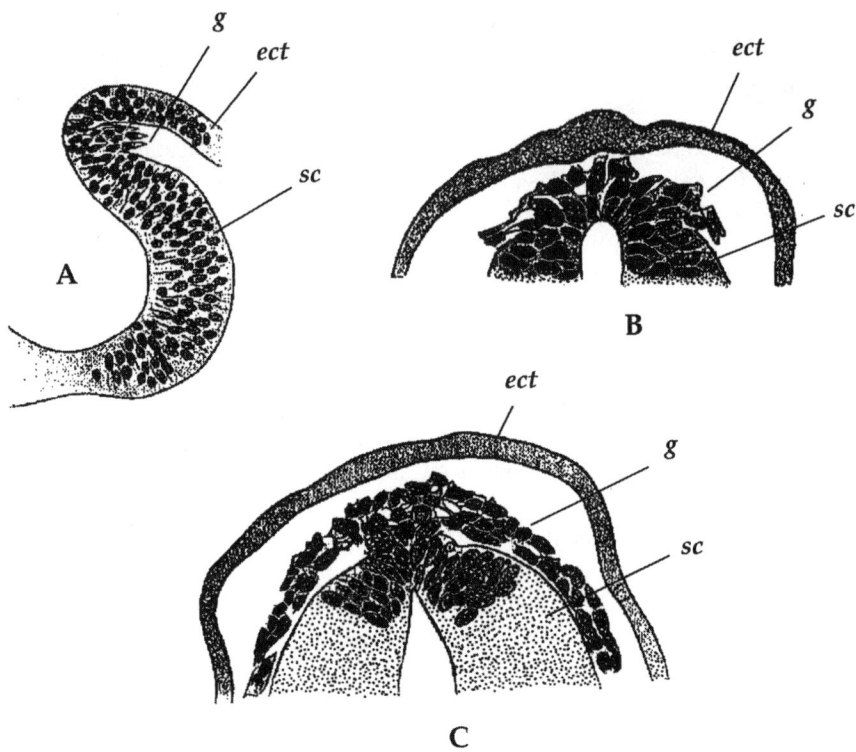

Fig. 6.1 The neural crest origin of spinal ganglia as seen in sections of neural folds of a chicken embryo (**A**) and of the electric ray (**B** and **C**). ect, ectoderm; g, ganglion; sc, spinal cord. Modified from Volume II of Text-Book of Embryology, by Kerr (1919)

NC was removed at the time of neural fold closure. Extirpating the *ventral* portion of the neural tube results in larvae lacking motor nerves but has no effect on spinal ganglia or sensory nerves; that is, the NC was shown to be a derivative of the *dorsal* neural tube. Further confirmation was provided by Raven, who transplanted NC between embryos of different species of amphibians, using such species-specific characters as differential cell or nuclear size to distinguish host from donor cells.[4]

A major source of the early confusion over whether ganglia arose from the neural crest, the dorsal neural tube, or adjacent ectoderm, was the observation that although the NC always arises at the lateral borders of neural ectoderm, the NC arises at different stages of neurulation in different vertebrates:

- NCCs do not migrate away from the trunk neural tube in sharks until late in neurulation, after the neural tube has invaginated and separated from the epidermal ectoderm (see Fig. 3.1 in Box 3.1).

- Neural crest appears at the open neural plate stage in amphibians and rodents, but at the closed-neural-fold stage in birds.
- In many teleost fish, neurulation is by cavitation of a neural keel, not by invagination of a flat neural plate (see Fig. 3.1 in Box 3.1).

Further complexity exists, not the least of which, is that the most caudal embryonic region arises by secondary neurulation without segregation into germ layers, often appearing as a tail bud (see Chapter 2). Furthermore, TNCC migration in teleost embryos is **between** the somites rather than through the rostral half of each somite as occurs in tetrapods. Using a combination of HNK-1 staining, SEM and DiI injection in the most complete analysis of the development of dorsal root ganglia in any teleost, the Mozambique tilapia, Laudel and Lim (1993) demonstrated that dorsal root ganglia contain sensory cells and motor fibers, and arise from NCCs that have migrated between the neural tube and the somites.

Spinal ganglia do not self-differentiate but require influences from other cells and tissues, primarily the somites. This conclusion goes back at least to the studies by Detwiler in the 1930s, in which removing somites led to localized loss of spinal ganglia and spinal nerves, while adding somites led to supernumerary ganglia and nerves.

An NC origin for the **cranial ganglia** in chicken embryos was one of the bases on which Wilhelm His identified the *Zwischenstrang* in 1868. Acceptance of the NC origin of cranial ganglia did not come quite as readily as acceptance of the NC origin of spinal ganglia. Some maintained a NC, others a **placodal origin**. As it turned out, both were and are correct. As most extensively shown in the research of Leon Stone and C. L. Yntema, many cranial ganglia receive contributions from NC *and* placodal cells (Fig. 6.2; and see Fig. 2.16), although cranial sensory ganglia in the North American sea lamprey may be entirely placodal in origin (McCauley and Bronner-Fraser, 2003). If confirmed with studies from other lamprey species, a NC contribution to placodes may be a gnathostome innovation.

This section now discusses the nature and origin of and contributions that arise from placodal ectoderm before moving on to discuss the autonomic nervous system.

Placodal Ectoderm

The NC appears at the neural–epidermal border, whether that border is at the normal site *in vivo* or created after ectopic induction of a neural tube within epidermal ectoderm (Figs. 6.3 and 6.4).

Epidermal sensory placodes (hereinafter, placodes), which also arise at the border of the neural plate (Fig. 6.3) give rise to cranial sense organs such as the nose and ear and represent an important source of neural tissue, especially for the central ganglia of the cranial nerves (Figs. 6.3 and 6.4) and in teleosts and amphibians for the lateral line, neuromasts, and mechanosensory systems (Box 6.1). Placodes develop as thickenings of the head ectoderm, either adjacent to the NC or from the

Fig. 6.2 The neural crest (**A**) and successive stages in the development of the roots and ganglia of the spinal nerves (**B** and **C**) in a shark embryo. (**A**) Origin of the neural crest (pr) by proliferation from the dorsal surface of the neural tube. (**B**) Initial formation of the root of the spinal nerve (pr). (**C**) An older embryo to show the position of the posterior root (pr) and spinal ganglion (sp.g) of the spinal nerve (n). Other abbreviations are: ch, notochord; ao, aorta; nc, neural canal; sc, somatic mesoderm; sp, splanchnic mesoderm; x, subnotochordal rod. Modified from Balfour (1881)

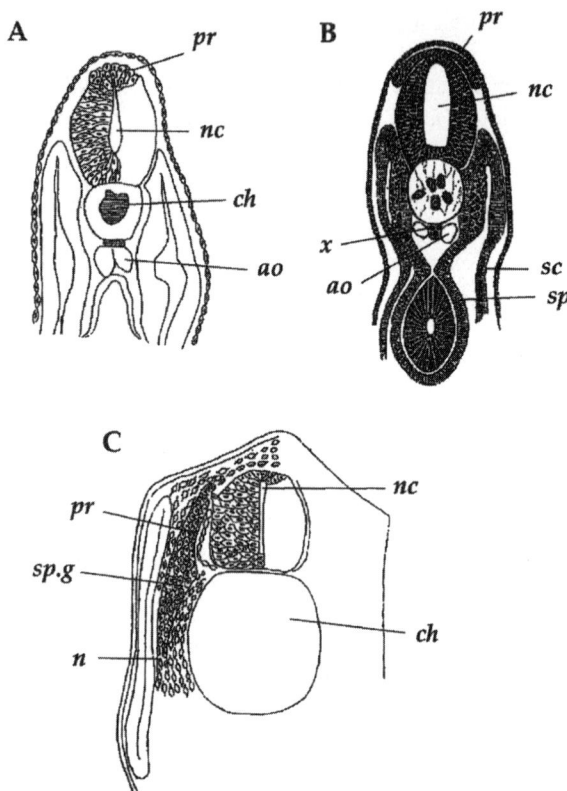

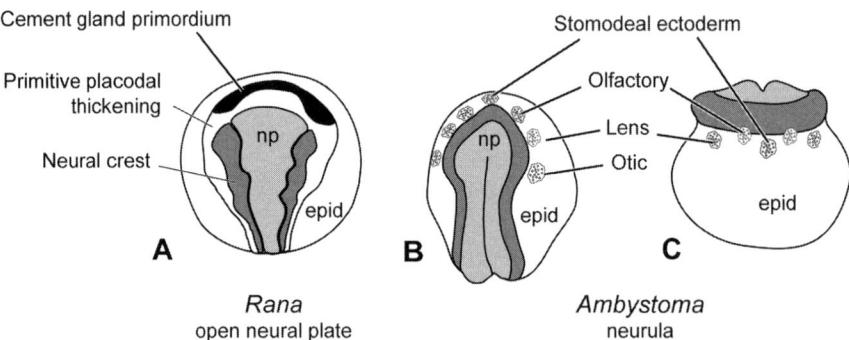

Fig. 6.3 Fate maps of cranial placodes in a typical frog (*Rana*, **A**) and in *Ambystoma* (**B** and **C**) to show their relationships to the neural crest, neural plate (np), and epidermal ectoderm (epid). **A** and **B** as seen in dorsal view with anterior to the *top*, **C** in rostral view. Adapted from Baker and Bronner-Fraser (2001)

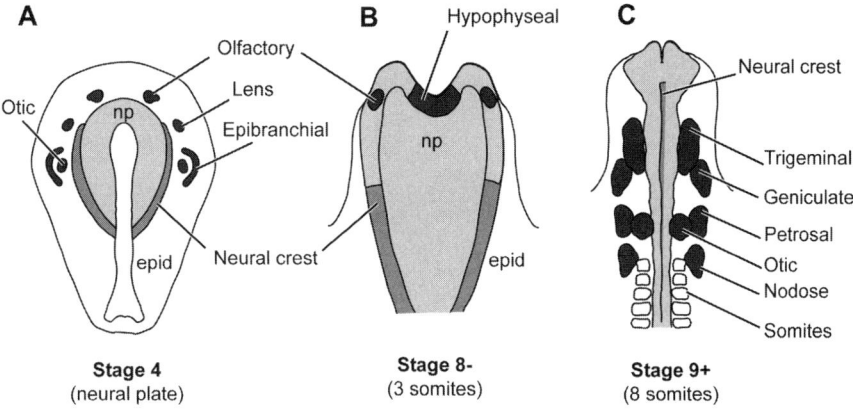

Fig. 6.4 Fate maps of cranial placodes in chicken embryos at three H.H. stages (**A**, **B**, and **C**), all shown in dorsal view with anterior to the *top*. Both the rostral–caudal positioning of individual placodes and their relationship to the neural crest, neural plate (np), and epidermal ectoderm (epid) can be seen. Adapted from Baker and Bronner-Fraser (2001)

neural folds themselves (Figs. 6.3, 6.4 and 6.5). In axolotls and in mice, perhaps in other groups, placodes arise from *lateral* neural fold ectoderm, while NC arises from *median* neural folds.

Box 6.1 The lateral line system

In amphibians and in fish, neuronal cells form a specialized system of nerves, the **lateral line nerves** and specialized electro- and mechanoreceptive organs, **neuromasts**, associated with the lateral line nerves. Both elements of the lateral line system, which arise from placodal ectoderm, have been lost from all amniotes. In fish, neuromasts are found either superficially in the epidermis or deep and associated with bony canals of the lateral line system. All amphibian neuromasts are superficial.[a] Here I discuss the lateral line system of teleost fish.

Bony and cartilaginous fish have an extensive lateral line system. Eight basic patterns, derived by heterochrony from an ancestral pattern, have been proposed in groups such as cichlids, whose evolutionary history is well resolved.

Neuromasts and lateral line nerves arise from ectodermal placodes.[b] NCCs pattern neuromasts but have not definitively been shown to give rise to neuromasts. On the basis of labeling zebrafish NCCs with DiI, however, it has been claimed that neuromasts have a dual origin, arising from NC *and* from non-NC ectoderm. In the absence of histological evidence and because of the difficulty of labeling NC in the neural folds without labeling placodal ectoderm

(Box 2.1), this claim remains unsubstantiated. As discussed in the text of this chapter, placodal ectoderm in some species arises from the lateral neural folds, NC arising from medial neural folds. In such species, placodal ectoderm may be induced by NC rather than forming from NC. Axolotl lateral lines and neuromasts express *Msx2* and *Dlx3*, molecular markers that could be used to track the effects of NCCs on neuromast origins, differentiation and morphogenesis.[c]

In a recent study of the origin of neurogenic placodes and cranial sensory ganglia in the lesser-spotted dogfish, *Tbx3*, was identified as a specific marker for lateral line ganglia in this species. *Ngn1* is required for development of lateral line and cranial sensory ganglionic precursors in zebrafish.[d]

As described in salmon and other genera, preosteogenic cells derived from the NC accumulate under the sensory organs (neuromasts) of the lateral line system. Circumstantial evidence suggests that developing neuromasts induce scales or dermal bone in teleosts and cartilage in sharks. Once formed in teleosts, lateral line bone contains a canal for the lateral line nerve and pores for the neuromasts. Canal neuromasts are distinctive; in the goldfish, for instance, they are 50–100 μm in diameter. Superficial neuromasts are <50 μm in diameter. Surprisingly, in two species of hexagrammid fish of the genus *Hexagrammis*, some trunk lateral lines lack neuromasts. Although amphibians possess neuromasts, they lack the canal neuromasts associated with bone in fish.[e]

---

[a] For the induction and development of placodes in anurans and urodeles, see Schlosser and Northcutt (2000, 2001) and Schlosser (2002a,b, 2006*).

[b] The story is more complicated than this; secondary neuromasts that develop postembryonically in the posterior lateral line in fish, arise from new waves of precursors and not by budding from the embryonic lateral line (Ledent, 2002).

[c] See Collazo *et al.* (1994) for the DiI-labeling study, S. C. Smith *et al.* (1990*) and Northcutt (1996) for associations between placodal and NC ectoderm, and Metscher *et al.* (1997) for *Msx2* and *Dlx3*. The role of the lateral line system, especially neuromasts, in specifying pigment patterns in anuran and urodele amphibian tadpoles is discussed in Chapter 5.

[d] See O'Neill *et al.* (2007) for *Tbx3* and Andermann *et al.* (2002) for *Ngn1*. A neurogenic gene also has been cloned from amphioxus and the earliest known marker for amphioxus neuroectoderm, appearing as two segmental bands in early neurulae and in the dorsal neural tube of mid-stage neurulae. Subsequent *AmphiNg* is expressed in epidermal chemosensory cells, and in the midgut along with an insulin-like peptide in a region that L. Z. Holland *et al.* (2000) surmise could be a homolog of the pancreas.

[e] See Holmgren (1940) and Wonsettler and Webb (1997) for lateral line patterns in sharks, rays, and bony fish, and Webb and Northcutt (1997) for neuromasts in nonteleost bony fish, in which multiple canal neuromasts between pore positions is the evolutionarily primitive condition. See Tardy and Webb (2003) and Webb and Shirey (2003) for lateral line canal and neuromast development in zebrafish, and Wilson *et al.* (2007) for Cad4, lateral line, and neuromast development in zebrafish. For the possible induction of bone or cartilage by organs of the lateral line, see Hall and Hanken (1985) and Webb and Noden (1993*).

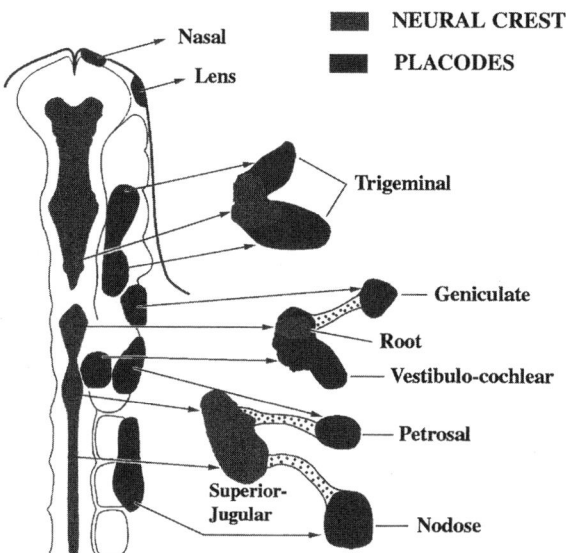

Fig. 6.5 A diagrammatic representation of the location of neurogenic placodal ectoderm (*black*) adjacent to the neural tube and of neural crest (*gray*) that contributes to placodes in a chicken embryo as seen from the dorsal side. Only the placodes on the *right-hand side* are shown. The most rostral placodes are the nasal and lens. The cranial sensory ganglia are placodal (geniculate, vestibulo-cochlear, petrosal, nodose ganglia), neural crest (root, superior jugular ganglia), or of mixed placodal and neural crest origin (Trigeminal ganglion). Note that ganglia can have a proximal component that is neural crest (superior jugular) and distal components that are placodal (petrosal as distal cranial ganglion IX; nodose as distal cranial ganglion X). Adapted from Webb and Noden (1993)

Placodes have been proposed to form in association with weak competence at the neural plate boundary, although the precise relationship between placodes and NC remains an active area of research. Lateral-line placodes in the Mexican axolotl are set aside at the late neural fold stage but can be induced as late as the early tail bud stage, the temporal window of induction paralleling loss of competence for induction by the ectoderm. The same group of experimenters examined development of direct-developing Puerto Rican frog, *Eleutherodactylus coqui*, which lacks neuromasts or ganglia associated with the lateral line. The loss is evident early in development; no lateral line placodes arise. Transplantation between coqui and axolotl embryos demonstrated that coqui ectoderm lacks the competence to respond to inductive signals. Competence for induction emerges as an important mechanism regulating placode formation.[5]

All placodes except the adenohypophyseal and lens placode give rise to neurons; some give rise to other cell types as well (Table 6.1). The earliest jawed vertebrates are thought to have had seven pairs of cephalic dorsolateral placodes. Fish and many amphibians have six dorsolateral lateral-line placodes, from which the lateral lines and lateral-line nerves arise (Box 6.1); the lateral-line system has been secondarily

Table 6.1 Overview of the major placodes, the sense organs to which they contribute, and the cell types that arise from them

Placode	Derivatives	Cell types
Olfactory	Nasal epithelia	Olfactory neurons, supporting cells olfactory epithelium
	Brain	Gonadotropin-releasing hormone neuroendocrine (GnRH) cells, glia
Hypophyseal	Adenohypophysis (anterior pituitary)	Endocrine cells
Lens	Lens of eye	Lens cells
Trigeminal	Trigeminal ganglia	Primary sensory neurons of the ganglion of the Vth nerve
Otic[a]	Inner ear	Sensory hair cells, supporting cells, sensory neurons of the ganglion of the VIII nerve
Lateral line[a]	Neuromasts	Mechanoreceptor and electroreceptor
	Ganglia of lateral line nerves	neurons, hair cells, supporting cells,
Epibranchial[b]	Ganglia of cranial nerves VII (the facial), IX (the glossopharyngeal) and X (the vagal)	Sensory neurons
	Taste buds,	Afferent neurons (innervation)
	Heart	Afferent innervation for receptors[c]
	Lungs, Intestine	Afferent neurons[d]

[a.] Together these make up the dorsolateral series of placodes.
[b] The three epibranchial placodes are the geniculate, petrosal, and nodosal.
[c] These neurons are involved in the transmission of information on heart rate and blood pressure.
[d] These neurons are involved in the transmission of information on distension of the gut epithelium and stimuli in bronchii of the lungs.

lost from all amniotes. The seven classically recognized classes of dorsolateral pla-codes (otic and lateral-line placodes), along with the six ventrolateral placodes, are set out in Table 6.2, as are the sense organs to which each placode contributes and the cell type(s) derived from placodal ectoderm in each sense organ.[6]

The adenohypophyseal placode (from which the anterior pituitary forms) is the most rostral, succeeded as one moves caudally by the hypophyseal, lens, trigeminal,

Table 6.2 Gene markers for individual placodes expressed early in placode specification[a]

Placode	Gene
Epibranchial	*Pax2, Sox2, Sox3, Phox2a*
Lens	*Pax6, Sox2, Sox3*
Olfactory	*Sox2[b], Sox3[b], Pax6, Dlx3, Dlx5*
Otic	*Pax8[b], Pax2, Dlx3, Dlx5, Dlx7, Sox2, Sox3, Spalt4*
Hypophyseal	*Six3, Pitx1*
Trigeminal	*Pax3[b], Sox2, Sox3, Pax3*
Lateral line	*Eya1, Sox2, Sox3*

[a] These genes are the first markers to be expressed in the development of the particular placode (Source: Baker and Bronner-Fraser (2001*), Schlosser (2005*), Barenbaum and Bronner-Fraser (2007*)).

and otic placodes (Figs. 6.3, 6.4, and 6.5 and Table 6.2). A century-old debate over whether the first cranial nerve (Nerve 0) arises from the olfactory placode is discussed in Box 6.2.

Box 6.2 The origin of cranial nerve 0

For more than a century, the most rostral cranial nerve, the *nervus terminalis*, traditionally named *nerve 0*, has been regarded as originating from the olfactory placode. A population of neurons within nerve 0 expresses gonadotropin-releasing hormone (GnRH). Known as GnRH neurons or luteinizing hormone releasing (LhRH) neurons, defective migration of these neurons is suspected to be the basis of Kallman syndrome in humans, in which the olfactory nerves fail to develop (Carstens, 2002).

Earlier studies on mouse, chick, amphibian, and lungfish are consistent with an olfactory placodal origin for these GnRH-positive neurons, which have been interpreted as arising from the medial aspect of the olfactory placodes, migrating to the CNS along the vomeronasal and terminal nerves. A challenge to accepted wisdom comes from a study tracing the origin of GnRH neurons in zebrafish, in which a caudal subset of GnRH neurons that contribute to the terminal nerve were traced back to the NC, and evidence provided for the origin within the adenohypophyseal placode of the rostral subset of neurons that migrate into the hypothalamus.[a]

---

[a] See the study by Whitlock *et al.* (2003) and a commentary on and discussion of it by von Bartheld and Baker (2004). See Box 4.2 and L. Z. Holland and N. D. Holland (2001*) for homologies of the adenohypophysis.

The **lens placode** (Figs. 6.3, 6.4, and 6.5), which segregates from head ectoderm in 30-h-old chicken embryos, shares an intimate association with the subjacent NCCs and associated extracellular matrix (ECM). Meier's data (1978*) convinced him that the placode/NC association was so intimate that the two must interact developmentally. One difference between the two is that delamination of cells from placodal ectoderm in chicken embryos does not involve the epithelial-to-mesenchymal transformation required for NCCs to delaminate (Graham *et al.*, 2007).

Placodal Markers and Specification of Placodal Ectoderm

Until recently, one difficulty with assessing placodal development was the lack of specific markers. Thanks to more recent work we have genetic markers for all placodes.[7] All placodes with the exception of the maxillomandibular trigeminal placode express one or more members of the *Pax* gene family:

- *Pax8* in the otic placode,
- *Pax3* in the ophthalmic trigeminal placode,

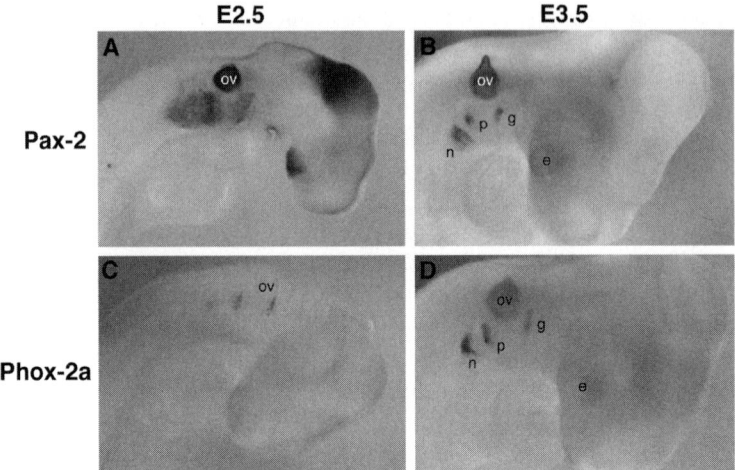

Fig. 6.6 Chicken embryos of 2.5 and 3.5 days of gestation to show otic and epibranchial placodes. (**A**) At 2.5 days *Pax2* is expressed in the otic vesicle (ov) and ventral epibranchial placodal ectoderm. (**B**) At 3.5 days, *Pax2* expression continues in the otic vesicle and labels the three epibranchial placodes, the geniculate (g), petrosal (p), and nodose (n). (**C**) At 2.5 days, *Phox2a* is expressed in a small number of neurons in the three epibranchial placodes but not in the otic vesicle. (**D**) At 3.5 days, *Phox2a* is expressed in the three epibranchial placodes. Figure kindly supplied by Clare Baker

- *Pax6* in lens and olfactory placodes, and
- *Pax2* in the otic and epibranchial placodes in all gnathostomes analyzed (Table 6.2; Fig. 6.6).

As summarized in Table 6.2, a combination of Pax and other markers (particularly members of the *Sox* gene families) enables early placode precursors to be differentiated from one another, even though individual markers may be expressed in more than one placode and function in other cell lineages. (Figure 6.6 shows *Pax2* in otic and epibranchial placodes and *Phox2a* in epibranchial but not in the otic placode in chicken embryos.) Nevertheless, judicious application of these markers has expanded our understanding of placodal origins, diversification, and relationships enormously in the past decade. One study on the formation of the otic placodes in chicken embryos and two studies on induction and specification of the otic placode are used as examples: one in chicken embryos and another in mice.

The first detailed fate map of otic placodal ectoderm in chicken embryos was produced by Streit in 2002. *Pax2* was used as the marker, although not all *Pax2*-positive cells contributed to the placode, a caution to be keep in mind when using a single marker that is not uniquely expressed in one subset of cells. Initially, future otic placodal cells are not adjacent but scattered in the ectoderm with future neural, NC, epithelial, and other placodal cells. Extensive cell movement brings otic placodal cells to the midline of the embryo adjacent to r5–6, where some cells from the neural folds contribute to the placode.

Using *Pax2*, *Sox3*, *Bmp7*, and Notch as markers, Groves and Bronner-Fraser (2000) localized the ectoderm from which the otic placode will arise in chicken embryos at the 4–5-somite stages, showed that the ectoderm was specified as otic by the 4–6-somite stages, that induction of the placode continued progressively to the 10-somite stage, when the otic placode was committed (committed meaning the earliest stage at which the placode could develop independently), and that the placode formed between the 12- and 14-somite stages. In a subsequent study, Martin and Groves (2006) showed that preplacodal ectoderm expressed some but not all otic placode markers in response to Fgf.

Pax2 is strongly expressed in the otic placode ectoderm in mouse embryos and weakly or not at all in the adjacent epidermal ectoderm. *Wnt* gene expression is enhanced in a subset of cells within the *Pax2*-positive placodal ectoderm. Wnts function through β-catenin. Inactivating β-catenin enhances the field of epidermal ectodermal cells at the expense of the extent of the otic placodal domain. Activating β-catenin produces the opposite result, the expanded otic placodal ectodermal expressing exclusively dorsal otocyst markers (Ohyama *et al.*, 2006).

Many more genes than included in Table 6.2 are expressed at later stages of placode formation; Figure 5 (pp. 319–321) in Schlosser (2006*) contains a detailed list of genes expressed at neural plate and tail bud stages in *Xenopus* embryos.

Following early studies showing that placodal ectoderm lies at the border between cranial neural plate/neural crest and epidermal ectoderm, we now know that multiple placodes can and do arise from individual regions within placodal ectoderm. As an example, a continuous region of thickened placodal ectoderm in Atlantic cod embryos gives rise to the lens, otic, and lateral-line placodes, and to the ectodermal lining of the opercular cavity (Miyake *et al.*, 1997). Such spatial patterning (regionalization) of placodal ectoderm may begin as early as gastrulation in some taxa.

The Panplacodal Domain

Perhaps the most important findings from recent studies is the recognition of a **pre- or panplacodal domain** at the border of the cranial neural tube within which individual placodes are induced at different times during development in response to different combinations of inductive molecules, competence to form individual placodes changing in space and over time.

The panplacodal domain is specified by members of two major families of transcription factors, the *Six* family (orthologous to *sine oculis* in *Drosophila*) and the *Eya* family (orthologous to the *Drosophila* gene *eyes absent*) are important, especially *Six1* and *Eya1*. Single copies of both genes are found through the metazoans, indicative of the ancient evolutionary origin of these transcription factors (see Figure 2 in Schlosser, 2007). As evaluated by Andrea Streit and Gerhard Schlosser who have spearheaded analysis of specification of the panplacodal domain, activation of *Eya* genes and members of the *Six1/2* and *Six4/5* families,—the 'panplacodal genes'—is under the control of Bmps, Fgfs, and Wnts; Fgfs promoting and Bmps and Wnts suppressing *Six*/*Eya*.[8] Target genes of *Six1* and *Eya1* are less well understood.

Table 6.3 Embryonic primordia providing inductive signals for individual placodes[a]

Placode	Primordium—gene
Hypophyseal	Diencephalon—Shh from midline tissues for induction; *Bmp4* (possibly aided by foregut endoderm) for patterning
Epibranchial	Pharyngeal arch endoderm—*Bmp7; Fgfs*
Lens	Optic cup—*Bmp4, Fgfs* to induce; Bmp7 to maintain
Olfactory	Anterior, mesendoderm, aided by forebrain
Otic	Mesendoderm—*Fgf3* (depending on species), other *Fgfs, Wnts*
Trigeminal	Neural tube
Lateral line	Mesoderm, neural plate

[a] Source: Baker and Bronner-Fraser (2001*).

Induction of Individual Placodes

The nature and source of the signals responsible for the specification of individual placodes within the panplacodal domain is also being revealed. Signals from a variety of developing embryonic regions/tissues induce placode formation, the most well known of which are outlined in Table 6.3.

Details of these inductions and of how individual placodes arise from common ectodermal fields continue to emerge. For example, zebrafish epibranchial and otic placodes are induced by a combination of signaling molecules, *Fgf3* and *Fgf8* (probably from the hindbrain) and *Foxi1,* the latter acting via *Pax8;* placodal initiation is prevented in the *hearsay* mutant, a mutation of *Foxi1,* which is the earliest marker for otic placodes in zebrafish. However, other signals may operate *in vivo* and, once formed, placodes evoked by the same signals come under independent control; chicken *Sox3* must be downregulated for epibranchial placodes to form from *Sox3*-positive preplacodal domains. If *Dlx3* and *Dlx4b* are knocked out in zebrafish, epibranchial placode formation is unaffected, although otic placodes fail to form. *Dlx3* and *Dlx7* also have been shown to act in concert to initiate otic and epibranchial placodal development in zebra fish.[9]

Such results are consistent with (i) shared signals for placode induction, (ii) what may be considerable redundancy of cascades of signals not yet sufficiently understood, and (iii) divergent signals for later development. Supposedly, we are entering a decade of intensive research on placodal ectoderm.

Induction of individual placodes in the context of progressive restriction of a panplacodal has just been elegantly demonstrated by Xu and colleagues (2008) using DiI and DiO labeling to fate map the ophthalmic and maxillomandibular trigeminal placodes in chicken embryos of 0–16 pairs of somites (24–48 h of incubation). The three major conclusions are that:

- precursors for the two placodes arise from a partially overlapping territory that is initially distributed as a large region of ectoderm that includes precursors of the geniculate and otic placodes and of epidermal ectoderm;

- no clear demarcation between the two placodes could be discerned, even at the latest stage examined when neurogenesis is underway in both placodes;
- some cells at the border expressed both the ophthalmic placodal marker *Pax3* and the maxillomandibular placodal marker, *Neurogenin1*, consistent with some cells responding to inductive signals that initiate both placodes.

The Autonomic Nervous System

As with the origin of cranial ganglia (but for different reasons), authoritative figures differed as to whether the NC contributed to the autonomic nervous system; as discussed below, enteric and visceral ganglia of the sympathetic nervous system were more convincingly shown to arise from NC.[10]

In part, the differences were technical; extirpation and transplantation studies gave less reproducible results than did vital staining. The clearest interpretation of the early studies is that *NC and ventral neural tube both* contribute cells to the autonomic nervous system (see Hörstadius, 1950*). Resolution of the origin of the autonomic nervous system from dorsal neural tube (NC) had to await studies on other vertebrates, notably chicken embryos. That said, the ventral neural tube has reappeared in a series of studies claiming that cells emigrating from the ventral neural tube form derivatives regarded as the exclusive property of NCCs (see Box 6.3).

Box 6.3 The ventral neural tube: A ventral neural crest?

NCCs are derivatives of the dorsal neural tube. The ventral neural tube gives rise to neurons. However, as indicated in the text, cells of the ventral neural tube can be switched to NCCs, if they are grafted into the migration pathway taken by NCCs.

Evidence from a series of studies in chicken embryos leads us to conclude that the ventral neural tube normally contributes chondrocytes to Meckel's cartilage and to the quadrate (Sohal *et al.*, 1999), two skeletal elements that have been regarded as exclusively of NC origin.

The evidence comes from studies in which ventral neural tube cells of chicken embryos were tagged with a viral LacZ marker and then injected into the lumen of the rostral hindbrain. The injected cells moved out of the lumen of the hindbrain. Unlike NCCs, these cells were HNK-1 negative and so could not be confused with migrating NCCs. Furthermore, because they had been marked and injected 2 days after the completion of CNCC migration from the dorsal neural tube, dangers of confusing the migrating marked ventral neural tube cells with NCCs were minimized. Subsequently, marked ventral neural tube cells were found as perichondrial cells and as chondrocytes in Meckel's cartilage and in the quadrate. Consequently, they were readily distinguishable from the HNK-1-positive NC-derived cells that comprise the majority of the chondrocytes in the cartilage.

This may not be the only situation in which non-NCCs contribute to skeletal elements thought to be entirely of NC origin. In their study, fate mapping the CNC of mouse embryos, Chai *et al.* (2000) observed a considerable number of nonlabeled (i.e., non-NC) cells in the dental mesenchyme forming dental papillae and teeth and in Meckel's cartilage. They considered the ventral neural tube as a possible source of these cells, but could not rule out a subpopulation of NCCs not labeled with their *Wnt1-Cre/R26R* marker.

According to other studies by Sohal and colleagues, ventral neural tube cells are even more versatile, forming smooth muscle cells of craniofacial arteries and veins, hepatocytes in the liver, and contributing to the trigeminal ganglion, vestibulocochlear nerve, and otic vesicle. *En route* to contributing to the liver, cells derived from the ventral hindbrain of 2-day-old embryos were found in craniofacial muscles at 7 days and to migrate from the site of attachment of the vagus nerve, through the gut, into the smooth muscle of the stomach and intestine to form hepatocytes in the liver. DiI labeling and the expression of the homeobox gene *Islet-1* were used to show that cells delaminated from the neural tube at site of attachment of the trigeminal nerve, migrated into the trigeminal ganglion, and then into the mesenchyme of the first pharyngeal arch.[a]

---

[a] See Ali *et al.* (2003) for the study on the otic vesicle and for references to the earlier studies, and Erickson and Weston (1999) for a critique of the possibility of accidentally labeling other cells in such studies.

We know from the evidence outlined in Chapter 3 that migrating NCCs require signals to differentiation into particular cell types. *Bmp4* and *Bmp7* within the dorsal aorta of chick embryos trigger autonomic neuronal differentiation from migrating NCCs (Reissmann *et al.*, 1996); ectopic expression of these growth factors *in ovo*, or maintenance of NCCs with *Bmp4* or *Bmp7 in vitro*, triggers autonomic differentiation.

The suspected dual NC and placodal origin of the autonomic nervous system also applies to the origin of *chromaffin cells of the adrenal g* land and to *ganglia associated with the aorta*. Interestingly, lampreys lack chromaffin cells and peripheral sympathetic neurons, perhaps an indication that these two cell types have had a long evolutionary linkage within a single NC population that arose in jawed vertebrates.

During the 1920s and 1930s, a series of experimental studies performed on chicken embryos established the NC origin of the sympathetic ganglia. Although not all early studies reported comparable results, subsequent experimental studies affirmed that these ganglia are indeed NC in origin. Enteric and visceral ganglia, chromaffin cells of the adrenal gland, and pigment cells were also shown to be NC in origin, and trunk, vagal, and sacral NCCs were identified (see section below, and see Chapter 10 for defects affecting adrenal and ganglionic NC derivatives).[11]

As with avian embryos (above), lysinated rhodamine dextran and DiI have been used with effect to show that single TNCCs are multipotential and can form neurons of the neural tube and contribute to the dorsal root ganglia, the sympathoadrenergic system, Schwann, and pigment cells (Serbedzija et al., 1991; Serbedzija and McMahon, 1977*).

Schwann Cells

As neuronal support cells, Schwann cells—named after the co-founder of the cell theory, Theodor Schwann—provide an insulating sheath around neurons of the peripheral nervous system; each Schwann cell sheathing a single neuron and aiding in the conductance of electrical signals through that neuron.

Peripheral motor nerves arise from neural ectoderm. Evidence for the NC origin of the Schwann cells that sheath peripheral motor nerves goes back to the extirpation and tissue culture studies of Harrison and Müller and Ingvar in the early 20th century; Harrison's were the first tissue culture experiments performed with any tissue. Subsequent vital dye staining of *Ambystoma* embryos confirmed the NC origin of Schwann cells, which we now know arise from vagal and sacral NC, although in birds the CarNC also provides Schwann cells for cranial nerve XII.[12]

Clonal cell culture has shown that Schwann cells share lineages with other NC derivatives—sensory and sympathetic neurons, pigment cells, and neurons of the central nervous system (Fig. 3.17). Schwann cells are often found in embryomas; malignant Schwannomas are composed of abnormal Schwann cells (see Chapter 10).

Glial Cells

Glial cells (neuroglia) also are supporting cells for neurons, in their case for neurons in nerves and in the brain. Two types of glial cells are known.

(1) **Oligodendrocytes** secrete myelin that forms a sheath (the myelin sheath) around the axons of neurons of the CNS, providing an insulating function and aiding the conduction of electrical signals along the nerve. Schwann cells (above) provide the same function to neurons of the peripheral nervous system, one Schwann cell per neuron. A single oligodendrocyte, however, can provide the sheath for dozens of neurons.

(2) **Astrocytes** (astroglia) are found in the brain where they provide support (especially to the vascular cells that form the blood–brain barrier), metabolic products, and aid in the transmission of electrical impulses.

Recent studies demonstrate that glial cells arise from multipotential NCC precursors: The NCCs that delaminate early in zebrafish embryos differentiate as glial cells, pigment cells, and as neurons of the dorsal root ganglia. The CNC of early quail embryos contains two classes of multipotential precursors, one that gives rise to glia, neurons, and chondrocytes, and a smaller population of cells that gives rise to these three cell types plus chromatophores (Baroffio *et al.*, 1991, and see Fig. 3.17). Stabilization of β-catenin by Wnt (the canonical Wnt pathway; see Fig. 2.5) inhibits glial and neuronal differentiation as a side effect of enhancing chromatophore differentiation.[13]

As discussed in Chapter 2, *Sox10* is essential for glial cell differentiation. Although *Sox10* is expressed in premigratory NCCs along the entire neural axis, it is rapidly downregulated in the earliest stages of the differentiation of many NCCs, but not in glial precursors. *Sox10* continues to be expressed at high levels throughout glial differentiation in embryos and in adults, where, at least in rats, it functions to maintain glial and neuronal potential in a subset of bipotential NCCs.

The differentiation of NCCs into glia and the maintenance of the glial phenotype require continuous expression of *Sox10* in glial cells. *Sox10* maintains the expression of Er3, a component of the Neuregulin receptor that regulates the expression of *neuregulin1*. *Sox10* is also involved in upregulating *Notch1*, which plays a role in specifying glial cell fate and rendering the cells unable to enter the neuronal lineage—or vice versa—it is unclear which occurs first. $Sox10^{-/-}$ mice lack all peripheral glia, autonomic and enteric neurons, and melanocytes because of increased cell death in premigratory TNCCs; *Neuregulin* serves as a postmitotic survival signal in $Sox10^{+/+}$ and $Sox10^{+/-}$ mice.[14]

The requirement of *Sox10* for initiation and maintenance of glial cell fate is a consequence of the inhibition of autonomic neuronal cell fates by *Sox10*. Specification of NCCs as sensory neurons—which occurs in part via *neurogenin* transcription factors such as *Ngn1*— is reduced in *Sox10* mutants, although whether by direct or indirect action has to be determined. Two key transcription factors involved in the specification of sympathetic neurons, *Mouse achaete-scute homologue 1 (Mash1)* and *Phox2B*, the expression of which inhibits sympathoadrenal differentiation, are activated by *Sox10*. Therefore, *Sox10* serves as a master switch; 'global' inhibition of a switch into the autonomic neuronal lineage establishes a default in which the glial lineage is promoted.[15]

Multipotency, lineage specification, and phenotype specification are intimately related and under considerable shared control. Shared potency is defined by receptors shared by NCCs early in their development (and so intrinsic?). Specification reflects later activation of intrinsic switch genes, such as *Sox10* and *Sox10*-mediated transcription factors and receptors, although there is considerable redundancy among members of subfamilies such as SoxE. To take a recent example, deleting *Sox9* from mouse glial (oligodendrocyte) cell precursors does not prevent their terminal differentiation. Deleting *Sox10* at the same time results in reduced numbers of glial cells because of induced apoptosis, although migration is also perturbed. *Sox9* and *Sox10* both elicit their actions by regulating the expression of Pdgf-α, a receptor that regulates survival and migration of oligodendrocyte precursors. Identification of these upstream transcriptional and downstream signaling targets is but the first step along a road to identifying target genes and gene cascades, only a few of which are known.[16]

Vagal and Sacral Neural Crest

The VNC contributes enteric ganglia to the entire length of the gut, the sacral neural crest (SNC) predominantly to the hindgut; lysinated rhodamine dextran and DiI have been used with effect to label individual murine NCCs, follow their migration, and demonstrate that enteric ganglia are SNC in origin. Simpson *et al.* (2007) developed and verified (with organ culture of embryonic chicken gut containing quail NCCs) a mathematical model demonstrating the key role of cell proliferation at the front of the wave of migrating NCCs in the invasion of the gut by NCCs destined to form enteric ganglia.[17] Human syndromes associated with failure of development of all or parts of the enteric nervous system are discussed in Chapters 9 and 10, with some emphasis on Hirschsprung disease.

Because they come from different levels of the neural tube lack (somite levels 1–7 and caudal to somite 18, respectively), and because VNCCs migrate rostrocaudally and SNCCs caudorostrally, avian sacral and vagal NCCs normally do not interact. SNCCs can invade other regions of the gut, as demonstrated by transplanting chicken SNCCs more anteriorly. In chicken embryos, VNCCs migrate ventrally and so they reach the developing gut several hours before SNCCs, because: the developing anterior intestinal portal is a barrier to SNCC migration; the sacral environment permits cells from the vagal level to enter the gut; because enteric neurons require environmental cues from the hindgut but VNCCs do not.[18]

In a recent screen of mutagenized zebrafish, Kuhlman and Eisen (2007) identified four mutations with specific affects on enteric neurons, and nine mutations with pleiotropic effects on enteric neurons and other NCC derivatives such as melanophores or craniofacial mesenchyme, and/or mesodermal derivatives such as somites and fins.

Vagal and sacral NCCs express *Sox10*, *EdnRB* (the gene for endothelin receptor B) and the proto-oncogene *Ret,* which is expressed at a fourfold higher level in vagal than sacral crest. Overexpressing *Ret* in SNCCs extends the regions of the developing gut to which SNCCs contribute. As itappears to shift these NCCs toward a vagal fate, *Ret* is a potential regulator of vagal–sacral lineages *in vivo.* As discussed in Chapter 9, Hirschsprung disease in humans reflects mutations in *RET and EDNRB* genes. In an interesting study using chick embryos, the same ECM signals were shown to evoke different cell types in cranial (mesencephalic) and trunk (sacral) NCCs, in part because of differential upregulation of *Hox* genes.[19]

Rohon–Béard Neurons

Giant, transient, mechanosensory ganglionic cells known as Rohon–Béard (R–B) neurons form a network associated with the dorsal neural tube in some amphibian embryos and tadpoles, and in the embryos and larvae of all cartilaginous and bony fish, described, for example, in the common skate, Mozambique tilapia, and sturgeon. Although called Rohon–Béard (R–B) neurons after Rohon (1884) and Béard

(1896), who described them most fully, they were reported first by Balfour (1878) in elasmobranchs.

Functionally, R–B neurons remain with the dorsal nervous system. Their cell processes (neurites) project to the hindbrain and to the skin as they function in mediating external sensations to the CNS. Ontogenetically, R–B neurons are transitory and confined to larvae; they are replaced by later developing neurons of dorsal root ganglia in amphibian and fish embryos but persist into adulthood in lampreys. As noted in Chapter 4, the dorsal cells of Rhode may serve a similar function in adult amphioxus but their relationship to R–B neurons is untested.[20]

Apoptosis Removes R–B Neurons

Why and how R–B neurons disappear from larvae has not been thoroughly investigated. Cell death certainly is involved; Béard reported the death of R–B neurons in skate embryos in his original 1896 description.

All zebrafish R–B neurons are eliminated by programmed cell death (**apoptosis**) by 5 days after fertilization. The nerve growth factor, neurotrophin 3 (Nt3), and electrical activity associated with Na^+ currents, both regulate apoptosis. Blocking Nt3 increases rates of apoptosis, while adding Nt3 enhances the rate at which R–B neurons undergo cell death, two results consistent with neural activity playing a role in the death of these neurons. Three markers of apoptosis in zebrafish R–B neurons are reduced to similar levels in the *macho* (*mao*) phenotypic mutation (which is unresponsive to tactile stimulation) or if Na+ currents are blocked with the drug tricaine.[21]

Neural Crest Origin and Relationships to Other Neurons

Béard hypothesized that amphibian Rohon–Béard neurons arise from the NC; they do not arise if TNC is removed. However, because R–B neurons and NC originate from separate lineages that can be identified as early as the 512-cell stage in *Xenopus*, their origin from the NC has been questioned. Separate lineages, combined with origination at the gastrula stage before NC induction, are certainly inconsistent with how we think of NC lineages arising. Lateral neural ectoderm has been proposed as the site of origin of R–B neurons (see Box 3.3).

R–B neurons arise at the border between neural and epidermal ectoderm (the neural plate border), they can arise independently of NCCs, but they can arise from a common precursor of NCCs. A very recent study by Rossi and colleagues (2008) has identified Bmp4 as required for the induction of R–B neurons in *Xenopus*. Their approach was to transplant medial neural plate into lateral ectoderm between pigmented and albino embryos to establish an ectopic neural plate border. The R–B neurons arose at the ectopic border, originating from both donor (neural) and host (epidermal ectodermal) cells. Levels of Bmp4 lower than required to induce neural ectoderm were sufficient to induce R–B neurons in neural ectoderm *in vitro*, consistent with the model discussed in Chapter 2 of intermediate levels of Bmp4 operating at the neural plate border. The Bmp-inhibitor *Noggin* inhibited R–B formation.

HNK-1 was used to identify neurons within the dorsal neural tube of lampreys that were taken to be homologs of R–B neurons. Although HNK-1-positive cells are found in lamprey hearts, and HNK-1-positive mesenchymal cells are found in the dorsal fins of ammocoete larvae, and although independent evidence demonstrates that NCCs contribute to the dorsal fin in lampreys, HNK-1 alone is not a sufficient marker to allow us to conclude that cells are NC in origin.[22]

Amphioxus contains a population of primary intramedullary sensory neurons known as **Retzius bipolar cells** that either synapse with epidermal secondary neurons or terminate within the epidermis. As R–B neurons are primary intramedullary sensory neurons, amphioxus Retzius bipolar cells have been proposed as homologs of vertebrate R–B neurons, the only major difference being that R–B neurons do not synapse with secondary neurons in the epidermis, although their neurites do ramify extensively within the epidermis (Fritzsch and Northcutt, 1993*). If this homology is correct, then Retzius bipolar cells are precursors of NCCs or represent a parallel evolution.

The presence of R–B neurons in amphibians and fish (and lampreys?) and what may be their homologs in amphioxus have been used to propose that R–B neurons might be evolutionary precursors of the large proprioceptive neurons of the mesencephalic trigeminal nucleus (MTN) that innervates the lower jaw (see Box 3.3). Given that the two lineages lie adjacent to one another in the blastula, that placodal and NC ectoderm lie in close proximity, and that some placodal ectoderm arises from the neural folds (see Figs. 2.10 and 2.12), a placodal origin cannot be ruled out; the extirpation studies could have removed ectoderm other than NC from the neural folds. Further study of these interesting cells is clearly merited.[23]

Genetic Control of R–B Neurons

Delta/Notch: Precursors of R–B neurons express *Delta* genes, which are ligands for Notch signaling. Misexpressing *DeltaA* leads to the loss of all R–B neurons, while mutants with decreased Delta signaling have excess R–B neurons and decreased TNCCs, results consistent with two cell populations or early segregation of a single population. Notch signaling inhibits the specification of R–B neurons in premigratory TNCCs, a result consistent with the inhibitory role of Notch over neuronal gene expression, although Notch signaling also leads to apoptosis in preneurogenic cells.[24]

Ap2:Under regulation by Wnt genes, *Ap2* transcription factors regulate the expression of *Snail2* in *Xenopus* embryosand so play maintaining-induced NCCs. Once NCCs are induced, *Ap2* is required for their segregation and survival (see Fig. 2.4).

Eliminating the genes *ap2a* and *ap2c* from zebrafish, which encode the transcription factors Ap2α and Ap2γ, eliminates all NCCs but leaves R–B neurons in almost normal numbers. Two interpretations of these results are that R–B neurons do not arise from the NC (and so are not affected) and/or that there is a separate transcriptional control over R–B neurons and NCCs, in which case the results are not informative about origins.

Midkine-b: A recent study in which expression of the heparin-binding growth factor Midkine-b (Mdkb) was altered in zebrafish embryos is consistent with Mdkb establishing the boundary between R–B neurons and NCCs. (Mdkb acts downstream of retinoic acid.) Overexpressing Mdkb increases the numbers of NCCs *and* R–B neurons. Morpholino knockdown of Mdkb dramatically reduces the NCC population *and* results in loss of all R–B neurons. Both findings are consistent with the origin of both cell types from a boundary region, where specification can be altered to match local needs in individual embryos.[25] Consequently, it may be more appropriate to regard this region as a **NC–RB neuron field** than to ask whether it is a NC field that can form R–B neurons or an R–B neuronal field than can form NCCs. Establishing more precise relationships between the two cell types will require more comparative analyses conducted within a phylogenetic context.

Notes

1. Readers interested in further details of the origin of neuronal cells from the amphibian NC will find that Chapters 2 and 4 of Hörstadius (1950*), comprising some 40% of his monograph, provide an excellent evaluation and synthesis of the earlier work. See Le Douarin (1986) for early studies on the neuronal NC derivatives in avian embryos. Reviews include Bronner-Fraser (1995) and Bronner-Fraser and Fraser (1997).
2. Because of cell surface differences, HNK-1-positive adrenergic subpopulations can be isolated using fluorescence-activated cell sorting (Maxwell and Forbes, 1991*).
3. For early studies on NC contribution to dorsal root ganglia, see the literature discussed in Hall and Hörstadius (1988*).
4. See the studies by Detwiler, Harrison, Raven, and DuShane cited in Hall (1999a*) for vital dye staining and experimental confirmation of the NC origin of spinal ganglia.
5. See Schlosser and Northcutt (2001) for induction of the lateral-line placodes in axolotls, Schlosser et al. (1999) for the studies on coqui, and Schlosser (2002a,b, 2006*, 2007*) for reviews.
6. See Le Douarin et al. (1986), Northcutt (1992, 1996), Webb and Noden (1993*), Northcutt et al. (1994, 1995), Osumi-Yamashita et al. (1994), and Northcutt and Barlow (1998) for overviews of placode development. See M. M. Smith and Hall (1990), S. C. Smith et al. (1994), Parichy (1996), Graham and Begbie (2000), Shimeld and Holland (2000), Baker and Bronner-Fraser (2001), and O'Neill et al. (2007) for discussions of the developmental and evolutionary links between placodal and NC ectoderm. It appears that the NC does not contribute cells to the derivatives of the more caudal epibranchial placodes in the North American sea lamprey (McCauley and Bronner-Fraser, 2003).
7. For recent overviews of our understanding of placodes, see Graham and Begbie (2000), Baker and Bronner Fraser (2001*), Streit (2004*), Litsiou et al. (2005), Schlosser (2006*, 2007*), and Lassiter et al. (2007). For the first fine-grained mapping of ophthalmic and maxillo-mandibular trigeminal placodes, see Xu et al. (2008).

8. See Streit (2004*) and Schlosser (2007*) for summaries and analyses of co-option of genes into placodal ectoderm and of their regulation of panplacodal ectoderm.
9. See Liu *et al.* (2003) and Nechiporuk *et al.* (2007) for *Fgf3* and *Fgf8*, Sun *et al.* (2007) for *Dlx3* and *Dlx4b*, Solomon and Fritz (2002) for *Dlx3* and *Dlx7*, Solomon *et al.* (2003) for the *hearsay* mutant, and Abu-Elmagd *et al.* (2001) for *Sox3*. A recent issue of the *Int J Devel Biol* (2007, 51[6/7]: 427–687) is devoted to ear development, including studies on otic placodes.
10. Yntema and Hammond (1947) provide a comprehensive review of the early studies on the autonomic and sympathetic nervous systems.
11. For the NC origin of autonomic ganglia, see Yntema and Hammond (1945) and the literature summarized in Hall and Hörstadius (1988*), Le Douarin (1982), Anderson (1997), Hall (1999a*), and Le Douarin and Kalcheim (1999*). For enteric and visceral ganglia, see Yntema and Hammond (1945) and Peters-van der Sanden *et al.* (1993).
12. See Harrison (1910) and Müller and Ingvar (1921, 1923) for studies on the NC origin of Schwann cells, and Hörstadius (1950*) for a discussion of early studies that gave contrary results.
13. See Raible and Eisen (1996) for migration and lineage restriction in zebrafish TNC, and Raible and Ragland (2005) for the role played by *Wnt* signaling.
14. See Paratore *et al.* (2001) for Sox10 and *Neuregulin* in mice.
15. See Carney *et al.* (2006) for *Sox10, neurogenin1* and specification of neuronal cell fate, and Kelsh (2006*) for an overview of the functions of *Sox10*. All sensory neurons of dorsal root ganglia require *Ngn2* or *Ngn1* to mediate the first and second waves of neurogenesis, respectively, the neurogenins being regulated by the canonical Wnt pathway (see Fig. 2.5).
16. See Hong and Saint-Jeannet (2005) and Kelsh (2006*) for literature on the various roles of *Sox10,* and Finzsch *et al.* (2008) for the study on Pdgfα.
17. An even more recent study not only using an entirely different approach, but also using chicken embryos, shows that leading edge CNCCs migrating from r4 into the pharyngeal arches out-proliferate trailing NCCs by 3–1 (Kulesa *et al.*, 2008).
18. See Pomeranz *et al.* (1993) and Serbedzija *et al.* (1991*) for the SNC, Burns *et al.* (2000, 2002) for lack of interaction between VNCCs and SNCCs of chicken embryos during migration, Hearn and Newgreen (2000) for lumbo-sacral and VNCC interactions, and Erickson and Goins (2000) for the ability of SNCCs to invade anterior gut. These cells are multipotential; they can form neuronal or glial cells when cultured on laminin but are normally inhibited from forming catecholaminergic neurons (Pomeranz *et al.* 1993).
19. See Delalande *et al.* (2008) for *ret*, and Abzhanov *et al.* (2004) for the CNC–SNC study.
20. See Balfour (1878), Rohon (1884), and Béard (1892, 1896) for the discovery and descriptions of Rohon–Béard neurons, Chibon (1974*) for their NC origin, Laudel and Lim (1993) for their development in the Mozambique tilapia, *Oreochromis mossambicus*, and Kuratani *et al.* (2000) for their development in sturgeon embryos and regression as DRG develop.
21. See Williams *et al.* (2000*), Cole and Ross (2001), and Svoboda *et al.* (2001*) for removal of Rohon–Béard neurons from zebrafish, and Hunter *et al.* (2001) for R–B neuron origins and the mesencephalic trigeminal nucleus. Both the trigeminal placode and R–B neurons provide the extensive network of epidermal neurites seen in sturgeon larvae (Kuratani *et al.*, 2000).
22. See Langille and Hall (1988b) and McCauley and Bronner-Fraser (2003) for NCC contributions to the dorsal fin in lampreys.
23. See M. Jacobson and Moody (1984) for lineage analysis of Rohon–Béard neurons, Lamborghini (1980) for their origin during gastrulation, and Rao and Jacobson (2005*) for a general discussion.
24. See Cornell and Eisen (2000, 2005*) for Delta/Notch functioning in zebrafish, and Yeo and Gautier (2004) for Notch and apoptosis of neurogenic cells.
25. See Li and Cornell (2007) for R–B neuron survival in zebrafish, Luo *et al.* (2003) and Saint-Jeannet (2006) for *Xenopus*, and Liedtke and Winkler (2008) for Midkine-b.

Chapter 7
Cartilage Cells and Skeletal Systems

Skeletal cells—cartilage and bone cells (chondrocytes and osteoblasts)—arise from embryonic mesenchyme, which may be neural crest (NC) or mesodermal in origin. This chapter concerns with those cartilage cells and the craniofacial and viscerocranial skeletal systems that arise from the NC. In most vertebrates these cartilaginous skeletons are replaced by bone. The cells that build the major parts of the teeth are of two types: odontoblasts, which deposit dentine, and arise from NC-derived mesenchyme, and ameloblasts, which deposit enamel, and arise from ectoderm (see Chapter 8).

The following major points concerning the NC origins of mesenchyme and of the craniofacial skeleton have been made so far:

- A century ago, the entrenched germ-layer theory made it almost impossible to accept a NC (ectodermal) origin of head mesenchyme and cartilage.
- Julia Platt (1893) demonstrated that the cartilages of the craniofacial and pharyngeal-arch skeletons of the mudpuppy arise from the NC, an origin confirmed in the Mexican axolotl by Hörstadius and Sellman (1941) and by de Beer (1947).
- HNK-1 labels both NC- and mesoderm-derived mesenchyme.
- Frizzled genes (which encode *Wnt* receptors) and *Snail2* are expressed in NCCs and in both NC- and mesoderm-derived mesenchyme.
- *Bmp2* is expressed in distinct fields in mesenchyme of NC origin, but neither in somatic nor in prechordal mesoderm.
- Bmps are involved in the regulation of those mesenchymal NCCs that form craniofacial skeletal and cardiac structures.
- Neural ectoderm, migrating NCCs providing the mesenchyme for each pharyngeal arch, and the mesenchyme and endoderm of each arch, express a combination of *Hox* genes specific to each arch, as beautifully demonstrated in the formation of the middle ear ossicles (Box 7.1).
- Pharyngeal endoderm is a major inducer of chondrogenesis from craniofacial mesenchyme and plays an important role in patterning the craniofacial skeleton.
- The long-standing theory of the origin of jaws from an anterior gill arch is not necessarily proven.

B.K. Hall, *The Neural Crest and Neural Crest Cells in Vertebrate Development and Evolution*, DOI 10.1007/978-0-387-09846-3_7,

Box 7.1 Middle ear ossicles

Of a number of distinctive skeletal features (synapomorphies) that delineate mammals from other tetrapods, perhaps the most well known are the three NC-derived middle ear ossicles, the **incus** (anvil), **malleus** (hammer), and **stapes** (stirrup), which together amplify and transmit sound from the external environment to the inner ear. The evolution of these ossicles from what once were reptilian jawbones is one of the most remarkable examples of the evolutionary transformation of developmental and adult structures.

Reptilian jawbones are the evolutionary precursors (homologs) of middle ear ossicles. Although morphologically distinct from mammalian middle ear ossicles, certain of the lower jawbones can be identified as homologs because their transformation to ossicles can be traced, both in the fossil record and in the development of the middle ear ossicles in marsupials.

The mammalian lower jaw consists of a single bone, the dentary. Reptilian lower jaws are comprised of five to seven membrane bones. Some of these bones were lost in the evolution to mammals. Others transformed into middle ear ossicles. The articulation of the jaw to the skull in the ancestors of mammals (therapsids) was between the articular and the quadrate, an articulation retained in extant reptiles, including birds.

Two of the middle ear ossicles—the incus and malleus—arose by transformation of the quadrate and articular bones, respectively, both of which arise in the first pharyngeal (mandibular) arch. The obvious corollary is that the articulation of the jaws to the skull in mammals involves different bones than it does in reptiles, a dentary–squamosal joint replacing the articular–quadrate joint. The stapes arose not from a reptilian jawbone but from a bone, the columella, that arises in the second pharyngeal (hyoid) arch. This dual origin from two pharyngeal arches goes a long to explaining why middle ear ossicles are so susceptible to perturbation during embryonic development and why defects do not always affect all three ossicles equally; malleus and incus but not stapes are reduced in *Ap-2* mutant mice and eliminated in *Hoxa2* knockout mice (see Chapter 10).

In mouse embryos, developing middle ear ossicles lie at the junction of the first and second pharyngeal arches (see Fig. 10.5), a position assumed to be typical of placental mammals. This is not the case for marsupials. From embryological evidence, we know that newborn short-tailed opossums have a jaw articulation that is neither typically mammalian nor typically reptilian. The incus and malleus arise on the side of the jaw and only later transform into functional ear ossicles. Connection with the stapes in the second arch is facilitated because development of the mandibular arch is faster than the development of the hyoid arch. This developing region constitutes what in Box 10.1 is discussed as a developmental field of action, a field over which *Hoxa2* exerts considerable control (see Box 10.2).

New fossil finds add to our understanding of the transformation at the level of the skeletal elements but, more importantly, at the level of developmental mechanisms that drove the evolutionary transformation. A newly described Mesozoic crown mammal, *Yanoconodon allini* from the Mesozoic, represents a transitional stage to the definitive mammalian condition and the condition seen in the embryonic development of extant mammals. In this species, well-differentiated middle ear ossicles are joined to the mandible by an ossified Meckel's cartilage. We expect to discover transitional stages, albeit only after prolonged effort. Another discovery—a dentary of the oldest known monotreme, *Teinolophos trusleri*, from the Early Cretaceous—has both angular and articular/prearticular (homologs of the therian mammalian ectotympanic and malleus, respectively) elements in a trough along the inner surface. The marsupial and placental mammalian elements are already on their way to forming middle ear ossicles, which means that they and monotremes evolved their middle ear ossicles independently.[a]

---

[a] For further details and literature on the evolution of middle ear ossicle, see Hall (2005a,b*) and Hall and Hallgrímsson (2008*). See Z.-X. Luo *et al.* (2007) for *Yanoconodon allini*, and Rich *et al.* (2005) for *Teinolophos trusleri*.

The evolutionary origins of cartilage, the pharyngeal skeleton, and the jaws were discussed in Chapter 4. This chapter examines the evidence for the NC origin of cranial mesenchyme and of the craniofacial skeleton during embryonic development albeit in a comparative and evolutionary context.[1]

In discussing his study on the NC origin of pharyngeal-arch cartilages in the Mexican axolotl (see below), de Beer (1947) raised the possibility that the vertebrate pharyngeal skeleton had always developed from NCCs. de Beer emphasized the importance of knowing the source of the pharyngeal skeleton in cyclostomes (hagfish and lampreys) and sharks, which were then regarded as 'lower', 'primitive', or 'ancestral' (see Fig. 4.12). As discussed in Chapter 4 and as illustrated in Fig. 4.13, lampreys occupy a crucial phylogenetic position, [2] even though, as discussed below, neither lamprey nor hagfish craniofacial skeletons are typically vertebrate. To understand the evolutionary origin of the craniofacial and visceral skeletons of jawed vertebrates, it is appropriate to begin with a discussion of the developmental origin of the cranial and pharyngeal-arch skeletons in hagfish and lampreys.

Pharyngeal Skeletons of Hagfish

We knew little about the NC in hagfish (Box 7.2) until a recent study of a shallow-water species, the Japanese hagfish, documented cells delaminating from the neural tube. These putative NCCs express *Pax6*, *Pax3/7*, *Sox9*, and *SoxEa*, but not the NCC marker gene, *Snail1*, all of which are expressed in NCCs in vertebrates (Ota *et al.*, 2007). However, we still do not have direct evidence that the hagfish cranial skeleton is NC in origin. Hagfish cartilage is certainly atypical in comparison with the cartilage of jawed vertebrates or even in lampreys, although both agnathans react positively to a human antibody against the N-terminus of type-II collagen, the 'cartilage-type' collagen found in the cartilages of all jawed vertebrates (Zhang and Cohn, 2006).

Box 7.2 The neural crest in hagfish embryos

Hagfish are structurally diverse but not structurally conservative; the Pacific hagfish, *Eptatretus stouti*, has a well-developed lateral line system but other 'hags,' such as the Atlantic hagfish, *Myxine glutinosa*, lack any component of a lateral line system. This diversity means that the small amount of information on hagfish embryonic development may or may not be representative of all species.[a]

The major study of hagfish embryonic development is based on a collection of live embryos of the Californian hagfish, *Eptatretus stouti* (formerly *Bdellostoma stouti*), made in the summer and fall of 1896 by Bashford Dean in Montery Bay off Pacific Grove, California—where Julia Platt (Box 7.3) was elected mayor in 1931. In his publication, Dean (1899) did not comment on the NC. [b]

There is some information on hagfish NC in a paper by Conel (1942) but based on embryos collected by Dean. Much of this paper is devoted to the NC in sharks, but in one figure and two paragraphs of text, Conel reported that the NC in *Eptatretus stouti* invaginates from the dorsal neural folds as a hollow sac connected to epidermal ectoderm above and to the dorsal surface of the neural tube below, with a lumen that is continuous with the cavity of the neural tube. This pouch or ganglionic vesicle, as he termed it, was continuous along the developing spinal cord but segmental, with thin intervening intersegmental connections. Conel (1942) claimed that cells associated with these pouches gave rise to the segmentally arranged dorsal root cranial ganglia of cranial and spinal nerves.

Invagination of NC as a sac or vesicle is not seen in any vertebrate, and so led many of us to regard hagfish as potentially distinct from vertebrates in this aspect, as indeed they are in other aspects (see text). Very recently, however, Kinya Ota in Shigeru Kuratani's laboratory obtained seven live embryos of the Japanese shallow water (inshore) hagfish, *Eptatretus burgeri*. Seven seems a paltry number, but as the first live embryos obtained in 110 years, these seven represent the beginning of what should be a new era of investigation on hagfish NC and hagfish embryology in general. Already we can say that the ganglionic vesicle described by Conel is a fixation or shrinkage artifact. Ota *et al.* (2007) saw no evidence of such a vesicle unless fixation was inadequate. In contrast, they identified putative NCCs in a strand alongside and apparently migrating from the dorsal neural tube (see their Fig. 3.f). Segmentally arranged populations of cells were identified in the trunk. Although not *Snail1*- or HNK-1 positive, these cells express *Pax6*, *Pax3/7*, *Sox9*, and *SoxEa*, all of which are NCC markers. The next batch of embryos is eagerly awaited.

[a] See Wicht and Northcutt (1995) and Braun and Northcutt (1997) for the study on the Pacific hagfish and Ota and Kuratani (2006*, 2007*) for comprehensive reviews of the history of research on hagfish (and lamprey) embryology.

[b] Franz Doflein of Munich University, who was also in California in 1896, obtained embryos from the same fisherman who collected for Bashford Dean. Von Kupffer used these embryos in his studies on head development in *Eptatretus* (Kupffer, 1906*). Dean may also have collected embryos of *Eptatretus burgeri* in Japan in 1900 and 1901 (Conel, 1931, pp. 97–98). If he did, they remain unstudied and their location unknown.

The lingual skeleton of the Atlantic hagfish, *Myxine glutinosa*, has two types of cartilage. One, unlike any other vertebrate cartilage, is similar to some invertebrate cartilages in consisting of hypertrophic chondrocytes filled with cytoplasmic filaments; Glenda Wright and her colleagues regard it as the most primitive of all the agnathan cartilages.

The second type, although histologically similar to vertebrate cartilage, possesses an ECM protein, *myxine*, that differs from collagen, elastin, and lamprin (which is found only in lamprey cartilages, and then only in some; see below) and also from the major matrix protein of the other form of hagfish cartilage.[3]

It has not been determined whether either of these cartilages is NC in origin and/or whether either is homologous with lamprey or gnathostome pharyngeal arch cartilages; recall, from the discussion in Chapter 4, that the pharyngeal-arch skeleton of extant cephalochordates is mesodermal in origin.

If hagfish pharyngeal arches are homologous with the pharyngeal arches of lampreys and jawed vertebrates (which are NC in origin), we would expect hagfish pharyngeal arch skeleton to be NC in origin. A non-NC origin for hagfish pharyngeal cartilages could postdate the hagfish/lamprey split, and either require a secondary reversion from a NC to a mesodermal origin or represent a separate evolutionary path, depending on the phylogenetic relationships of hags, lampreys, and jawed vertebrates.

But, what about lampreys, whose embryos are more accessible and for which more data are available?

Pharyngeal Skeletons of Lamprey

Cartilages

The cartilages of lampreys are structurally and biochemically distinct from one another, from hagfish cartilages, and from the cartilages of other vertebrates. Although type II collagen is a *sine qua non* of cartilage, the only sites within the cartilages of the North American sea lamprey where type II collagen and the gene *Col2a1* are found is in the perichondrium and in the most superficial ECM. The bulk of the cartilaginous ECM consists of branched fibrils and fibers, which are 150–400 Å in diameter and composed of a structural protein, lamprin, unique to lampreys; the amino acid composition of lamprin is unlike type II collagen in having only traces of hydroxyproline.[4] Adding to the complexity, the North American sea lamprey has three distinct types of cartilage:

- annular and neurocranial cartilages, which are based on lamprin and are NC in origin;
- branchial basket and pericardial cartilages, which are based on a second protein with affinities to elastin and which are also NC in origin (Fig. 7.1);
- mucocartilages, which are histologically distinct from the remainder of the skeleton and are not NC in origin (Wright *et al.*, 2001*).[5]

Fortunately, we have experimental evidence for the embryological origin of lamprey cranial and pharyngeal skeletons. Although this evidence was obtained after that for the NC origin of amphibian cranial and pharyngeal skeletons, the relevance of the lamprey studies to the origin of jaws (see Chapter 4) makes it sensible to discuss them first. Damas (1951) concluded that the head cartilages of the European river lamprey were NC in origin. After perturbing embryos and noting that defective pharyngeal chondrogenesis was preceded by a failure of contact between pharyngeal endoderm and mesenchyme, Damas concluded that, as in amphibian embryos, interactions between NC-derived mesenchymal cells and pharyngeal endoderm were required to initiate chondrogenesis in lamprey embryos.

In the 1950s, David Newth performed a series of extirpations of neural folds from European brook and river lamprey embryos, along with transplantations of the neural folds to amphibian embryos to determine the origin of the head cartilages; his results are discussed in some detail in Hall (1999a*). In his final study, Newth (1956) found that NCCs differentiated into chondroblasts and formed cartilages if they were grafted into the branchial region of crested newt or palmate newt neurula-stage embryos but not if they were grafted into the flank, implying the requirement of a pharyngeal endoderm–NCC interaction as in amphibians.

No further extirpations of lamprey NC were reported until Langille and Hall (1988b, 1989) removed 50-μm regions of CNC from early neurula-stage embryos (shown as I–VII in Fig. 7.2) and followed development to the ammocoete larval stage. Depending on the region of NC extirpated, the trabeculae and pharyngeal

Fig. 7.1 The distinctive histology of lamprey pharyngeal arch cartilages (*arrows*) can be seen in these low (A, × 265) and high (B, × 460) magnification sagittal sections through a stage-17 larval North American sea lamprey. The trabecular cartilage (t) adjacent to the notochord (n) and beneath the developing neural tube (nt) is also evident. From a specimen kindly provided by Robert Langille

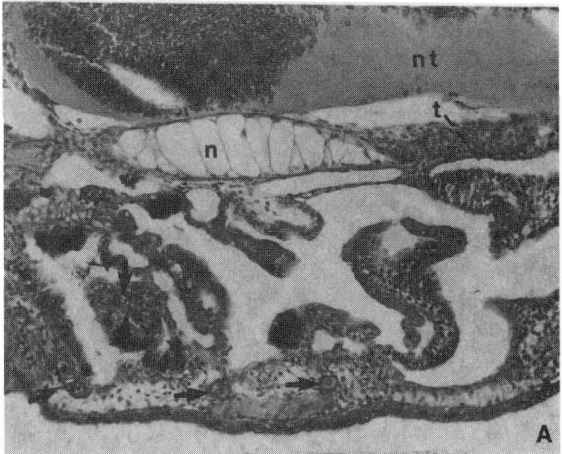

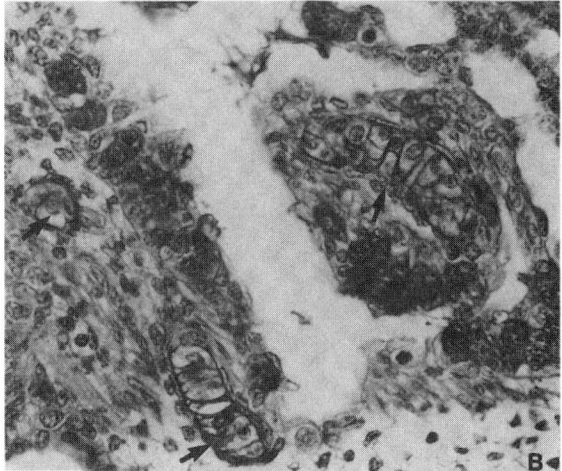

arch cartilages were entirely absent or substantially reduced, results consistent with their origin from NCCs. Parachordal cartilages, cartilage of the otic capsule, and the cranial mucocartilages developed in all larvae, results consistent with a mesodermal origin of these cartilages.[⊕]

Regional extirpations of the NC allowed Langille and Hall to identify two non-chondrogenic and five chondrogenic regions within the lamprey CNC (Fig. 7.2). The rostral prosencephalon (equivalent to the transverse neural folds in amphibians) and the CNC caudal to the sixth pair of somites are nonchondrogenic (Table 7.1). If similar to amphibians and birds, the rostral prosencephalon would lack NCCs, and

---

[⊕] The otic capsule is mesodermal in origin in amphibians but chimeric in birds, receiving cells from NC and mesoderm.

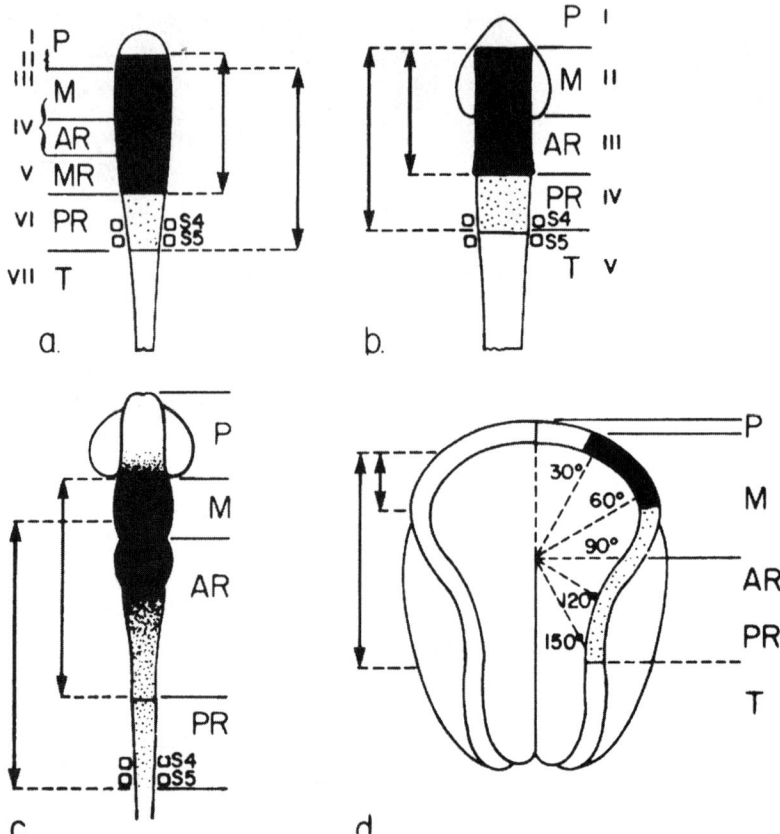

Fig. 7.2 Regionalization of the CNC showing the cranio-caudal extent of the neural crest contribution to the chondrocranial skeleton (*black*; *double-headed arrows*) and to the visceral skeleton (*stippled*; *double-headed arrows*) in (**a**) the North American sea lamprey, (**b**) a teleost, the Japanese medaka, (**c**) a bird, the chicken, and (**d**) representative urodele amphibians, the Mexican axolotl and the Spanish ribbed newt, as seen in dorsal views of the developing neural tubes. The skeletogenic CNC extends from mid-prosencephalon caudal to the level of somites 4 or 5 (S4, S5) in all species. P, prosencephalon; M, mesencephalon; AR, anterior rhombencephalon; MR, mid-rhombencephalon; PR, posterior rhombencephalon; T, trunk neural crest cells. Numbers I–VII in (a) and I–V in (b) represent boundaries of regions excised to generate the fate maps for lamprey and Japanese medaka skeletogenic neural crests. The angles from the midline in (d) represent sectors of the neural crest extirpated from Spanish ribbed newt embryos by Chibon (1974*), which has projected onto the fate map generated forMexican axolotl embryos by Hörstadius and Sellman (1946)

the caudal crest would contain nonchondrogenic TNCCs. The chondrogenic CNC extends from the caudal region of the prosencephalon caudally to the fifth pair of somites (regions II–VI in Tables 4.1 and 4.2), and is rostrocaudally patterned, the trabecular cartilages that form the floor of the skull arising from regions II to V in Table 7.1, the pharyngeal arch cartilages (Fig. 7.1) from regions III to VI in Table 7.2 (and see Figs. 7.1 and 7.2).

Table 7.1 Percentages of larvae with missing or deficient trabeculae and/or branchial arches following removal of neural crest from embryos of the lamprey *Petromyzon marinus*[a]

Region removed[b]	Trabeculae	Branchial arches
I anterior prosencephalon	0	0
II posterior prosencephalon	78	0
III anterior mesencephalon	50	70
IV posterior mesencephalon/anterior rhombencephalon	25	75
V mid-rhombencephalon	75	100
VI somites 4–5	0	70
VII somite 6	0	0

[a]Based on data in Langille and Hall (1988b).
[b]See Fig. 4.3 for the seven, 250 μm regions of neural crest extirpated. $N = 4$–10 specimens/region removed. The levels of the neural tube and/or adjacent somites corresponding to each region are indicated.

Table 7.2 Percentages of larvae showing missing or deficient pharyngeal arches 1–7 following removal of 250 μm regions of neural crest from embryos of the sea lamprey, *Petromyzon marinus*, to illustrate the rostrocaudal level of the cranial neural crest contributing cells to the pharyngeal arch cartilages[a]

Region removed[b]		Pharyngeal arches[c]		
		1–3	4 & 5	6 & 7
I	anterior prosencephalon	0	0	0
II	posterior prosencephalon	0	0	0
III	anterior mesencephalon	71	28	0
IV	posterior mesencephalon/anterior rhombencephalon	33	66	0
V	mid-rhombencephalon	50	50	25
VI	somites 4–5	14	51	71
VII	somite 6	0	0	0

[a]Based on data in Langille and Hall (1988b).
[b]The levels of the neural tube and adjacent somites corresponding to each region are indicated. See Fig. 5.2 for the seven regions. $N = 3$–7 specimens/region removed.
[c]branchial arch 1 is the most rostral, 7 the most caudal. The arches are grouped as rostral (1–3), mid (4 and 5), and caudal (6 and 7).

This rostrocaudal patterning of the skeletogenic CNC in lampreys is remarkably congruent with the patterning in jawed vertebrates (Fig. 7.2). Furthermore, the rostrocaudal contribution of caudal CNC to the pharyngeal arches is shared with a teleost fish, the Japanese medaka (regions III–VI in Table 7.3). We can conclude that: **the more caudal the pharyngeal arch, the more caudal the CNC from which it arises**, a regionalization of the skeletogenic CNC established early and strongly conserved throughout vertebrate evolution; see Langille and Hall (1989) for further discussion.

More recently, McCauley and Bronner-Fraser (2003) used DiI injection to follow the fate of NCCs in the North American sea lamprey. Their study confirmed the

Table 7.3 The elements of the cartilaginous skeleton of the Japanese medaka, *Oryzias latipes*, that arise from the cranial neural crest, shown in relation to the region of the neural crest from which they arise[a]

Region of the neural crest	Cartilaginous element
Prosencephalon	none
Mesencephalon	Orbital cartilage, Meckel's cartilage, quadrate, trabeculae, ethmoid plate, lamina orbitonasalis, pterygoid process, epiphyseal cartilage, basihyal, symplectic
Preotic (anterior to mid-rhombencephalon)	Orbital cartilage, epiphyseal cartilage, basihyal, hyomandibula
Postotic (rhombencephalon caudal to the level of the 4th pair of somites)	ceratohyal, hyomandibula, basibranchial, hypobranchials I–IV, ceratobranchials I–V, epi- and pharyngobranchials IV
Caudal to level of 4th pair of somites	none

[a]Based on data in Langille and Hall (1987, 1988a). The polar cartilage, hypophyseal, acrochordal, anterior and posterior basicranial commisures, basilar plate, otic capsule, occipital arch, and tectum synoticum receive no contribution from NCCs and are mesodermal in origin.

conservation of pathways of migration and reported an important new finding; the more caudal pharyngeal arches are populated by NCCs that migrated rostrally *and* caudally from the CNC, a pattern not seen in any jawed vertebrates. Consequently, NCCs from a single level of the neural tube contribute to more than one pharyngeal arch and in so doing mix with NCCs from other axial levels; the strict *Hox* gene code of axial contribution to a single arch may be a feature of jawed vertebrates.

Patterns of gene expression in the lamprey NC are consistent with regional specification of the CNC and/or the origin of cells/tissues from the NC. In the Japanese lamprey, *Lampetra japonica:*

- *Pax6* (*LjPax6*) is expressed in the eyes, brain, nasohypophyseal placode, and oral ectoderm;
- *LjPax9* in the endoderm, pharyngeal pouches, hyoid arch, nasohypophyseal plate, and in the mesenchyme but not the muscle of the velum;
- *LjDlx1/6* in craniofacial and pharyngeal mesenchyme and in the forebrain; and
- *LjEmx* —a homeobox gene orthologous to the gene *empty spiracles* (*ems*) in *Drosophila*)—is expressed initially in the neural tube and then in craniofacial mesenchyme.

Two SoxE genes (*SoxE1* and *SoxE2*) in the North American sea lamprey are expressed in migrating NC-derived mesenchyme and in chondrocytes of the posterior pharyngeal arches; *SoxE1* is required for posterior but not for mandibular arch skeletal development. A third Sox gene, *SoxE3*, is expressed in the mandibular arch and in perichondria of the developing cartilage.[6]

Vitamin A

The neural tube and NCCs of jawed vertebrates are sensitive to excess vitamin A (retinoic acid); hypervitaminosis-A results in defects in cells of NC origin. Treating lamprey embryos with vitamin A results in a dose- and stage-dependent loss of the pharynx and truncation of the rostral neural tube, two results consistent with sensitivity of NCC and/or pharyngeal endoderm to vitamin A having predated the separation of agnathan and jawed vertebrate (see Chapter 10).

Amphibian Craniofacial Skeletons

Newth's studies on lamprey embryos were preceded by studies using amphibian embryos to demonstrate the NC origin of craniofacial mesenchyme and much of the craniofacial and viscerocranial skeleton, beginning with Julia Platt in 1893 (Box 7.3). Sven Hörstadius devoted over one-third of the text and over half the figures in his 1950 book and a monographic 170-page paper published with Sven Sellman in 1946 to the contribution of NC to the axolotl cranial skeleton.

Box 7.3 Julia Platt (1857–1935)

American-born Julia Platt was a remarkable woman.[a] Educated at Harvard, Radcliffe College, Chicago, Freiburg, and Munich, she worked with many of the leaders of the day, including C. O. Whitman, E. B. Wilson, and Richard Hertwig. One of the first women to receive a Ph.D. degree from a German university—Freiburg in 1889, where August Weismann was the principal referee for her thesis—Platt was one of the first women neuroscientists, and one of the first two women to join the American Society of Morphology (later the American Society of Zoologists (ASZ), now the Society for Integrative and Comparative Biology (SICB)). The Julia Platt Club, a forum for presenting the results of research in evolutionary morphology, had its inaugural meeting in Boston in January 1998.

Platt's outstanding training was followed by the publication of 12 papers in the decade between 1889 and 1899: one on axial segmentation in chicken embryos, three on the development of head cavities, one on amphioxus, six on differentiation in the mudpuppy (including the ectodermal origin of head cartilages and development of the lateral lines), and one on the specific gravity of protozoa and tadpoles.

Platt was outspoken and critical of her male colleagues, justifiably so if she thought they were arguing from weak factual bases. Such attributes did not sit well with those men deciding academic appointments. Despite a

---

yearlong search in the United States and Europe, she failed to obtain an academic position. Instead, she turned to civic work in Pacific Grove, California. In 1931, aged 74, Julia Platt became that city's first female mayor.

---

[a] Zottoli and Seyfarth (1994) is the only in-depth analysis of the life and work of Julia Platt.

Extirpating and Transplanting Amphibian Neural Crest

Twenty-four years have elapsed after Platt's 1893 study before Landacre's analysis, based on histological sections of normal development, demonstrated that NC and not placodal ectoderm is the source of the visceral cartilages in Jefferson's salamander. Landacre (1921) also produced the first fate map of those parts of the skeleton arising from the NC—the anterior trabeculae cranii and the pharyngeal arch skeleton, except for the second basibranchial, which Landacre determined to be mesodermal in origin.

Leon Stone (Box 7.4), who was the first to publish studies based on extirpating selected regions of urodele and anuran NC (using embryos of North American salamanders and pickerel frogs), confirmed Landacre's descriptive study of the NC origin of the skeletal cartilages and produced a more refined fate map of the CNC (Fig. 7.3). Raven (1931, 1936) went beyond Stone's approaches by grafting **trunk neural crest** from Mexican axolotl embryos in place of CNC of *Triturus* embryos; the transplanted TNC failed to produce the cartilages that would have arisen from CNC.

Box 7.4 Leon Stansfield Stone (1893–1980)

Leon Stone was educated at Lafayette College and at Yale University, where he spent a 42-year career in the Anatomy Department, retiring in 1961 as Bronson Professor of Comparative Anatomy Emeritus. Stone pioneered cinematography and time-lapse techniques in experiments on regeneration of the amphibian iris, lens, and retina, and on eye development in fetal (pouch) possums.

In Stone's pioneered studies, he extirpated regions of the neural tube in the North American spotted salamander and the pickerel frog, established patterns of NCC migration, refined Landacre's fate map of the CNC (see Fig. 3.1 in Box 3.1), and established that many cranial ganglia receive contributions from the NC and from placodal ectoderm (see Figs. 2.16 and 6.5).

E. S. Crelin of Yale University, who was a Ph.D. student of Stone's (Stone himself having been a student of Ross Harrison), told me that after the publication of the 1922 paper mapping the NC, Stone visited Germany, where the heretical notion of cartilage arising from an ectodermal derivative led to him being dubbed 'Neural Crest Stone'.

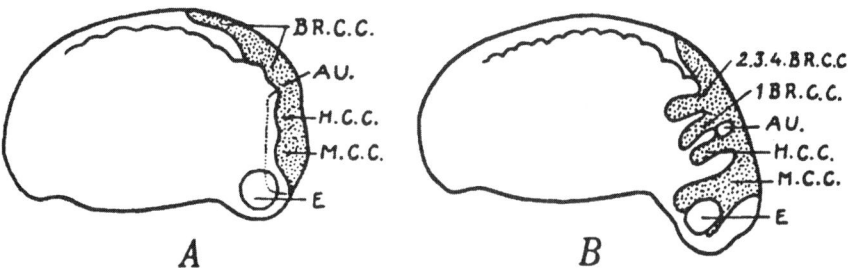

Fig. 7.3 Stone's (1926) fate map showing the origin of the craniofacial mesenchyme from premigratory (**A**) and late migratory (**B**) CNC in embryos of the North American spotted salamander. Distinctive mandibular (M.C.C.), pre- and postotic, and branchial (pharyngeal) arch populations (BR.C.C.) are evident in **B**, the mandibular population in two streams rostral and caudal to the eye. AU, auditory vesicle; 1–4 BR.C.C., mesenchyme of branchial arches 1–4; E, optic vesicle; H.C.C., hyoid arch mesenchyme; M.C.C., mandibular mesenchyme. Reproduced from Hörstadius (1950) with the permission of Dagmar Hörstadius Ågren

In 1929, Ross Harrison of Yale University published his classic experiments on the intrinsic determination of organ rudiment size. When he grafted eye primordia between embryos of two species of salamanders of different sizes, growth of the eye followed the pattern set by the donor, not the host.[7] Harrison transplanted hindbrain neural folds—including premigratory NC—from North American spotted salamander to the equivalent positions in tiger salamander or Mexican axolotl embryos. Two days later, after the NCCs had migrated into association with the endoderm of the developing pharynx, Harrison transplanted the combined pharyngeal endoderm and associated grafted NCCs to an equivalent position in another embryo (Fig. 7.4). Taking advantage of his finding that the skeletal elements of the pharyngeal arches of these species are distinctive in size, Harrison confirmed the origin of the pharyngeal skeleton from NCCs of the donor embryo.

Inherent size specificity within the skeletogenic NC was shown in Wagner's classic 1949 study, in which he grafted NC from the yellow-bellied toad into the equivalent position in the neural folds of alpine newt embryos. By grafting neural folds from only one side—an extension of the approach used by Harrison—Wagner was able to create chimeric embryos in which patterning of the skull and pharyngeal skeletal on the grafted side could be compared with the intact side. As the grafted portions of the skull and jaws were toad in type and size and the host structures

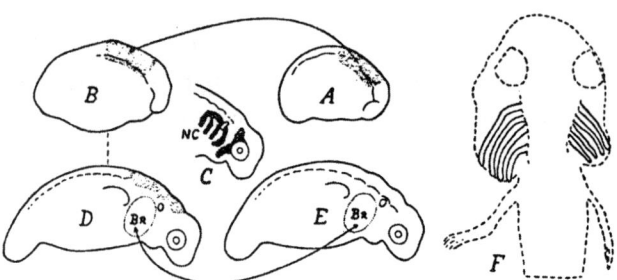

Fig. 7.4 Transplanting pharyngeal arch neural crest and adjacent ectoderm between Mexican axolotl and tiger salamander embryos(**A** to **B**) results in migration of the grafted neural crest into the visceral arches of the donor embryo (NC in **C**). The branchial region (BR) was transplanted to the normal postmigration site 2 days later (**D** to **E**). Harrison used size differences between graft and donor species (**F**) to demonstrate the neural crest origin of the visceral arch cartilages and that the size of these elements was an intrinsic property of the neural crest cells. Reproduced from Hörstadius (1950) with the permission of Dagmar Hörstadius Ågren

newt in size, we can conclude that transplanted NCCs express patterns of morphogenesis and growth reflecting species-specific intrinsic properties of the skeletogenic cells. [8]

As reported in their classic papers in the 1940s, Hörstadius and Sellman performed an extensive series of experimental studies on cranial development in the Mexican axolotl, which form the bulk of Chapter 3 in Hörstadius (1950*). In this pioneering study, Hörstadius and Sellman mapped the rostrocaudal extent of the NC that produces the trabeculae and the mandibular, hyoid, and pharyngeal arch skeletons (Fig. 7.5), dividing the urodele NC into three regions:

- NC rostral to the mid-prosencephalon (the transverse neural folds), a region that does not produce skeletal tissues;
- CNC caudal to the mid-prosencephalon, which gives rise to the trabeculae and pharyngeal arch skeleton; and
- TNC, which does not produce skeletal tissues.

Hörstadius and Sellman further subdivided the cranial skeletogenic crest (Fig.7.5) and demonstrated intrinsic morphogenetic specificity along the rostrocaudal extent of the NC with respect to the type of skeletal elements that formed. Crest destined to form trabeculae did not form visceral cartilages when transplanted to a more caudal region of the neural folds, a region from which the NC that forms pharyngeal arch cartilages normally arises. Conversely, crest destined to form pharyngeal arch cartilages did not form trabeculae after being transplanted more rostrally to the position from which trabeculae normally arise. The same regionalization holds true for lampreys (discussed above), the Japanese medaka, and chickens (both discussed below) and is a fundamental feature of the organization of the NC in vertebrate embryos (Fig. 7.1).

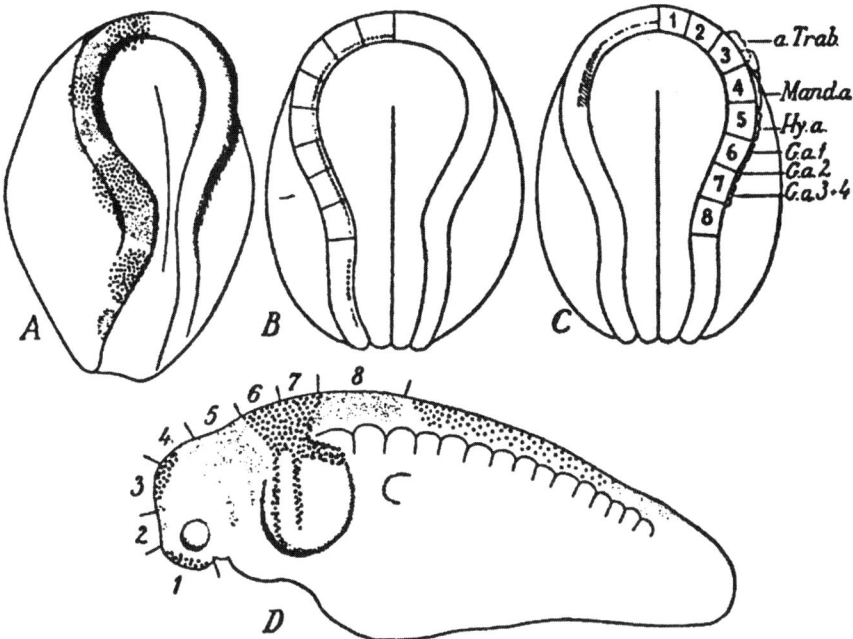

Fig. 7.5 By applying two different vital dyes to the neural crest before cell migration—neutral *red* and Nile *blue*, shown in coarse and fine stippling in **A** —Hörstadius and Sellman mapped the rostral–caudal extent of the skeletogenic CNC in Mexican axolotl embryos into eight regions, shown in **B** and **C**. Regions 1, 2, and 8 (rostral to the mid-prosencephalon and the TNC, respectively) are nonskeletogenic. Regions 3–7 show the rostrocaudal patterning of the skeletogenic neural crest, which gives rise to the anterior trabeculae (a. Trab.), mandibular, hyoid, and gill arch skeletons (Mand.a, Hy.a, G.a.1–4). **D** shows the pattern of migration of populations of neural crest cells into the same eight regions. Reproduced from Hörstadius (1950) with the permission of Dagmar Hörstadius Ågren

Epithelial–Mesenchymal Interaction Required to Initiate Chondrogenesis

Balinsky (1940) removed the entire endoderm (entoderm in his terminology) from late-neurula-stage embryos of the European common salamander. In his words:

> The absence of entoderm produces the greatest effect in the development of the branchial regions...The entodermless embryos possess no external gills, no cartilaginous branchial skeleton, no visceral arches at all. (p. 197)

Although Balinsky discussed the inductive role of the endoderm on eye and heart development—the endodermless embryos showed 'a total absence of the heart'—he did not draw the same conclusions for the skeleton.

Hörstadius and Sellman obtained presumptive evidence for inductive tissue interactions in the initiation of various head cartilages, including chondrification of the otic capsule (a cartilage that develops from head mesoderm under the influence of the otic vesicle), and differentiation of NC-derived pharyngeal arch cartilages under the influence of pharyngeal endoderm.[9] Further, they extended Raven's studies on the differences between CNC and TNC by demonstrating that CNC grafted into the trunk fails to chondrify unless pharyngeal endoderm is included in the graft, in which case gill-like openings and associated pharyngeal cartilages develop.

Wagner (1949) observed that migrating CNCCs made multiple contacts with the endoderm and that the endoderm responds to these contacts by dividing and thickening, providing perhaps the first indication of reciprocal interaction between NC and endoderm. Hall and Coffin-Collins (1990) showed the same phenomenon in chicken embryos, a mitogenic role of CNCCs that merits closer study, especially with respect to the developmental and evolutionary links between NC and placodes (see Chapter 4) and the epithelial–mesenchymal interactions required to initiate the differentiation of NCCs.

Amphibian NCCs neither preferentially migrate toward pharyngeal endoderm *in vitro* nor do they differentiate as chondroblasts if NC and pharyngeal endoderm are placed near, but not in contact with, one another. Evidently, pharyngeal endoderm does not exert its influence by a diffusible factor; direct contact with NCCs is required to initiate chondrogenesis.[⊗] Once chondrogenesis is initiated, however, NCCs can transmit over distances of some 300 μm the ability to chondrify to other NCCs, apparently without direct cell-to-cell contact. As discussed in Chapter 2, such a self-generating transfer of cell fate (lateral induction) transfers neural induction to adjacent cells.[10]

This epigenetic view of the differentiation of cartilage from NCCs has been confirmed repeatedly and extended to other types of NCCs. As discussed in Box 7.4, epithelial–mesenchymal interactions elicit the differentiation of cartilage (and bone) from NCCs in all vertebrates examined.

Cascades of Interactions in Amphibian Craniofacial Development

Interactions involving the NC and adjacent embryonic layers such as the pharyngeal endoderm initiate a cascade of interactions integrating associated tissues and organs as functional units in anurans and in urodeles. Embryonic origins and cascades of interactions can be tricky to unravel and separate. Taste buds, the last sensory receptors to arise during development, provide a nice exemplar before tackling the craniofacial region as a whole.

[⊗] The story of cartilage induction is more complex than a simple pharyngeal endodermal induction. While only pharyngeal endoderm is inductive in alpine newts, pharyngeal endoderm and stomodaeal ectoderm induce in the North American spotted salamander, and pharyngeal endoderm and notochord induce in the Spanish ribbed newt.

Color Plate 1

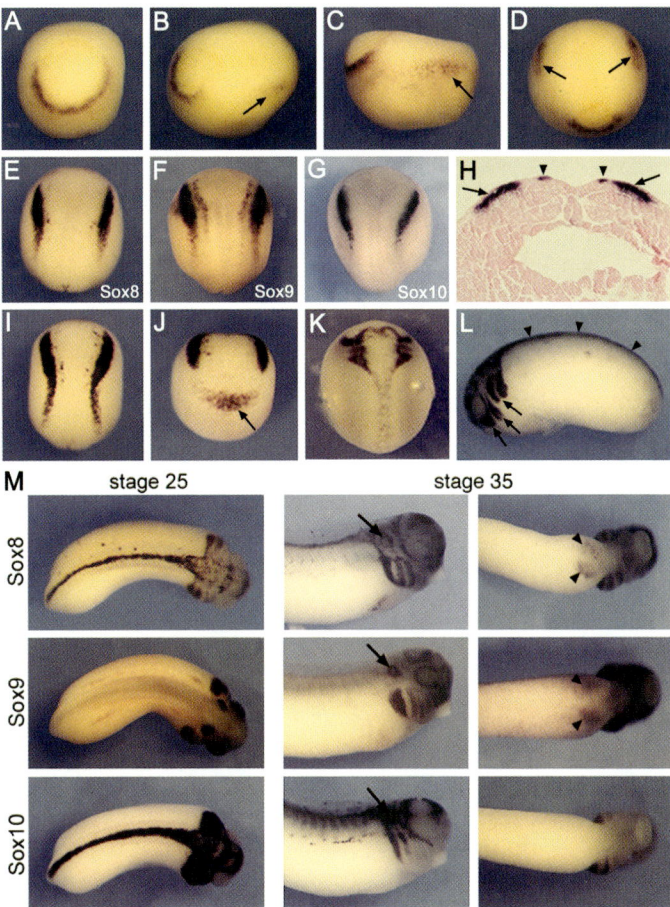

Fig. 2.6 Expression of *Sox8*, *Sox9*, and *Sox10* in NCCs and in NCC derivatives in *Xenopus* embryos. *Sox8* is expressed around the blastopore in gastrulae (**A**, **B**, **C** and **D**) and then lateral to the neural plate (*arrows* in **B**, **C**, and **D**). In slightly later embryos, *Sox8* (**E**), *Sox9* (**F**), and *Sox10* (**G**) are expressed in the neural folds, the site of the future NC. Panel (**H**) shows *Sox 8* expression in both medial (*arrowheads*) and lateral (*arrows*) NC, shown here in a transverse histological section. (**I** and **J**) slightly later stage in neurulation showing the extent of expression of *Sox8* in the NC and expression in the future cement gland (*arrow* in **J**). With closure of the neural tube — shown dorsally in **K** and laterally in **L** — Sox8 is expressed in migrating CNCCs (*arrows* in **K**) and in premigratory TNCCs (*arrowheads* in **L**). The nine panels in (**M**) compare expression of the three *Sox* genes at two tail bud stages (25 and 35). Note co-expression in CNCCs but downregulation of *Sox9* in TNCCs. The three genes are expressed in the otic vesicle (*arrows*). Sox8 and Sox9 (but not *Sox10*) are expressed in the primordium of the pancreas (*arrowheads*). Figure kindly provided by Jean-Pierre Saint-Jeannet

Color Plate 2

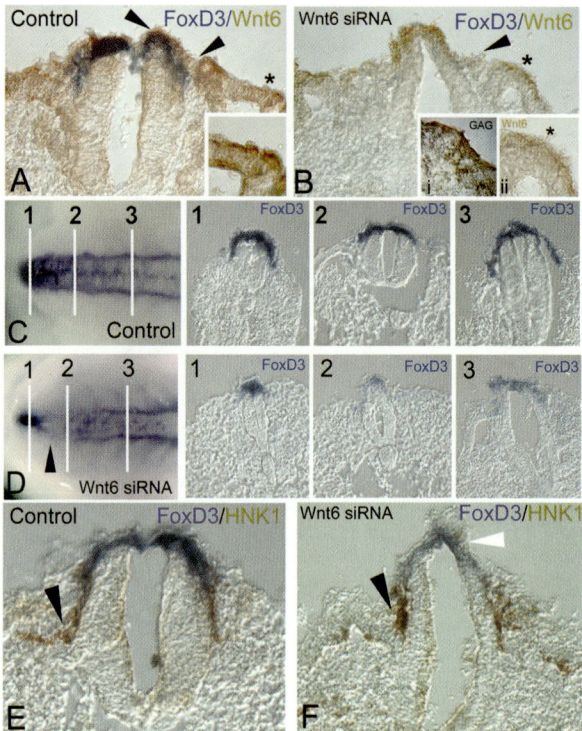

Fig. 2.7 *Wnt6* and NC induction depicted in cross-sections of the neural tubes of H.H. stage 18 (3-day) chicken embryos using FoxD3 protein and HNK-1 as NC markers. (**A**) Wnt6 (*brown*) and FoxD3 (*blue*) expression in neural ectoderm (*arrowheads*) in a control embryo. The area marked * is shown in the insert. (**B**) Reduced Wnt6 (*brown*) expression and absence of FoxD3 (*blue*) expression in a Wnt6 siRNA-treated embryo. (**C and D**) Reduced expression of FoxD3 (*blue*) at three rostrocaudal levels of the dorsal neural tube in a Wnt6 siRNA-treated embryo (**D**) when compared with control (**C**). (**E**) FoxD3 (*blue*) expression in the dorsal neural tube of a control embryo (**E**) is reduced significantly in Wnt6 siRNA-treated embryo (**D**, *white arrowhead*). Figure kindly supplied by Imelda McGonnell

Color Plate 3

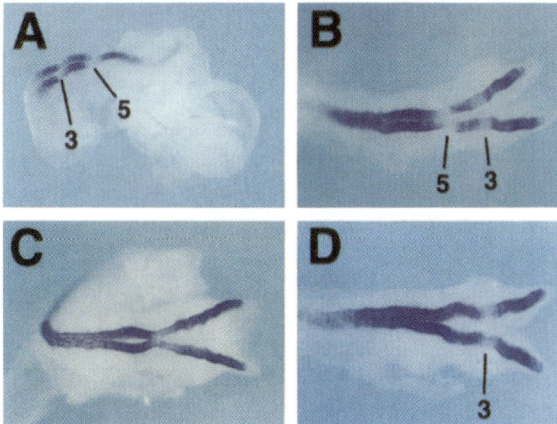

Fig. 2.14 Expression of *Msx3* in 8–9-day-old mouse embryos. (**A**) *Msx3* is expressed strongly in rhombomeres 1, 2, and 4 and in the spinal cord, and weakly in r3 in this 7-somite embryo seen in lateral view with anterior to the *left* and r3 and r5 identified. (**B**) This 10-somite embryo, seen in dorsal view with anterior to the *right*, shows weak expression of *Msx3* in r3 and lack of expression in r5. (**C**) There is uniform expression of *Msx3* throughout the hindbrain and spinal cord in this 18-somite embryo seen in dorsal view (anterior to the *right*). (**D**) The gap in expression in r5 seen in the normal embryos (**B**) is not seen in this 10-somite embryo carrying the *Kreisler* (*Krmlkr*) mutation. Kreisler codes for a transcription factor that regulates rhombomere segment identity through *Hox* genes. Indeed, r5 may not have developed in this embryo. Reprinted from a figure kindly provided by Paul Sharpe from *Mechanisms of Develop*, Volume 55, Shimeld *et al.* (1996). Copyright © (1996) with permission from Elsevier Science

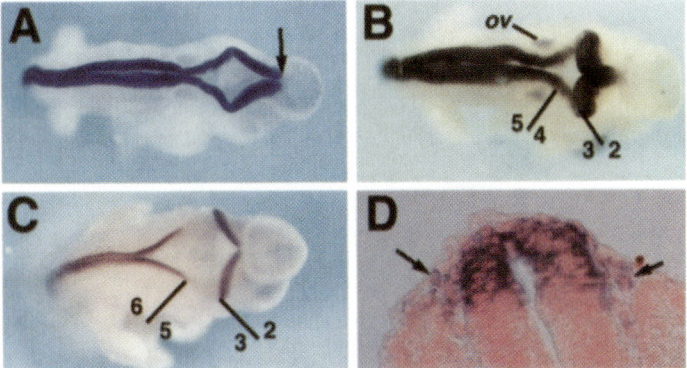

Fig. 2.15 Expression of *Msx3* in 9.5–11.5-day-old mouse embryos seen in dorsal view with anterior to the *right* (**A, B,** and **C**) and in histological cross-section (**D**). (**A**) Expression is strong in hindbrain and spinal cord at 9.5 days. The arrow marks the hindbrain–midbrain boundary, expression being negative in the midbrain. (**B**) Expression is similar at 10.5 days of gestation. 2, 3, 4, and 5, rhombomeres 2, 3, 4 and 5; OV, the otic vesicle, which displays nonspecific trapping of the antibody. (**C**) At 11.5 days of gestation, expression is restricted dorsally and is absent from rhombomeres 3–5. (**D**) A transverse section of the neural tube of an embryo of 9.5 days of gestation shows *Msx3* expression in the dorsal neural tube and in NCCs adjacent to the neural tube. Reprinted from a figure kindly provided by Paul Sharpe from *Mech Develop*, Volume 55, Shimeld *et al.* (1996). Copyright © (1996) with permission from Elsevier Science

Color Plate 4

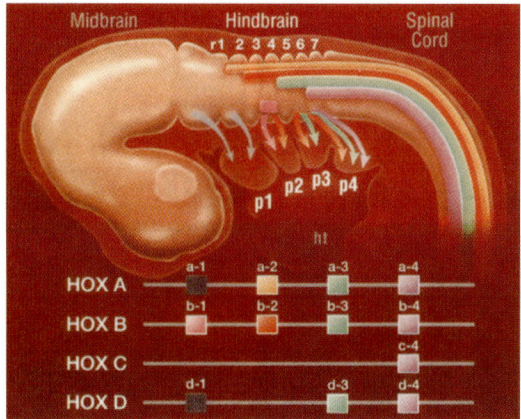

Fig. 2.18 *Hox*-gene expression in the rhombomeres of the hindbrain (r1–r7) and in the pharyngeal arches (p1–p4) is shown in this reconstruction of a mouse embryo of 9.5 days of gestation. *Colored bars* in the neural tube and *colored arrows* in migrating neural crest cells represent expression domains of *HoxA–HoxD*, which are also shown in the panel at the *bottom*. Some genes, such as *Hoxa2*, are expressed in the hindbrain but not in migrating neural crest cells. Reproduced from Manley and Capecchi (1995), with the permission of Company of Biologists Ltd.

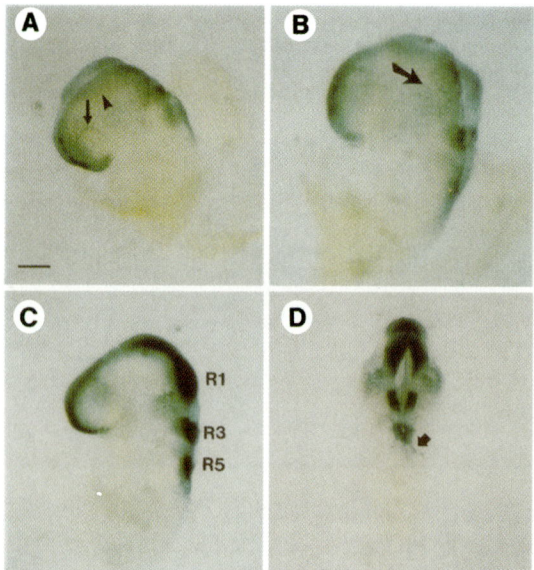

Fig. 4.7 Expression of *Pax7* in CNC as seen in lateral (**A, B,** and **C**) or dorsal (**D**) views of whole mounts of 8.5- (**A** and **B**) and 9.5- (**C** and **D**) day-old mouse embryos. (**A**) Neural crest cells surrounding the optic vesicle (*arrow*) and at the level of the midbrain (*arrowhead*) are highlighted. (**B**) The arrow indicates neural crest cells migrating from the hindbrain. (**C**) Rhombomeres 1, 3, and 5 are identified and strongly express *Pax7*. (**D**) This dorsal view highlights expression in alternating rhombomeres. R5 is marked by the *arrow*. Bar = 200 μm (**A** and **C**); 240 μm (**B**); and 160 μm (**D**). Reproduced from the color original in Mansouri *et al.* (1996), with the permission of Company of Biologists Ltd., from a figure kindly provided by A. Mansouri

Color Plate 5

Fig. 5.4 An adult Mexican salamander to show the typical pattern of pigmentation. As metamorphosis does not normally occur adults retain the external gills and the dorsal median fin. Compare with the metamorphosed adult in Fig. 5.6. Figure kindly supplied by Lennart Olsson

Fig. 5.6 An adult Mexican axolotl that was induced to metamorphose retains the larval pigment pattern, loses the external gills and median dorsal fin, and has a differently shaped head than the normal unmetamorphosed adult shown in Fig. 5.4. Figure kindly supplied by Lennart Olsson

Fig. 5.7 The variation in patterning (horizontal stripes, vertical stripes, spots, and albino shown) and in color found within a single genus (*Danio*) is shown in these images of six species of zebrafish, which are, clockwise from the *top left*, the zebrafish, *Danio rerio*; pearl danio, *D. albolineatus*; *D. sp. Nov.*; *D. choprae*, *D. feegradei*; and the spotted danio, *D. nigrofasciatus*. Reprinted with permission from the Annual Review of Ecology, Evolution, and Systematics, Volume 38 © 2007 by Annual Reviews www.annualreviews.org. Images kindly provided by David Parichy

Color Plate 6

Fig. 5.8 A close-up view of the body of a *leopard* (*leo*) mutant zebrafish showing the lines of pigmented spots made up of melanophores (*blue–black*) that replace the horizontal stripes seen in wild-type zebrafish (Fig. 5.7). Individual melanophores and the pigment-free zone around each spot are clearly visible, as are the *yellow* xanthophores and the iridescent iridophores. Image kindly provided by Michelle Connolly

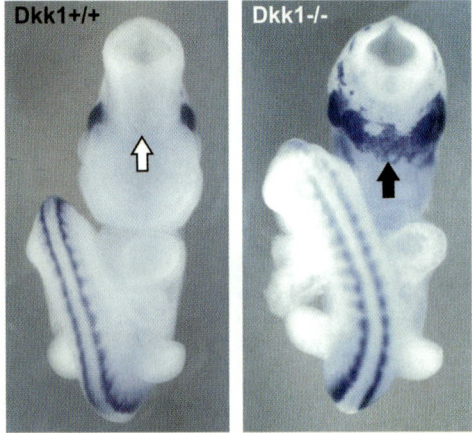

Fig. 7.7 Wild-type (*left*) and a *Dkk1* mutant mouse (*right*) at 9.5 days of gestation, viewed from the anterior, to show that *Sox10* is not expressed in the anterior neural folds of the wild type (*white arrow*) but is expressed ectopically in the anterior neural folds of the *Dkk* mutant (*black arrow*). Figure kindly provided by Roberto Mayor. Reproduced from *Dev Biol* (2007) 309:208–221 with the permission of the publisher Elsevier Limited

Color Plate 7

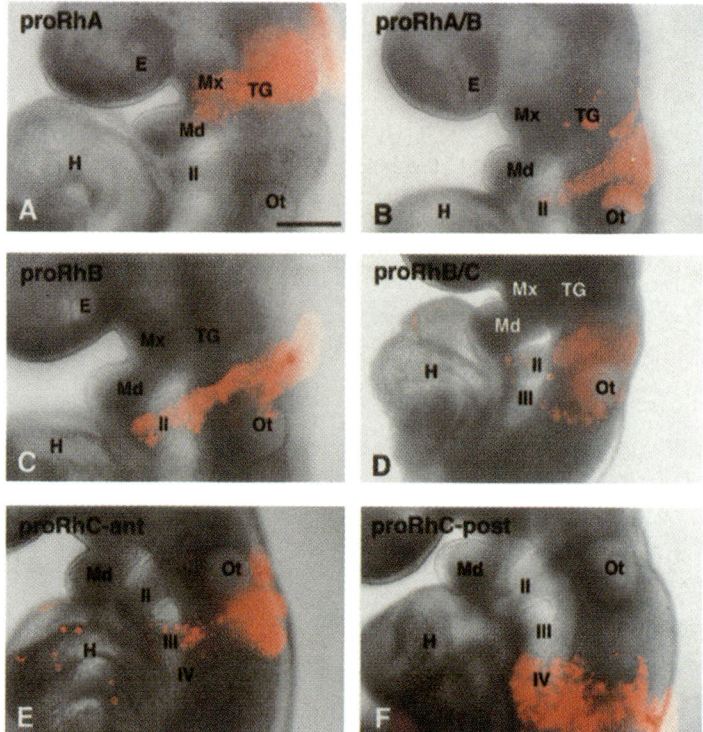

Fig. 7.19 The segmental migration of neural crest cells (*red*) into individual pharyngeal arches as demonstrated following injection of DiI either into pro-rhombomeres **A, B,** or **C** of mouse hindbrains (seen in **A, C, E,** and **F**) or into the boundary between rhombomeres **A/B** or **B/C** (shown in **B** and **D**). II–IV, second to fourth pharyngeal arches; (**E**) eye primordium; (**H**) heart; Md, mandibular prominence; Mx maxillary prominence; Ot, otic vesicle; TG, trigeminal ganglion. Bar = 200 μm. Reproduced from Osumi-Yamashita *et al.* (1996), with the permission of Blackwell Science Pty. Ltd., from a figure kindly provided by N. Osumi-Yamashita

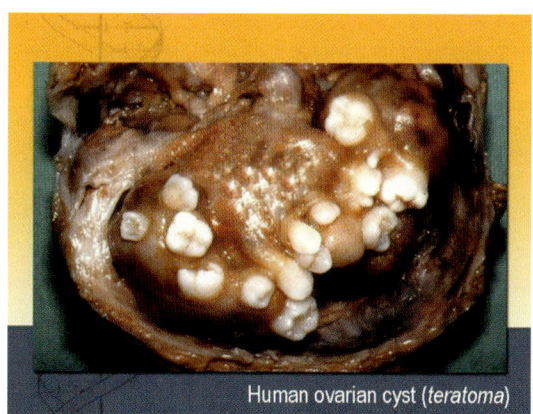

Fig. 8.1 Extraordinarily complete teeth, the largest of which are 10 mm across and molariform with multiple cusps. These teeth developed in a teratoma in a human ovary. Image supplied by Paul Sharpe courtesy of the Anatomy Museum, Guy's Hospital, London

Color Plate 8

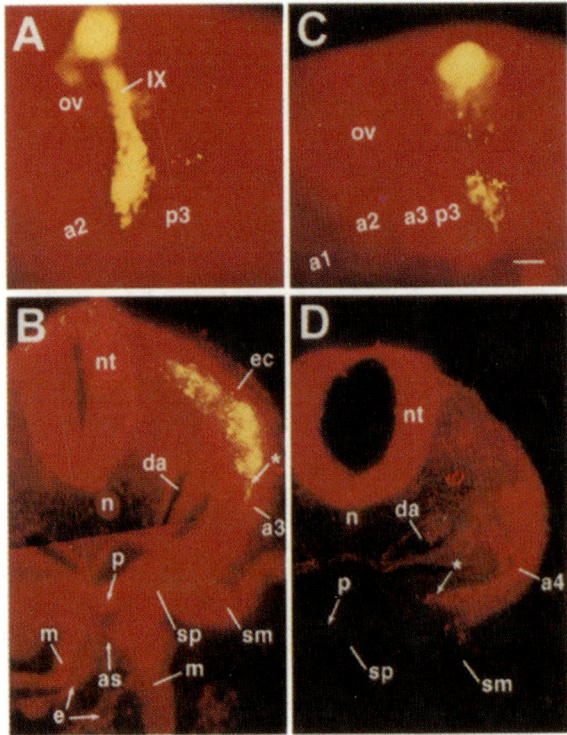

Fig. 8.4 Migration of neural crest cells to visceral arches 3 and 4 is shown in these sections of chicken embryos in which the neural crest was labeled with DiI, shown as *yellow* fluorescence. (**A**) Labeled cells migrating from r7 into arch 3 and forming the tract of the glossopharyngeal nerve IX. (**B**) In this transverse section of a chicken embryo of H.H. stage 14, DiI-labeled neural crest cells (ec) that arose from r7 can be seen migrating toward the third pharyngeal arch (a3). The leading edge of the neural crest cells (*) lies lateral to the pharynx (p). (**C**) Labeled neural crest cells from the level of somite 2 migrate toward the fourth visceral arch. (**D**) A cross-section of (**C**) showing the ventral location (*) of the leading edge of the migrating cells. Other abbreviations are: a1–a4 and p3, pharyngeal arches; da, dorsal aorta; e, endocardium; m, myocardium; n, notochord; nt, neural tube; ov, otic vesicle; p, pharynx; sm, somatic mesoderm; sp, splanchnic mesoderm. Reproduced from Suzuki and Kirby (1997) from a figure kindly supplied by Margaret Kirby. Reprinted by permission of Academic Press, Inc.

Color Plate 9

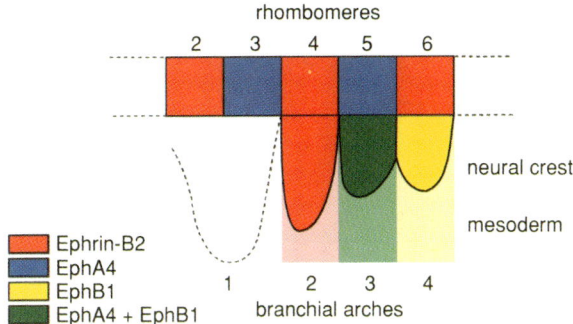

Fig. 10.6 Expression of EphrinB2 (*red*), EphA (*blue*) and EphB1 (*yellow*) in rhombomeres 2–6, in migrating neural crest cells and in cranial mesoderm, is highly regionalized with respect to the four pharyngeal arches. EphrinB2 is expressed in r4 and in the second pharyngeal arch. EphB1 is expressed in r6 and the fourth pharyngeal arch. The boundary of EphrinB2 is shown in *red*, the boundary of cells expressing EphA4 plus EphB1 in *green*. Receptor–ligand interactions at the boundary between pharyngeal arches 2 and 3 restricts cells that are migrating from r5 to the third pharyngeal arch and prevents intermingling of second and third arch cells. Reproduced from A. Smith *et al.* (1997) with the permission of Current Biology Ltd., from a figure kindly supplied by David Wilkinson

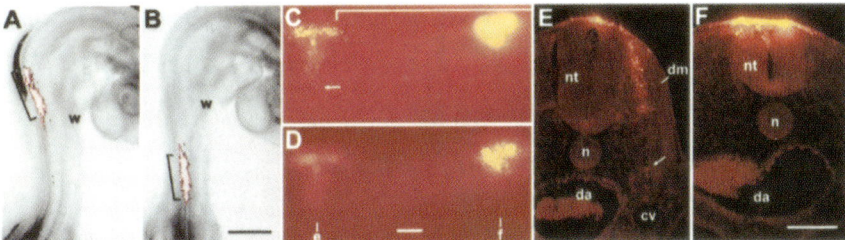

Fig. 10.7 Ablating the neural crest followed by injection of DiI demonstrates absence of regulation; no labeled cells are seen either in the normal migration pathway (illustrated in Fig. 7.16) or at the final site. (**A, B**) Trunk neural crest was ablated (*square bracket*) either adjacent to the most caudal somites (somites 12–17 in **A**) or to the most cranial unsegmented mesoderm (*square bracket* in **B**) from 17- or 20-somite-stage embryos (**A** and **B**, respectively). Multiple sites in the neural tube in the ablated region were then injected with DiI. In neither case can neural crest cells be seen migrating from the ablated region, signifying lack of regulation. w, wing bud. Bar = 1 mm. (**C, D**) This 17-somite embryo is viewed in two focal planes. The *bracket marks* the ablated neural crest. The *yellow* fluorescence marks sites of DiI injection either immediately rostral to the ablated region (the *left* zone of fluorescence, seen in focus in **C**) or within the ablated region (the *right* zone of fluorescence, seen in focus in **D**). Neural crest cells are migrating from the rostral neural crest (*arrow* in **C**) but do not deviate into the ablated area. As expected, no cells migrate from the neural tube in the ablated region. e and f mark the position of the transverse sections shown in **E** and **F**. Bar = 100 μm. (**E**) Transverse section rostral to the ablated region marked e in (**D**). DiI-labeled cells are migrating medial to the dermamyotome (dm) and in the primary sympathetic trunk (*arrow*). cv, cardinal vein; da, descending aorta; n, notochord; nt, neural tube. (**F**) Transverse section through the ablated region marked as f in (**D**). DiI-labeled cells are concentrated adjacent to the dorsal neural tube and not seen in the normal migration pathway; compare with **E**. Abbreviations as in **E**. Bar = 100 μm for **E** and **F**. Reproduced from Suzuki and Kirby (1997) from a figure kindly supplied by Margaret Kirby. Reprinted by permission of Academic Press

It was not clear whether taste buds of the tongue developed from, or were induced by, NCCs until a combination of transplantation and organ culture by Barlow and Northcutt (1997) demonstrated that:

- taste buds arise from local pharyngeal endoderm and not from NC or placodal ectoderm,
- NC does not induce taste buds,
- axial mesoderm signals to pharyngeal endoderm to evoke taste buds, and
- taste buds do not require innervation to develop.[11]

The cranial ganglia of cranial nerves VII, IX, and X associated with innervation of the taste buds are of placodal and NC origin. In the Mexican axolotl, gustatory (taste) neurons of these cranial ganglia that innervate the taste buds are placodal but the nongustatory (touch, pain) neurons that innervate the oral epithelium are NC in origin, distinctive function mirroring embryological origin.[12]

Anurans: Signaling pathways involved in induction of the NC in *Xenopus* are discussed in Chapter 2, and see Fig. 2.9.

The painted frog was used in a series of experiments to investigate the role played by the NC in the formation of the tissues and structures associated with the developing mouth. The experimental procedure was to associate NC with one or more adjacent regions of neural-fold-stage embryos, and then to graft the combined tissues either into the flanks of similarly aged embryos or into the blastocoele of younger embryos. Supernumerary mouths formed in these ectopic sites. Formation of the mouth requires interaction between stomodaeal endoderm and ectoderm, neural folds, and NCCs. Cusimano-Carollo (1972) interpreted her results as evidence for a cascade of interactions emanating from the pharyngeal endoderm as follows:

- The endoderm from regions equivalent to the 0–90° sectors mapped by Chibon (Fig. 7.2d; and see below) acts inductively on rostral NCCs to initiate the differentiation of the supra- and infrarostral cartilages (the cartilages of the larval mouth).
- At the same time, pharyngeal endoderm acts inductively on future stomodaeal ectoderm.
- Once differentiated, the cartilages act upon the ectoderm, enabling it to invaginate to form the horny teeth, horny beak, and oral papillae of the developing larval mouth.

However, some anurans such as *Xenopus laevis* lack suprarostrals and have a single midline infrarostral and keratinized mouthparts. Urodeles also lack these cartilages. Consequently, the cascade seen in the painted frog cannot operate in all amphibians.

Urodeles: The only experimental studies on urodele embryos that parallel those using the painted frog are transplantation studies using Spanish ribbed and alpine newt embryos. As in the painted frog, all structures associated with the mouth form

only if NC, prechordal and lateral head mesoderm, stomodaeal endoderm, and stomodaeal ectoderm are present.

The earliest ossifications in newts are dermal and associated with tooth-bearing
bones. Formation of the vomers and of the palatine bones depends on the prior
differentiation of the cartilaginous trabeculae, while formation of the dentary and
the splenials depends on the prior differentiation of Meckel's cartilage, indicating temporal (and perhaps inductive) relationships between chondrogenesis and
osteogenesis.[13]

Mutations affect cascades of interactions and have been most studied in zebrafish
(see section below). Unfortunately, amphibians, about whose NC so much is known,
are not ideal animals for genetic screening or for generating mutations. Nevertheless, one mutation has revealed elements of inductive cascades in the axolotl.
The *premature death* (*p*) mutation in Mexican axolotls affects the subpopulation of
NCCs that produce the craniofacial cartilages and contribute to the heart. The NC
cartilages fail to form, although other NC derivatives develop normally. The site of
action of mutant genes as epithelial or mesenchymal can be determined by establishing reciprocal tissue recombinations between mesenchyme and known inductive epithelia from wild-type and mutant embryos (Fig. 7.6). *p* Endoderm induces
heart and cartilage development from mesenchyme of wild-type embryos, but *p*
mesenchyme cannot respond to inductive influences from wild-type pharyngeal

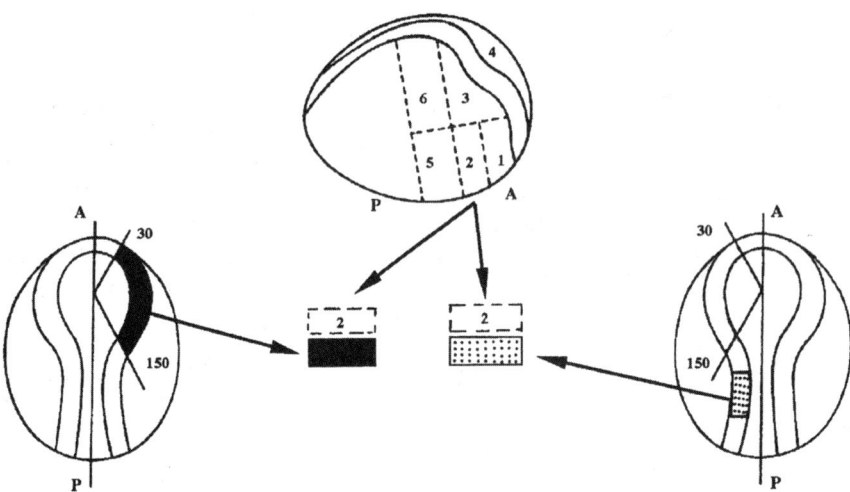

Fig. 7.6 Cranial neural crest (*black*) or trunk neural crest (*stippled*) from wild-type or mutant
amphibian embryos (shown here at the neurula stage, viewed from the dorsal surface) can be
recombined with an epithelium (shown as region 2 of 6 regions of inductively active endoderm
in the lateral view of a neurula) and maintained in organ culture to test inductive activity of the
epithelium, and to test responsiveness of particular populations of neural crest cells. The region
bracketed between 30 and 150° represents the chondrogenic neural crest (see also Fig. 7.3d). A,
anterior; P, posterior. Modified from Hall (1987)

endoderm. Competence of the mesenchyme to respond to induction is defective in *p* mutants (Graveson and Armstrong, 1996).

Labeling Amphibian CNCCs

The next advance over transplantation came in the 1960s when ³H-thymidine-labeled neural folds were grafted into unlabeled host embryos and the fate of radioactively labeled migrating cells followed using autoradiography. One of the most detailed analyses of the distribution of the skeletogenic CNC using this approach is the series of studies initiated by Chibon (1964) using embryos of the Spanish ribbed newt into which ³H-thymidine-labeled NCCs were grafted. The resulting fate map for urodele skeletogenic CNC forms the basis of our understanding of the regionalization of urodele (and anuran) NC.

Urodele Fate Maps: Chibon divided the cranial neural folds into sectors or wedges, marked out in 30° arcs from the midline of neurula-stage embryos (Fig. 7.2d), which can be compared with the regions used in Langille and Hall's analysis of lamprey NC, summarized in Fig. 7.2a. He grafted individual labeled sectors into unlabeled embryos and determined their skeletogenic potential as outlined below.

(i) The *most rostral 30° sector* —the transverse neural folds—provided no skeletal cells, supporting the findings of Hörstadius and Sellman for the Mexican axolotl and consistent with Langille and Hall for the lamprey. Seufert and Hall (1990) later demonstrated that the transverse neural folds of *Xenopus* embryos lack NCCs and are unable to form cartilage, even when challenged with inductively active epithelia. The mechanism by which the most rostral neural folds are prevented from forming NCCs in zebrafish and mice has been shown in a most elegant study by Carmona-Fontaine *et al.* (2007) to involve inhibition of the canonical Wnt/β-catenin pathway by the protein Dickkopf1 (Dkk1) secreted by prechordal mesoderm, and concomitant ectopic expression of *Sox10* in the anterior neural folds (Fig. 7.7 [Color Plate 6]).

(ii) *The next most caudal sector* (30–60°) provided the cells for the most rostral skeletal elements in the head, the rostral trabeculae. More caudal skeletal elements arise from more caudal sectors of CNC, as in lampreys, and so represent an ancient pattern predating the agnathan–gnathostome split (Fig. 7.6d).

(iii) Palatoquadrate, posterior trabeculae, and the basal plate of the skull arise from the *next most-caudal sector* (60–90°);

(iv) Meckel's and hyoid arch cartilages form from the 70 to 100° sector;

(v) hyobranchial and rostral pharyngeal arch cartilages from the sector at 100 to 120°, and

(vi) caudal pharyngeal arch cartilages from the *most caudal CNCCs* in sector 120 to 150°.

This precise fate mapping confirmed Raven's conclusions of distinct chondrogenic cranial and nonchondrogenic trunk neural crests in the Mexican axolotl and

Fig. 7.7 Wild-type (*left*) and a *Dkk1* mutant mouse (*right*) at 9.5 days of gestation, viewed from the anterior, to show that *Sox10* is not expressed in the anterior neural folds of the wild type (*white arrow*) but is expressed ectopically in the anterior neural folds of the *Dkk* mutant (*black arrow*). Figure kindly provided by Roberto Mayor. Reproduced from *Dev Biol* (2007) 309:208–221 with the permission of the publisher Elsevier Limited (see Color Plate 6)

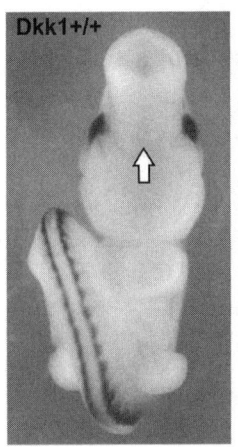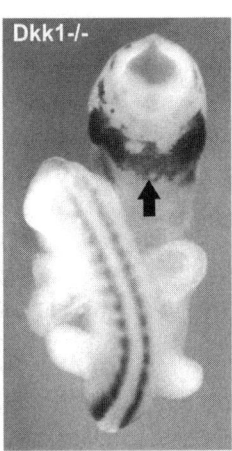

Triturus and confirmed Hörstadius and Sellman's study demonstrating a regionalized skeletogenic neural crest in axolotls.[14]

Anuran Fate Maps: Another approach to labeling cells is to use two species whose cells can be distinguished from one another because of a naturally occurring marker. To this end, Sadaghiani and Thiébaud (1987) transplanted NC between *Xenopus laevis* and *X. borealis*. In addition to confirming the patterns of NCC migrations described by Stone for the axolotl (Fig. 7.8), they demonstrated three streams of migrating cells that contribute cells to the craniofacial skeleton:

- **a mandibular stream** from the midbrain and whose cells form the Meckel's cartilage, and the quadrate, ethmoid, and trabeculae;
- **a hyoid stream** from the hindbrain and whose cells form the ceratohyal; and
- **a branchial stream**, from which the pharyngeal arch (gill) cartilages arise.

This map of separate streams of anuran CNCCs compares very closely with the map of urodele chondrogenic regions obtained by Chibon. The chondrogenic CNC of the oriental fire-bellied toad has since been mapped and is essentially similar (Fig. 7.8). **Indeed, CNCC migration in three streams is conserved from lampreys to placental mammals** (Fig. 7.9) but perhaps not in marsupials (Box 7.5). Given the apparent conservation of migrating subpopulations of NCCs, Gass and Hall (2007) have argued that each subpopulation represents a module, in the sense of the model proposed by Atchley and Hall (1991). This model, which is based on modular cell condensations, was elaborated with mammalian NC-derived skeletal elements as the paradigmatic example. As modules operating individually (the condensation for a phalange in a digit) or through interactions (the cell lineages that form the mammalian dentary) populations of specified cells generate the phenotype during development and are the cellular basis for modification of the phenotype in evolution (1991).

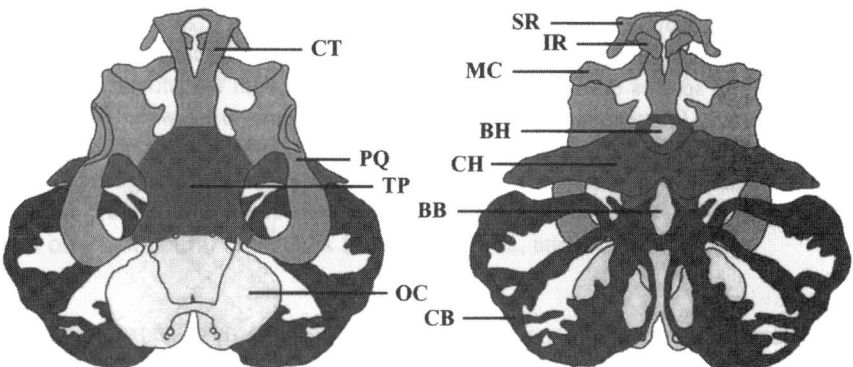

Fig. 7.8 The contribution of populations of migrating CNCCs to the cartilaginous larval skull of the Oriental fire-bellied toad as determined by vital dye labeling and as seen in dorsal (*left*) and ventral (*right*) views. The mandibular stream (*pale gray*) gives rise to the trabecular horn (CT), palato-quadrate (PQ), suprarostral (SR), infrarostral (IR), and Meckel's cartilage (MC). The hyoid stream (*medium gray*) gives rise to the trabecular plate (TP) and ceratohyal (CH). The branchial streams (*darkest shading*) give rise to the ceratobranchials (CB). Cranial mesoderm (*lightest shading*) gives rise to the otic capsule (OC), basibranchial (BB), and basihyal (BH). Reproduced from Olsson and Hanken. Copyright © (1996). From a figure kindly supplied by Lennart Olsson. Reprinted by permission of Wiley-Liss Inc., a subsidiary of John Wiley & Sons, Inc.

Timing of Migration

Although the *number* of streams is highly conserved across the vertebrates (Fig. 7.9), variation in the time at which NCCs migrate—and therefore in the stage of development of the neural folds/tube when migration occurs—differs between the major groups of vertebrates but not within groups. A recent comparison of the timing of NCC migration in two closely related species of ranid frogs, however, revealed substantial temporal differences that in a phylogenetic context we regard as heterochronies. In the European common frog, *Rana temporaria*, mandibular, hyoid, and branchial streams are migrating and have separated when the neural plate is still wide open (i.e., the neural folds have not started to close), whereas in the black-striped frog, *Rana* (*Sylvirana*) *nigrovittata*, migration is initiated after the neural folds have fused (Mitgutsch *et al.*, 2008). Clearly, more comparative studies of this kind are required.

Although migration of CNCCs in three streams is conservative (even when timing varies), there are subtleties of cell populations and cell lineages within the NC that we do not yet fully understand. In *Xenopus* and in *Gastrotheca riobambae*, the egg-brooding marsupial frog, prominent streams of NCCs migrate from r5 (Fig. 7.10). This is an unusual pattern if compared with chicken and mouse embryos, in which r5 (and r3) contain NCCs, but the cells undergo apoptosis before migration can be initiated. The marsupial frog also has an unusually large-yolked egg (even for an amphibian), upon which develops a large flat epiblast (as in avian embryos), which goes on to form an embryo with large external gills. Del Pino and colleagues

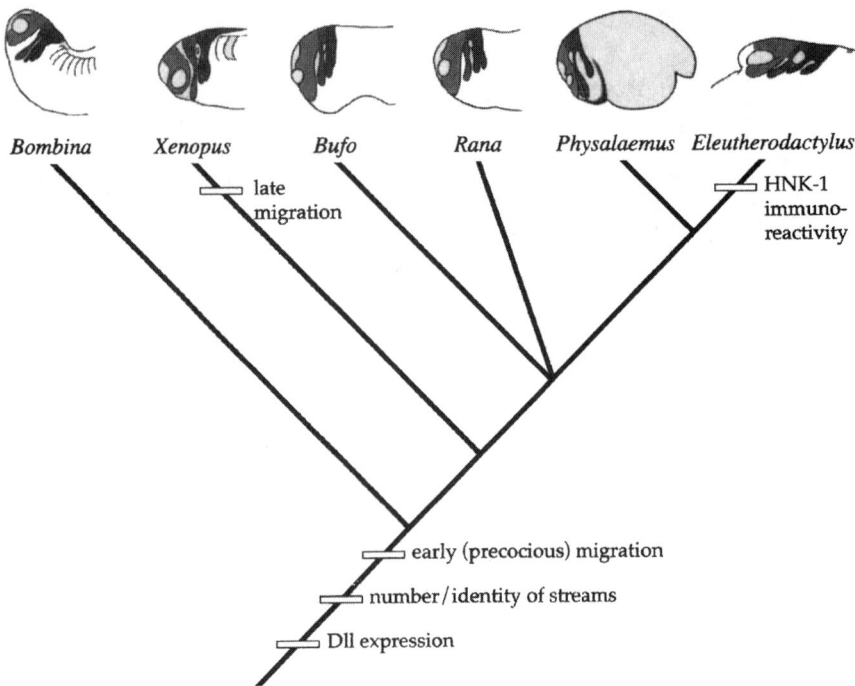

Fig. 7.9 The three major streams of migrating neural crest cells (mandibular, pre- and postotic; lightest to darkest shading respectively) are highly conserved evolutionarily in amphibians with tadpoles and in the direct-developing species *Eleutherodactylus*. Early migration is the primitive condition. Late migration as seen in *Xenopus* is derived. *Dlx* is expressed in all species examined so far, while reactivity with the antibody HNK-1 is a derived feature in Puerto Rican coqui, convergent with HNK-1 expression in amniotes. The Tungara frog from Cost Rica, which like coqui is a leptodactylidid frog, has a small-yolked egg, produces a tadpole and has neural crest cells that do not react with HNK-1. I thank Lennart Olsson for providing this figure and the data on which it is based

Box 7.5 Marsupial CNCCs

All marsupials are immature at birth, comparable in their development to a 12-day-old mouse embryo or a 10-week-old human embryo. In dramatic contrast to placental mammals, the phase of intrauterine development in marsupials is short (2.5–10 days, depending on the species) and followed by a prolonged period of postnatal development attached to the mother, as much as 100 days in some species. Further, lactation can continue intermittently for a year or more after birth, while weaning varies from around 60–295 days after birth, depending on the species.

While we have not mapped the skeletogenic CNC for any marsupial (or monotreme), Kathleen Smith laid the modern basis for analyses of marsupials when she compared the craniofacial development of four species of marsupials with the development of five eutherian mammals. By emphasizing the developing central nervous system and its relationship to skeletal and muscle development, Smith demonstrated that marsupial CNS development is delayed relative to skeleto-muscular development. Within the marsupial CNS (and in contrast to placental mammals), development of the olfactory placodes and the craniofacial region is advanced relative to forebrain development. Neurogenesis and a short gestation period are rate limiting in marsupial development.[a]

Smith went on to examine NC development in embryos of the short-tailed opossum, *Monodelphis domestica*. Her major findings—given here with the contrasting situation in placental mammal embryos or other vertebrates—are:

- Early migration of NCCs from an open neural plate in opossum embryos stands in contrast to migration from the closed neural plate in placental mammals.
- A single stream of CNCCs populates the mandibular, maxillary, *and* frontonasal processes, in contrast to the three streams seen in all other vertebrates, including lampreys.
- The early accumulation of NCCs in the facial region of opossum embryos allows for the development of a much larger first arch and accelerated development of craniofacial structures in comparison with developing placental mammals.
- Delayed development of the central nervous system, especially of the forebrain, delays the ossification of cranial bones in comparison with the timing of ossification in placental mammals.[b]

---

[a] A draft of the genome of the short-tailed opossum is now available (Mikkelsen *et al.*, 2007).
[b] See Vaglia and Smith (2003) and K. K. Smith (2006*) for the origin and migration of NCCs in marsupial embryos.

(2007) discuss whether this unusual mode of development is a requirement of a species with such large external gills or whether it is a general trait associated with the development of the anuran pharyngeal skeleton.

What about **direct developing frogs**, which also have large-yolked eggs and modified gastrulation but which show various degrees of loss of larval skeletal elements?

CNCC migration begins *before neural tube closure* in a number of amphibian species, including the direct-developing Puerto Rican frog, *Eleutherodactylus coqui*. In such species, which have eliminated the tadpole stage from the life cycle, most of the cranial cartilages found in the tadpoles of other species (indirect developers)

Fig. 7.10 A dorsal view of an early embryo of the egg-brooding marsupial frog, *Gastrotheca riobambae* showing three major streams of migrating CNCCs delaminating from alternate rhombomeres, and the prominence of the mandibular stream in this species. Modified from Del Pino and Medina (1998), published by UBC Press, Leioa, Vizcaya, Spain

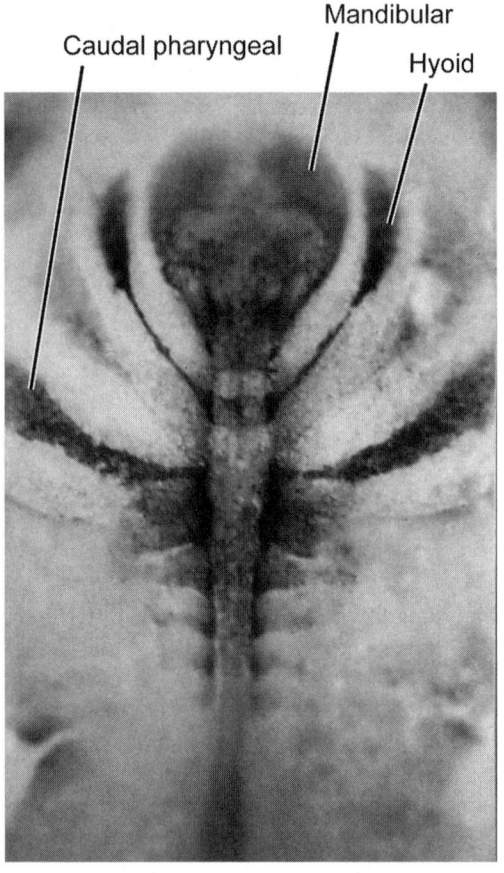

fail to form. The skeleton that forms is either typically adult or typical of the mid-metamorphic stages of species with tadpoles and metamorphosis; direct-developers, of course, do not metamorphose. Despite these fundamental changes in skeletogenesis, cranial crest migration in coqui is surprisingly conserved when compared with *Bombina*, *Xenopus,* and *Rana*, three species with tadpoles in their life cycles (Fig. 7.9). The only difference appears to be enhancement of the mandibular stream, as we saw above in the marsupial frog.[15]

Dlx is expressed in the mandibular stream of NCCs in *Xenopus*, patterns the vertebrate forebrain, and patterns the more rostral portions of the developing brain. Indeed, *Dlx* is expressed in the NCCs of all amphibians with tadpoles examined so far (Fig. 7.9).

Is *Dlx* expressed in coqui? The four *Dlx* genes cloned from coqui are expressed in a pattern similar to the patterns of expression of *Dlx* in species with a tadpole in their life cycle (indirect developers).[16]

Mapping CNCCs in Fish

Elasmobranchs

Claims for a NC contribution to the cranial mesenchyme in sharks were made, independently, in the late 19th century, by Kastschenko and Goronowitsch; Fig. 7.11 illustrates populations of migrating CNCCs in an embryo of the winter skate.

Holmgren (1940, 1943) provided substantial evidence from descriptive studies for a NC origin of much of the head skeleton in the velvet belly lantern shark, spiny and lesser-spotted dogfish, thornback, and rays.

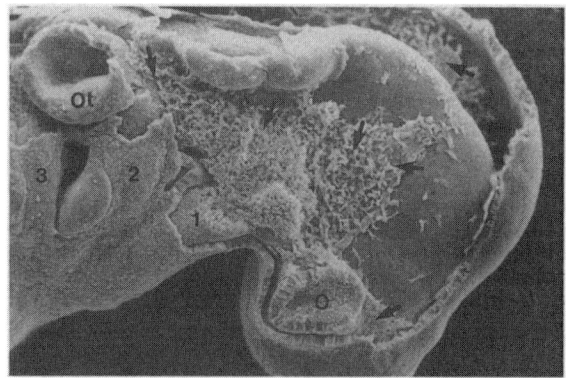
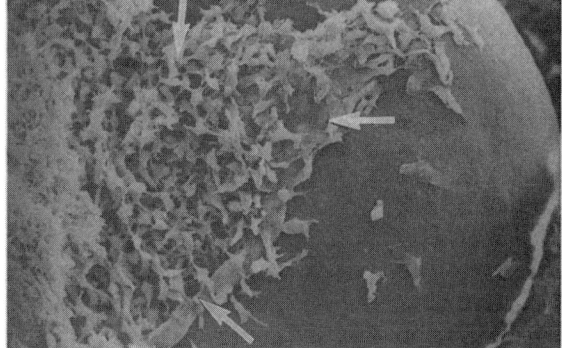

Fig. 7.11 Scanning electron micrographs at lower (*top*) and higher (*bottom*) magnifications of the developing head of an 84–85 somite-stage (60-day-old) embryo of the winter skate, collected off the coast of Nova Scotia. The ectoderm was removed to reveal populations of migrating neural crest cells (*arrows*). O, optic vesicle; Ot, otic vesicle; 1–3, the first (mandibular), second (hyoid), and third pharyngeal arches. Figure kindly supplied by Tom Miyake

Teleosts

In four studies published between 1888 and 1893, Kastschenko and Goronowitsch concluded that the NC contributes to the cranial mesenchyme in sharks and teleosts, and that dentine of the teeth of sharks and the common carp originate from NC. [17] Although, as discussed above, CNCC migration occurs in streams, each crest segment is restricted (determined) to form individual cell lineages of a single pharyngeal arch and its skeleton, a determination that is set before migration is initiated (Raible and Eisen, 1996).

No fate map of the CNC of any teleost was available until a study on the Japanese medaka provided a detailed description of the development of the cranial and pharyngeal skeleton and an experimental investigation of the effects on skeletal development of removing 100- to 150-μm-long regions from the CNC (Fig. 7.2b). Langille and Hall (1987, 1988a) determined that the skeletogenic NC extends from the midbrain caudally to the hindbrain, and that the most rostral (prosencephalic) neural tube and NC caudal to the fifth pair of somites is not chondrogenic, a pattern seen in lampreys, amphibians (the transverse folds above) and other vertebrates (Fig. 7.2). The chondrocranium, the skeleton of the hyoid and pharyngeal arches, [18] and the bulk of the rostral neurocranium were shown to originate from NCCs. Table 7.3 contains a full list of cartilages with a NC origin in the Japanese medaka and the regions of the CNC from which they arise.

Regional populations of NCCs, including mandibular and hyoid populations have been described in Mississippi paddlefish and white sturgeon and equated with similar populations in other vertebrates (see Bolker, 2004).

Zebrafish Mutants and Pharyngeal Arch Development

Zebrafish are ideal animals for saturation mutagenesis, a technique that opened wide a window onto a vista revealing many mutants affecting the NC or NC derivatives.

On the basis of screening for morphological changes in the phenotype, over 100 mutations affecting the pharyngeal arches and some 50 affecting craniofacial development were described in the first round of mutagenesis.[19] Using molecular markers, Henion *et al.* (1996) produced parthenogenetic diploid embryos and screened them for pleiotropic mutants. Mutations involving CNCCs and TNCCs were identified, including mutants lacking all NC cartilages and others in which cartilages were abnormal or in which only pharyngeal cartilages failed to develop. The affects on different pharyngeal arches of several genes revealed by mutant screening are outlined below; see Box 3.2 for mutations of *Edn1* and lower jaw development.

Distalless (Dlx) and the First and Second Pharyngeal Arches

In vertebrates, *Dlx* genes play at least two important roles: patterning the fore- and hindbrain and establishing the dorsoventral patterning of the pharyngeal endoderm (see Chapter 2).

In zebrafish, as in other vertebrates, *Dlx* exerts specific actions on the craniofacial cartilages of the first and second pharyngeal arches (which arise from hindbrain NC) but has little effect on midbrain-derived chondrogenic crest cells (Ellies *et al.*, 1997). *Dlx1a* (and other *Dlx* genes) is upregulated by *Dlx2a*, which plays a key role in specifying and maintaining CNCCs and their chondrogenic and ganglionic derivatives. Mutations in mice demonstrate that *Dlx1* and *Dlx2* play overlapping roles in patterning skeletal and soft tissues of the first and second pharyngeal arches.[20]

Chameleon (Con) and the Caudal Pharyngeal Arches

Con embryos, which lack functional Dispatched 1 (Disp1) protein, have reduced caudal pharyngeal arches and abnormally patterned hyoid arches. Disp1 is involved in the release of *Shh*, indicating that Shh is involved in patterning zebrafish pharyngeal arches.

Chinless (Chn) and the Absence of NCC from all Pharyngeal Arches

Chn disrupts skeletal fate and the interactions between NCCs and muscle progenitors; all seven pharyngeal arches lack NC-derived cartilages as a primary defect and lack mesoderm-derived muscles as a secondary defect. However, cartilaginous and muscle precursors *are* present as condensations of cells, indicating that *Chn* affects the differentiation and not the origin of these NCCs (Schilling *et al.*, 1996a). *Chn* affects NC and mesoderm, and therefore may:

- act independently on NC-derived chondrogenesis and mesoderm-derived myogenesis;
- act downstream of NCC initiation and specification, perhaps inhibiting the epithelial–mesenchymal interactions required for NCC differentiation; or
- have secondary effects on mesodermal muscle progenitor cells because of defective NC and therefore defective mesodermal interactions.

Given our present knowledge, the last seems the most likely.

Ninja and the Growth of Pharyngeal Arch Cartilages

Ninja is an induced recessive lethal mutation with reduced expression of *crestin* in the rostral neural tube and hindbrain. *Crestin*, a member of a family of retroelements (i.e., mobile genetic elements containing a reverse transcriptase gene), is expressed in all premigratory and migratory NCCs in zebrafish, but is downregulated with differentiation (Luo *et al.*, 2001).

Although NC-derived precursors of the pharyngeal arches (including the jaws) form in *Ninja* mutants, the mandibular and hyoid skeletons are small, and the cerato-branchials fail to form. Interestingly and unusually for a mutation affecting CNCCs, the number of vagal or sacral NCCs that form enteric neurons is also reduced and those that form fail to initiate neurogenesis, most likely because of pleiotropy rather than as secondary consequence of CNCC loss.

Skeletogenic NCCs in Reptiles

If we exclude birds, which cladistic phylogenies nest within the Reptilia, we know little about the skeletogenic CNC in nonavian reptilian embryos.

CNCCs

Alligators: Mark Ferguson (1985) used American alligator embryos to show that, as in the snapping turtle, NCC migration starts from the midbrain as two cell populations. One is more rostral, invading the mandibular processes, the other more caudal, populating the maxillary processes. A subsequent stream of cells from the prosencephalon provides the mesenchyme of the medial and lateral nasal processes. Ferguson also provided experimental evidence for CNCC contribution to the mandibular processes when he administered 5-fluoro-2′-deoxyuridine (FUDR, which blocks DNA synthesis and so cell division) to alligator embryos to coincide with the migration of either mandibular or maxillary streams of cells. Subsequently, these embryos displayed either greatly reduced lower jaws (depletion of the mandibular stream) or deficient palates (depletion of the maxillary stream).

Turtles: In what is now a classic paper, Meier and Packard (1984) used scanning electron microscopy to visualize migrating CNCCs in the snapping turtle. NCCs first emerge from the midbrain, then as two streams from the rostral and caudal hindbrain. These three streams then fuse to form a continuous mass of cranial mesenchyme.[21]

Evidence that the mandibular and pharyngeal skeletons arise from NCCs and that the skeletogenic NC is regionalized as in other vertebrates (Fig. 7.2) was obtained in a study in which NCCs were extirpated from snapping turtle embryos (Toerien, 1965). Extirpating cranial neural folds rostral to the optic region at the 4–6 somite stage resulted in embryos that lacked Meckel's and quadrate cartilages, evidence of their origin in the NC. Extirpating postoptic neural folds caudal to the level of the second pair of somites resulted in embryos lacking hyoid arch cartilages and the columella (presumptive evidence for their NC origin), and with abnormal tympanic membranes, the latter indicating either a dual origin from NC and mesoderm or a role for NCC in tympanic membrane formation.

The development of cartilages found in the hearts of Spanish terrapin, *Mauremys leprosa*, have been studied and may be NC derived (see Chapter 8), as are the cartilages in the hearts of birds.[22]

TNCCs

We can make some interesting comments about skeletogenic TNCCs in turtles, not with respect to cartilage but concerning bone from TNCCs.

In recent studies, Scott Gilbert has sought the embryological origin of the shell in the red-eared slider turtle, *Trachemys scripta* (see Gilbert *et al.*, 2007 for an overview).

The shell consists of a dorsal carapace and a ventral plastron, each composed of many dermal bones. The ribs and bony elements that form in the dermis form the plastron. The carapace arises from dermal ossifications. Demonstration of a TNCC origin of the bones would add significantly to our knowledge of the evolutionary pathways taken by NCCs during tetrapod evolution.

Although no evidence of a NCC contribution to the carapace or plastron is available from NCC labeling, cells of the plastron cross-reacted with polyspecific antibodies against four proteins found in migrating NCCs in other tetrapods—HNK-1, PdgfRα, FoxD3, and p75. One of the bones of the carapace—the nuchal bone—cross-reacts with HNK-1 and PdgfRα. Cells surrounding the gastralia ('stomach bones') also react with HNK-1. Although none of these four proteins is an exclusive NCC marker, the combination of reactivity in the plastron is suggestive. One difficulty is that in other tetrapods these gene products are expressed in migrating NCCs and expression is lost with differentiation. In turtles we see expression in osteoblasts.

Avian CNCCs

Some of the earliest work on the NC origin of mesenchyme and/or cartilage was undertaken on avian embryos. In studies published in the 1890s, Goronowitsch made one of the first claims for the NC origin of cranial mesenchyme.

³H-thymidine Labeling

Seventy years later, the research programs pursued by Jim Weston and Mac Johnston for their PhD degrees and published in 1963 and 1966, respectively, ushered in an exponentially increasing interest in, and acquisition of knowledge about, the avian NC. These papers, along with those on the amphibian NC by Chibon (discussed above), are benchmark studies.

Johnston labeled a set of chicken embryos with ³H-thymidine, removed cranial neural folds from embryos at stages between head fold and four pairs of somites (24–28 h of incubation) and grafted the labeled neural folds into the equivalent position in nonisotopically labeled embryos of the same age. Host embryos were fixed over the ensuing 8 days and autoradiographs prepared. Johnston: (i) demonstrated that the neural folds contained NCCs; (ii) mapped the migration pathways taken by these CNCCs as they populate the maxillary, mandibular, and frontonasal processes; and (iii) identified ³H-thymidine-labeled cells in much of the developing connective tissue of the head and in some of the developing cartilages in the oldest embryos he examined—Meckel's cartilage of the lower jaw, the cartilage of the cranial base, and the hyoid cartilages.

Quail/Chicken Chimeras

There are limitations to isotopic markers such as ^{3}H-thymidine:

- The only cells labeled are those that are synthesizing DNA when the ^{3}H-thymidine is applied.
- Label is diluted with each wave of DNA synthesis/cell division and can only be followed for comparatively short periods, unless the cell cycle is prolonged and/or proliferation slow, as is so in some species.
- Labeled DNA from necrotic cells may be 'picked up' by unlabeled cells.

Nevertheless, ^{3}H-thymidine remained the best marker until Catherine Le Lièvre from France began to graft Japanese quail NCCs into chicken embryos and used the characteristic pattern of the packing of the heterochromatin within quail nuclei to distinguished quail from chicken cells. It rapidly became clear that quail cells grafted into chicken embryos are incorporated to produce normal, albeit chimeric,[⊕] embryos, and that labeled cells can be identified readily and followed into embryonic and even adult life. The subsequent development of a quail nonchicken perinuclear antigen (QCPN) [23] enhanced visualization of quail cells grafted into chicken hosts, in both whole mounts and serial histological sections.

Quail/chicken chimeras can survive past hatching, although 'spinal cord chimeras', in which a portion of the chicken TNC is replaced by the equivalent quail region, do show some breakdown in tolerance after hatching. Interestingly, the immune response begins in peripheral nerve ganglia (which are derived from the grafted NC), rather than centrally within the spinal cord (Kinutani *et al.*, 1989).

The Chondrogenic CNC

Figure 7.12 shows the extent of the CNC and the boundary between cranial and trunk neural crest in chicken embryos.

Le Lièvre and Le Douarin produced the most detailed and first complete map of the CNC for any species. Another extensive set of data on the craniofacial derivatives of the avian NC comes from the elegant studies carried out by Drew Noden at Cornell University. Initially Noden grafted ^{3}H-thymidine-labeled NCCs into unlabeled host embryonic chickens to analyze the migratory pathways and migratory behavior of CNCCs. Then he switched to quail/chicken chimeras.[24]

CNCC migration in chicken embryos begins from the midbrain at the five-somite stage. These cells form all the NC-derived mesenchyme of the first pharyngeal arch and portion of the mesenchyme of the second. The remaining mesenchyme of the

---

[⊕] Although called chimeras, the term is not strictly correctly applied if a chimera is an organism with large contributions from each species, often half from one species, half from another. These grafted embryos are NC chimeras, not chimerical animals. Nevertheless, the term quail/chicken chimera is usually used as shorthand for a chicken embryo/adult containing quail cells.

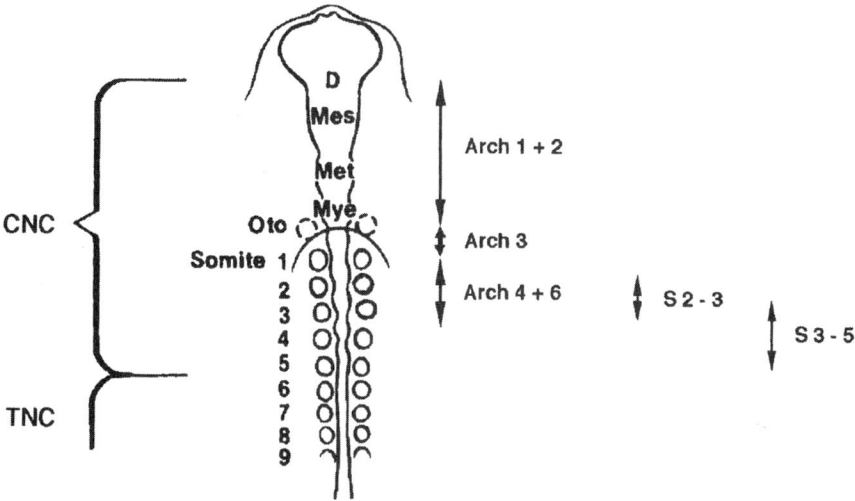

Fig. 7.12 This dorsal view of a chicken embryo shows the location of the CNC (CNC) and the boundary between CNC and TNC. Contribution of neural crest mesenchyme to each of the six visceral arches and to postotic regions adjacent to somites 2–5 is shown. D, diencephalon; Mes, mesencephalon; Met, metencephalon; Mye, myelencephalon; Oto, otocyst. Modified from Nishibatake, Kirby and Van Mierop (1987)

second arch and of pharyngeal arches 3 and 4 is derived from rhombencephalic NC, as are the calcitonin-producing parafollicular cells of the ultimobranchial bodies of the thyroid gland. Noden demonstrated that the pathways of migration are not irreversibly fixed before delamination commences; patterns of migration are normal after regions of the NC are exchanged.

All the cartilages and bones of the cranial and pharyngeal skeletons of chicken embryos are derived from NC, with the exception of the occipitals, sphenoid bones, and basal plate cartilage (which are mesodermal) and the otic capsular cartilage and frontal bones (which receive contributions from NC and mesoderm). Noden extended the fate maps of the craniofacial skeleton produced by Le Lièvre and Le Douarin, mapping the contribution of the NC to the trigeminal[25] and ciliary ganglia (see Figs. 2.14 and 2.19), head dermis, and craniofacial connective tissue. He also determined the embryological origins of the cephalic and cervical musculature and blood vessels and that dorsal iris muscles are NC but ventral iris muscles mesodermal in origin.

Cell Lineages in CNCCs

Clonal culture of migrating CNCCs and of cells from the caudal pharyngeal arches of avian embryos allowed Bronner-Fraser (1987) to reveal unexpected heterogeneity; posterior pharyngeal arches contain four clones of postmigratory NCCs

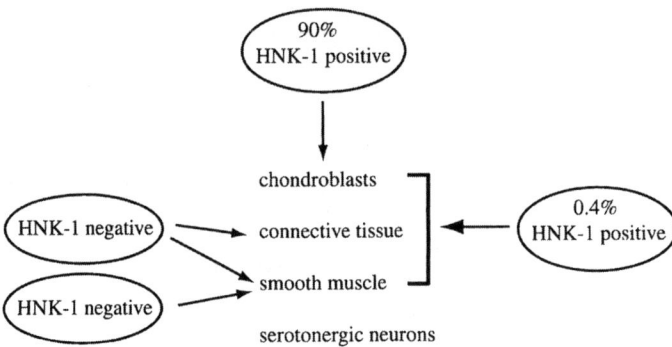

Fig. 7.13 Four lineages of neural crest cells present in the mesencephalon/metencephalon of chicken embryos are revealed following clonal cell culture. The percentages refer to percentages of cells that are HNK-1 positive in the two lineages containing HNK-1-positive cells. The more restricted lineages are HNK-1 negative. Based on data in Ito and Sieber-Blum (1993)

(Fig. 7.13).[26] Pharyngeal arch cell lineages are identified by whether they are HNK-1 positive or negative and by the cells types arising from them in clonal cell culture.

HNK-1 is a marker of migrating NCCs (see Fig. 2.3) *and* is required for epithelial —> mesenchymal transformation and initiation of migration of NCCs, which can be blocked *in vivo* and *in vitro* by perturbing avian NC with an antibody against HNK-1. NCCs accumulate beside the neural tube and ectopically *within* the lumen of the neural tube, migration having been both blocked and redirected medially. Addition of HNK-1 allows these cells to detach from the lumen.

Mesencephalic/metencephalic quail NCCs, cloned on a 3T3 feeder layer for some 7–10 days, display diverse developmental potentials ranging from pluripotency (see Table 1.2) to committed lineages. Ito and Sieber-Blum (1993) identified four clonal lineages (Fig. 7.13):

- a lineage, 90% of whose cells are HNK-1 positive, from which arise connective-tissue fibroblasts, chondroblasts, serotonin-containing neurons and smooth muscle cells (NC-derived smooth muscles are found in the pharyngeal arteries and in the muscles that activate feathers of the head and neck);
- a lineage, only 0.4% of whose cells are HNK-1 positive, from which arise connective tissue, chondroblasts, and smooth muscle, but not neurons;
- an HNK-1-negative lineage that forms connective tissue and smooth muscle; and
- another HNK-1-negative lineage that forms only smooth muscle.[27]

Shigetani *et al.* (1995) used DiI-labeling of postotic NCCs to identify what they call the **circumpharyngeal crest**, which contains a population of cells that emerge from postotic rhombomeres (r5 caudal to the level of the boundary of somites 3–4) and migrate along a dorsolateral pathway before subdividing into populations destined for individual pharyngeal arches (Figs. 7.14 and 7.15). Circumpharyngeal crest

A

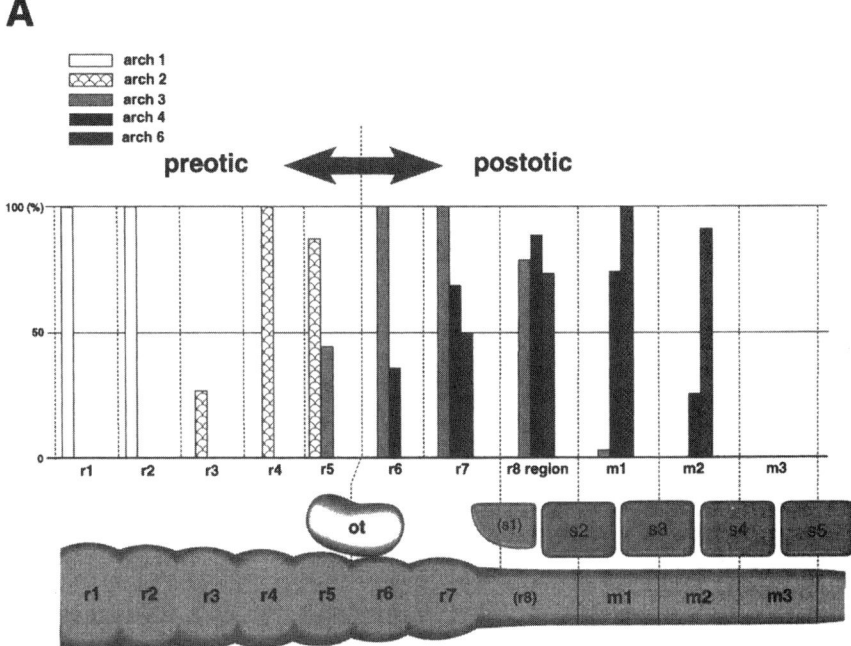

B

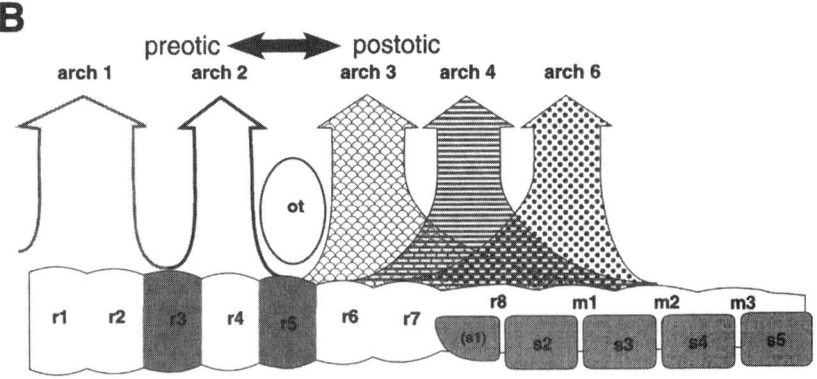

Fig. 7.14 The origin and distribution of pharyngeal arch crest in embryonic chickens as determined from DiI-labeling. (**A**) Pre- and postotic crest (ot, otocyst) is shown with reference to rhombomeres 1–8 (r1–r8), myelomeres 1–3 (m1–m3), and somites 1–5 (s1–s5). The vertical axis shows the percentage of embryos showing labeled cells in each pharyngeal arch. (**B**) A schematic representation of the origin of pharyngeal arch cells along the neural axis to show the registration between rhombomeres and arches in the preotic crest (e.g.,r1 and r2 contribute to arch 1; r4 to arch 2), and the absence of registration in the postotic region (cells from r6 caudal to s3 contribute to arch 3, for example). Reproduced from Shigetani *et al.* (1995), with the permission of Blackwell Scientific Pty. Ltd., from a figure kindly provided by Shigeru Kuratani

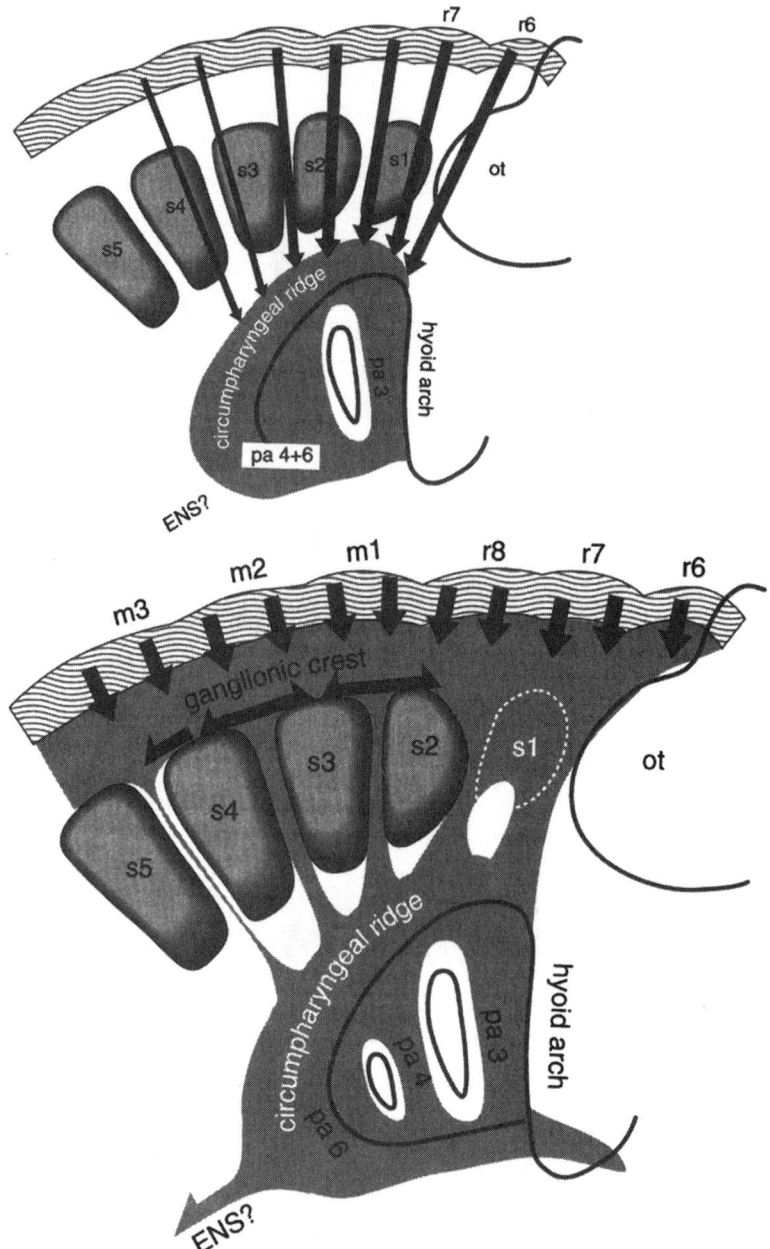

Fig. 7.15 Patterns of early (*top*) and late (*bottom*) migration of the circumpharyngeal crest into pharyngeal arches 3, 4, and 6 (pa 3–pa 6) as deduced from DiI-labeling of premigratory neural crest cells. *Arrows* in the *top figure* show early migrating cells as a single population. Later migrating cells (*short arrows* in the *bottom figure*) form the ganglionic crest. Enteric neuroblasts (ENS?) originate from the postotic population of neural crest cells. m1–m3, myelomeres 1–3; ot, otocyst; r6–r8, rhombomeres 6–8; s1–s5, somites 1–5. Reproduced from Shigetani *et al.* (1995), with the permission of Blackwell Scientific Pty. Ltd., from a figure kindly provided by Shigeru Kuratani

cells in populations adjacent to three or more somites contribute to each pharyngeal arch (Fig. 7.15), meaning that strict segmental identity between neural tube and pharyngeal arches is not maintained. Like Noden, these authors demonstrated that the pattern of migration is intrinsic; it is preserved if cells are transplanted more caudally along the neural tube.

Noden established an important attribute of NCCs, namely, that mesenchyme derived from NC does not mix with mesoderm-derived mesenchyme, with the important consequence that a **neural crest/mesoderm boundary** can be identified in the developing chicken head. Extrapolation of this boundary to other vertebrates, including humans, reinforces the importance of studies on embryological origins, with the underlying but unproven presumption that the NC–mesoderm boundary is conserved across the vertebrates. Coupling Noden's findings with studies on epithelial–mesenchymal interactions undertaken at the same time by Hall and Thorogood led to the conclusion that (i) the fate of NCCs is influenced by their position and (ii) differentiation is elicited by selective interaction with embryonic epithelia (see Box 3.4).

CNCCs and Muscle Patterns

By emphasizing serial evaginations of the pharynx, a metameric paraxial mesoderm (somitomeres), and NC-derived mesenchyme are unique to and present in all vertebrates, Noden showed that NCCs impose patterns onto developing muscles. Subsequently, Koentges and Lumsden (1996) discovered a previously unsuspected compartmentalization of avian rhombomeric NCCs in which muscles attach to NCCs with the same rhombomeric origin, the connective tissue sheaths of the muscles arising from NCCs derived from that level. In a urodele and an anuran (the Mexican axolotl and the oriental fire-bellied toad, respectively), NCC have been shown to 'position' the visceral muscles, the connective tissue of which are also NC in origin.[28]

McGonnell and colleagues (2001) identified a population of caudally migrating CNCCs that arise from the postotic neural tube to migrate along the ventral surface of the somites. These cells come to lie on the clavicle at the site of attachment of the cleidohyoid muscle; that is, CNCCs influence skeletomuscular patterning of an element in the shoulder girdle.

Postotic NCCs have been studied with single-cell resolution in mouse embryos in an extremely detailed analysis of interconnections between NCC-forming skeletal elements and mesodermal cells producing muscles of the neck and shoulder regions. In a study that may force us to rethink how we approach skeletal/muscle origins/interactions, Matsuoka and colleagues (2005) identify what they term 'cryptic cell boundaries' influencing both NC and mesodermal cells. If confirmed, their analysis (i) shifts the NCC–mesodermal boundary to the shoulder girdle, (ii) gives the primacy in patterning to these boundaries, rather than to NCCs, and (iii) places the shoulder girdle developmentally within the 'head'. Even more recently, Rinon *et al.* (2007) demonstrated that, in the absence of CNCCs, myoblasts arise, remain proliferative, and fail to differentiate normally, resulting in altered and abnormal patterning of muscles *and* of the craniofacial skeleton. Clearly, we need to learn more about mesoderm–NCC interactions.[29]

Mapping the Mouse CNC

Much of the interest in the NC from the 1970s onward was prompted by the knowledge that many craniofacial defects and inherited conditions in humans involve tissues known to be NC in origin in nonmammalian vertebrates. Defects involving pigment cells (albinism), the craniofacial skeleton (cleft lip and palate, asymmetrical facial growth, first arch syndromes), adrenal glands (medullary carcinoma), or sympathetic neurons of spinal ganglia (neuroblastomas) are discussed in Chapters 9 and 10. Indeed, so linked are embryos, defects, and the genes underlying the defects that Pierce titled his 1985 paper on the topic, 'Carcinoma is to embryology as mutation is to genetics'.[30]

Patterns of NC origin seen in other vertebrates should also be seen in mammals, including humans; hence the extrapolation of the NC–mesoderm boundary from chickens to humans and the use of nonmammalian species as models for humans. Until a couple of decades ago, we had to rely on the interpretation of static evidence, usually from histological sections, for our understanding of mammalian embryos. In the 1990s, however, there was what amounts to an explosion of publications on the origin, migration, mapping, and differentiative capabilities of mammalian (mostly mouse) NCCs. A synopsis of these mammalian studies follows, with some emphasis on developing techniques that have permitted mammalian NCCs to be mapped.

The technical problems of embryo survival after *ex utero* surgery or in long-term whole embryo cultures may be mitigated by an experimental technique developed by Muneoka and colleagues (1986). Rodent embryos lying within the amniotic cavity are exteriorized onto the abdominal surface, operated on, or injected, and then returned to the abdominal cavity, where they continue to develop *ex utero* for the remainder of the gestation period. Serbedzija *et al.* (1992) used this technique in combination with DiI labeling and *in vitro* culture to determine the timing of the onset of CNCC migration from murine neural tubes. Migration was initiated from the rostral hindbrain at the 11-somite stage, from mid- and caudal hindbrain at 14 somites, and from the forebrain at 16 somites (see Table 2.1). This pattern, whereby migration does not begin from the rostral forebrain and spreads caudally but rather begins more caudally, is typical of nonmammalian vertebrates.

Tan and Morriss-Kay, and Smits-van Prooije and colleagues, labeled embryos with ^{3}H-thymidine and/or wheat germ agglutinin conjugated to gold particles, removed the neural folds and grafted them into the equivalent position in an unlabeled embryo that was allowed to develop in culture for up to 3 days (the maximum for *in vitro* survival of whole-rodent embryos of these ages). Using immunohistochemistry to visualize the wheat germ agglutinin-labeled cells, these authors describe the migration routes and final locations of CNCCs. NCCs in rodents migrate well below the superficial ectoderm; NCCs in avian embryos primarily migrate subectodermally. The NCCs did not originate from the forebrain but did arise from midbrain caudally to the postotic caudal hindbrain, confirming the cytological studies of sectioned embryos discussed earlier.[31]

Because cultured whole embryos do not survive long enough to follow the fate of labeled cells, Tan and Morriss-Kay (1986) removed mandibular arches or other regions from cultured embryos and grafted them into the anterior chamber of the eyes of adult rats, where they continued to develop and differentiate. In this way, the effective life of labeled NCCs was extended greatly. Labeled cartilage developed in such grafts, confirming its origin from the NC.

Tam and his colleagues have investigated whether NC and cranial paraxial mesoderm are co-distributed in mouse embryos. They use a combination of NCC grafting—labeling with a fluorescent dye (DiI, DiO; Fig. 7.16) or with wheat germ agglutinin conjugated to gold particles, Fig. 7.17)—followed by 48 h of *in vitro* culture. Their experiments show that mesoderm and NC arising at the same axial levels have common destinations and share similar patterns of regionalization. Hence, mesoderm from somitomeres I, III, IV, and VI contributes to the same craniofacial tissues as does NC adjacent to these somitomeres (forebrain, caudal midbrain, and rostral–caudal hindbrain), indicative of global segmental patterning as seen in birds (Figs. 7.18 and 7.19[Color Plate 7]). Using this method of labeling followed by organ culture, Trainor and colleagues (2002) identified three streams of NCCs migrating from the hindbrain (as we saw is the case in other vertebrates), far fewer NCCs emerging from r3 and r5, and demonstrated that NCC could migrate from both odd- and even-numbered rhombomeres.[32]

The subsequent behavior of mesodermal- and NC-derived mesenchyme differs in craniofacial and pharyngeal arch regions. Mesenchyme from both sources mixes extensively in the periocular, periotic, and cervical regions but segregates within the pharyngeal arches, the NC-derived mesenchyme toward the periphery, and the

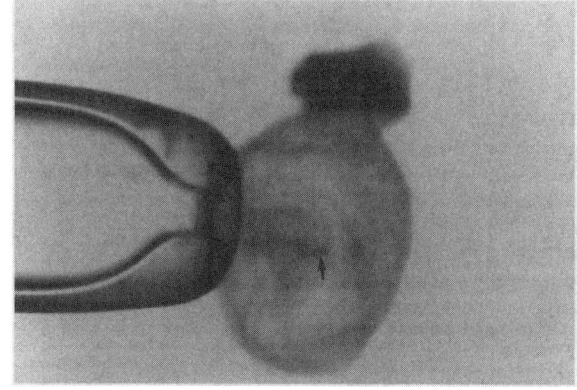

Fig. 7.16 An 8.5-day-old (5-somite-stage) mouse embryo supported on the end of a micropipette. DiI was injected into the rostral hindbrain (*arrow*). Reproduced from Trainor and Tam (1995), with the permission of Company of Biologists Ltd., from a figure kindly provided by Patrick Tam

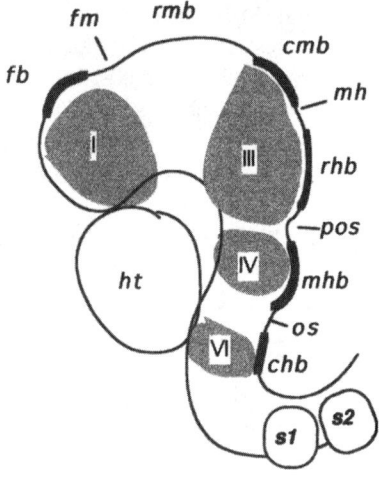

Fig. 7.17 The cephalic region of an 8.5-day-old (5-somite-stage) mouse embryo showing the location of somitomeres I, III, IV, and VI (*gray*) and the regions of the developing brain that correspond in position to those somitomeres. *Thick lines* indicate labeled regions of neural crest whose fate is shown in Fig. 7.20. The brain regions, from rostral to caudal, are fb, forebrain; fm, forebrain–midbrain junction; rmb, rostral midbrain; cmb, caudal midbrain; mh, midbrain–hindbrain junction; rhb, rostral hindbrain; pos, preotic sulcus; mhb, middle hindbrain; os, otic sulcus and chb, caudal hindbrain. ht, heart; s1 and s2, somites 1 and 2. Reproduced from Trainor and Tam (1995), with the permission of Company of Biologists Ltd., from a figure kindly provided by Patrick Tam

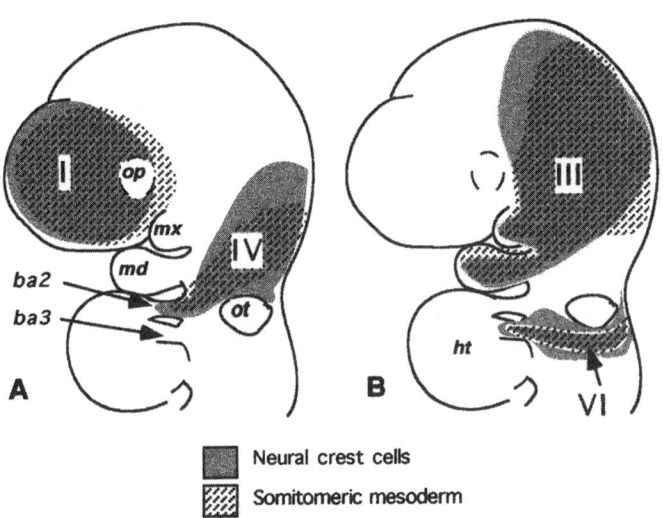

Fig. 7.18 The distribution of CNCCs and somitomeric mesoderm as seen in a 10.5-day-old embryo labeled at 8.5 days (see Fig. 7.19) and cultured for 2 days. (**A**) The distribution of neural crest cells from forebrain and middle midbrain correlates closely with the distribution of cells from somitomeres I and IV. (**B**) The distribution of neural crest cells from caudal midbrain and from rostral and caudal hindbrain correlates closely with the distribution of cells from somitomeres III and VI. ba 2, ba 3, second and third branchial arches; ht, heart; md, mandibular process; max, maxillary prominence; op, optic vesicle; ot, otic vesicle. Reproduced from Trainor and Tam (1995), with the permission of Company of Biologists Ltd., from a figure kindly provided by Patrick Tam

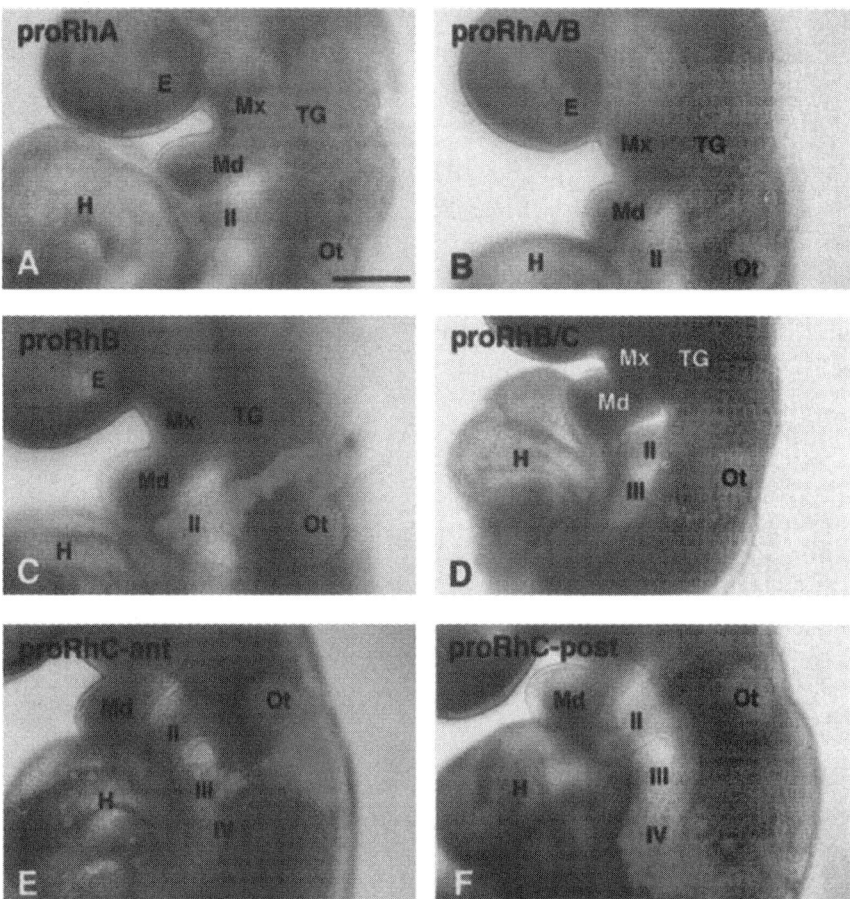

Fig. 7.19 The segmental migration of neural crest cells (*red*) into individual pharyngeal arches as demonstrated following injection of DiI either into pro-rhombomeres **A, B,** or **C** of mouse hindbrains (seen in **A, C, E,** and **F**) or into the boundary between rhombomeres **A/B** or **B/C** (shown in **B** and **D**). II–IV, second to fourth pharyngeal arches; (**E**) eye primordium; (**H**) heart; Md, mandibular prominence; Mx maxillary prominence; Ot, otic vesicle; TG, trigeminal ganglion. Bar = 200 μm. Reproduced from Osumi-Yamashita *et al.* (1996), with the permission of Blackwell Science Pty. Ltd., from a figure kindly provided by N. Osumi-Yamashita (see Color Plate 7)

mesodermal mesenchyme toward the center (as was previously shown for avian embryos). The timing of these events was laid out in a pair of papers by Chan and Tam.

Chan and Tam (1986) removed the neural plates or neural folds with their resident populations of NCCs from 0- to 4-somite (7.5- to 9.5-day-old) rat embryos and grafted them under the kidney capsule. Cartilage, bone, and hair follicles developed in these renal grafts. The capacity to form skeletal tissues was lost at the 5-somite stage, reflecting either prior delamination of the skeletogenic cells or lack of an appropriate epithelial environment to elicit differentiation. The latter is more likely, especially given how murine NCCs disperse into the first pharyngeal arch. These NCCs migrate superficially, rather than more deeply, retaining an association with the overlying epithelium. Nichols (1986) noted that this affinity of epithelial and NCCs immediately preceded the onset of the epithelial–mesenchymal interaction, which Hall (1980) had shown to be a prerequisite in mice for the initiation of mandibular chondrogenesis and osteogenesis in mice.

The NC contributes mesenchyme throughout the mouse head. Chan and Tam (1988) followed the migration of midbrain NC from their initial delamination through breaks in the basement membrane. Midbrain cells were traced into craniofacial and periocular mesenchyme, mandibular and maxillary arches, and trigeminal ganglia, all comparable locations to those seen in avian and amphibian embryos. A rostral–caudal regionalization of the CNC was evident:

- Cells of the maxillary processes originate from midbrain and rostral hindbrain.
- Cells of the mandibular processes originate from rostral hindbrain and preotic caudal hindbrain.
- Cells of the hyoid arch arise from pre- and postotic caudal hindbrain.

Injecting DiI into fore- or midbrain murine crest cells allowed Osumi-Yamashita et al. (1994) to demonstrate that the rostral edge of the prosencephalon gives rise to the head ectoderm, including the nasal placode, Rathke's pouch, and the oral epithelium, as it does in chicken embryos (Fig. 7.19 [color]). The lateral edges of the prosencephalon provide the NC-derived mesenchyme of the frontonasal mass, while the midbrain provides NC mesenchyme to the first pharyngeal arch.

Osumi-Yamashita et al. (1996) adapted their injection methods to analyze the contributions made by NCCs to later stages of murine embryogenesis. They labeled specific populations of NCCs with fluorescent dyes and cultured whole embryos for as long as possible, after which regions such as the mandibular arches were established in organ culture. In this way they followed CNC into developing teeth as late as the cap stage (see Chapter 8). Focal injections of DiI into defined regions of mouse neural tubes was used to great advantage to produce fine-grained maps of the NC. By injecting DiI into pro-rhombomeres of the hindbrain, NCC lineages and the segmental contribution of hindbrain lineages to the pharyngeal arches were identified (Figs. 7.16 and 7.19 [Color Plate 7]). Injecting DiI into pre- and postotic NC demonstrated that only even-numbered rhombomeres in the preotic crest produce NCCs. They found it difficult to recognize even- and odd-numbered rhombomeres in the postotic NC, which coincides with the head–trunk interface and the transition from cranial to TNCCs.[33]

The development of a two-component (Cre/lox) genetic system using two mouse lines, one in which Cre recombinase is expressed under the control of the *Wnt1* promoter as a transgene, the other using the ubiquitous ROSA26 conditional reporter as a substrate, restricts expression of the *Wnt1* transgene to migrating NCCs. Using this approach, Chai et al. (2000) followed CNCC migration and craniofacial development during early embryonic development. A summary of the distribution of CNCCs between 9.5 and 13.5 days of gestation is outlined in Table 7.4. CNCCs populate the craniofacial processes, mandibular and maxillary arches, nerve ganglia, and Meckel's cartilage; contributions to the teeth and bone are discussed in Chapter 8. Surprisingly, at 13.5 days of gestation they found non-NC-derived cells associated with Meckel's cartilage. At 17.5 days these cells were localized to the posterior portion of Meckel's cartilage, the region undergoing ossification. Two possibilities for the nature of these cells are that they are (i) osteogenic cells brought in via the vascular system or (ii) degenerating chondrocytes. Either cell type could be NC in origin but no longer expressing the marker, or from a non-NC source.

In another study using *Wnt1-Cre/R26R* transgenic mice, Jiang *et al.* (2002) explored the origins of the skull in considerable depth, with special emphasis on the coronal and sagittal sutures, major sites of bone growth. The coronal suture unites the frontal and parietal bones, the sagittal suture unites left and right parietal bones. In chicken embryos, these bones occupy a region of transition from NC to mesodermal origin. In the transgenic mouse embryos, they were also found to lie at the interface between NC-derived (frontal) and mesoderm-derived (parietal) bones of the skull, with NCCs forming the intervening mesenchyme. The coronal suture is laid out at 9.5 days of gestation as the caudal boundary of the frontonasal mesenchyme (cf. Table 7.4). With growth of the brain and expansion of the associated osteogenic mesenchyme, a layer of NCCs forms the covering of the brain, the meninges. Reduced development of the *NC*-derived meninges in retinoic acid-treated embryos inhibits ossification of the *mesoderm* parietal bones (consistent with an inductive interaction) but does not inhibit ossification of the frontal bone (con-

Table 7.4 Localization of CNCCs and their derivatives in the craniofacial region in *Wnt1-Cre/R26R* transgenic mice between 9 and 13.5 days of gestation[a]

Day of gestation	Theiler stage	Number of pairs of somites	Labeled cells/tissues
9.5	15	21–29	frontonasal processes first and second pharyngeal arches spinal dorsal root ganglia
10.5	17	35–39	frontonasal, mandibular, and maxillary processes second pharyngeal arch, third arch primordium surrounding the olfactory pit ganglia of the trigeminal, facial, glossopharyngeal, and vagus nerves
12.5	20.5		Meckel's cartilage condensation
13.5	21.5		non-NC cells also within Meckel's cartilage

[a]Based on Chai *et al.* (2000). See also Table 2.1 for stages of NCC delamination/migration in relation to stages of development as determined from histological analysis.

sistent with autonomous development of the frontal or that an inductive interaction occurred earlier in development).

Notes

1. Much of this evidence is discussed in detail in the first edition of this book, which should be consulted for a fuller treatment. This chapter summarizes the older evidence and updates it with more recent studies.
2. For discussions of lamprey and hagfish relationships, see Hall and Hanken (1985), Maisey (1986), Langille and Hall (1988b), Forey (1995), Janvier (1996, 2007), Kuratani et al. (1998, 2001, 2002), Donoghue (2002), and Kuratani (2004, 2005).
3. See Wright et al. (2001*), Cole and Hall (2004a,b*), and Hall (2005b*, 2007a) for overviews on agnathan and invertebrate cartilages.
4. Seventy-four hydroxyprolines/1,000 residues in type II collagen compares with 1–22 in North American sea lamprey cartilages, the number of amino acid/1,000 residues depending on the cartilage; 1 in the annular cartilage, 3 in the neurocranium, 17 in the pericardial, and 22 in the branchial basket.
5. See Wright et al. (2001*) for lamprey cartilages, and Cole and Hall (2004a,b*) and Hall (2005b*, 2007a) for invertebrate cartilages.
6. See Ogasawara et al. (2000), Myojin et al. (2001), Murakami et al. (2001), and McCauley and Bronner-Fraser (2006) for the lamprey genes. Neidert et al. (2001) clones four Dlx genes from North American sea lamprey, Petromyzon marinus, and visualized their differential expression in forebrain, CNC, pharyngeal arches and placodal ectoderm.
7. See Harrison (1929), Stone (1932), and studies by Twitty discussed in Hall (1999a,b*) on the determination of the size of organ primordia; and Yntema (1955*), Holtfreter (1968), and Hall (2005b*) for literature on nasal capsule size.
8. For discussions of intrinsic patterning of the NC, see Lumsden (1988) and Hall and Hörstadius (1988*). See Thorogood (1993*) for an insightful analysis of the (not necessarily congruent) studies on regionalization of the skeletogenic NC in amphibians and birds. See Shimamura et al. (1995) for an overview of various theories seeking to explain the rostrocaudal organization of the neural tube, including those of Wilhelm His.
9. We have evidence for inductive interaction between pharyngeal endoderm and pharyngeal-arch-destined mesenchyme for more species of urodele and anuran amphibians than for any other vertebrate group. Urodeles studied include the alpine newt, crested newt, Mexican axolotl, North American spotted salamander, and Spanish ribbed newt. Anurans include the edible frog, painted frog, South African clawed-toed toad, tiger salamander, and the wood frog.
10. See Holtfreter (1968), Corsin (1975), Hall (1980, 2000a,c, 2005b*), Graveson and Armstrong (1996), Seufert and Hall (1990), and Epperlein et al. (2000a, 2007a*) for the amphibian studies and for reviews.
11. See Barlow and Northcutt (1997) and Northcutt and Barlow (1998*) for studies on the development of amphibian taste buds, and Barlow (1999*, 2001) for reviews.
12. See Harlow and Barlow (2007) for these two types of neurons. Another group of NC cells that may be involved in sensing touch, although this is unproven, are Merkel cells, which are found in the dermis of birds and in the epidermis of other vertebrates, often associated with sensory nerve endings (Grim and Halata, 2000). An association with APUD cells (Chapter 9) and a neuroendocrine function has also been suggested.
13. See Seufert and Hall (1990) for studies with Xenopus. The rostral cartilages in Xenopus are homologous to the suprarostrals in other taxa, although some homologize the rostral cartilages to the ethmoid trabecular plate. Endochondral bones are poorly developed in urodeles and develop immediately before metamorphosis.

14. See Chibon (1964, 1974*) for the transplantation studies and Hall (1987, 1999a*) and Thorogood (1993*) for discussions of them.

15. See Krotoski *et al.* (1988) for crest cell migration in *Xenopus* spp., Olsson and Hanken (1996), Hanken *et al.* (1997), Olsson *et al.* (2002), and Hall (2003b) for patterns of migration of NCCs in coqui and comparisons with other amphibians, and Kerney *et al.* (2007) for the less dramatic loss in the direct-developing Sri Lankan frog, *Philautus silus* and for an overview of larval skeletal loss with direct development.

16. See Fang and Elinson (1996) for patterns of *Dlx* gene expression in *E. coqui*, and Elinson (1990) and Callery *et al.* (2001) for the study of direct-developing frogs for understanding the evolution of ontogeny.

17. See de Beer (1947) and Hall (1999a*) for the pioneering studies by Kastschenko and Goronowitsch.

18. The pharyngeal skeleton in the swordtail also has been shown to be NC in origin (Sadaghiani *et al.*, 1994*).

19. For the initial reports on targeted mutagenesis of the zebrafish genome, see Neuhass *et al.* (1996), Piotrowski *et al.* (1996), Schilling *et al.* (1996b), and Schilling (1997). See Schilling *et al.* (1996b), Piotrowski and Nüsslein-Volhard (2000), and Yelick and Schilling (2002) for jaw and pharyngeal arch mutants, Malicki *et al.* (1996) for mutations affecting ear development, and Kelsh *et al.* (1996) for mutations affecting pigmentation.

20. See Ellies *et al.* (1997) and Sperber *et al.* (2008) for *Dlx1a* and *Dlx2a* in zebrafish CNCCs and pharyngeal chondrogenesis.

21. See Hall and Hallgrímsson (2008) for a discussion of the phylogeny of avian and nonavian reptiles and of how the term reptiles (or Reptilia) is used nowadays.

22. See López *et al.* (2000, 2003) for cardiac cartilages and the likely NC origin.

23. QCPN is available from the Developmental Studies Hybridoma Bank, University of Iowa, Iowa City, IA. 52242 (http://dshb.biology.uiowa.edu/).

24. For these pioneering studies mapping the avian NC, see Le Lièvre (1971a,b, 1974, 1976, 1978), Le Douarin (1974*, 1982*, 1988), Le Lièvre and Le Douarin (1974, 1975), and Le Douarin *et al.* (1993). For pathways of migration of NCCs in avian embryos, see Noden (1983a*, 1984a*).

25. After NCCs have migrated, the ventral neural tube also contributes cells to the trigeminal ganglion, while neural tube–ectoderm interactions are required for placode formation (Noden, 1984a*).

26. See also Ito and Sieber-Blum (1993) and Ito *et al.* (1993) for clonal analyses.

27. See Baroffio *et al.* (1988), Ito and Sieber-Blum (1993), and Ekanayake and Hall (1994) for clonal culture of NCCs.

28. See Ericsson *et al.* (2004) and Olsson *et al.* (2001) for the studies on the axolotl and anuran, respectively. Olsson and colleagues refer this patterning role of NC-derived connective tissue as 'cryptic segmentation'.

29. For segregation of individual muscles and the patterning of muscles by NC-derived connective tissue, see Noden (1983a,b, 1984b, 1987). For detailed fate mapping of avian cephalic mesoderm, see Noden (1987*), Couly *et al.* (1992, 1993), and Huang *et al.* (1997). The claim by Matsuoka *et al.* (2005) that post-otic NCCs form primary endochondral bone associated with the clavicle is complicated by the formation of secondary cartilage on the mouse clavicle and its subsequent (but secondary) endochondral ossification. They use this evidence to conclude that intramembranous and endochondral patterns of ossification cannot be simply equated with NC and mesodermal origin, respectively. This view of the dichotomy is too broad; Meckel's cartilage, which is undoubtedly of NC origin, undergoes endochondral ossification in many vertebrates.

30. For NC and craniofacial defects, see Opitz and Gorlin (1988), Pierce (1985), Cohen and Baum (1997), and Chapter 10 herein.

31. Whether the forebrain is segmented remains controversial; see Northcutt (1995), N. D. Holland *et al.* (1996), and Creuzet *et al.* (2006) for reviews. See Tan and Morriss-Kay

(1986*), Smits-van Prooije *et al.* (1988*), Tam (1989), and Chan and Tam (1988) and Osumi-Yamashita and Eto (1990) for pioneering studies using labeled rodent NC.

32. See Tam (1989) and Trainor and Tam (1995) for mesoderm and NC patterning.

33. Ferguson and Graham (2004) examined the head–trunk interface in chicken and murine embryos using as their criteria: coincidence of the occipital–cervical boundary with somite 5/6; formation of transient DRG opposite somite 4/5; and absence of *Hoxb3* expression and presence of CNC differentiative capabilities rostral to somite levels 3/4. Furthermore, and as discussed in Chapter 10, the most rostral extent of expression of *Hoxd4* is at the first cervical somite. Expressing *Hoxd4* more rostrally, results in homeotic transformation of the occipital bones into cervical vertebrae.

Chapter 8
Teeth and Hearts: The Odontogenic and Cardiac Neural Crests

Teeth

Remember that teeth are composite organs constructed from two mineralized tissues; **enamel**, which is deposited by ameloblasts that are epithelial in origin, and **dentine**, which is deposited by odontoblasts that originate from the neural crest (NC). Enamel and dentine are deposited and mineralized by ameloblasts and odontoblasts, respectively. Additional cell and tissue types in mammalian teeth, such as pulp, cementum, the periodontal ligament, and alveolar bone are also NC in origin.[1]

The following points concerning the NC origin of odontoblasts and dentine have been made so far:

- In studies published between 1888 and 1893, Kastschenko and Goronowitsch concluded that dentine of the teeth of sharks and of the common carp originate from NC.
- Platt (between 1893 and 1897) and de Beer (1947) demonstrated that the odontoblasts in the mudpuppy and Mexican axolotl arise from neural crest cells (NCCs).
- Clonal cell lines from melanophore tumors obtained from goldfish and from the Nibe croaker form odontoblasts.
- Bmp2 and Bmp4 play important roles in tooth development.
- *Patch* mutant mice display tooth defects because of defective glycosaminoglycans in the ECM, reflecting a deletion in the gene encoding PdgfRα.

The following points concerning the evolutionary history of dentine and/or of teeth have been made so far:

- The developmental loss of *Bmp4* was a factor in the evolutionary loss of teeth in toothless teleost fish (see Box 3.5).
- Loss of ectodermal competence is associated with the evolutionary loss of teeth in birds.

B.K. Hall, *The Neural Crest and Neural Crest Cells in Vertebrate Development and Evolution*, DOI 10.1007/978-0-387-09846-3_8,

The Odontogenic Neural Crest

Although it has long been known that amphibian teeth are derivatives of the NC, and although there is circumstantial evidence for a NC origin for fish teeth, the first experimental analysis of the NC origin of mammalian teeth was not undertaken until Lumsden's 1988 report. The technique was to excise cranial neural crest (CNC) (or trunk neural crest (TNC); see below) from 6- to 12-somite-stage mouse embryos (the early 9th day of gestation, when the neural folds are elevating and beginning to converge; see Table 2.1), recombine the NC with epithelium from the mandibular arches or limb buds from 9- or 10-day-old embryos, and graft these epithelial–NC recombinations into the anterior chamber of the eye. Lumsden determined that:

- CNC grafted alone forms cartilage.
- CNC grafted with mandibular or limb bud epithelia forms cartilage, as well as perichondrial and membrane bone.
- CNC grafted with mandibular epithelium forms teeth.

The teeth were comprised of dentine and enamel deposited by odontoblasts and ameloblasts, respectively, demonstrating the differentiation of epithelial and NCCs. Cartilage, bone and dentine development in these grafts demonstrates the NC origin of these three tissue types and provides indirect information on the requirement for epithelial–mesenchymal interactions in the differentiation of bone and dentine (see Box 3. for bone and Box 8.1 for teeth). Similar conclusions can be drawn from the presence of cartilage, bone, and dentine—the latter associated with enamel to form easily recognizable teeth—in teratomas that form after rodent embryonic ectoderm is grafted under the kidney capsule or that arise spontaneously in humans, as in the extraordinary example shown in Fig. 8.1[Color Plate 7].[2]

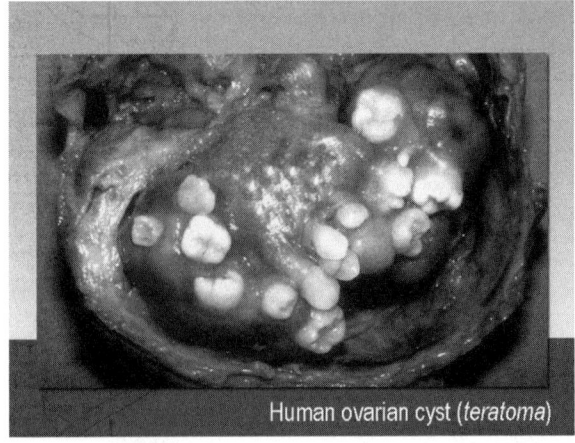

Fig. 8.1 Extraordinarily complete teeth, the largest of which are 10 mm across and molariform with multiple cusps. These teeth developed in a teratoma in a human ovary. Image supplied by Paul Sharpe courtesy of the Anatomy Museum, Guy's Hospital, London (see Color Plate 7)

Human ovarian cyst (*teratoma*)

Box 8.1 Reciprocal interactions and tooth development

One of the first indications that teeth arise following interactions between epithelial and NC-derived mesenchymal cells came from transplantation studies between amphibian embryos undertaken by Holtfreter and then by Sellman in the 1930s and 1940s. Two clear results from these studies were that teeth only form if endoderm is transplanted along with the NCCs and that removing endoderm is sufficient to prevent tooth formation *in vivo*.

More recently, because of their relevance to human tooth development, abnormalities, repair, and/or replacement, attention has turned to mammalian teeth: one example of an abnormality is the formation of supernumerary teeth in the autosomal-dominant skeletal syndrome, *Cleidocranial dysplasia,* which results from a mutation in *RUNX2*, a gene with a major role in regulating bone and tooth formation (see Box 10.3). As most completely understood in mouse embryos, mammalian teeth develop as the result of an extensive series of precisely timed, coordinated, and reciprocal epithelial–mesenchymal interactions that regulate cell division and cell differentiation in enamel-forming epithelial and dentine-forming mesenchymal components.

NCCs migrate to the first and second pharyngeal arches. Subsequently, as the mouth cavity begins to form, these NC-derived mesenchymal cells interact with the buccal epithelium to form dental papillae, within which odontoblasts differentiate. The first of a series of reciprocal interactions between mesenchyme and epithelium initiates the epithelial enamel organs, within which ameloblasts differentiate. Subsequent interactions between ameloblasts and odontoblasts direct further differentiation, the deposition of pre-enamel and pre-dentine and the mineralization of these organic matrices as enamel and dentine. The essential interactions[a] in this cascade (with inductions shown as —>) are:

- dental mesenchyme —> oral epithelium to proliferate
- oral epithelium —> dental mesenchyme to proliferate to form a dental papilla
- the dental papilla —> oral epithelium to form an enamel organ containing preameloblasts
- preameloblasts —> dental papilla cells to differentiate as preodontoblasts, which transform into odontoblasts
- preodontoblasts —> preameloblasts of the enamel organ to differentiate into ameloblasts, which synthesize and deposit pre-enamel
- ameloblasts —> odontoblasts to synthesize and deposit pre-dentine
- the organic matrices of pre-enamel and pre-dentine mineralize to form enamel and dentine.

Maintaining the specificity of murine teeth—whether incisors (incisiform) or molars (molariform)—is controlled by NC-derived odontoblasts late in embryonic development. However, early in development, tooth type is specified by the epithelium component, specifically by the site along the mandibular arch from which the epithelium is derived. Furthermore, as discussed in this chapter, mandibular epithelium can evoke teeth from the most rostral murine TNC if the two tissues are brought into association.[b]

---

[a] The molecular basis of tooth induction, which is now quite well understood, is outside the scope of this review; for summaries see Järvinen *et al.* (2006), Mitsiadis and Smith (2006), Kettunen *et al.* (2007), and Pummila *et al.* (2007).

[b] See Lumsden (1988), Imai *et al.* (1996), and Hall (2005b*) for the origin of odontoblast precursors in the NC and for patterning of teeth by oral ectoderm.

Making use of the two-component Cre/lox genetic system introduced in Chapter 7, Chai *et al.* (2000) followed the migration and differentiation of CNCCs from 9.5-day-old embryos to 6-week-old adults. Dental mesenchyme of the dental papilla, derivatives of the dental papilla (odontoblasts, dentine matrix, pulp, cementum, and fibroblasts of the periodontal ligament), along with mandibular and Meckelian chondrocytes, and the articular disc of the temporomandibular joint, all arise from CNCCs. To their surprise, and as summarized in Box 6.3, apparent non-neural-crest cells contributed to the dental papilla and to Meckel's cartilage (see Chapter 7 for the latter).

Lumsden's study went considerably further than showing that mandibular epithelium evokes tooth formation from CNC-derived odontogenic mesenchyme. He showed that mandibular epithelium can evoke teeth from the most **rostral murine TNC**, provided that the two tissues are brought into association so that they can interact.[3]

Teeth but not Cartilage from Trunk NCCs

As initially determined by Chibon in his experiments on the NC of the Spanish ribbed newt, and as extended by studies in other urodele amphibians, anurans, and birds, the NC consists of a *chondrogenic cranial* and a *nonchondrogenic trunk* neural crest (see Chapter 7). This difference in chondrogenic potential has been extrapolated to bone and dentine (teeth) with the conclusion that the **CNC is skeletogenic and the TNC nonskeletogenic**.

In the classic study using chicken embryos, TNCCs grafted in place of CNC migrated with CNCCs and formed 'mesectodermal' derivatives (connective tissue, dermis, and muscle) but neither cartilage nor bone (Nakamura and Ayer-Le Lièvre, 1982). TNC was defined on the basis of the fate maps of chondrogenic and nonchondrogenic NC, but chondrogenesis does not equate with skeletogenesis.

A similar approach with axolotl embryos is inappropriate because transplanted TNCCs do not migrate away from the CNC. A different approach has to be used. In a classic study with axolotl embryos, the most rostral TNC (defined as trunk on the basis of the fate map of the chondrogenic NC) was associated with embryonic pharyngeal endoderm (the *in vivo* inducer of teeth in the axolotl) and shown to form teeth but not cartilage (Fig. 8.2, and see Fig. 7.6). At the very least, a fate map of chondrogenic NC is not a map of odontogenic NC, and cells identified as the most rostral TNC form teeth.[4] The same is true for the most rostral TNC of mouse embryos.

Lumsden (1988) found that cells derived from the most rostral TNC of mouse embryos participate in tooth formation when combined with mandibular epithelium. TNCCs formed the dentine; mandibular epithelial cells formed the enamel. Each tooth developed in association with a follicle, periodontal ligament, and alveolar bone (bone of attachment), although no other bone and no cartilage develops. Lumsden concluded that the bone of attachment develops from the dental papilla, a finding supported by studies in which tooth germs from older embryos are allowed to develop ectopically.[5]

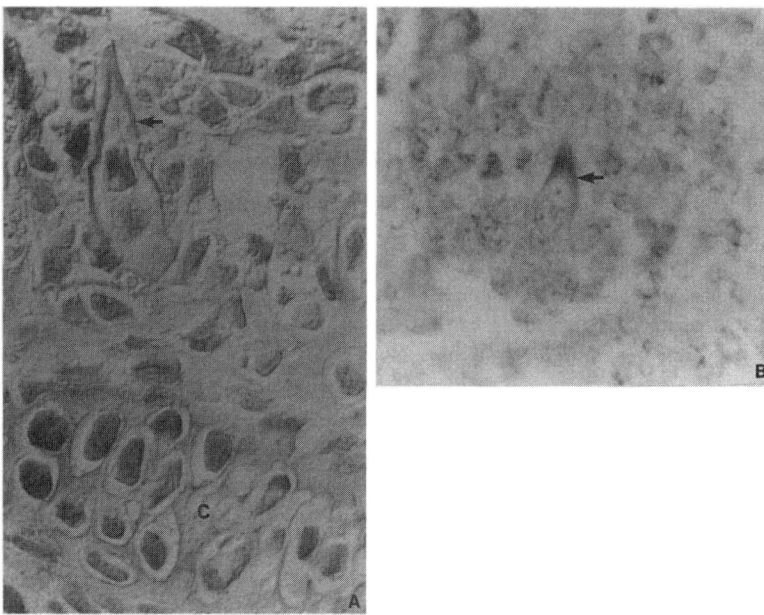

Fig. 8.2 (**A**) A tooth (*arrow*) is shown adjacent to Meckel's cartilage (**C**) from the *lower* jaw of the Mexican axolotl in a figure kindly provided by Moya Smith. The tooth and Meckel's cartilage are derived from CNC. (**B**) Culturing the rostral TNC with embryonic endoderm elicits development of teeth (*arrow*) but not cartilage. See text for details

Cartilage from TNCCs?

McGonnell and Graham (2002) used two approaches to investigate whether TNCCs can form cartilage.

In the first, TNC adjacent to future cervical and thoracic somites in H.H. stages 10–15 chicken embryos (10–24 somites) was established in organ culture. In the other, Japanese quail NCC were dispersed, reaggregated, and implanted into the first pharyngeal arch of chicken embryos. Development of chondroblasts and osteoblasts was reported to occur from the TNCCs maintained *in vitro*, and the quail cells were reported to contribute to chicken skeletal elements *in ovo*.

Although they interpreted these results as indicating that avian TNC is skeletogenic, they utilized only the most rostral TNCCs. Their conclusion, while it pertains to these rostral cells (and parallels the work on axolotls and mice), does not establish skeletogenic potential for more caudal avian TNCCs. Indeed, Lwigale and colleagues (2004), who sought evidence for the skeletogenic ability of chicken and Japanese quail NCCs in relation to their position along the rostral–caudal axis, found that CarNCCs produced little cartilage and TNCCs none. Ability to form cartilage and sensory neurons was rostrocaudally limited, but not by *Hox* genes.

Very recently, Calloni *et al.* (2007) reported chondrogenesis from TNCCs isolated from 20- to 25-somite-stage quail embryos and maintained in clonal culture in the presence of *Shh*. The incidence of chondrogenesis in the presence of *Shh* was 0.25% and 10% from TNCCs and CNCCs, respectively. Incidence in the absence of *Shh* was estimated as 0.04% and 4%, respectively. The evidence for chondrogenesis is compelling, although it is not clear from their methodology whether CarNC was included in the region removed, an inclusion that could bring the results into line with those reported by Lwigale *et al.* (2004). This may be moot, if, as the authors suggest, the cells are multipotential NCCs with neural and mesenchymal potential, and lie upstream of any NCC progenitors previously isolated.

A recent study raises the possibility that murine TNCCs *may* have the potential to form chondroblasts, but in the past we had not have found the right conditions to evoke that potential. It is being said so because Ido and Ito (2006) obtained chondrogenesis after they isolated TNCCs from the region adjacent to the *last* six somites of mouse embryos with 21–29 somites (9.5 days gestation).

Their approach was to culture the TNCCs in the presence of Fgf2, which is known to alter expression patterns of *Hox9* genes and to influence *Id2*, a CNCC marker and a negative regulator of basic helix–loop–helix gene products. Either a latent chondrogenic potential in TNCCs was revealed or the Fgf2 may have converted TNCCs into stem cells; another member of this helix–loop–helix gene family, *Id3*, a c-Myc target (see Fig. 2.4), is required for NCC formation in *Xenopus*; embryos in which *Id3* has been knocked out using a morpholino lack CNCCs—dorsal neural tube cells are switched from a NCC fate to CNS precursors—while overexpression prolongs the time NCCs spend before initiating differentiation (Light *et al.*, 2005). See the last section of Chapter 10 for a discussion of NCCs as stem cells.

Origination of Dentine and Bone

Why is the most rostral TNC odontogenic? An evolutionary approach sheds light on this question.

In their scenario for the origin of the NC-based skeleton, Northcutt and Gans (1983) and Gans and Northcutt (1983) presented arguments for cartilaginous capsules around the sense organs and for a pharyngeal skeleton as the primary **endoskeletal** tissue.

As we saw from evidence documented in Chapter 4, Cambrian cephalochordates reveal this stage of prevertebrate evolution. If modern cephalochordates have retained the ancestral condition, however, the cartilages are mesodermal and not NC in origin. Evidence was outlined in Chapter 4 for the co-option into the newly evolving NC of genes regulating chondrogenesis from several tissues in the protovertebrate chordate. Neither bone nor dentine is present in any Cambrian cephalochordate; their originations were later evolutionary events in association with the evolution of enamel and an **exoskeleton** composed of odontodes or denticles, not too dissimilar from tooth primordia. Assignment of conodonts as craniates on the basis of the presence of bone and dentine in conodont elements (Box 8.2, Fig. 8.3; and see Fig. 4.1) has the potential to complicate this early-endoskeleton, late-exoskeleton scenario.

Box 8.2 Conodonts

Among fossilized skeletal remains are the scales of jawless ostracoderm 'fish' and conodont elements from the Upper Cambrian and Lower Ordovician.

Conodonts have a geological history extending from the Lower Cambrian to the Upper Triassic. They are represented as fossils by minute, tooth-like, phosphatic structures known as *conodont elements*, parts of an elaborate 'pharyngeal' feeding apparatus. Analysis of the histology of the elements that constitute conodont elements and the subsequent assignment of conodonts to the chordates raises some fascinating issues concerning the origin of bone and dentine with implications for our understanding of the origination of the vertebrate skeletal systems.[a]

Over the past 15 years, many researchers have come to recognize conodonts as chordates or even vertebrates, basing that recognition on the identification of bone, dentine, and enamel in conodont elements and the discovery that the animal containing conodont elements (of which only some 12 specimens are known) had a notochord, segmental muscle blocks, tail fin, and paired eyes.

Where conodonts fit within the chordates remains unresolved but has implications for our understanding of the evolution of the NC and of endoskeletal

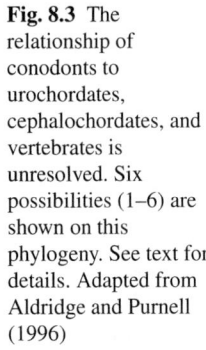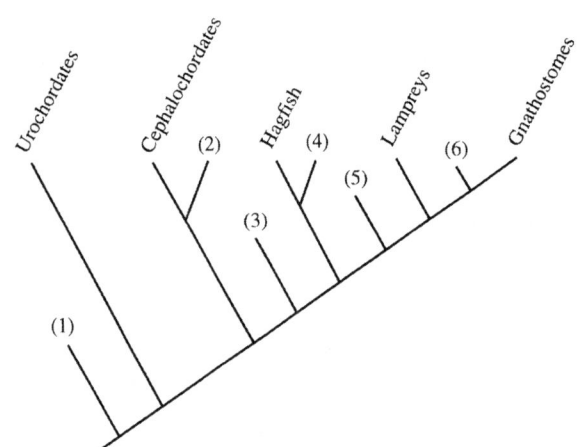

Fig. 8.3 The relationship of conodonts to urochordates, cephalochordates, and vertebrates is unresolved. Six possibilities (1–6) are shown on this phylogeny. See text for details. Adapted from Aldridge and Purnell (1996)

and exoskeletal tissues. Six possibilities are shown in Fig. 8.3. Locating con-odonts basal to the vertebrates (positions 1–4 in Fig. 8.3) requires that elements of the muscular and skeletal systems evolved independently in conodonts and vertebrates or that they were lost independently in hagfish (position 4). Locat-ing conodonts basal to position 5 requires secondary loss of skeletal tissues in lampreys.[b]

---

[a] For the earliest chordate skeletal remains, see M. M. Smith and Hall (1990, 1993) and Sansom *et al.* (1994, 1996, 1997) See Sansom *et al.* (1992) and Donoghue (1998) for the his-tology and growth of conodont elements. See Donoghue *et al.* (2000) and Donoghue (2002) for conodont elements and the conodont animal.

[b] Aldridge and Donoghue (1998) and Janvier (1996, 2007) summarize the evidence for rela-tionships between conodonts and hagfish and/or lampreys, concluding that placing conodonts as a sister group to lampreys best fitted the data available.

Gans and Northcutt outlined the following sequence for the origination of dentine and bone:

- the elaboration of dentine and its availability for use in electroreception because of hydroxyapatite rather than calcite in the dentinal matrix;
- the appearance of layers of bone and its utilization, as an armor for protection, and as a Ca^{++}-storage reservoir; and
- the elaboration of a more generalized bony skeleton.

Bone arising from NCCs is (i) endoskeletal where it replaces or encases pri-mary cartilage and (ii) exoskeletal (dermal bone) where it forms by direct ossifi-cation without going through a cartilaginous phase. Therefore, the appearance of bone in the endoskeleton was a later evolutionary event than was the formation of cartilage. The deposition of bone in the endoskeleton, either around cartilage (perichondral ossification) or within (and replacing) cartilage (endochondral ossification) required the prior evolution of cartilage, and so is evolutionarily secondary.

The origin of the exoskeleton was a later evolutionary event than was the origin of the endoskeleton. Mesoderm- and NC-derived bone supporting the skull—mesoderm and NC both contribute to cranial bones in modern-day vertebrates—and mesoderm-derived trunk appendicular (fin and limb) and axial (vertebral) cartilage and bone also were later evolutionary events.[6]

As discussed in Chapters 4 and 7, the pharyngeal skeleton in all vertebrates examined is derived from NC not mesodermal in origin. Given that cartilage preceded bone in the endoskeleton, a cartilaginous pharyngeal skeleton derived from NC was the original vertebrate endoskeleton, genes necessary for chondrogenesis having been co-opted from genes present in mesoderm, ectoderm, and endoderm of a proto-vertebrate (see Chapter 4). The early association of NC-derived skeletal tissues with the pharynx established what was thought to be a dichotomy between a skeletogenic cranial and a nonskeletogenic trunk NC. As we saw in the preceding section, however, we need to recognize, not a skeletogenic NC, but chondrogenic, osteogenic, and odontogenic lineages of NCCs, or even chondrogenic, osteogenic, and odontogenic NCs.

Many early jawless vertebrates possessed an extensive dermal armor formed from the fusion of individual odontodes, extended along the trunk, to the tip of the tail in some taxa. Teeth are remnants of this once-extensive exoskeleton, the basic building block of which was an odontode of dentine, enamel, and basal bone. One interpretation of such an extensive dermal armor is that the entire TNC was capable of forming exoskeleton in those early forms.

Scale formation, including the development of dentine in scales, is widespread along the trunk in some modern and many fossil fish. The fin skeleton also includes fin rays that are NC in origin (see Box 5.1). Dermal denticles in the little skate, *Leucoraja erinacea*, appear first on the tail and only much later along the fins, developing in a rostrocaudal gradient (Miyake *et al.,* 1999). As seen in the dermal denticles of sharks, skates, and stingrays, and in the bony skin ossicles of some teleost fish, retention of odontogenic ability in the TNC should not be totally surprising. Nevertheless, it is.

One implication of the loss of the body exoskeleton during the evolution of most vertebrates is a parallel loss of skeletal-forming TNCCs. We expect those extant vertebrates that retain an extensive exoskeleton, especially in the trunk, to have retained an odontogenic and osteogenic TNC. The alternative scenario would be of CNCCs migrating the length of the body to form denticles or scales.[7]

Hearts

The heart has long been traced to a precardiac field of mesoderm. Recently, however, with the discovery of a secondary mesodermal heart field and the discovery that NCCs are major contributors to the structure and function of the heart, heart development has become much more complicated.

The secondary heart field, which arises from mesoderm beneath the foregut, provides the myocardial cells and smooth muscle cells for the conotruncal region of the developing heart. Cells from the secondary heart field are essential for the outflow tract of the heart to elongate in order to bring the aorta, pulmonary arterial trunks, and ventricles into alignment. However, our major concern is with the cardiac neural crest (CarNC) and its contributions to the heart. As with the vertebrate head and exoskeleton (the 'new head' and the 'new skeleton'), the vertebrate heart, with its CarNCC contributions, can be regarded as a 'new heart,' part of a 'new cardiovascular system' composed of modular units, only some of which (the mesodermal components) existed in prevertebrate chordate ancestors.[8]

The following points concerning relationships between the NC and the developing heart have been made so far:

- A normal component of the NC, the cardiac neural crest, contributes cells to the valves, septa, and major vessels of the heart.
- *Bmp2* and *Bmp4* are key players in the determination of those mesenchymal NCCs that contribute to the heart.
- The CarNC in birds produces few cartilage cells.
- *Fgf8* is required to pattern the zebrafish heart.
- Deleting *Sox10* from mouse embryos results in defects in heart septation and of the outflow tracts.
- Heart defects are found in the mouse mutant, *Patch*, in which the gene for PdgfRα is deleted.
- Some of the genes that regulate heart development in vertebrates are present in the genome of *Ciona intestinalis*.

Indirect Effects of Cranial NCCs on Heart Function

CNCCs have an *indirect* effect on heart and blood vessel development through the extracellular matrices they deposit and within which the heart develops. Ablating CNCCs affects heart development because loss of these NC derivatives alters the environment in which the heart develops by altering blood flow or the tissue milieu. Although specific, such effects are indirect and independent of any direct contribution NCCs may make to the heart (Hutson and Kirby, 2003*).

Direct Effects of Cranial NCCs on Heart Function

A second but more *direct* role for NCCs was revealed in studies in which ablation of the CarNC provided evidence of a requirement for NCCs for:

- general electrical conductance within the heart (Gurjarpadhye *et al.*, 2007),

- normal functioning of the pacemaker, and
- proper alignment of the specialized and heart muscle cells of the bundle of His.

The bundle of His consists of an aggregation of heart muscle cells that transmit electrical impulses from the atrio-ventricular node to Purkinje fibers in the ventricles, which function as a secondary pacemaker regulating heart beat. The sino-atrial node is the primary pacemaker. Neither the bundle of His nor the pacemaker are derived from NCCs.

A direct role of the NC in heart development, most well studied in avian embryos, is the provision of NCCs from the cardiac neural crest to form essential elements of the heart, including smooth muscles.

The Avian Cardiac Neural Crest

Smooth muscle: **Smooth muscle cells** are found in the heart and in the muscles that activate the feathers of the head and neck.

Smooth muscle cells arise from unipotential or multipotential progenitors. As noted in Chapter 7 (see Fig. 7.15), clonal culture of mesencephalic/metencephalic quail NCCs reveals a lineage of HNK-I-positive cells capable of forming smooth muscle cells, fibroblasts, chondroblasts, and serotonin-containing neurons. A unipotential lineage that forms only smooth muscle can also be isolated. In rat embryos, the appearance of HNK-1-positive cells correlates with the appearance of conductive tissue within the heart, and HNK-1-positive cells have been reported in lamprey hearts. It has not been determined whether the rodent or lamprey cells are NC in origin.[9]

Consistent with the studies of quail NCCs, murine NCCs can be switched into a smooth muscle lineage under the influence of Tgfβ (and into glial or autonomic neuronal pathways under the influence of glial growth factor (Ggf) and Bmp2/4, respectively, (Anderson, 1997)).

Migration through the pharyngeal arches: we now know that NCCs produce the smooth muscles at the base of the aortic arches in chicken embryos. Evidence of this direct contribution of CNCCs to the heart from a CarNC and mapping of the cardiac crest came from an extensive series of studies transplanting diI-labeled and/or quail NCCs into chicken embryos. In the early 1980s, Margaret Kirby and her colleagues established that NCCs delaminating from r6 to r8 caudally to the level of the third somite, migrate through pharyngeal arches 3, 4, and 6, and colonize the outflow region (**outflow tract**) of the developing heart. Outgrowth of branches of cranial nerves IX, X, and XII occurs at the same time (Fig. 8.4 [Color Plate 8]).

The CarNC in chicken embryos provides a unique set of proteins to the pharyngeal region. Five proteins localized in migrating CarNCCs are eliminated if the CarNC is removed. Interestingly, several stages later, the normal protein profile is restored, indicating regulation of CarNCCs, a process discussed toward the end of the Chapter 10. Cells from the cardiac neural crest adjacent to somites 1–7 produce

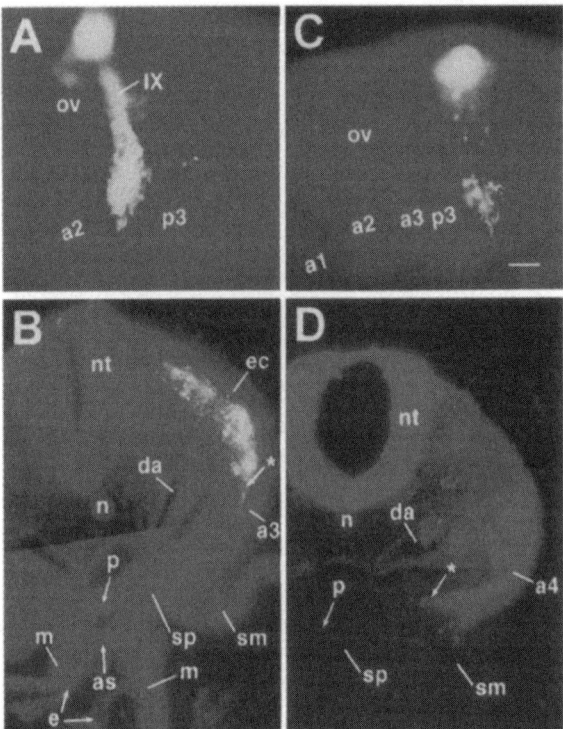

Fig. 8.4 Migration of neural crest cells to visceral arches 3 and 4 is shown in these sections of chicken embryos in which the neural crest was labeled with DiI, shown as *yellow* fluorescence. (**A**) Labeled cells migrating from r7 into arch 3 and forming the tract of the glossopharyngeal nerve IX. (**B**) In this transverse section of a chicken embryo of H.H. stage 14, DiI-labeled neural crest cells (ec) that arose from r7 can be seen migrating toward the third pharyngeal arch (a3). The leading edge of the neural crest cells (*) lies lateral to the pharynx (p). (**C**) Labeled neural crest cells from the level of somite 2 migrate toward the fourth visceral arch. (**D**) A cross-section of (**C**) showing the ventral location (*) of the leading edge of the migrating cells. Other abbreviations are: a1–a4 and p3; pharyngeal arches; da, dorsal aorta; e, endocardium; m, myocardium; n, notochord; nt, neural tube; ov, otic vesicle; p, pharynx; sm, somatic mesoderm; sp, splanchnic mesoderm. Reproduced from Suzuki and Kirby (1997) from a figure kindly supplied by Margaret Kirby. Reprinted by permission of Academic Press, Inc. (see Color Plate 8)

cardiac ganglia, while cells adjacent to somites 1–3 produce mesenchymal cells. Hypobranchial muscles and Schwann cells for cranial nerve XII also arise from the CarNC.

Unlike most other NCC populations, cardiac NCCs in chicken embryos migrate and invade the heart primordium as two streams of cells, each with different potentials:

• Early migrating cells leave the neural tube between H.H. stages 9 and 10 and migrate to the aortico-pulmonary septum and pharyngeal arch arteries.

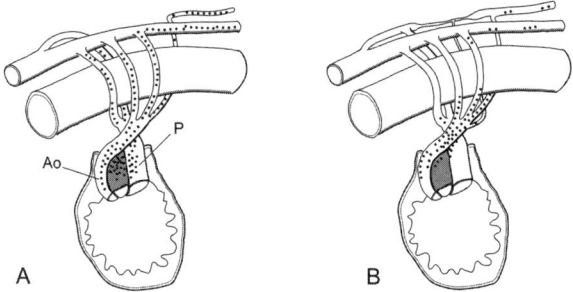

Fig. 8.5 The CarNC of chicken embryos consist of an early migrating population (**A**) that contributes cells to the proximal and distal regions of the arteries of the pharyngeal arches, to the pulmonary trunk (P) and to the aorticopulmonary septum (Ao) and a late-migrating population (**B**) that contributes cells to the distal portions of the arteries of the pharyngeal arches. Modified from Boot *et al.* (2003)

- Later migrating cells leave the neural tube at H.H. stages 12 and migrate only to (are restricted to?) the proximal portion of the pharyngeal arch arteries (Fig. 8.5).

Grafting these cells allowed Boot and colleagues (2003) to demonstrate the intrinsic potential of each population. Transplanting late-migrating CarNCCs into younger hosts showed that cell fate is not restricted; late CarNCCs can differentiate as mesenchyme and contribute to the aortico-pulmonary septum.

Once in the outflow tract, CarNCCs form:

- mesenchyme of the endocardial cushions,
- the semilunar valves,
- the middle layer (tunica media) of the aortic arches and their branches,
- much of the aortico-pulmonary septum, which separates the aorta from the pulmonary artery, and
- much of the conotruncal septa of the truncus arteriosis in the outflow tract of the heart.[10]

The aortico-pulmonary septum: The **aortico-pulmonary septum** is a CarNC-derived permanent feature of the heart that separates pulmonary from nonpulmonary blood flow; no smooth muscle α-actin is deposited where smooth muscle cells of the aortico-pulmonary septum should form in chicken embryos from which the CarNC has been removed.

Some CarNCCs in the aortico-pulmonary septum may play a more indirect role; Poelmann *et al.* (1998) identified a sub-population of CarNCCs in chicken embryos whose removal from the septum by apoptosis may play a role in initiating cardiac myogenesis. Such heterogeneity has implications for the regulation of CarNC, a topic discussed below.

Deleting the cardiac neural crest: As you might expect given the range of cell types they produced, deleting chicken CarNC produces:

- persistent truncus arteriosis,
- malformations of the aortico-pulmonary septum,
- transposition of the major heart vessels,
- high ventricular septal defects,
- single outflow vessels emerging from the right ventricle or over the ventricular septum, and
- abnormal outflow of blood from the heart, including increased velocity of blood flow through the dorsal aorta, and lower blood pressure.

Specific defects can be induced by controlling the time when NCCs are deleted or by deleting the cells adjacent to only one of somites 1, 2, or 3. Deleting more caudal NCCs (adjacent to somites 10–20) removes sympathetic cardiac innervation, resulting in sympathetically aneural hearts. Unexpectedly, deleting the CarNC was found to be associated with intrinsic myocardial defects, evidenced as a defective contraction of heart muscle accompanied by ventricular dilatation, supporting the role for NCCs in regulating the functioning of heart muscle discussed above.[11]

Cardiac Cartilages

The presence of cartilages in the hearts of species in all vertebrate groups other than amphibians immediately leads one to suspect pathology, infection, parasitism, or trauma. But it is not so.

All individuals in some species (chickens and Japanese quail, for example) have cardiac cartilages or bones as constitutive elements. The NC origin has been proposed and, in birds, demonstrated. Cartilage is a constitutive feature of the proximal aorta, pulmonary trunk, and semilunar valves of chick embryos from H.H. 37 onward. The NC origin of these cardiac cartilages has been demonstrated in chick and quail, expanding the list of NC-derived cartilages, and demonstrating that cardiac crest can contribute skeletal elements to the trunk.[12]

The development of cardiac cartilages in the Spanish terrapin, *Mauremys leprosa*, has been extensively studied (López *et al.*, 2003). First evident as a mesenchymal condensation (as is true for cartilages outside the heart as shown in Fig. 3.13) along the aortico-pulmonary and ventricular horizontal septa, the differentiation of chondroblasts and the deposition of a type II collagen-positive ECM produces a hyaline cartilage that extends along both septa, which, you will recall, are NC in origin. Anytime up to 18 months after birth, a second cartilage develops in the fibrous cushion that supports portion of the semilunar valve.

Cartilage is less frequent in teleosts and mammals; 11% of Syrian (golden) hamsters, *Mesocricetus auratus*, have cartilage along the fibrous attachment pulmonary valves. Habeck (1990) observed islands of cartilage or bone at the margins of the carotid bodies in 0.5% of 1,395 rats. He thought (as it turns out, correctly) that these arise from pharyngeal arch mesenchyme, although Habeck thought the mesenchyme was mesodermal in origin. The NC origin is more likely, as shown in birds, and, as proposed by Blanco *et al.* (2001) to explain the presence of cartilaginous nodules in the bulbus arteriosus of four species of teleosts fish.

Cardiac Neural Crest in Fish and Amphibians

Much less research has been undertaken on the presence and contributions of a CarNC in fish or amphibians. A population of NCCs in zebrafish embryos, located more rostrally than the CarNC in chicken embryos, migrates to the developing heart, where the cells differentiate into cardiac muscle cells. These cardiomyocytes also appear to play a role in heart morphogenesis—specifically for the looping that transforms the straight heart tube into an S-shaped organ, and for contraction of heart muscle (see Stoller and Epstein, 2005* for literature).

Rhombencephalic NCCs in Mexican axolotl embryos migrate to the aortic arches and to the developing truncus arteriosus but not to the atria or ventricles (as do CarNCCs in chicken embryos).

Cardiac Neural Crest in Mammals

Using a combination of DiI labeling and embryo culture, Fukiishi and Morriss-Kay (1992) demonstrated a CarNC, the cells of which contribute to the heart after migrating through the pharyngeal arches in rat embryos. The first map of the CarNC in human embryos was published by O'Rahilly and Müller (2007); CarNCCs migrate from r6 and r7 to the third pharyngeal arch and then to the truncus arteriosus at Carnegie stage 12.

A number of molecular have been used to follow the migration of CarNCCs through the pharynx or to determine their arrival in the cardiac primordia; *pro-α2 (V)* collagen transcripts are expressed in craniofacial mesenchyme, skeletal precursors, and heart valves of murine embryos.

As identified using enhancer regions within the *Pax3* promoter to drive expression of Cre-recombinase, cells from the CarNC of mouse embryos migrate through pharyngeal arches 3, 4, and 6 to colonize the outflow tract of the developing heart. This and a subsequent study using a 5.5-Kb region of the *Wnt1* promoter to drive expression of Cre-recombinase in NCCs revealed labeled cells invading the pharyngeal arches and associating with the aortic arches before invading the outflow tract of the heart (10.5 days gestation) and contributing cells to the aortico-pulmonary septum, the aortic valve, ductus arteriosis, and endocardial cushions. It is not clear whether the CarNC contributes cells to the mesodermal linings of the heart (the epi- and myocardia) and/or influences conductance of electrical signals through, or the contraction of, cardiac myocytes.[13]

As discussed in Chapter 4, heart defects are found in the mouse mutant, *Patch*, the phenotype of which reflects a deletion of the gene encoding PdgfRα. Loss of the receptor results in aortic arch and craniofacial defects because of the requirement for PdgfRα on late stages of NCC differentiation. Recent analyses have demonstrated a role for the second Pdgf receptor, PdgfRβ, in recruiting CarNCCs into the conotruncal regions for the proper development of the aortico-pulmonary and ventricular septum.[14]

A genetic approach has now been used to follow defects in the outflow tract of developing mouse hearts after ablating the CarNC at specific times, beginning at 7 days of gestation. Semaphorins, which direct axon guidance and patterning (see Chapter 3), direct CarNCCs during their migration through the pharyngeal arches; Src, which functions as a tyrosine kinase substrate, is required for development of the CarNCCs in the outflow tract (Fagiani *et al.*, 2007; High and Epstein, 2008). Bmp2 and the Notch signaling pathway are involved in septation of the outflow tract, the epithelial —> mesenchymal transformation associated with valve development, and proliferation of smooth muscle cells in the aortic arches. This is an important signaling pathway; over 20 *Notch* mutant mice show cardiovascular defects.[15]

Cardiac Defects

Sox genes: The major region of expression of *Sox9* after NCC migration is in the mesenchyme of the pharyngeal arches and in derivatives of the CarNC; *Sox9* has a general role in regulating the specification and differentiation of chondrocytes; *Sox9* and *Sox10* regulating the synthesis of the *Col2a1* gene and so type II collagen CNCCs (see Fig. 4.10). Depleting *Sox9* in the murine CarNC eliminates the ability of those cells to form cartilage; Sox9⁻ cells arise but cannot form the condensations necessary for the initiation of chondrogenesis.[16]

In mouse and zebrafish embryos, *Sox9* and *Sox10* display nonoverlapping functions in cranial and cardiac and in trunk and vagal crest, functions that include regulating NCC formation. Conditional inactivation of *Sox9* in mouse embryos leads to impaired heart development—especially evident in the heart valve formation—because of impaired expansion of CarNCCs and impaired production and deposition of ECM proteins. *Sox4* is expressed in those NCCs that contribute to the outflow tract of the heart; *Sox4*- and *Sox11*-deficient mice fail to form the septum required for the semilunar valves to form.

Deleting *Sox 10* or *Sox11* from mouse embryos results in defects of the outflow tracts because of failure to form the septum required for the semilunar valves to form. Neonates die immediately after birth, with craniofacial defects (including cleft lip and palate) and defects in heart septation and the outflow tracts (as seen in *Sox4*-deficient mice), indicating involvement of *Sox11* with proper differentiation of cranial and CarNCCs (Hong and Saint-Jeannet, 2005*).

Pax3: Because of important functions at the condensation stage of organogenesis (see Fig. 3.14), *Pax* family genes are key players in the development of many organs, in part because *Pax3* regulates *FoxD3*, a winged helix transcription factor required for the specification and migration of NCCs (see Fig. 2.7). The *mother superior* mutation removes *FoxD3* activity in zebrafish, dramatically reducing the numbers of NCC precursors available to contribute to the craniofacial skeletal, neuronal, and melanophore lineages. Perhaps even more dramatic is the recent demonstration that deleting *FoxD3* from mouse embryos results in perinatal death of newborns in which cranial neural crest-derived tissues are either missing or extremely reduced in size, the peripheral nervous system consists only of small DRG and cranial nerves, and the enteric nervous system fails to develop at all. On the other hand, CarNC, which is reduced early in embryogenesis, is compensated for (by regulation? see Chapter 10) by mid-gestation when most CarNC derivatives are normal.[17]

The *Splotch* mutant results from an inversion of intron 3 in *Pax3*, resulting in the production of abnormally spliced mRNA transcripts. Mice homozygous for the *Splotch* gene have defective aortic arches, thymus, thyroid, and parathyroid glands (see Chapter 10).

Sp2H, one of the six mutant or induced alleles of the *Splotch* locus, is associated with persistent truncus arteriosus and related conotruncal heart defects. The transient expression of *Pax3* in developing heart primordia at 10.5 days of gestation is not seen in embryos carrying the *Splotch* mutation. From their studies with *Sp2H* mutant mice, Conway and colleagues (1997, 2000) demonstrated that *Pax3* functions autonomously in expanding one subpopulation of CarNCCs, a population that is reduced in Sp^{2H} embryos, but which migrates normally in a wild-type environment.

Gap junction proteins and cardiac defects: Deletions or deficiencies of NCCs in mouse embryos produce specific cardiac defects, such as conotruncal anomalies, aortic coarctation, or bicuspid aortic valves, providing animal models for the DiGeorge syndrome, a human syndrome with involvement of CarNCCs (see Chapter 9). Heart defects arising from deletion of CarNC are associated with the underdevelopment of thymus and parathyroid glands, two organs that receive contributions from the NC (see Table 2.1 and Box 9.3). The parathyroid arises in the third pharyngeal arch but can be traced to an epipharyngeal placode earlier in development (Manley and Capecchi, 1995).

Cell communication is altered in conjunction with heart defects and may explain the etiology of some heart defects. The gap junction protein **connexin 43** (Cx43) is expressed in the dorsal neural tube and in subpopulations of NCCs, including those of the CarNC. Overexpressing *Cx43* in mice results in delayed/defective neural tube closure and conotruncal heart defects (Ewart *et al.*, 1997).

Some congenital heart defects are mediated via disruption of Cx43 and a second gap-junction protein Zonula occludens1 (Zo1), which mediates interactions between migrating NCCs, Cx43 and actin. The mediation occurs especially in response to homocysteine, which with a second amino acid, methionine, is an integral component of the metabolic pathways for purine synthesis and for gene regulation involving histone methylation. Serum concentrations of homocysteine are elevated in folic acid or vitamin B_{12} insufficiency, resulting in higher risk for cardiovascular and congenital defects involving NCCs. Of the genes whose transcription is altered following homocysteine-induced inhibition of migrating CarNCC, 30% have roles in cell migration and adhesion and 15% in proliferation and apoptosis.[18]

This brief discussion of cardiac defects arising from defective CarNCCs provides an introduction to the more extended discussion of NC-based defects in the next two chapters.

Notes

1. See Smith and Hall (1990, 1993) and Hall (2005b*) for NCC contributions to teeth.
2. See Svajger et al. (1981) and Hall (2005b*) for teratomas derived from grafts of embryonic ectoderm.
3. See Lumsden (1988), Imai et al. (1996), and Hall (2005b*) for the origin of odontoblast precursors in the NC and for patterning of teeth by oral ectoderm, and Järvinen et al. (2006), Mitsiadis and Smith (2006), Kettunen et al. (2007), and Pummila et al. (2007) for recent studies on the genetic basis of murine tooth development via epithelial–mesenchymal interactions.
4. See M. M. Smith and Hall (1993) and Graveson et al. (1997) for tooth-forming TNC.
5. See Lumsden (1988), M. M. Smith and Hall (1990) and Atchley and Hall (1991) for the development of teeth in ectopic sites and for the origin of alveolar bone from the dental follicle.
6. For the evolutionary origins and relationships of vertebrate skeletal tissues, see Hall (1987, 2000a,b*, 2007a*), M. M. Smith and Hall (1990, 1993), Langille and Hall (1993), Sansom et al. (1997), and Hall and Witten (2007).
7. See M. M. Smith and Hall (1990, 1993), Graveson et al. (1997), and Hall (1999a*, 2000c) for further elaboration of tooth formation from TNCCs.
8. See Opitz and Clark (2000*) and Hutson and Kirby (2003*) for overviews of the CarNC. Atchley and Hall (1991) and Schlosser (2002c) for modules as fundamental units of morphological evolution. Gass and Hall (2007) present the arguments for regarding subpopulations of NCCs as developmental and evolutionary modules.
9. See Ito and Sieber-Blum (1993) for the clonal cultures, and Nakagawa et al. (1993) and Hirata et al. (1997) for HNK-1-positive cells in the hearts of rats and lampreys.
10. See Ito and Sieber-Blum (1993*) and Ito et al. (1993) for the clonal analyses, and see Le Lièvre and Le Douarin (1975), Kuratani and Kirby (1992), Kirby et al. (1993), Hutson and Kirby (2003*), and Stoller and Epstein (2005*) for studies on NC contributions to the aortic arches and to the heart.
11. For deletion of NC and compensation for that deletion, see Kirby et al. (1993) and Hutson and Kirby (2003*). For regulation of the heart field in zebrafish, see Serbedzija et al. (1998).
12. See López et al. (2000, 2003) and Hall (2005b*) for literature on cardiac cartilages in birds.
13. See Jiang et al. (2000) and Stoller and Epstein (2005*) for the mouse studies and for discussions of the possibility that CarNC may contribute cells to the epicardium and/or the myocardium and whether such contributions might be mouse strain specific. See Porras and Brown (2008) for the ablation studies.
14. See Tallqvist and Soriano (2003) for PdgfRα, and Richarte et al. (2007) for PdgfRβ in cardiac development and for discussions of the roles of the two Pdgf receptors.
15. See Fagiani et al. (2007), High and Epstein (2008), and Varadkar et al. (2008) for these signaling pathways in CarNCC development.
16. See Mori-Akiyama et al. (2003) for Sox9 and chondrogenesis, especially of CarNCCs, Suzuki et al. (2006) for Sox9, Sox 10 and type II collagen, Hong and Saint-Jeannet (2005*) and Saint-Jeannet (2006*) for a recent review, and Lincoln et al. (2007) for inactivation of Sox9 in mouse embryos.
17. See Lister et al. (2006) for FoxD3 and Montero-Balaguer et al. (2006) for the mother superior mutation in zebrafish, and Teng et al. (2008) for the mouse study. Dottori et al. (2001) found FoxD3 expressed in premigratory and migrating NCCs in chicken embryos and that

FoxD3 functioned upstream of *Pax3*, increased expression of FoxD3 leading to enhanced NCCs and decreased numbers of interneurons.

18. See Boot *et al.* (2006) for Cx43 levels in migrating NCCs, Rosenquist *et al.* (2007) for gene transcription by CarNC in response to homocysteine and for folic acid, homocysteine, and defects of the CarNC, and see Weston *et al.* (2007) for links between histone methylation and cardiac defects.

Part III
Abnormal Development and the Neural Crest

There is a long history of using abnormal development to provide clues to the control of normal development. Mutants, chromosomal abnormalities, drug-induced phenotypes, or phenotypes resulting from the application of abnormal levels of compounds natural to the body (hormones and vitamins as two examples) all have provided important information on developmental mechanisms and how those mechanisms can go awry.

Tumors of neural crest origin (neurocristopathies) and birth defects both serve the same purpose for us, illuminating the stages of NC formation and NCC development that are especially susceptible to change. Research into such associations does not stop at their classification but rather leads to increased understanding of the mechanisms that cause the defects and, ultimately, toward improved treatment or even prevention.

Embryos have the means to protect themselves against some agents that would otherwise lead development astray. Regulation—the ability of NCCs to compensate for lost cells—is one such mechanism. Another is the maintenance of stem cells into adult life. Regulation and NCCs as stem cells brings us full circle in our discussion of the properties and potential of NCCs.

Chapter 9
Neurocristopathies

Bolande introduced the term **neurocristopathy** in 1974 for syndromes, tumors, and/or dysmorphologies involving neural crest cells (NCCs). The shared embryonic origin evoked by the name provides the rationale for Bolande's classification and brings us a full circle to germ layers and to the germ-layer theory discussed in Chapter 2.

The following points concerning neurocristopathies were made in earlier chapters, especially Chapter 5:

- Neurocristopathies are defined as syndromes involving one or more types of NCCs.
- The recognition of neurocristopathies by developmental biologists, clinicians, and medical geneticists acknowledges germ-layer origin as a common link between affected cells, tissues, and organs.
- NCCs or tumors derived from NCCs can respond to tumor-promoting agents by transforming into a different cell type (e.g., neurons to chromatophores).
- *SOX10* has been implicated in the etiology of two neurocristopathies, Hirschsprung disease (aganglionic megacolon), and Waardenburg–Shah syndrome (hypopigmentation; see Chapter 4).

Antiquity

Interest, indeed fascination with developmental abnormalities, stretches back at least to the Neolithic period: A sculpture of a two-headed person dated 6500 BCE was excavated from southern Turkey; cuneiform characters on seventh century BCE clay tablets discovered in the Royal Library at Nineveh record 62 different human malformations known to the Babylonians.

Syndromology and Neural Tube Defects

Syndromes involving neural tube defects (Table 9.1) are second only to heart defects (recall the involvement of NCCs in heart development discussed in

B.K. Hall, *The Neural Crest and Neural Crest Cells in Vertebrate Development and Evolution*, DOI 10.1007/978-0-387-09846-3_9,
© Springer Science+Business Media, LLC 2009

Table 9.1 A survey of the major neurocristopathies

Tumors

Pheochromocytoma: A tumor of the chromaffin tissue of the adrenal medulla that can arise as
    early as 5 months of age in humans. High concentrations of vanillinemandelic acid are
    characteristic.Synonym: Endocrine neoplasia III, Multiple.

Neuroblastoma: A common, malignant embryoma involving the adrenal medulla and ganglia of
    the autonomic nervous system. Displays a very high incidence of spontaneous regression. [The
    autonomic nervous system can also function abnormally in defects other than neuroblastomas.
    Prader–Willi syndrome is one example.]

Medullary carcinoma of the thyroid: A tumor of parafollicular calcitonin-forming 'C' cells of
    the thyroid, representing 7% of all thyroid tumors. Synonyms: endocrine neoplasia II,
    Multiple, hypercalcitoninemia.

Carcinoid tumors: Tumors of the bowel and gastrointestinal tract involving enterochromaffin
    cells.

Nonchromaffin paraganglioma (chemodectoma) of the middle ear: A tumor involving ganglia
    of the neck and the middle ear. Over 90% of patients suffer hearing loss. Associated with
    erosion of bone.

Hirschsprung disease: Absence or reduction of ganglia of the colon (aganglionic megacolon)
    1:5,000 to 1:8,000 live births; 80% of cases are in males. Synonyms: aganglionic megacolon,
    colon aganglionosis. Cause: mutation in *Sox10*.

Neuroectodermal pigmented tumor: A rare tumor of the maxilla consisting of clusters of
    neuroblast-like cells, with or without melanin. Can arise as early as the first year of life. High
    urinary excretion of vanillinemandelic acid. Results in destruction of bone and displacement
    of teeth. Synonym: Melanotic progonoma.

Clear cell sarcoma: A sarcoma of tendons and aponeuroses.

Tumor syndromes

von Recklinghausen neurofibromatosis: Multiple neuroid tumors of the skin and abnormal
    pigmentation. An autosomal dominant; 1:3,000 live births. Six or more *café-au-lait* spots
    (areas of pigmentation) greater than 1.5 cm diameter are diagnostic for the syndrome. Neural
    neoplasia occurs throughout the nervous system. Skeletal abnormalities and scoliosis may be
    associated.

Sipple syndrome: An association of pheochromocytoma and medullary thyroid carcinoma. An
    autosomal dominant, displaying elevated levels of calcitonin, catecholamine in excess, and
    ectopic production of ACTH. Synonyms: Medullary thyroid carcinoma and pheochromo-
    cytoma syndrome.

Multiple mucosal neuroma syndrome: Mucosal tumors (neurofibromas) of the tongue, lips,
    and eyelids; medullary carcinoma of the thyroid, sometimes with neural involvement.

Wermer-Zollinger-Ellison syndrome: Neoplasia of endocrine glands (pituitary, adrenal,
    parathyroid, pancreas, thyroid). A rare autosomal dominant. Synonym: Multiple endocrine
    adenomatosis.

Neurocutaneous melanosis: Giant pigmented nevi (birth marks) of the skin

Malformations

Mandibulofacial dysostosis: Hypoplastic zygomatic arch, orbital and supraorbital ridges,
    dysplastic ears, micrognathia, defective ear ossicles; high incidence of cleft palate. An
    autosomal dominant, largely treatable with corrective surgery. Synonym: Treacher Collins
    Syndrome.

Otocephaly: Absence of lower jaw, low-set ears which may be fused in the midline;
    microstomia; ear ossicles, temporal bone, palate, and maxilla may be deformed. Absence of
    mesenchyme derived from the neural crest. Associated with heart, pharyngeal, and limb
    anomalies. Synonyms: Agnathia, microstomia, and synotia.

Table 9.1 (continued)

CHARGE association: **C**oloboma of the iris, **H**eart defects, **A**tresia of choanae, **R**etardation of physical and mental development, **G**enital anomalies, and **E**ar anomalies and/or deafness (CHARGE).

Others

Albinism: A very heterogeneous group of conditions. Absence of skin pigmentation because of lack of melanoblasts. Synthesis of tyrosinase inhibited. An autosomal dominant. Synonym: Piebaldness.

Waardenburg syndrome: An autosomal-dominant exonic mutation in the HuP2-paired domain as an intragenic deletion in *Pax* 3. Maps to the distal arm of chromosome 3. Incidence 1:4,000 live births. Pigment defects of the hair (white forelock) and skin. May be associated with deafness. Increased incidence of cleft palate. Ocular hypertelorism. Due to abnormal migration or survival of melanoblasts.[a]

Waardenburg–Shah syndrome: A dominantly inherited condition with major effects on two populations of NCCs — pigment cells (hypopigmentation) and enteric ganglia (aganglionic megacolon or Hirschsprung disease) with associated deafness and eye involvement. Waardenburg–Shah syndrome combines the features of Waardenburg syndrome and Hirschsprung disease. Basis: A mutation in *SOX10* that acts cell autonomously in NCCs.

[a] Waardenburg syndrome is named after Petrus Johannes Waardenburg (1886–1979), the Dutch ophthalmologist and pioneer in the application of genetics to ophthalmology. In 1932, the year he wrote his classic textbook, *The Human Eye and its Genetic Disorders*, Waardenburg was one of the first to suggest that Down's syndrome might be a chromosomal aberration.

Chapter 8) as a source of perinatal mortality in humans, accounting for some 15% of perinatal deaths in the UK alone, with a worldwide prevalence of 0.5–0.6 individuals per 10,000 births. Syndromology is not dead; new syndromes are described every year and websites updated weekly.[1]

Folic acid, which is required for the synthesis of nucleotides and methionine, and for methylation, is effective in preventing as many as 70% of human neural tube defects, including spina bifida, to which defective NCCs contribute. Because of a relationship between ultraviolet light and folic acid, melanin also may protect against neural tube defects. Cardiac defects arise in mice following knockout of *Folr-1*, the gene for the folate-binding protein, but can be alleviated by giving folic acid maternally.

For women who have already produced a child with a neural tube defect, a folic acid supplement of 4 mg/day, taken before a subsequent pregnancy, reduces the risk of a recurrent neural tube defect by 87%. Caution is still required, however, in attributing all improvements to folic acid. An analysis of the incidences of anencephaly and spina bifida in Atlanta (USA) before and after the enrichment of grains with folic acid was inconclusive as to whether folic acid had an additional effect or merely enhanced already declining incidences of both defects.[2]

The Utility of the Germ-Layer Theory

Because abnormalities are often expressed as a syndrome, with defective development of more than one cell type or tissue, and because so many different cell types arise from the NC, it was inevitable that NC origin would be seen as the common 'explanation' for syndromes involving combinations of neuronal pigment, craniofa-cial skeletal, heart, adrenomedullary, and/or other cell types derived from the neu-ral crest (NC). A typical syndrome, described almost 50 years ago, a *malignant embryoma* with pulmonary metastases, was removed from the chest of a 61-year-old male. Ganglia, connective tissue, cartilage, Schwann cells, and pigment cells (all NC derivatives) were all present.

Does grouping NC defects as neurocristopathies merely classify them on the basis of an outmoded germ-layer theory or does this classification help us under-stand the etiology of the defects? Because NCCs share so many characteristics—commonality of embryonic origin, bipotentiality, migratory capability, and activa-tion by growth and other epigenetic factors—we can invoke the explanatory powers of NC origin, just as pathologists and developmental geneticists correlate develop-mental effects/defects with commonality of origin from a single germ layer. The pervasiveness and utility of germ-layer thinking is illustrated by the willingness of clinicians to classify malformations on the basis of germ layer of origin. We use embryopathy for defective development involving the entire embryo and neu-rocristopathy for defective development of the NC or its derivatives. It is less com-mon to see the terms mesoderm-, ectoderm-, or endodermopathy, although partic-ular syndromes are classified based on the involvement of derivatives from only one germ layer. Thus, osteochondromas affect mesodermal tissue; ectodermal dys-plasias affect ectodermal derivatives, while juvenile colonic polyposis affects cells of endodermal origin.[3]

Identifying commonality of origin from a single germ layer means that the likely time, location, and sometimes the cause(s) of defective development can be iden-tified. Analyses of neoplasias of mixed developmental origin can reveal important information about the control of NCCs and/or reveal defects affecting pathways in a fundamental biological process, such as cell division shared by cells from different germ layers. *Multiple endocrine neoplasia type 1 (MEN1)* in humans is character-ized by endocrine neoplasia of glands that are NC in origin (the parathyroid) and glands that are not (the pituitary, which arises from the adenohypophyseal placode and neural ectoderm [see Fig. 6.5], and the pancreas, which arises from endoderm). The primary cause of *MEN1* is a loss-of-function mutation in the nuclear tumor suppressor protein, menin, which interferes with the functioning of the cell cycle inhibitor, p27; a recent study showed that inactivation of menin in murine NCCs results in cleft palate, cranial bone defects, and perinatal death because of a decrease in p27 (Engleka *et al.*, 2007).

The developing pancreas of rat embryos is colonized by two waves of NC pre-cursors. An initial wave colonizes the bowel and then acquires the ability to migrate, in a second wave, into the pancreas. Although murine pancreatic cells can produce neurite outgrowths and although the neurites contain neurofilament protein, pancre-atic cells are endodermal in origin. Endoderm and NC therefore share the ability to form neurite outgrowths, just as mesoderm and NC share the ability to generate mesenchyme.[4] Neither commonality of cell type nor commonality of metabolism need imply commonality of developmental origin, a caution to keep in mind when invoking the 'explanatory' powers of germ layer of origin.

Types of Neurocristopathies

Neurocristopathies may involve the adrenal medulla, endocrine organs of NC origin (*MEN1* above), pigment cells, ganglia or Schwann cells of the autonomic nervous system, the heart, craniofacial skeletal, and/or mesenchymal NCCs. Some well-studied classes of neurocristopathies, with the affected NC-derived tissues in parentheses, are:

- neuroblastomas (adrenal medulla, autonomic ganglia),
- mandibulofacial dysostosis, first arch syndrome, Treacher–Collins syndrome (facial skeleton, ears), and
- otocephaly (ear defects, heart anomalies, and loss of the lower jaws).

Because Schwann cells accompany axons of peripheral nerves into the limbs, limb abnormalities may accompany neurocristopathies as a secondary consequence.

   To Bolande, a neurocristopathy could arise following a defect at any stage of NC or NCC development—migration, proliferation, cell-to-cell interaction,

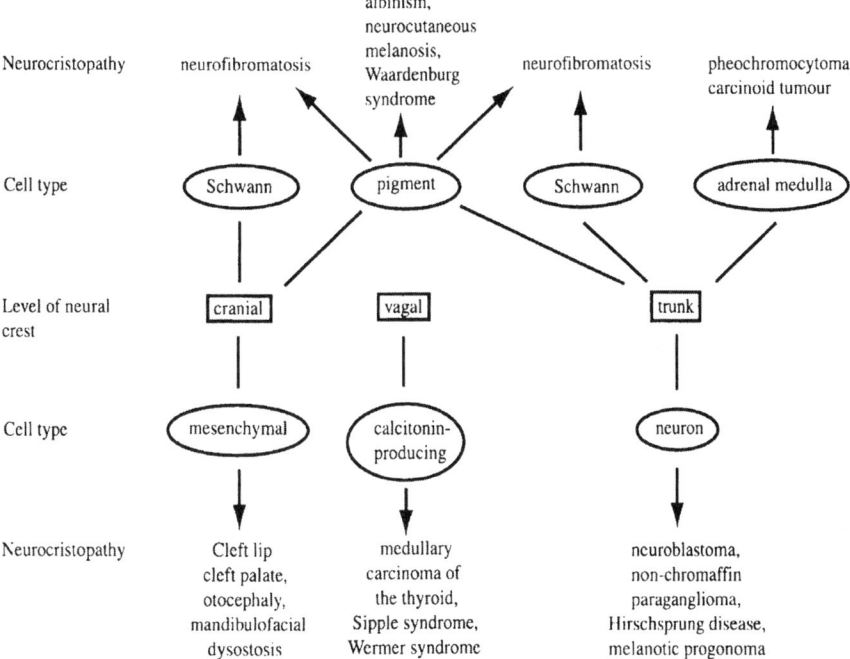

Fig. 9.1 This flow chart illustrates the origin of neurocristopathies from particular cell types (*circled*) arising from cranial, vagal, or trunk regions of the neural crest. For example, abnormalities in Schwann cells from cranial or trunk crest are associated with production of neurofibromatosis; abnormalities of mesenchymal cells from CNC are associated with cleft lip, cleft palate, otocephaly and mandibulofacial dysostosis, and so forth

differentiation, or growth. He divided neurocristopathies into **simple** (a single, usually localized pathological condition) and **complex** (usually multifocal syndromes for associations of several simple neurocristopathies). Here, neurocristopathies have been grouped into:

- **tumors**, in which only one tissue is affected;
- **tumor syndromes**, where several tissues are involved;
- **malformations** (mandibulofacial dysostosis, otocephaly, and the CHARGE syndrome); and
- a **miscellaneous** group of others (Table 9.1).

In Fig. 9.1, the neurocristopathies are grouped according to the rostrocaudal region of the NC involved and according to the type of NCC from which they arise. Many neurocristopathies reflect disruption of the two basic mechanisms, cell migration and cell proliferation. Table 9.2 shows how the neurocristopathies listed in Table 9.1 fit into these two categories of disrupted NCC migration or proliferation. The primary gene defects in some two dozen human syndromes are summarized in Table 9.3.

Syndromes can be genetically complex, partly but not entirely because phenotypic expression is dependent on genetic background. The *first arch* (*far*) locus in mice, which produces a classic first arch syndrome of NC defects, is partly dominant in ICR/Bc mice but recessive in BaLB/c mice (Harris and Juriloff, 2007*). Such a shift in dominance allows two syndromes to arise from the same mutation.

The remainder of the chapter introduces five neurocristopathies that reflect changes in a variety of NCC derivative and for which we have information concerning their mechanism of origin at several levels. They are the CHARGE syndrome,

Table 9.2 A summary of the neurocristopathies according to whether they are based in migration or proliferation defects of neural crest cells[a]

Migration defects	Proliferation defects
Defects of the anterior chamber of the eye	Pheochromocytomas
Cardiac aorticopulmonary septation defects	Neuroblastoma
CHARGE association	Medullary carcinoma of the thyroid
Cleft lip ± cleft palate	Carcinoid tumor
Cleft palate	Melanotic progonoma
DiGeorge syndrome	Neurofibromatosis
Frontonasal dysplasia	Wermer (Sipple) syndrome
Goldenhar syndrome	Neurocutaneous melanosis
Hirschsprung disease	
Waardenburg syndrome	
Isotretinoin embryopathy	

[a]Migration defects produce malformations, proliferation defects produce dysplasias. This arrangement follows M. C. Jones (1990), who may be consulted for supporting literature.

Table 9.3 Human syndrome, their chromosome location, and (where known) the gene and protein product associated with the syndrome[a]

Syndrome	Chromosome location	Gene symbol	Protein
Achondroplasia	4p16.3	*FGFR3*	Fibroblast growth factor receptor 3
Alagille syndrome 1	20p12	*JAG1*	Jagged1 transmembrane protein
Apert	10q26	*FGFR2*	Fibroblast growth factor receptor 2
Atrial septal defect with Atrioventricular conduction defects	5q34	*NKX2E*	NK2 transcription factor
Campomelic dysplasia	17q24.3–q25.1	SOX9	SRY (sex-determining region Y)-box 9
CHARGE	8q12.1	*CHD7*	Chromodomain helicase DNA binding protein 7
Cleidocranial dysplasia	6p21	*RUNX2*	Runt-related transcription factor 2
Craniosynostosis	10q26	*FGFR2*	Fibroblast growth factor receptor 3
Crouzon syndrome	10q26	*FGFR2*	Fibroblast growth factor receptor 3
Di George	22q11.2	*TBX1, CRKL*	250-kB Microdeletion of 22q11.2 containing TBOX gene 1 (*TBX1*) and *the v-crk sarcoma virus CT10 oncogene homolog [avian]-like* (CRKL)
Down	Trisomy 21		Additional chromosome 21
Ectodermal dysplasia	2q11–q13	*EDAR*	Ectodysplasin 1, anhidrotic receptor
Goldenhar[b]	?	?	
Hirschsprung[c]	5p13.1–p12	*GDNF*	Glial cell line derived-neurotrophic factor
Holoprosencephaly	7q36	*SHH*	Sonic hedgehog
Kallman syndrome xx	8p12	*FGFR1*	Fibroblast growth factor receptor 3 (10% of cases)
Gonadotropin-releasing hormone 1	4q13.2	*GNRH*	
Neurofibromatosis, type 1[d]	19q11.2	*NF1*	Neurofibromin
Neurofibromatosis, type 2	22q12.2	*NF2*	Neurofibromatosis 2
Proteus	10	PTEN	*Phosphatase and tensin homolog* tumor-suppressor gene
Treacher Collins[e]	5q32–q33.1	TCOF1	*Treacher–Collins Franceschetti syndrome 1* gene coding for nucleolar phosphoprotein Treacle

Table 9.3 (continued)

Syndrome	Chromosome location	Gene symbol	Protein
Waardenburg	2q35	*PAX3*	Paired box homeotic gene 3
Waardenburg–Shah	22q13	*SOX10*	SRY (sex-determining region Y)-box 10
Wermer (Sipple) syndrome	8p12–11.2	*WRN*	Werner gene producing a helicase

[a] Only those syndromes discussed or mentioned in the text are included.
[b] Also known as Oculo-Auriculo-Vertebral/OAV syndrome.
[c] Several other forms of Hirschprung disease are known (Table 9.1), one based on changes in *EDNRB* (endothelin receptor B) located on chromosome 13q22, another based on changes in *RET*, the RET transforming sequence, located on chromosome 10q11.2.
[d] Also known as von Recklinghausen neurofibromatosis.
[e] Also known as mandibulofacial dysostosis and Franceschetti–Zwahlen–Klein syndrome.

neuroblastomas, von Recklinghausen Neurofibromatosis, APUDomas, and DiGeorge Syndrome.

CHARGE Syndrome

The diversity of tissues involved in a neurocristopathy is illustrated in the CHARGE syndrome, an acronym from **C**oloboma of the iris, **H**eart defects, **A**tresia of choanae, **R**etardation of physical and mental development, **G**enital anomalies, and **E**ar anomalies and/or deafness (Table 9.1). The acronym, which was coined by Pagon *et al.* (1981), now applies to several hundred affected individuals, the prevalence being as high as one in 10,000 live births. Heart, nasal and oral cavities, brain, reproductive organs, ears, and the overall growth are affected; all affected individuals have hearing loss and developmental delays, and almost 90% have defective eye tissues. NCCs make a substantial contribution to ocular tissues, many of which are under the regulation of *PAX6*.

PAX6 *and* CHD7

Eye defects in *Pax6*–heterozygote mice (*Pax6$^{+/-}$*) can be traced back to the abnormal deployment of NCCs into the developing eyes. In early chicken embryos, progression of *Pax6* activity along the developing neural tube is controlled by caudal regression of *Fgf8*, a temporal signaling system that helps to set up the caudal and rostral patterning of the neural tube (Box 9.1). It is unclear whether defective *PAX6* in humans with the CHARGE syndrome can be traced to a similar early stage.[5]

Box 9.1 Fgf8, patterning, and symmetry in zebrafish

Studies with zebrafish embryos demonstrate that *Fgf8* is required to establish the correct patterning—whether symmetrical or asymmetrical —of the brain, craniofacial and pharyngeal skeleton, heart, and alimentary canal.

In chicken, but not zebrafish embryos, *Fgf8* functions to regulate cell death among NCCs, as well as regulating proliferation and migration. In part, signaling in chicken and zebrafish is indirect, *Fgf8* acting via cranial mesoderm, brain, and pharyngeal endoderm, *Fgf8* (and *Fgf3*) being required for the formation and patterning of pharyngeal endoderm in zebrafish (Crump *et al.*, 2004).[a]

Fgf8 regulates the asymmetrical expression of Nodal, a molecule involved in establishing left–right asymmetry in Hensen's node early in the development of many if not all amniotes by virtue of Nodal's asymmetrical distribution in the lateral plate mesoderm. Referred to as the **Nodal cassette**, a signaling cascade that is highly conserved across the vertebrates, *Fgf8* functions to establish left–right embryonic axis by regulating enhancers of symmetry, such as *Pitx2*, and inhibitors of *Nodal* signaling and enhancers of L–R asymmetry such as *Lefty1* and *Lefty2*.[b]

Fgf8 is also required for the formation of Kupffer's vesicle, which is the zebrafish equivalent of Hensen's node. Left–right laterality of zebrafish pharyngeal skeletons is not established in the absence of Kupffer's vesicle, indicating a patterning role for Kupffer's vesicle equivalent to that played by Hensen's node in tetrapods. To that end, it is notable that Hensen's node *and* Kupffer's vesicle contain ciliated cells and that directed ciliary movements provide the cellular basis for asymmetry.[c]

---

[a] Reduced functioning of *Fgf8* in zebrafish results in craniofacial defects and in alteration of left–right craniofacial symmetry (Albertson and Yelick, 2007).
[b] Inhibiting *Nodal* signaling by Lefty2 or Cerberus (Cerb-S) promotes neuronal differentiation in human embryonic stem cells (J. R. Smith *et al.*, 2008).
[c] See Albertson and Yelick (2005*) and Kreiling *et al.* (2007) for the zebrafish studies.

Insofar as one can associate a syndrome with a single causative gene, the causative gene for CHARGE syndrome is *Chromodomain helicase DNA-binding protein7* (*CHD7*), a gene ubiquitously expressed in a number of human tissues. Indeed, a heterozygous mutation of *CHD7* occurs in some 60% of individuals with the syndrome. To obtain a model for embryonic expression in humans, the chicken ortholog, *Chd7*, was sequenced. *Chd7* is a pan-neuronal marker in early embryos (2–3.5 days incubation) when the neural tube and placodes are laid down and specified. *Chd7* is expressed in olfactory, optic, and otic placodes, in the pharyngeal arches, but not in nonneuronal ectoderm, a pattern consistent with the range of tissues affected in individual humans with CHARGE syndrome.[6]

Neuroblastomas

Malignant neoplasia is rare in humans, the estimated occurrence being one in 30,000 live births. Half the malignant neoplasias in human fetuses or newborns are primary leukemias or Wilm's tumor of the kidney. The other half consists of *neuroblastomas*, tumors of the adrenal medulla and of the autonomic nervous system (Table 9.1). Neuroblastomas are neurocristopathies because their neoplastic cells are exclusively NC in origin, although not all neuroblastomas are identical; cell lines derived from different neuroblastomas respond differently to Tgfβ with respect to proliferation, morphology, and the synthesis of ECM products (Rogers *et al.*, 1994*).

Malignant tumors of NC origin are somewhat more common in children than in adults, although the vast majority of potential neuroblastomas never appear clinically and so are not diagnosed. A high incidence (40 times expected) of microscopic neuroblastomas is reported in the adrenal glands of infants below 3 months of age. Such tumors could remain dormant, regress, or continue to progress to become tumors of the sympathetic nervous system such as malignant ganglioneuromas or ganglioneuroblastomas, or neuroendocrine tumor of the chromaffin cells of the adrenal gland (pheochromocytomas).

A survey of 90,000 surgical biopsies of malignant tumors undertaken in the mid-1970s revealed ten skin tumors, each containing two or three neoplastic tissues whose normal counterparts are derived from the NC. These were (with the normal counterpart in brackets) malignant melanomas (melanocytes), gangliomas (neurons), malignant Schwannomas (Schwann cells), and chondromas (chondroblasts).[7] Transmission electron microscopic and/or enzymatic analysis can also provide evidence of NC origin, as in a human soft tissue sarcoma that displayed Schwann cell (myelin sheaths and myelinated axons) and fibroblast differentiation (cross-banded collagen), or a clear cell sarcoma of tendons and aponeuroses.

Analysis of cell lines can provide evidence of the NC origin of neuroblastomas and of the ability of cells derived from neuroblastomas to modulate their phenotype. La-N-5, a human neuroblastoma cell line, differentiates into cholinergic neurons if challenged with retinoic acid. Because neuronal, Schwann, and melanocytic cells develop *in vitro*, these cell lines have been interpreted as recapitulating the development of NCCs, although reversion to NCCs is not complete.

The Neoplastic State

Neoplastic cells are identified by the production of tumor-specific products. They are neither embryonic cells expressing some abnormal products, nor are they abnormal cells expressing some embryonic products (although neoplastic cells do express fetal antigens). Neoplasia represents an alternate differentiation fate, and not regression to a prior state.

Neuroblastomas often consist of cells that are morphologically identical to migrating NCCs. They appear to be mesenchymal cells that have failed to differentiate. Macrophage inhibitory factor (MIF), which is highly expressed in neuroblastomas, is expressed in NCC and is a chemoattractant for NCCs. Such cells often undergo spontaneous regression to nonmalignant cells forming a ganglioneuroma or a neurofibroma, rather than to normal neurons or chromaffin cells. An intriguing question is obviously why they do not (cannot?) transform to normal cells. The answer lies partly in the neoplastic state as a stable differentiated state that differs from embryonic cells and from normal fully differentiated cells. In the transformation to a neuroblastoma, an NCC expresses a new pathway of differentiation that is only partially reversible; the cell can go back along the neoplastic pathway but not as far back as when the decision to become neoplastic was made.[8]

Diagnosis

Techniques used in nuclear medicine facilitate the detection and treatment of neuroblastomas.

Neuroblastomas and pheochromocytomas can be detected with [135]iodine metaiodobenzyl-guanidine. Pheochromocytomas and melanotic neuroblastoma (Table 9.1) are associated with abnormal synthesis of catecholamine. Diagnosis is facilitated by measuring the urinary level of vanillinemandelic acid (VMA), which is a metabolite of catecholamine. Mean urinary concentrations of 2.68 VMA μg/mg creatine (range 0.5–4.8 μg) for tumor-free individuals and 15.7 μg/mg (range 4.5–50) for those with pheochromocytoma, are typical.

The 7S Ngf receptor is lost from NCCs as they differentiate; a monoclonal antibody against the receptor has been used to separate multipotential NC stem cells that form neurons or Schwann cells but does not isolate differentiated neurons or Schwann cells. The cells of some tumors such as human melanomas, however, regain the Ngf receptor. Reappearance of the receptor on tumor cells is one line of evidence of the NC origin of the tumors. Further evidence comes from monoclonal antibodies generated against neuroblastomas. These antibodies cross-react with ganglia, with satellite and chromaffin cells, and can be used to screen for neuroblastomas.[9]

All the neurofibromas removed by Riopelle and Riccardi (1987) from individuals with von Recklinghausen neurofibromatosis (see below) and established in culture-released Ngf, implicating the local production of Ngf in the development of neuronal elements within the tumors. However, neither Ngf, nor indeed any other growth or humoral factor, modulates, let alone alleviates, neurofibromatosis; the evidence of primary humoral imbalance in neurofibromatosis is tenuous.

RaLP

RaLP is a new member of the Src family of tyrosine kinase substrates, which maintain cell proliferation, survival, and apoptosis, as well as mitochondrial membrane permeability. *Src*, the canonical member of this family, is required for development of the CarNCCs in the outflow tract of the heart.

RaLP encodes a 69-kDa protein only found in melanomas in adults. *RaLP* has now been shown by Fagiani and colleagues (2007) to be expressed in nine human melanoma cell lines. Notable findings in this study are:

- *RaLP* is expressed at low levels in normal melanocytes.[10]
- *RaLP*-enhanced expression is restricted to invasive and metastatic melanoma cells (expression is high when melanoma cells become migratory and begin to invade other tissues).
- Silencing *RaLP* expression in metastatic melanoma cells using RNAi reduces tumorogenesis *in vivo*.
- Ectopic expressing of *RaLP* in *RaLP*-negative melanocytes or nonmetastatic melanoma cells provides a substrate that activates two receptors (IGR1R and EGFR), resulting in the *initiation* of Ras/mitogen-activated protein kinase (MAPK) signaling and the initiation of cell migration.[11]
- In contrast, silencing *RaLP* in *RaLP*-positive melanoma cells inhibits migration *without affecting* MAPK signaling.

The conclusions—that *RaLP* is a marker for melanomas, is involved in regulating migration/metastasis, and acts via Ras-dependent- and independent pathways—have important implications for the next phase of research on melanomas, including pharmacological approaches.

Model Systems

Model systems for simple neurocristopathies such as NCC tumors are available. C-1300 mouse neuroblastoma maintained *in vivo* or as a cell line continues to produce axon-like extensions (typical of neurons) and to synthesize catecholamines (typical of adrenergic neurons). Neural crest tumor cell lines contain lectin receptors and respond to lectins by increased proliferation; that is, the balance between proliferation and cytodifferentiation, so important in neoplasia, can be manipulated *in vitro*.[12]

Direct experimental evidence of response to tumor-promoting agents of NCCs or of tumors derived from NCCs comes from studies using TNCCs or cell lines established from tumors (neuroblastoma, melanoma, or pheochromocytoma) with an NC origin. Such treated NCCs divide more rapidly, delay pigmentation, and block the adrenergic phenotype, showing that tumor promoters can redirect the differentiation of NCCs.[13]

von Recklinghausen Neurofibromatosis

Neurofibromatosis is one of the most common, if not the most common, tumor involving neural and pigment cells. With an occurrence of 1 in 2,500 to 1 in 3,000 live births (Table 9.1), the US alone has 100,000 affected individuals. Scientific interest in neurofibromatosis was reawakened in the 1980s with the inauguration of a National Neurofibromastosis Foundation in the US, a new journal, and publication of the proceedings of two workshops and a comprehensive treatise on the syndrome.[14]

Included in the proceedings of the 1981 conference edited by Riccardi and Mulvihill is a translation of the classic account of neurofibromatosis from the 1882 treatise by the eminent German pathologist, Friedrich Daniel von Recklinghausen—a clinical description of two individuals with neurofibromatosis and the first clear demarcation of the syndrome. Although he identified neurofibromatosis as a distinct entity, von Recklinghausen did not describe the first individuals or name the condition, which is now known as *von Recklinghausen neurofibromatosis* or *type 1 neurofibromatosis* (NF1).

An estimated mutation rate of 1 mutation/10,000 gametes/generation ranks *von Recklinghausen* neurofibromatosis as one of the two human genes with the highest mutation rates. The other is the X-linked gene for Duchenne muscular dystrophy. The high mutation rate in the muscular dystrophy gene may be attributed to the large size of the gene, which spans at least 33 Kb of genomic DNA. The high mutation rate of genes affecting neurofibromatosis, however, probably requires a more complex explanation, reflecting:

(a) heterogeneity of the neurofibromatoses, which are categorized into at least four types (central, Schwannoma, segmental, and other variant neurofibromatosis);

(b) the finding that individual syndromes involving more than one set of genes are often pooled in single analyses; and

(c) misidentification of other syndromes as von Recklinghausen neurofibromatosis.

The lack of homogeneity in individuals reliably diagnosed with von Recklinghausen neurofibromatosis is consistent with low hereditability and single locus mutations producing the majority of new cases. The *neurofibromin1* (*NF1*) gene was mapped to the long arm (pericentromeric region) of chromosome 17 (17q11.2) in 1987 and cloned in 1990. Over 1,000 mutations of the gene have now been associated with type 1 neurofibromatosis. Other neurofibromatoses are genetically distinct; central or bilateral acoustic neurofibromatosis (NF2), for example, results from mutation of *neurofibromin 2* (*NF2*), which maps to chromosome 22.

NF1, a tumor suppressor gene, encodes a 2,818-amino-acid-long protein neurofibromin that is expressed in neurons and Schwann cells. NF1 regulates cell division via the ras pathway. The gene is complex; embedded within its introns are several other genes transcribed in the opposite orientation. Targeted disruption of *NF1* produces syndromes, involving heart malformations and hyperplasia of sympathetic ganglia.[15]

The avian neurofibromatosis-1 gene (*aNf1*), cloned as a 432-bp cDNA, has 82% similarity to the human gene at the nucleic-acid level and 93% similarity to the human protein sequence. A probe to the 12.6-Kb transcript of the chicken gene showed that the gene is expressed from H.H. stage 11 to adult life. While mRNA is expressed ubiquitously, the protein is most highly expressed in a subset of migrating NCCs, and subsequently, in a subset of NC derivatives.

Involvement of Non-Neural-Crest Cells

Non-NCCs are involved in neurofibromatosis; both secondary involvement of mesodermal derivatives—scoliosis in 53% of affected individuals and tibial pseudarthroses in 19% of the same individuals—and cardiac malformations. Malignancies not NC in origin—leukemia, stomach cancer, and Wilm's tumor of the kidney—are also associated with neurofibromatosis.

Investigators have gone to considerable lengths to explain the involvement of non-NC derivatives in neurocristopathies, explanations including:

- *genetic pleiotropy* —the same gene is affected in more than one tissue—the most likely explanation;
- *displaced migration and metaplasia* —NCCs migrate into and differentiate ectopically within tissues that normally lack them;
- *epigenetic*, because the microenvironment surrounding more than one cell type is affected; or
- the production of a *stimulatory factor* (*s*) that acts on another cell type.

John Merrick—the 'Elephant Man'

von Recklinghausen neurofibromatosis received much publicity because of a single case, Joseph Merrick (1862–1890), the 'Elephant Man', the subject of magazine articles, a Broadway play, a motion film, and at least four popular books.

Joseph Merrick was diagnosed as having the most severe expression of von Recklinghausen neurofibromatosis, characterized by grossly abnormal skin (hence the name Elephant Man) and severe skeletal abnormalities, including overgrowth and exostoses of the calvarium (which is NC in origin) and hemihypertrophy of such mesoderm-derived long bones as the femur and radius.

The difficulty with the diagnosis of von Recklinghausen neurofibromatosis in this case is the lack of *café-au-lait* spots, which are diagnostic for the syndrome (Table 9.1), and the severity of the skeletal overgrowth and hypertrophy, especially since the majority of the affected skeletal elements were mesodermal, not NC in origin.⊕ Furthermore, no other individual diagnosed with von Recklinghausen

---

⊕ The phenotype of human bone cells maintained *in vitro* can be correlated with their embryological origin. Comparison of osteogenic cells from the mandible (which arise from NC) with cells from the iliac crest (which are mesodermal in origin) of four individuals showed that mRNA for *Fgf2* and *IgfII* increased in mandibular cells, while mRNA for alkaline phosphatase decreased (Kasperk *et al.*, 1995). mRNA for Tgfβ; increases in iliac crest cells, which also divide more slowly than do mandibular cells.

neurofibromatosis has such skeletal involvement. The skeletal defects are secondary to a primary defect in the NC; neurofibromas produce a bone-matrix-stimulating factor(s) that could account for the skeletal overgrowth and hemihypertrophy.

The solution to this dilemma turned out to be devilishly simple. Joseph Merrick did not have von Recklinghausen neurofibromatosis. Rather, he exhibited an even more rare condition, **Proteus syndrome**, of which only 200 individuals have been diagnosed worldwide, and of which Joseph Merrick's is the most severe ever.[16]

Proteus syndrome is characterized by macrocephaly, lipomas, hyperostosis of the skull, hypertrophy of the long bones and thickening of the skin and dermis, including plantar hyperplasia. Although Proteus syndrome is now known not to be a neurocristopathy, its genetic basis is unresolved. The tumor-suppressor gene, *phosphatase and tensin homologue* (*PTEN*) on chromosome 10, which is mutated in some individuals with multiple advanced cancers, may be involved (Loffeld *et al.*, 2006).

Animal Models and Mutations

A difficulty in determining the mechanism of pathogenesis of neurofibromatosis was that no mammalian animal models were available until the late 1980s, a surprising situation given the plethora of mutants affecting NC derivatives. Neurofibromas are occasionally seen in cattle, horses, dogs, and birds, but in these animals they arise following a viral infection and not as an inherited defect. Furthermore, these neurofibromas do not involve pigment cell or skeletal defects.

Although some mouse mutants exhibit pleiotropic effects on several NC derivatives (pigment cells, nerve cells, and cells of the craniofacial skeleton), none mimic neurofibromatosis. One mutant discussed earlier, *Patch*, results in abnormal pigmentation, craniofacial defects, and defects of the thymus, heart, and teeth. In contrast to situations in which mutations in different genes produce the same phenotype, interactions between genes can produce syndromes not seen with either gene alone. Thus, mice that are double mutant for *Patch* and *Undulated* display an extreme form of spina bifida; single mutants, on the other hand, show no evidence of even mild spina bifida.[17]

Transgenic mouse models are available in which Schwannomas and facial bone tumors can be investigated. One is associated with osteogenic sarcomas of the facial bones with extension into the periodontal ligament—both tissues that originate in the NC—making this a good model for neurocristopathies. In 1987, Hinrichs and colleagues described a transgenic mouse model utilizing tumors induced by the *TAT* gene of human T-lymphotropic virus type 1, which closely resemble tumors in von Recklinghausen neurofibromatosis.

The same phenotype can result from the function of two different genes or genetic pathways if the two genes act in different ways. For example, mutations of the loci *Dominant Spotting* and *Steel* result in similar defects in melanocytes. *Dominant Spotting* affects NCCs directly. *Steel* affects the ECM, altering collagen and glycosaminoglycans, and thereby suppressing melanocyte development. Steel factor (c-kit ligand, stem cell factor, mast cell growth factor), a receptor and the product of the *Steel* locus, along with its ligand, the c-kit receptor tyrosine kinase, is required for the survival but not for the differentiation of melanocyte precursors. c-Kit receptor tyrosine kinase is required in a subpopulation of cells to facilitate melanocyte differentiation during fin regeneration in adult zebrafish (Rawls and Johnson, 2001).

c-Kit signaling, which is regulated by *Sox10* and *Mitf* in mice and by *Mitfa* in zebrafish, is required for the migration of melanocytes in mice and of melanophores in zebrafish; in the *Mitfa* zebrafish mutant, *Nacre*, only melanocyte formation is inhibited. Induction of *Sox10* in zebrafish NCCs in response to *Wnt* signaling may be required to make NCCs competent to produce *Mitf* as a response to a second Wnt signal in a pathway that is required to specify NCCs as melanophores. Very recently, Miller *et al.* (2007) provided important insights into the evolution of regulation across the vertebrates when they showed that parallel evolution of *cis*-regulation of *Kit ligand* (*Kitlg*), the gene for c-kit ligand, elicits light coloration in marine and freshwater species of threespine sticklebacks and in humans.[18]

APUDomas

The acronym APUD—for cells that exhibit *Amine Precursor Uptake and Decarboxylation* —was coined in 1969 for:

- polypeptide-hormone-synthesizing cells of the gut,
- calcitonin-producing cells of the ultimobranchial bodies of the thyroid gland,
- cells of the pancreas and adrenal gland, and
- cells of the carotid body (a chemoreceptive organ).

These cells share the uptake and decarboxylation pathway and are all NC derivatives. Like neurocristopathies, the APUD concept seeks to explain a suite of characters on the basis of commonality of developmental origin.

APUD cell types—there may be as many as 40—decarboxylate the catecholamine precursor, 3, 4-dihydroxyphenylalanine (known as DOPA) or the serotonin precursor, 5-hydroxytryptophan. Either amine can be visualized using a fluorescent or radioactive label, facilitating cytochemical visualization of APUD cells. The amines and/or peptides function as hormones or as neurotransmitters for what is termed a **diffuse neurendocrine system** involved in regulating the autonomic nervous system, the somatic division of the nervous system, and nonendocrine cells, including other APUD cells. Indeed, injecting a neurotransmitter such as serotonin (Box 9.2) into newborn mice leads to the production of tumors involving APUD cells (APUDomas), while exposing newborn mice to adenosine causes dysplasia of NC and multiple NC tumors, including tumors containing cartilage and bone surrounded by melanocytes. Neoplastic changes can be elicited experimentally in APUD cells of NC origin by administering steroid hormones or endotoxin.[19]

Box 9.2 Serotonin and NCC migration

Serotonin (5-hydroxytryptamine, 5-HT) may play a role in regulating migration *and* morphogenesis of NCCs.

Serotonin transporter messenger RNA is expressed in rat NCCs and in NC-derived tissues, such as autonomic ganglia, tooth primordia, adrenal medulla, chondrocytes, and neuroepithelial cells in the skin, heart, intestine, and lung. Transporter mRNA is expressed by sensory ganglionic neurons and by neuroepithelial cells, both of which are targets for outgrowing sensory neurons.

Craniofacial Development

Serotonin inhibits the migration of murine NCCs maintained in Boyden chambers; the 5-HT2B serotonin receptor is expressed on the surfaces of NCCs, and on enteric neurons and myoblasts of the small intestine and stomach (Fiorica-Howells *et al.*, 2000). Serotonin is expressed transiently in murine embryonic facial epithelia (palatal, tongue, nasal septum, maxillary, mandibular), at times (especially days 10 and 11 postconception) that correlate with morphogenesis.

Shuey *et al.* (1992) proposed that serotonin is a regulator of craniofacial morphogenesis. Site-specific malformations occur if mouse embryos are exposed to serotonin-uptake inhibitors. The mechanisms involve decreased proliferation and increased cell death of mesenchyme located five or six cell diameters away from the epithelium and increased proliferation in subepithelial mesenchyme. Shuey and colleagues concluded that serotonin regulates the epithelial–mesenchymal interactions that initiate craniofacial differentiation and morphogenesis. Injecting serotonin into newborn mice elicits death of NCCs and the production of APUDomas (this chapter), again implicating serotonin and serotonin receptors on NCCs in development and dysmorphogenesis. 5-HT2B receptors cloned from mice and humans, along with receptors 5-HT2A and 5-HT2C, modulate mitogenic signals from serotonin. Inhibiting 5-HT2B inhibits NCC migration and elicits their death by apoptosis (Choi *et al.*, 1997).

Heart Development

Serotonin also affects morphogenesis of the cardiac cushion in murine embryos, in part because of its action on the NC derivatives; although mesoderm-derived mesenchymal cells also contribute to the cushion.

Inactivating the gene *5-HT2B*, which codes for the 5-HT2B serotonin receptor, leads to heart defects that cause embryonic and neonatal death, indicating that the 5-HT2B receptor is an important regulator of cardiac development, as it is of craniofacial development. Other neurotransmitters may regulate NCC differentiation; inhibiting norepinephrine with uptake inhibitors inhibits the adrenergic phenotype in clonal cultures of pluripotential TNCCs.[a]

---

[a] See Sieber-Blum (1990*), Moiseiwitsch and Lauder (1995), and Yavarone *et al.* (1993) for serotonin distribution and function and Canan *et al.* (2000) and Snarr *et al.* (2007) for cardiac development.

Publication of the APUD concept led Nicole Le Douarin and her collaborators to investigate the NC origin of APUD cells. Fontaine (1979) grafted fragments of the thyroid rudiment, pharyngeal pouch, or pouch endoderm/mesenchyme in various combinations from 18 to 45 somite (9–11 days) mouse embryos onto the chorioallantoic membrane of embryonic chickens. Calcitonin-synthesizing cells (C-cells) first appear in pharyngeal pouch (4th pharyngeal arch) mesenchyme, then in the endoderm and finally in the thyroid. Thus, C-cells are NC in origin in mice as they are in birds.

Approximately half of the 40 APUD cell types are NC or neurectodermal in origin, including calcitonin-synthesizing cells and cells of the carotid body, adrenal gland, and aortic paraganglia. APUD cells that are not NC in origin include the endocrine islet cells of the pancreas and the enterochromaffin cells of the gut and respiratory tract.[20]

Hirschsprung Disease

Hirschsprung disease (aganglionic megacolon), with an incidence of 1 per 5,500–7,000 live births in the United States—possibly as high as 1 per 1,500 live births in some countries, with an incidence 4 times higher in males than females—is a potentially fatal disease in which the caudal part of the intestine (the large intestine or colon) lacks enteric ganglia; the small intestine may also lack ganglia, producing what is known as *short-segment disease*. Effects of Hirschsprung disease are specific to VNCCs (caudal hindbrain), the source of the enteric ganglia, and in which *SOX10* functions cell autonomously. In a mathematical model developed by Landman *et al.* (2007) to investigate development of the enteric nervous system and the etiology of Hirschsprung disease, directional migration of NCCs is driven by cell proliferation, which takes place only in a small region at the forefront of the invading cell population.

VNCCs colonize the gastrointestinal tract progressively in a rostrocaudal wave. In humans, NC-derived neuroblasts are normally present in the small intestine by 7 weeks of gestation and in the colon by 12 weeks. As in *Lethal Spotted* (see Chapter 3) and *Piebald Lethal* mouse mutants, it appears that NCC migration is not impaired in Hirschsprung disease. Rather, and again as in mouse mutants, the environment of the intestine cannot support the differentiation of enteric neurons.

Although cardiac and skeletogenic NCCs are unaffected, some 20% of individuals with Hirschsprung disease also have another syndrome—typically Down syndrome, Waardenburg–Shah syndrome (see Chapter 4) or a neurocristopathy—that does affect other NCC populations. Hirschsprung disease is most often associated with a mutation in *SOX10* (Brooks *et al.*, 2005), although defects in 11 genes have been associated with the syndrome.[21]

The genetic basis of Hirschsprung disease is not straightforward, however. Interactions between two proteins coded for by mutated forms of two genes have been identified in some affected individuals. One is a mutation in the *RET* proto-oncogene, resulting in loss of function of the RET receptor on those NC-derived neuroblasts that normally would differentiate into enteric ganglia. RET interacts with the G-protein-coupled receptor, endothelin receptor B (EDNRB), encoded by the gene *EDNRB*, which is required to connect enteric neurons to the intestine. Although other genes (*SOX10, EDN3, ECE1, NTN, SIP1*) are associated with Hirschsprung disease, the disease is not manifest without mutations in both *RET* and *EDNRB* genes. Mutations in the RET gene are also associated with neuroblastomas and with Down syndrome, which is found in 6% of individuals with Hirschsprung disease.[22]

In this disease, species specificity makes it difficult to extrapolate results from rodents to human; *EdnRB*-deficient enteric NCCs in rat embryos migrate normally into *EdnRB*-deficient intestines and differentiate into enteric neurons. *Ret* transduces signals from Gdnf (glial cell-line-derived neurotrophic factor) to regulate survival, proliferation, migration, and differentiation of cells derived from rodent enteric NC. The genes *Gfra*, which codes for the Ret co-receptor, and Edn3 and its receptor gene *EdnRB*, along with several transcription factors, including *Sox10, Pax3*, and *Phox2b*, also regulate enteric ganglionic precursors. The *EdnRB* pathway is required for normal migration of enteric neuronal and melanocyte precursors in chicken embryos, but not for specification or survival of either cell type.[23]

In a recent study by D'Autréaux and colleagues (2007), enteric neurons failed to develop in mice with a mutation in *Sip1*, a Smad1-interacting gene whose ortholog is associated with Hirschsprung disease in humans. Murine enteric ganglia require expression of *Hand2* but not *Hand1* (basic helix–loop–helix transcription factors) for terminal differentiation, but not for earlier stages of enteric neurogenesis or to colonize the gut, revealing some of the genetic complexity underlying enteric neuronal development in different species. Caution also has to be exercised when extrapolating consequences of mutation in or knockout of a mouse gene that is orthologous to the human gene; because protein sequences change adaptively in different species, null mutations in human and mouse orthologs often result in different phenotypes (Liao and Zhang, 2008).

DiGeorge Syndrome

DiGeorge syndrome (athymic agenesis) is the most complex of the neurocristopathies and has a frequency of 1 in 4,000 live births.

The heart, thymus, and parathyroid glands fail to develop or are severely underdeveloped, with accompanying craniofacial anomalies (see Box 9.3 for the thymus).[24] Abnormal development of the aortic arches and craniofacial malformations such as micrognathia, palatal, and ear defects characterize DiGeorge syndrome; 97% of the 161 individuals with DiGeorge syndrome examined by van Mierop and colleagues in the mid-1980s had heart defects. This association in the one syndrome of thymic, cardiac, and craniofacial crest derivatives, coupled with experimental evidence from chicken embryos, is consistent with a defective NC as the basis for defects. This claim is made even though DiGeorge syndrome is: (i) not only a NC defect (much of the heart is NC in origin); (ii) although many structures that arise from the NC are not involved in the syndrome; and (iii) although the involvement of NCCs may be secondary to primary defects in pharyngeal epithelia or mesodermal mesenchyme (see below). The rationale applied in regarding CHARGE syndrome as a neurocristopathy applies equally to grouping DiGeorge syndrome with the neurocristopathies.

Box 9.3 The thymic neural crest

Lymphoid stem cells are transported to the rudiment of the thymus via the developing vascular system. The thymic epithelium, a derivative of the pharyngeal endoderm, is stimulated to divide after encountering migrating NCCs. Thymic epithelial cells then stimulate the proliferation of lymphoid stem cells and mediate their differentiation into thymocytes in response to the winged-helix transcription factor gene *forkhead box N1* (*Foxn1*)[a] and the Tbox gene *Tbx1*. The NC also contributes connective tissue cells and endothelial cells to the blood vessels of thymus glands. Interestingly, NCCs from various rostrocaudal regions of the neural tube can participate in thymic development under experimental conditions, even if they do not during normal development.[b]

Hoxa3 mutant mice are athymic, a condition that may reflect *Hox3a* regulation of *Pax1*. The thymus is either underdeveloped or fails to form in mice in which Fgf8 has been knocked down. Absence or severe reduction of the thymus in humans is associated with the cardiac defects in the

---

DiGeorge syndrome. Such associations, recognized as 'neural crest defects' or neurocristopathies, are the topic of Chapter 9.[b]

---

[a] A mutation in *FOXN1* correlates with T-cell immunodeficiency in humans.
[b] See Le Douarin (1982) for NC contributions to the avian thymus, T. J. Wilson *et al.* (1992) for the microenvironment in which the avian thymus develops, Kuratani and Bockman (1991) for studies on axial levels, Manley and Capecchi (1998) for *Hox3a*, and Abu-Issa *et al.* (2002) for *Fgf8*.

Genetics

A major reason for the variability in DiGeorge syndrome lies in its genetic (chromosomal) basis and in the many functions of deleted genes. DiGeorge syndrome is the most frequent microdeletion syndrome in humans; that is, the most common syndrome based on the elimination of a region of a chromosome, in this case, deletion of a multigene 250-Kb region from chromosome 22; 90% of affected individuals share a deletion of approximately 30 genes from the 22q11.2 region of chromosome 22.

The phenotype of individuals with DiGeorge syndrome overlaps the phenotypes of individuals with velo-cardio-facial syndrome or with conotruncal anomaly. Eighty percent or more of individuals with these three syndromes have deletions of one allele of chromosome 22q11.2. Consequently, the three syndromes—DiGeorge, velo-cardio-facial, and conotruncal anomaly—have been grouped as the ***CATCH-22 syndrome*** for **C**ardiac defects, **A**bnormal facies, **T**hymic hypoplasia, **C**left palate, **H**ypocalcemia, associated with chromosome **22** microdeletion.

Genes Involved

DiGeorge syndrome is characterized by pleiotropic defects involving genes that regulate the degradation of proteins. Surprisingly—because it is not usually considered as a mechanism that determines cell fate—**protein degradation** may be involved in switching cell fate in early neural lineages; blocking the function of the *Xenopus* gene *Cullin1*, whose product is an E3 ubiquitin ligase involved in the stabilization of β-catenin, results in more cells being specified as NCCs and the additional cells forming melanophores rather than neurons of cranial ganglia; that is, *Cullin1*-mediated protein degradation determines the fate of cells generated in its presence (Voigt and Papalopulu, 2005).

Several mouse models produce phenotypes similar to DiGeorge syndrome— 'similar to' because some involve genes deleted in DiGeorge individuals, while others involve genes outside the deleted region but that modulate the severity and the penetrance of the syndrome. All the genes studied so far affect pharyngeal arch development, whether they lie within the microdeleted region or not. Two genes within the deleted region in the human genome, *TBX1* and *CRKL,* are potential central players. *Tbx1* is an ancient gene; *AmphiTbx1/10*, the ortholog of *Tbx1* and *Tbx2*, is expressed in amphioxus branchial arch endoderm and somitic mesoderm. Keep this in mind as we explore the role of *TBX1* in humans.

TBX1 and CRKL: The *TBOX* gene, *TBX1*, is located within the 22q11.2 region of human chromosomes 22. *TBX1* is expressed in epithelial and mesodermal cells of the pharyngeal arches but not in arch mesenchyme of NC origin. Mutations in *TBX1* have been identified in individuals with DiGeorge syndrome, implicating deletion of *TBX1* in the onset of the syndrome. Furthermore, the inactivation of *Tbx1* in mice is associated with a phenotype typical of DiGeorge syndrome.

Tbx1 also regulates development and remodeling of the arteries of the pharyngeal arches. Expression of *Tbx1* in mice, especially in pharyngeal epithelia, is affected by *Shh*, vascular endothelial growth factor (*Vegf*) and the Bmp antagonist, *Chordin* (see Box 4.1), revealing several potential pathways involving genes that influence onset of the syndrome but that lie outside the microdeleted region; targeted inactivation of *Chordin* in mice is associated with a syndrome equivalent to DiGeorge syndrome, and is embryonic lethal. A *cis*-regulator of *Shh*, located upstream of *Tbx1* and required to regulate *Tbx1* in pharyngeal endoderm, has a Fox (forkhead transcription factor) binding element, bringing the Fox proteins Foxc1, Foxc2, and Foxa2 into play in regulating *Tbx1*. *Tbx1* is also genetically linked to *Fgf8* (*Tbx1* regulates the expression of *Fgf8* in pharyngeal endoderm, and this source of *Fgf8* is required for normal development of the pharyngeal arches and the thymus in mice), and is linked to *Fgf10* (*Tbx1* can upregulate *Fgf10*); disrupting the *Shh* network is associated with a number of genetic disorders affecting NCCs, including holoprosencephaly (Box 10.1) and medulloblastoma.[25]

CRKL: Another candidate gene deleted from region 22q11.2 is *CRKL* (*v-crk sarcoma virus CT10 oncogene homolog [avian]-like*).

CRKL encodes a 39-kDa protein involved in the cellular response to growth factors and to other molecules involved in focal adhesion; the protein couples to the cell surface transmembrane protein CD 34. *Crkl* is linked to *Tgfβ* signaling in NCCs, and Tgfβ and Tgfβ-receptors both are essential for skeletogenesis from craniofacial NCCs; inactivating *alk2*, a Bmp type 1 receptor in mice, results in underdevelopment of the mandibles and cleft palate.[26]

Crkl regulates the expression of *Fgf8*, indicating a potential link back to *Tbx1*. Fgfs in pharyngeal endoderm pattern pharyngeal skeletal elements (Box 9.1) and haploinsufficiency of *Fgf8* in zebrafish results in craniofacial defects and in alteration of left–right craniofacial symmetry (Albertson and Yelick, 2007).

Hox genes: Analysis of animal models has revealed the potential involvement of *Hox* genes in DiGeorge syndrome.

Disrupting *Hoxa5* by gene targeting in mice produces homozygous (*Hoxa5*[−/−]) embryos that lack the thymus and parathyroid glands, and that have throat, heart, and craniofacial defects (Chisaka and Capecchi, 1991). These symptoms in mice parallel the DiGeorge syndrome in humans, and defects seen after ablating the CarNC (see Chapter 8).

Hoxa3 also plays a major role in the development of the thymus and thyroid, regulating the differentiation of mesenchymal NCCs (Manley and Capecchi, 1995, 1998). *Hoxa3* mutant mice are athymic (see Box 9.3) and have severely underdeveloped thyroid glands, defects that may be mediated through *Hoxa3* control of *Pax1*. *Hoxa3*, *Hoxb3*, and *Hoxd3* have partly overlapping functions in regulating migration of the ultimobranchial bodies that contribute calcitonin-synthesizing parafollicular cells to the thyroid.

Polycomb (*Pc-G*) response elements that lie within the regulatory regions of homeotic genes are important regulators of *Hox* genes in vertebrates, as indeed they are of homeotic genes in *Drosophila*. *Polycomb* response elements maintain heritable states of transcriptional activity through modification of chromatin structure. At least nine such elements are known in mice. Targeted disruption of *Rae28* (the murine ortholog of *Drosophila polyhomeotic*) results in a rostral shift of the anterior expression border of *Hox* genes and subsequent posterior skeletal transformations. Other abnormalities in NC derivatives include the eye, cleft palate, ectopic occipital bones, abnormal parathyroids, thymus, and heart defects.[27]

In summary, the ability to unite abnormalities in such apparently disparate structures as the heart, face, and thymus through primary (or secondary) involvement of the NC highlights the utility of searching for the developmental basis of human abnormalities as far back into development as the primary embryonic layers from which the affected cells arise. Research into such associations does not stop at their classification but rather leads to increased understanding of the mechanisms that cause the defects and, ultimately, toward improved treatment or even prevention. To that end, the final chapter revisits the development of NCCs in the context of craniofacial birth defects.

Notes

1. See Persaud *et al.* (1985) for depictions of malformations from antiquity, and Gorlin *et al.* (2001) for catalogs of birth defects. See Cohen (1989, 1990) for a ten-part overview on syndromology.
2. See Corcoran (1998) for neural tube defects resistant to folic acid, Besser *et al.* (2007) for the Atlanta study, Grosse and Collins (2007*) for folic acid and recurrent neural tube defects, Jablonski (2006*) and Beaudin and Stover (2007*) for ultraviolet light/melanin/folic acid and neural tube defects and for the metabolic action of folate metabolism on pyrimidine synthesis, and Zhu *et al.* (2007) for the *Folr-1* study.
3. See Opitz and Gorlin (1988) and Hall (1999a*) for germ layer of origin and the etiology of defects. See Cohen (1989) and Hall (1990) for types of classifications of syndromes, including the use of embryological criteria.
4. For the colonization of the pancreas by neural precursors, see Kirchgessner *et al.* (1992). For production of neurites by endodermal cells of the pancreas, see Teitelman (1990).
5. See Kanakubo *et al.* (2006*) and Bertrand *et al.* (2000) for the mouse and chicken studies, respectively.

6. For neurocristopathies, see Bolande (1974, 1981), Opitz and Gorlin (1988), M. C. Jones (1990), and Johnston and Bronsky (1995). For the CHARGE syndrome, see Pagon *et al.* (1981), Cohen (1989), and Aramaki *et al.* (2007).

7. See Pascualcastroviejo (1990) for tumors of the NC.

8. See Maclean and Hall (1987), Hodges and Rowlatt (1994), and Weinberg (2006) for distinctive features of neoplastic cells, and for developmental approaches to cancer, tumors, and syndromes.

9. See Stemple and Anderson (1993) for the monoclonal antibody against the Ngf receptor used to isolate stem cells, and Oppedal *et al.* (1987) for the monoclonal antibody against neuroblastomas.

10. Preliminary studies cited by Fagiani *et al.* (2007) demonstrate that *RaLP* is expressed at high levels at 7.5–10.5 days of gestation in the neural tube, NCC precursors of the peripheral nervous system, Schwann and chromaffin cells, but is not expressed at 12.5–16.5 days of gestation.

11. Demonstrations in the 1940s of the induction by saline solution of rostral neural tissues, eyes, and lens and olfactory placodes from ectoderm of the North American spotted salamander, a puzzling result, now are known to result from dephosphorylation of ERK and activation of the Ras/MAPK pathway (Hurtado and de Robertis, 2007).

12. *Xenopus* CNC is sensitive to a monoclonal antibody against a lectin that is specific to the premigratory stage of NC development; almost 20% of embryos given the antibody show major deformities of the lower jaw and reduced numbers of chondrocytes in the head.

13. See Nishihira *et al.* (1981) for the response of NCCs to tumor promoters.

14. For basic studies on neurofibromatosis, see Riccardi and Mulvihill (1981) and Riccardi and Eichner (1986). For an up-to-date web site, see http://www.ninds.nih.gov/disorders/ neurofibromatosis/neurofibromatosis.htm.

15. See the proceedings of the 9th Human Gene Mapping Workshop held in Paris (*J. Med. Genetics*, 1987, 24, 513–544) for chromosome mapping of the neurofibromatosis gene, and Ponder (1990) for an overview of its cloning. See Brannan *et al.* (1994) for chromosome mapping and targeted gene disruption, and Riccardi and Eichner (1986) for categories of neurofibromatosis.

16. See Biesecker *et al.* (1999) for diagnosis, and the Proteus Syndrome Foundation (http://www.proteus-syndrome.org/) for updated information.

17. See Morrison-Graham *et al.* (1992) and Soriano (1997) for *Patch.*

18. See Murphy *et al.* (1992) and Morrison-Graham and Weston (1993) for *Steel*, steel factor and *White spotting*, which also affects ECM encountered by NCCs. Soluble and transmembrane forms of Steel factor are known: soluble Steel factor disrupts the migration of melanocyte precursors; transmembrane Steel factor disrupts their survival (Wehrle-Haller and Weston, 1995, 1997). See Lister *et al.* (2001) and Raible and Ragland (2005*) for *Mitfa.*

19. See Pearse (1977) for APUD cell types, Nozue and Ono (1989) for serotonin-induced APUdomas and adenosine-induced NC tumors, and Atsumi *et al.* (1990) for endotoxin.

20. For the NC origin of calcitonin synthesizing and other APUD cells, see Pearse (1977), Le Douarin (1982), and references in Hall (1999a*) and Le Douarin and Kalcheim (1999*).

21. See Martucciello *et al.* (2007) for a summary of mutations in the RET gene, Cleves *et al.* (2007) for congenital defects associated with Down syndrome, and Reeves *et al.* (2001) for the interplay between genetic, environmental, and stochastic factors in Down syndrome and in mouse models of Down syndrome.

22. See Gershon (1999*) and Simpson *et al.* (2007*) for migration of enteric ganglionic precursors, Carrasquillo *et al.* (2002) for RET and EDNRB pathways, and Passarge (2002) and Brooks *et al.* (2005*) for models of Hirschsprung disease.

23. See Mosher *et al.* (2007) and Lee *et al.* (2003) for the studies in rat and chicken embryos, respectively.

24. See Wurdak *et al.* (2006*) for developmental changes in DiGeorge syndrome. See Thomas *et al.* (1998) for an animal model for CATCH-22.

25. See Garg *et al.* (2001) for *Tbx1* and *Shh* in murine pharyngeal arches, Cohen (2003) for an overview of the molecular and clinical aspects of *Shh* and the *hedgehog* signaling network,

Abu-Issa *et al.* (2002) for the role of pharyngeal ectodermal and endoderm *Fgf8* in murine craniofacial development, and Bachiller *et al.* (2003) for inactivation of *Chordin*.

26. The Bmp receptors *alk2* and *alk3* are also required for normal NCC-derived outflow tract and septal development and for mesoderm-derived myocardial development in mouse hearts. In a summary of the data on mouse mutants with cleft palate, Juriloff and Harris (2008) identified two interacting sets of signaling pathways, one including *Bmp4, BmpR1a,* and *Wnt9b,* the other including *Sox11.* See Dudas and Kaartinen (2005*) for the Tgfß receptors.

27. See Takihara *et al.* (1997) for targeted disruption of the polycomb element in mice, Schumacher and Magnuson (1997) for *Polycomb* response elements and transcriptional regulation in mammals (including nine genes identified in mice), and Yu *et al.* (1998) for knockout of *Mll* and subsequent dysplasia or hypoplasia of the pharyngeal arches.

Chapter 10
NCC Development Revisited in the Context of Birth Defects

Seventy percent of all known birth defects in humans involve abnormalities of the head and neck. Of these, 10% result from chromosomal abnormalities, 20% are single-gene defects producing complex syndromes, while the remaining 70% are polygenic or multifactorial.

Clinical geneticists recognize four classes of birth defects, summarized by Spranger *et al.* (1982) as:

1. **malformations,** which are inherited defective developmental processes
2. **deformations**, which are mechanical disruptions to embryos
3. **disruptions,** which are external interference with developmental processes, as seen after a teratogen is administered, and
4. **dysplasias, which are** neoplastic changes, such as those discussed in Chapter 9.

A fifth class, **syndromes**, is sometimes added, although some dysmorphologists and clinicians prefer to regard all four classes (including neurocristopathies; see Chapter 9) as syndromes, using the term in the generic sense of a group of changes that collectively characterize an abnormal condition or birth defect.

In an insightful analysis, Dufresne and Richtsmeier (1995) proposed a parallel classification to that outlined above, but based on response to surgery. Although based on clinical response, differential responses to surgery reflect the causes of craniofacial defects: a poor response reflects a growth disorder; an enhanced response reflects a defect in which growth processes are unaffected (Fig. 10.1). Dysplasias, for which response to surgery is often poor, reflect disruptions of growth in localized regions or regional changes in growth rate. Deformations, for which the response to surgery is usually good, reflect local influences, while malformations and disruptions fall somewhere between local and regional influences on growth, on the one hand, and poor-to-excellent esthetic or functional responses to surgery, on the other (Fig. 10.1).[1]

This chapter revisits the development of NCCs in the context of birth defects, with some emphasis on craniofacial malformations and disruptions. It:

- provides a general outline of the various *stages in the development of NCCs* and how disrupting processes operating at these stages can lead to craniofacial defects;

B.K. Hall, *The Neural Crest and Neural Crest Cells in Vertebrate Development and Evolution*, DOI 10.1007/978-0-387-09846-3_10, © Springer Science+Business Media, LLC 2009

Fig. 10.1 This classification of defects proposed by Dufresne and Richtsmeier (1995) is based on the response of individuals to surgery (poor —> excellent). Dysplasias and syndromes reflect regional influences on growth and poor response to surgery; deformations reflect local influences on growth and excellent response to surgery. See text for details

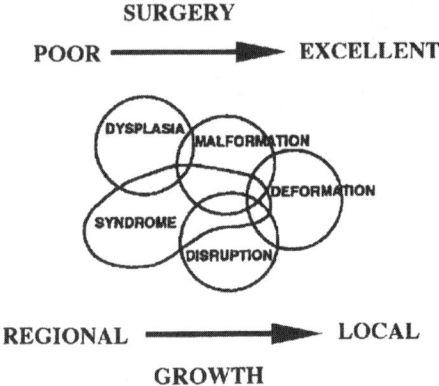

- considers defects that appear following the administration of *excess vitamin A* to pregnant mammals or to embryos maintained *in vitro*;
- discusses *mutations* affecting NCCs, some of which were introduced in previous chapters, especially Chapter 9;
- points out some elements of *species specificity* in mechanisms of abnormal craniofacial development; and finally,
- considers *regulation*, which is the ability of embryos or regions (fields; Box 10.1) within embryos to compensate for the effects of genetic or environmental changes that would otherwise disrupt development. Regulation brings us full circle to the controls over the normal development and differentiation of NCCs.

Box 10.1 Holoprosencephaly: A developmental field defect

Several types of holoprosencephaly have been described with various degrees of neural and facial anomalies. As a group, holoprosencephaly has an incidence among humans of 1.2 individuals per 10,000 live births. I say 'as a group' because holoprosencephaly does not have a single cause. A teratogen such as alcohol taken during pregnancy can result in holoprosencephaly in newborns, although some 40% of human cases involve a chromosomal defect, often trisomy of chromosome 13. Although variable in origin, all cases of holoprosencephaly can be traced back to abnormal division of the forebrain along the midline, leading to defective formation of the cerebral hemispheres and failure of the olfactory tracts and corpus callosum to form. Underdevelopment of the pituitary and a single central incisor in the maxilla also are characteristic (Cohen, 2006*).

Holoprosencephaly can be characterized as a **developmental field defect**, that is, a defect whose origin is in an integrated embryonic region rather than in a single developing element. Carstens (2002*) provided an insightful synthesis of molecular, developmental, and clinical data to make the case that the facial midline is a developmental field.

NCCs play important roles in the development of the midline developmental field. They control the expression of *Fgf8* in the rostral neural folds from which the prosencephalic organizer arises, and are involved in neural tube closure, patterning the prosencephalon and mesencephalon and restricting expression of *Shh* to the ventral prosencephalon.[a]

Some 17% of familial cases of holoprosencephaly result from a mutation in *SHH* or in the *SHH* signaling network, often involving the receptor PATCHED (PTCH), shown to be a lipoprotein receptor in *Drosophila*. Another protein, DISPATCHED (DISP1), which has 12 sterol-sensing transmembrane domains and regulates the release of *SHH*, has been implicated in holoprosencephaly. However, other mechanisms can cause different forms of holoprosencephaly. One type is based on abnormal functioning of the BMP receptors, BMPR1a and BMPR2b (Fernandes *et al.*, 2007).

Shh and Nkx

Multiple times and modes of action of *SHH* go some way toward explaining the considerable phenotypic variability among individuals with holoprosencephaly.

Shh is expressed in the midfacial region of developing mammalian embryos. Inhibiting *Shh* with cyclopamine at different developmental stages elicits a range of defects typical of holoprosencephaly, affects that can be interpreted as disrupting mid- and upper-facial fields earlier in development.[b]

Shh secreted by the ventral pharyngeal endoderm is required for lower jaw development in chicken embryos. Blocking *Shh* after the forebrain is patterned, but before *Shh* is expressed in the midfacial region, limits phenotypic changes to the face. Blocking *Shh* at later stages influences *Shh* expression and function in the pharyngeal endoderm and is associated with a broader range of phenotypic changes. *Shh*, derived from the forebrain in chicken embryos and from the ventral neural tube and oral ectoderm in zebrafish, is required for correct patterning of facial structures in both species. Studies using cyclopamine, forskolin, or ethanol to inhibit *Shh* show that not all the facial changes in zebrafish embryos can be attributed to direct inhibition of *Shh*.[c]

The Nkx family of homeodomain transcription factors function downstream of *Shh*. Both *Shh* and *Nkx2.1* are disrupted in embryonic mice in which holoprosencephaly is induced following maternal exposure to ethanol. An expanded zone of midline expression of *Shh* in the blind cave morph of the Mexican tetra, *Astyanax mexicanus*, disrupts patterning of the forebrain by expanding the expression of *Nkx2.1a*. This expanded expression is followed by enhanced proliferation in the basal diencephalon and hypothalamus, demonstrating the importance of the telencephalic midline organizer (field) in patterning the fore-brain and structures, which depend upon the forebrain.[d]

---

[a] See Creuzet et al. (2006) for the study with *Fgf8* and Harris and Juriloff (2007*) for an analysis of the 190 mouse mutants and strains with neural tube closure defects, in most of which the primary defect is failure of neural-fold elevation. See Opitz and Clark (2000) for the field theory in development and medical genetics and Shiota et al. (2007) for an analysis of embryogenesis in individuals with holoprosencephaly.

[b] *Xenopus* CNCCs are especially sensitive to the steroidal alkaloid cyclopamine, which inhibits *Shh* production; most craniofacial cartilages fail to develop. Other CNC derivatives are normal, as is the TNC and its derivatives, indicating dependence of mesenchymal CNCCs on *Shh* signaling (Milos et al., 1998).

[c] See Cordero et al. (2004) Marcucio et al. (2005), Wada et al. (2005), Eberhart et al. (2006), Britto et al. (2006), and Loucks et al. (2007) for experimental studies using chicken and zebrafish embryos and for the relationships between neural tube and oral ectodermal sources of *Shh*. See Le Douarin et al. (2007) for the relationship between *Shh* signaling, NCCs and the anterior neural tube in mediating brain and facial development.

[d] See Higashiyama et al. (2007) for *Nkx2.1* in mice, Menuet et al. (2007*) for *Astyanax mexicanus*, and Franz-Odendaal and Hall (2006) for selection on *Shh* in Mexican tetra. Mutations in *NKX2E* in humans are associated with atrial septal defects and atrioventricular conduction defects.

Susceptible Stages of Neural Crest Development

As with all developmental defects, the earlier in embryonic life a NC defect occurs, the more likely it is to be life-threatening, major, and/or to produce a structural abnormality. Craniofacial defects involving NCCs could arise from:

- **the total failure of NC development** because of failure of the inductive inter-actions that generate NCCs during neurulation (see Chapter 2). Such a failure would almost certainly be lethal or result in major structural abnormalities to the entire embryo.
- **abnormalities in NCC migration** because of defects in the NCCs themselves, in abnormal extracellular matrices, or in abnormal cells with which NCCs interact (see Chapter 3). Such defects result in major structural abnormalities to entire units such as the face, cranium, jaws, ears, heart, or peripheral nervous system.
- *Abnormalities in the differentiation* of NCCs because of intrinsic problems such as altered cell surface adhesion, defective gene networks/pathways/cascades, or defective inductive interactions (see Chapters 2, 3, 5, 6, 7, and 8). Individual NC derivatives such as bones, cartilages, teeth, or glands would either fail to form or form abnormally.
- *abnormalities in tissues* such as muscles, nerves, or blood vessels with which NCC derivatives interact (see Chapter 2). Because such disruptions occur later in

embryonic development than the other three, minor structural defects or defective function are the likely outcome.

Defective Migration

NCCs delaminate and migrate away from the neural tube by decreasing their adhesion to neural and epidermal epithelia, degrading the basal lamina, becoming mesenchymal, selectively adhering to specific components of the ECM, and moving along extracellular pathways (see Chapter 3). NCC migration could be abnormal if any of these steps was defective. For example, a delay in the onset of migration could result in NCCs failing to reach their final site, arriving in reduced numbers or arriving too late to undergo the interactions necessary for differentiation.[2]

Defective migration could occur, not because of an abnormality in the NCCs themselves but rather because of an abnormality in the extracellular environment through which the cells migrate or with which they interact. Extracellular spaces fail to form in a number of mutation-based craniofacial defects, with the consequence that NCC migration is prevented and development is defective. One example, albinism in the axolotl, was discussed in Chapter 5.[3]

Human craniofacial defects with a mutational basis are interpreted as resulting from varying degrees of defective CNCC migration. For example, abnormal migration of midbrain NCCs at Carnegie stages 10 and 11 leads to incomplete development of the chondrocranium, defective midline fusion, cyclopia, and holoprosencephaly (Box 10.1). Nonhuman animal models bear out such interpretations.

Treacher–Collins syndrome (mandibulofacial dysostosis; Franceschetti–Zwahlen–Klein syndrome), which affects 1 in 10,000 individuals, is characterized by reduced development of the lower jaw (micrognathia), absence or malformation of the ears, which are often set low on the head, downward slanting eyes, and sometimes cleft palate. The syndrome results from an autosomal-dominant mutation of the *Treacher–Collins–Franceschetti syndrome 1* gene, *TCOF1*, which encodes for the 144-KDa nucleolar phosphoprotein, Treacle. In a mouse model for Treacher–Collins syndrome, haploinsufficiency of *Tcof1* is associated with fewer migrating NCCs and subsequent craniofacial malformations. *Tcof1* and Treacle regulate the production of mature ribosomes and regulate NCC formation and proliferation cell autonomously (i.e., by acting within NCCs themselves).[4]

Of course, birth defects may have an environmental rather than a mutational basis. Birth defects in babies born to diabetic (hyperglycemic) mothers include craniofacial defects ranging from exencephaly and agnathia at one extreme to retardation of maxillary and mandibular growth at the other. The incidence of such defects is 3–5 times higher in the offspring of diabetic mothers than in babies born to nondiabetic mothers, but can be reduced by stimulating the maternal immune system[5] (Hrubec et al., 2006).

Defective craniofacial development can be induced experimentally by delaying NCC migration. High concentrations of glucose inhibit *in vitro* migration of CNCCs isolated from normal and diabetic rats and lead to outflow tract and pharyngeal-arch arterial defects in chicken embryos. Both results are consistent with glucose concentration as a contributing factor to defects arising in embryos born to diabetic mothers. Exposing murine embryos to ethanol disrupts migration to such an extent that NCCs migrate inward toward the lumen of the neural tube, rather than away from the tube.[6]

Defective Proliferation

Disrupting the normal rate of NCC division can result in craniofacial defects. Consequently, embryos are particularly sensitive to deviations from normal development at stages when growth is rapid or when growth rate is changing. Such times are **critical** or **sensitive periods**.[7]

Lectins, plant proteins that bind to specific sugar residues in membrane glycoproteins of animal cells, have been used to localize D-mannose, D-glucose, and N-acetyl-D-glucosamine in the zone of fusion of the neural folds in mice and rats. This regionally and temporally controlled fusion creates a sensitive period at which neural tube (and NC) defects can arise.

Although NCCs divide as they migrate, cell division increases once NCCs reach their final site. Such enhanced proliferation reflects increasing interactions among NCCs and between NCCs and epithelia and is critical for condensation and for condensed cells to differentiate (see Figs. 3.13 and 3.14).

Enhanced Cell Death

Inhibition of cell division and/or of migration, which can lead to extensive cell death, follows exposure of embryos to X-rays, steroid hormones, or alcohol.

Apoptosis (programmed cell death) has achieved much prominence over the past two decades as a mechanism operating during normal development to remove excess cells during differentiation and to mold and sculpt embryonic tissues during morphogenesis, temporally and spatially controlled waves of proliferation and apoptosis in the epibranchial placodes of the tree shrew, *Tupaia belangeri*, select precursor neurons for nodose, petrosal, and geniculate ganglia (Washausen *et al.*, 2005). Over a 100 years ago in a study of skate embryos, Béard described cell death as a normal part of the development of Rohon–Béard neurons (see Chapter 6). At the level of gene networks, *Bmp4* and *Msx1* are important regulators of apoptosis in NC induction (see Chapter 2), in rhombomeres and in the pharyngeal arches (discussed below).

Apoptosis in Rhombomeres: Bmp4 promotes the migration and differentiation of murine TNC and induces apoptosis in hindbrain NCCs. Apoptosis is specific and targeted. Cells from r3 and r5 (odd-numbered rhombomeres) undergo apoptosis as a result of interactions with the even-numbered rhombomeres r2 and r4. As a result of these interactions, *Msx2* and *Bmp4* are activated in the odd-numbered rhombomeres.

Msx2– transgenic mice die before birth with multiple craniofacial malformations. The balance between cell survival and apoptosis—which is regulated by

Msx2 —is disrupted in these transgenic embryos. Injecting Bmp2 or Bmp4 into the neural tube increases the expression of *Msx* genes. The expression of *Msx2* in rhombomeres that normally do not express *Msx2* (and from which NCCs normally delaminate) induces cell death and eliminates crest cell migration. Even-numbered rhombomeres appear to use Bmp4 as their signaling molecule; they contain Bmp4, and Bmp added to r3 and r5 upregulates *Bmp4* and *Msx2*, as it does in even-numbered rhombomeres.[8]

Apoptosis in Pharyngeal Arches: Later in pharyngeal-arch development, Bmp2 and Bmp4 selectively regulate apoptosis in CNC and TNC derivatives, targeting the first and second arches (especially the mandibular arch) and progenitor cells of sympathoadrenal neurons. Neuronal survival requires Fgf and Ngf, a growth factor dependence that can be preconditioned by pretreatment with Bmp2 and Bmp4. Early activation of these normal patterns of cell death or their expansion to cells that would normally not undergo apoptosis can contribute to defective development. As discussed below, vitamin A-induced craniofacial defects are mediated, in part, through enhanced cell death.[9]

Transcriptional regulation of *Hox* genes regulates apoptosis in the mandibular and maxillary arches. *Mll*, the mammalian ortholog of the *Drosophila gene Trx,* regulates *Hox* genes to prevent excess cell death in murine pharyngeal-arch mesenchyme. *Hox* gene expression in mouse embryos becomes dependent on *Mll* between 9 and 9.5 days of gestation, an interval that marks an important phase of chromatin reorganization (Yu *et al.*, 1998). *Hox* gene expression is abolished in *Mll*[-/-] embryos. Because of enhanced apoptosis in arch mesenchyme (but not in arch ectoderm), *Mll*[-/-] embryos have severely reduced mandibular arches and defective maxillary arches.

Defective Induction

As discussed in Box 3.4 and in Chapter 7, epithelial–mesenchymal interactions play essential roles in initiating the differentiation of NCCs. Consequently, abnormal interactions can lead to defects (Hall, 2005b*). The neurocristopathies discussed in the previous chapter are prime examples, as are many craniofacial defects involving the skeleton. Groups of craniofacial defects interpreted as resulting from defective secondary tissue interactions are von Recklinghausen neurofibromatosis, tumors and neurocristopathies (see Chapter 9), and facial dysmorphogenesis and holoprosencephaly (Box 10.1).

Craniofacial dysmorphogenesis associated with Down syndrome (trisomy 21) is a consequence of a reduction in the population of CNCCs that invade the first pharyngeal arch, resulting in smaller mandibular processes. Here the causative agent is chromosomal and genetic. *Ts65Dn*, a mouse mutant with Down syndrome features, is trisomic for approximately half of the genes found on the human chromosome 21. The NCC population and the size of the mandibular arch are reduced in *Ts65Dn* mice, because fewer NCCs are induced, divide, and proliferate. *Shh* has been implicated in playing a causal role by Roper *et al.* (2006).

In the following section, defects following the administration of vitamin A are used to illustrate the susceptibility of NCCs to external agents.

Vitamin A, Craniofacial Defects, and the Neural Crest

Vitamin A exists in several forms of retinoids, each with a different name—an alcoholic form, retinol; an aldehyde, retinal; and an acid, retinoic acid. Retinol and retinal are preformed versions of retinoic acid (vitamin A), the active form that affects animal development, metabolism, and gene transcription.

Direct Action In Vivo

At low concentrations, vitamin A is involved in determination of cell number and in the differentiation and dedifferentiation of NCCs.

Vitamin A is teratogenic, however, if administered to pregnant women at concentrations of 50,000–100,000 international units/day. The resulting abnormalities simulate inherited craniofacial defects (first-arch syndromes) that appear following a mutation. Administrating vitamin A to pregnant mammals of other species, or cultivating mammalian embryos with vitamin A, produces craniofacial defects similar to those seen in humans.

From 9.5 days onward in murine embryos retinoic-acid- and cellular retinoic-acid-binding protein (CraBP-I and -II) localize to the NC, NC derivatives (craniofacial mesenchyme, the pharyngeal arches, dorsal root ganglia, and enteric ganglia of the gut), to r2, r4, and r5, and r6, to somites, and to major blood vessels, limb bud mesenchyme, and flank and facial muscles (Fig. 10.2). Expression of the retinoid X receptor-γ nuclear receptor gene is even more widespread: in presomitic mesoderm at 8 days postconception; in the frontonasal processes, pharyngeal arches, limb and sclerotomal mesenchyme at 9.5–11.5 days postconception; in all precartilaginous mesenchyme at 12.5 days; and in all cartilages, keratinized epithelia, teeth, and vibrissae from 13.5 days onward. Such ubiquitous distribution helps explain why exogenous retinoids have such widespread actions on so many tissues. Although retinoic acid affects derivatives of both CNCs and TNCs, the effects on cranial derivatives are much greater and lead to more major anomalies, including absence of the mesenchyme of the third and fourth pharyngeal arches and arterial defects, along with placodal anomalies.[10]

The action of retinoic acid on NCCs is stage dependent, as demonstrated by Lee *et al.* (1995) in rat embryos to which retinoic acid was administered at 9.0 or 9.5 days of gestation, and in which cell migration and cell fate were assayed. Administering retinoic acid at 9 days decreases the size of the first pharyngeal arch and initiates ectopic migration of crest from r1 and r2 to the second arch (normally these cells migrate into the first arch). Administering vitamin A at 9.5 days elicits fusion of the first and second arches but not ectopic NCC migration.

In a subsequent study in pregnant rats, retinoic acid was inhibited by maternal administration of an antifungal agent at 9.5 days gestation. The resulting abnormal

Fig. 10.2 Injection of 14C-labeled retinoid to pregnant mice allows retinoid to be visualized during embryonic development, as can be seen in these sections through embryos between 8.5 and 11.5 days of gestation. At 8.5 and 9.5 days (**A, B**), radioactivity (shown in *white*) is seen in the neural plate (NP) and in early migrating neural crest cells (NC). At 10.5 and 11.5 days (**C, D**), radioactivity is seen in the frontonasal processes (FNP), the roof of the midbrain (MB), and in the neural tube (NT). Some radioactivity is also seen in the limb buds. Bar = 500 μm. Reproduced from Dencker *et al.* (1990), with the permission of Company of Biologists Ltd., from figures kindly supplied by L. Dencker

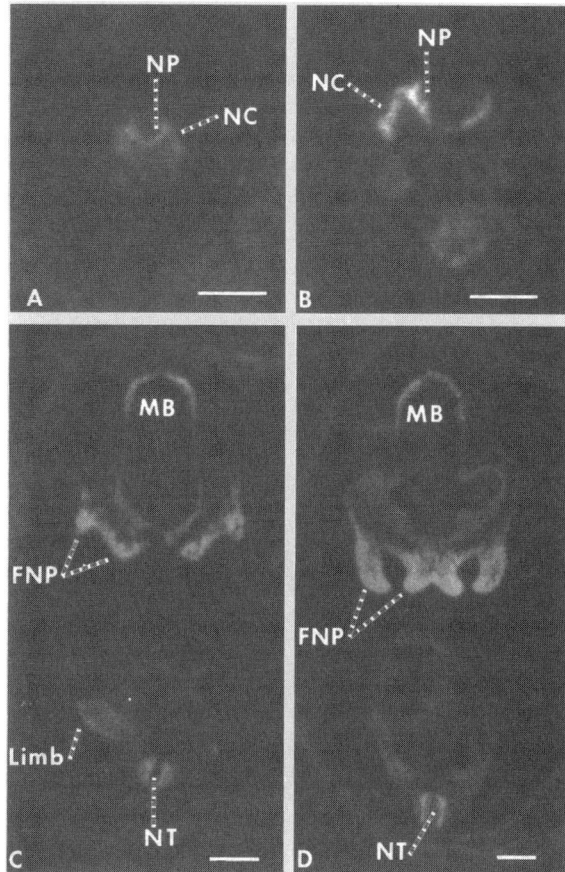

migration and condensation of NCCs in the first pharyngeal arch and the formation of ectopic maxillary cartilages were interpreted in light of the knowledge that the palatoquadrate and Meckel's cartilages arise from mandibular condensations of NCCs (see Fig. 10.5) and that the maxillary condensation produces rostral neurocranial cartilages. In chicken embryos, arch transformation occurs in response to enhanced expression of Bmps and retinoic acid (Box 10.2).[11]

Box 10.2 Hoxa2 knockout and second- to first-arch transformation

Hoxa2 acts upstream of *Sox9* and of the gene *Six2* (which encodes a homeodomain protein) and plays an important role in initiating chondrogenesis; *Hoxa2* is expressed in chondrogenic condensations of the second arch of wild-type mouse embryos, but is not expressed in second-arch condensations. What are the consequences of knocking out *Hoxa2*?

Knocking out *Hoxa2* in mice converts second arch to first-arch structures resulting in duplication of first-arch elements (Meckel's cartilage and the middle ear ossicle) and consequent loss of second-arch elements (stapes).[a] An additional and ectopic skeletal element(s) often also forms (Fig. 10.5; and see Box 7.1). With the transformation of the second to a first arch, *Hoxa2* is expressed ectopically in the new 'first arch'. Thus, in embryos that lack *Hoxa2*, the second arch is transformed into a first arch that produces first-arch skeletal elements.

Two approaches were used in obtaining these results. Gendron-Maguire *et al.* (1993) deleted exon 1, the first 32 base pairs of exon 2, the splice acceptor site, and all the introns from *Hoxa2*. Rijli *et al.* (1993) deleted exon 1, the first 72 base pairs of exon 2, and the translation initiation site. In both approaches, embryos showed abnormal development of the malleus and incus of the middle ear, mirror-image duplications of the malleus and incus, and mirror-image duplications of the tympanic bone of the skull. An ectopic squamosal bone was also formed. Stapes and stylohyal cartilages were missing, and an ectopic cartilage was formed in the embryos in both studies (Fig. 10.5). This developmental information on the role of *Hoxa2* has been coupled with knowledge of the developmental transformations involved in the evolution of the mammalian lower jaws and middle ear ossicles and given an evolutionary spin (Box 7.1).

---

[a] Knocking out *Hoxa2* in chicken embryos also results in conversion of the second to a first arch, while overexpressing *Hoxa2* converts the first to a second arch. *Hoxa2* 'selects' second-arch fate by modification of expression of Hoxa2 in the tissues surrounding arch NCCs (Grammatopoulos *et al.*, 2000).

Administering retinoic acid at a dose of 60 mg per kg body weight to female golden Syrian Hamsters on the eighth day of pregnancy results in defects of the facial skeleton in all surviving embryos. As early as 4 h after administration, NCCs display ultrastructural signs of cytological damage such as swollen mitochondria and dilated nuclear and endoplasmic reticular membranes. Dead NCCs were first seen 8 h after retinoic acid administration with maximal numbers at 24 h after administration. Later in development, surviving NCCs had not migrated as far as those in control embryos, producing local deficiencies in cell density and subsequent skeletal defects. This study demonstrates the early and extensive effects of retinoic acid on CNCC survival and migration and the consequent defects in the facial skeleton, although more recent analyses have revealed a primary role for pharyngeal endoderm in patterning the pharyngeal arches (Box 10.2) and pharyngeal endoderm as a (the?) target for retinoic acid (see Chapter 4).

Nonmammalian vertebrates also have been used in teratogenic studies with the aim of understanding mechanisms of action of vitamin A. Administering vitamin A to chicken or quail embryos does not disrupt dorsoventral patterning of the

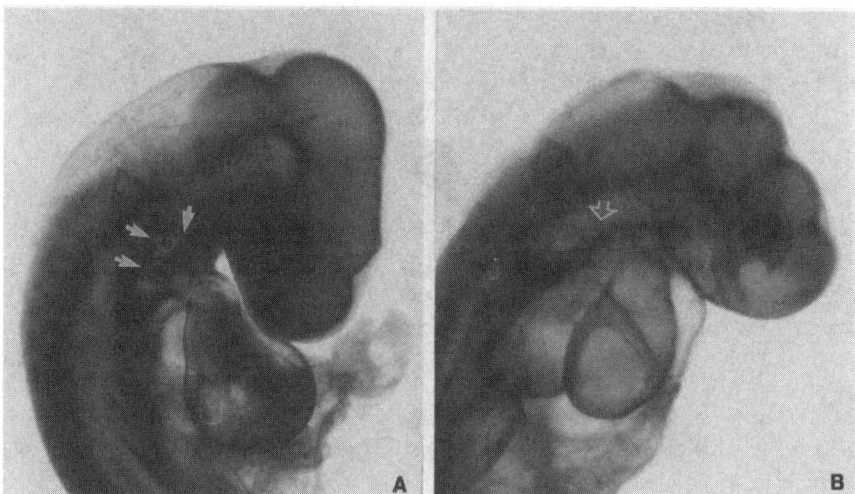

Fig. 10.3 A control chicken embryo (**A**) and embryo exposed to retinoic acid (**B**) to show the effect of retinoic acid on the pharyngeal arches, which are well developed in the control embryo (*arrows* in **A**) but severely underdeveloped in the embryo treated with retinoic acid (*arrow* in **B**). Also note the underdeveloped forebrain and displaced otic vesicle in the treated embryo. Reproduced from Moro-Balbás *et al.*, Retinoic acid induces changes in the rhombencephalic neural crest cells migration, and ECM composition in chicken embryos, *Teratology* 48:197–206, Copyright © (1993), from a figure kindly supplied by Jose Moro-Balbás. Reprinted by permission of Wiley-Liss Inc., a subsidiary of John Wiley & Sons, Inc.

neural tube, disrupts segmentation of r4–r8 and caudal hindbrain development, and prevents NCC migration, leading to craniofacial defects such as severe underdevelopment of the pharyngeal arches (Figs. 10.3 and 10.4). In another avian model, 13-*cis*-retinoic acid was injected into embryos between 2 and 6 days of incubation and the embryos examined at 14 days. Craniofacial and cardiovascular defects were both induced. Interestingly, defects affecting craniofacial mesenchyme are greatest if retinoic acid is injected *after* the onset of NCC migration, implying affects in this species on the localization and differentiation of NCCs rather than on their migration.

Zebrafish embryos are also sensitive to inhibition of retinoic acid and the disruption of genes whose products are involved in the metabolism of retinoic acid, chondrogenic NCCs being especially sensitive. Problems associated with species specificity are taken up later in the chapter.[12]

Craniofacial Defects

Vitamin A especially affects the facial processes and outgrowth of the facial skeleton because of its specific action on NC-derived mesenchyme in those processes. Varying degrees of severity occur after vitamin A is administered at doses between 50,000 and 100,000 international units/embryo, ranging from defects affecting the

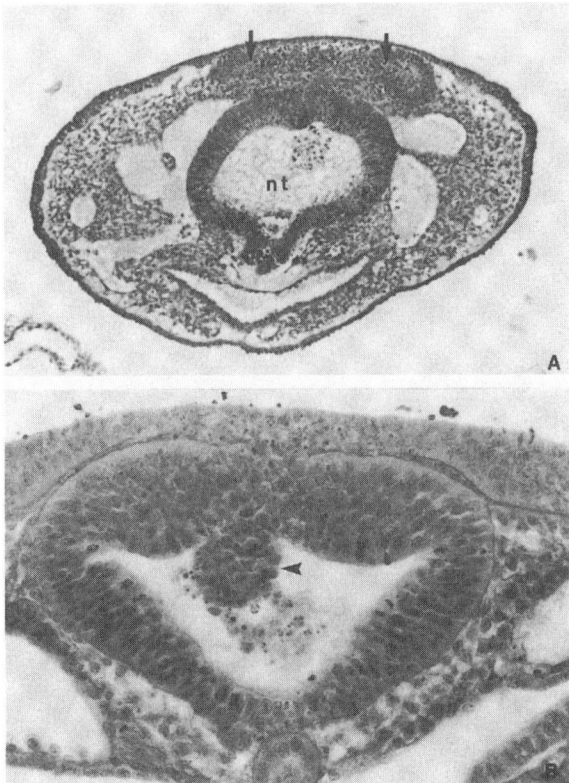

Fig. 10.4 Cross-sections through the neural tube of a chicken embryo to illustrate the effect of retinoic acid on the migration of NCCs. (**A**) is a transverse section through the rostral rhomben-cephalon. Neural crest cells (*arrows*) have accumulated above the neural tube (nt) and their normal migration having been inhibited. (**B**) is a transverse section through the caudal rhombencephalon. Neural crest cells (*arrowhead*) have migrated ectopically into the lumen of the neural tube. Repro-duced from Moro-Balbás *et al.*, Retinoic acid induces changes in the rhombencephalic neural crest cells migration, and ECM composition in chicken embryos, *Teratology* 48:197–206, Copyright © (1993), from a figure kindly supplied by Jose Moro-Balbás. Reprinted by permission of Wiley-Liss Inc., a subsidiary of John Wiley & Sons, Inc.

whole facial complex to defects affecting one side of the face only. The former defects are analogous to Treacher–Collins syndrome, the latter to hemifacial micro-somia or mandibulofacial dysostosis, in which the mandible, zygomatic arch, and middle ear ossicles are abnormal. Grant and colleagues (1997) showed that mouse embryos exposed to retinoic acid on day 9 do not display defects associated with the NC or with the frontonasal processes, indicating temporal specificity of retinoid action in combination with dosage effects.

In typical experiments with mammalian embryos, administration of vitamin A to pregnant mice results in elements of the ear and craniofacial skeleton failing to form, or the development of skeletal tissues in ectopic positions. For example,

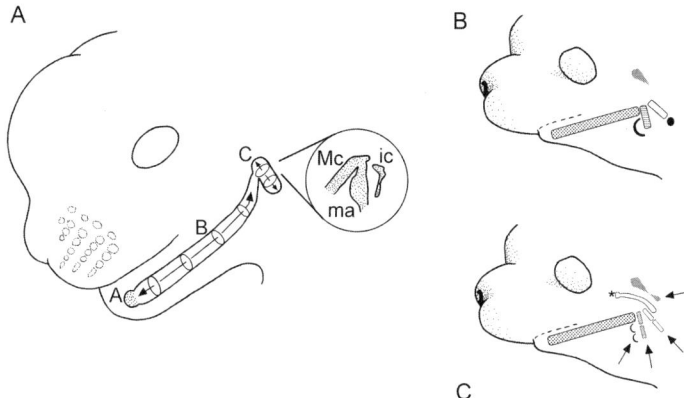

Fig. 10.5 Developmental origin of the middle ear ossicles in wild-type mouse embryos (**A, B**), and the affect of *Hoxa2* knock-out on their development (**C**). (**A**) The first (mandibular) arch prechondrogenic condensation gives rise to the symphyseal cartilage (**A**) that unites *left* and *right* lower jaws, and Meckel's cartilage (**B**) that forms the foundation for the *lower* jaw. The proximal region of the first-arch condensation (**C**), which lies at the boundary between the mandibular and hyoid arches, becomes part of Meckel's cartilage (Mc), and the malleus (ma). The incus (ic) originates from a separate condensation in the second arch. In **B** and **C** the *lower* jaw and *middle* ear ossicles of a wild-type mouse (**B**) are compared with those in a mouse in which *Hoxa2* has been knocked out (**C**). *Arrows* in **C** identify elements that are duplicated in the knock-out mouse; the squamosal bone (*top arrow*) and the three middle ear ossicles. The * identifies an ectopic cartilage that exists only as a condensation in the wild type. (**B, C**) based on data in Rijli *et al.* (1994). Figure prepared by Tom Miyake

cartilage does not develop in mammalian maxillary processes during normal development, although ectopic maxillary cartilages are found in rodent embryos treated with vitamin A. In fact, an entire craniofacial structure can develop in an abnormal position in rat embryos treated with vitamin A. Development of the otic capsule in the first rather than the second pharyngeal arch is a typical change.

With respect to ear defects, the tympanic ring (the embryonic precursor of the tympanic bone) is lost and the external acoustic meatus altered, a developmental coupling also seen in *Hoxa2* mutant mice, which have duplicated tympanic rings and an altered external acoustic meatus (Box 10.2); retinoic acid and *Hox* genes interact in initiating the developmental interactions between these two ear elements (see below).[13]

In an innovative test of how structures become integrated by shared developmental or functional attributes, and of when during ontogeny the switch from one to the other occurs, Miriam Zelditch and her colleagues examined ontogenetic variation in skull development in laboratory rats and in the cotton rat, *Sigmodon fulviventer*. For the cotton rat, Zelditch and Carmichael (1989) compared models based on embryological origin (neural crest vs. mesoderm), pharyngeal arch of origin (mandibular vs. maxillary), and functional integration (muscle function). Characters derived from the same pharyngeal arch covary early in ontogeny because shared embryological origin is a major determinate in the establishment of developmental fields of action (Box 10.1). Later in ontogeny, repatterning is explained by functional rather than by developmental integration.[14]

Mechanisms of Action

Abnormal migration of subpopulations of NCCs is the most ready explanation for several defects involving ectopic cartilages in the pharyngeal arches of mammalian embryos. Alternate explanations for defects such as *hemifacial microsomia* include local hemorrhaging; see the following section. It seems more likely, however, that hemorrhaging is a secondary consequence of primary defects in chondrogenic NCCs. Hemifacial microsomia shows us that mandible, zygomatic arch, and middle ear ossicles form a *developmental field* of action (see Boxes 7.1 and 10.1).[15]

Primate embryos, such as fetal pig-tailed macaques, exhibit defective mandibular development if vitamin A is administered maternally during NCC migration and craniofacial morphogenesis (20–44 days of gestation), establishing a link between nonhuman primates and human syndromes. The conclusion that vitamin A alters NCC migration is based on:

- administration of vitamin A during NCC migration evokes the syndromes, administration earlier or later does not;
- the accumulation of NCCs in abnormal embryonic locations after administering vitamin A; and
- the inhibition of NCC migration *in vitro* by vitamin A.[16]

As might be expected, given the multiple roles vitamin A plays in cell metabolism, vitamin A affects other important aspects of the development of NCCs:

- slowing the division of NCCs by increasing the length of the mitotic cycle,
- enhancing naturally occurring programmed cell death in the neuroepithelium,
- causing blebbing of the cytoplasm of NCCs,
- slowing the deposition of such ECM products as hyaluronan,
- inhibiting interactions between NCCs and ECM products (interactions that are required to initiate normal migration),
- acting directly on NC mesenchyme to inhibit the differentiation of membrane bones in chicken embryos, and
- being taken up by, and accumulating in, NCCs themselves, in a time- and site-specific manner.

Growth Factors: In part, retinoic acid may exert its actions through growth factors such as Tgfβ1 and Tgfβ2, both of which are expressed in the neuroepithelium of 8.5- to 10.5-day-old mouse embryos and downregulated by retinoic acid at the neural-plate stage of embryogenesis; downregulation is greatest in neuroepithelium and in the CNC.

Fgf3 may also be a downstream target of retinoic acid. In avian embryos, *Fgf3* is expressed in the epiblast, in rhombomeres (and then at rhombomere boundaries), in pharyngeal endoderm, and in the ectoderm of the second and third pharyngeal arches. Similarly, in murine embryos, Fgf3 is expressed in the brain, dorsal ectoderm, second-arch ectoderm and otic placodes and can be used as a marker for changes following retinoid administration. In zebrafish, reduction of *Fgf3* eliminates cartilage from pharyngeal arches 3–6, without affecting chondrogenesis in the other arches (see Box 9.1).

Correlations between *Fgf3* expression and craniofacial defects were used to demonstrate the dual identification of the fused pharyngeal arches, and that pro-rhombomere 2, the second pharyngeal arch, and the otic ectoderm represent a field with respect to response to the retinoids. This correlated response (field effect) arises even earlier in development. The gene for a basic domain leucine zipper (bZip) transcription factor from the Mafb family (*kreisler* in mice, *valentino* in zebrafish) regulates *Hox* gene expression and plays a role in subdividing pro-rhombomere 2 into r5 and r6. *Valentino*-mutant zebrafish and *kreisler*-mutant mice display similar phenotypic defects in pharyngeal-arch derivatives and in the inner ear, indicating that regional behavior can be traced to early patterning events in the neural tube.[17]

Indirect Effects

Vitamin A also affects NCCs indirectly because defects following vitamin A administration are likely to be based on more than one mechanism, depending, in part, on time of exposure and dose. For example, localized hematomas associated with vitamin-A-induced craniofacial defects in laboratory animals are often quite large and so prevent craniofacial mesenchyme from forming in sites such as the mandibular arch or developing ear, with the consequence that mandibular or ear structures fail to form.

Vitamin A acts via the epithelia surrounding developing craniofacial structures, especially the pharyngeal endoderm. Consequently, defective epithelia could compromise growth or differentiation of the subjacent mesenchyme, leading to craniofacial defects.

Evolutionary Origins of Sensitivity to Retinoic Acid

Reproducible defects and deletions in neurocranial and pharyngeal skeletal elements occur after mummichog and zebrafish are exposed to exogenous retinoic acid. The close correlation between defect and dose of retinoid administered, and the discovery that retinoic acid shifts the expression boundary of *Hoxd4* rostrally in a dose-dependent manner in the central nervous system and in the pharyngeal arches of the Japanese flounder, are two results consistent with a response conserved across the vertebrates. Response to retinoic acid, however, has not remained constant across the vertebrates. Pharyngeal-arch endoderm is not specified by retinoic acid in zebrafish, although the more lateral portion of the pharyngeal endoderm is sensitive to retinoic acid, and retinoic acid is required for proper patterning of the pharyngeal arches in all vertebrates.[18]

If neural tube sensitivity to vitamin A is a feature of all vertebrates, we would expect the response to be present in jawless vertebrates, and it is. Treating embryos of Japanese lampreys with all-*trans* retinoic acid results in rostral truncation of the nervous system and loss of the mouth, pharynx, esophagus, heart, and endostyle, that is, loss of many of the structures that characterize the vertebrate head (Kuratani *et al.*, 1998). Patterned effects of retinoic acid on neural and facial structures have an even more ancient evolutionary history; retinoids activate *Hox* genes to pattern the pharyngeal endoderm in amphioxus (see Chapter 4). An influence of retinoic acid on pharyngeal endoderm therefore precedes effects on NCCs.

Defects Following Disruption of the *Hox* Code

Evidence derived from normal embryos demonstrates that *Hox* genes pattern the pharyngeal region (see Fig. 2.17 and Chapter 2).

Hox genes are regulated by vitamin-A *Msx1*, which, as we have seen, plays multiple roles in craniofacial development, is expressed more strongly in avian embryos deficient in vitamin A. Expression of *Msx1* in premigratory NCCs is especially enhanced, although *Msx1* is also expressed ectopically, implicating vitamin A as a regulator of the *Msx1* pathway.

Hox genes have been disrupted in mouse development following homologous recombination in embryonic stem cells and introduction into the germ lines of mouse embryos. Although such experiments are elegant, functional overlap between *Hox* genes and the ability of one gene to compensate for the inactivation of another means that loss of function need not necessarily result in an altered phenotype, even if the eliminated gene is involved in the production of that phenotype. As discussed below, *Hoxa2, Hoxa3,* and *Hoxb4* interact in complex ways in the development of the facial skeleton in chicken embryos.[19]

Hoxa1 and *Hoxa2* play especially important roles in the NC and its mesenchymal derivatives and/or in the tissue interactions that induce them. Studies in which these *Hox* genes have been disrupted in zebrafish or in mice are discussed below. Disrupting a third *Hox* gene, *Hoxa3*, in mice produces a suite of defects in NC derivatives, with defects in the pharyngeal arches, thymus, thyroid, hyoid bone, and hyoid arch skeleton.

Hoxa1

Zebrafish: The retinoic acid phenotype was phenocopied in zebrafish by overexpressing *Hoxa1* (Alexandre *et al.*, 1996). Overexpression alters rostral hindbrain growth and changes the fate of those neural and NCCs that delaminate from r2. The resulting abnormalities include abnormal growth of the hindbrain and defective development of r2 NCCs. Older embryos show diminished growth or complete loss of the mandibles and an increase in second-arch structures; mandibular-destined cells are rerouted. Meckel's cartilage and the palatoquadrate of the mandibular (first pharyngeal) arch fail to form, while second-arch hyoid cartilages are enlarged and partially duplicated.[20]

Mice: *Hox* genes control first- and second-arch development and regulate developmental interactions between the two arches.

Only five rhombomeres develop in the hindbrain of mice mutant for *Hoxa1*; r5 is absent and r4 is reduced. Disrupting *Hoxa1* results in ear, hindbrain, and pharyngeal-arch defects, consistent with a role for *Hoxa1* in specific NCC populations. Changes in the palate and skull following allelic disruption of *Hoxa1* and/or *Hoxa2* can be compensated for and defects associated with *Hoxa1* restore normal palates in *Hoxa2* mutants.

Hoxa2

Homeotic transformations can be elicited if boundaries of *Hox* genes are expressed more rostrally. Disruption of *Hoxa2* induces defects in the pharyngeal region corresponding to the altered rostral expression; as discussed in Box 10.2, *Hoxa2*, regulates specification of the first pharyngeal (mandibular) in response to retinoic acid. In skeletogenic NCCs, *Hoxa2* is a target of the transcriptional activator protein *Ap2α*. Downstream targets of *Hoxa2* include *Sox9* and *Six2*.[21]

Neural crest mesenchyme derived from rhombomeres that normally populate first and second pharyngeal arches fails to produce normal second-arch derivatives, producing first-arch structures instead. The result is a homeotic transformation of the entire second-arch skeleton into a first-arch skeleton (Box 10.2).

Similarly, the normal rostral extent of expression of *Hoxd4* is at the first cervical somite. Following expression of *Hoxd4* more rostrally, the occipital bones (which lie rostral to the cervical vertebrae) are transformed into bones that mimic cervical vertebrae.[22]

Hoxa1, Hoxa2, *and* Hoxb1

Hoxa1 disrupts rhombomere organization in mouse hindbrains, while *Hoxb1* influences the fate of cells that emerge from r4; activity of *Hoxa1* is required to establish the anterior limits of expression of *Hoxb1* at the r3/r4 boundary (Barrow *et al.*, 2000). Mice with double mutations of *Hoxa1* and *Hoxb1* show defects either not seen in single mutants or expressed in mild form—patterning defects of r4, r5, and r6, major disruptions of middle and external ear development, loss of the second pharyngeal arches and all second-arch structures—resulting in major craniofacial defects.[23]

Gale and her colleagues used ectopic expression of retinoic acid in r4 of stage 10 rat embryos to demonstrate that retinoic acid alters the expression of zinc finger genes such as *Krox20* (the upstream regulator of *Hoxb2* in the hindbrain), eliminating expression of *Hoxb1* but leaving *Hoxa2* expression unaffected. One consequence is the development of facial ganglia more rostrally than normal. Retinoic acid modification of the *Hox* code of the hindbrain leads to r2 and r3 transforming into r4 and r5.[24]

Mutations and Birth Defects

Craniofacial defects similar to those following treatment of laboratory animals with excess vitamin A arise as the result of spontaneous mutations. Examples of mutant genes are *looptail* and *Patch* in mice and the mutations associated with neurocristopathies in humans (see Chapter 9).[25]

Looptail

Looptail (*Lp*) and *Patch* lead to premature differentiation; NCCs migrate along abnormal pathways because of precocious alterations in the extracellular matrices they encounter. Consequently, NCCs begin to differentiate before sufficient cells have accumulated for normal craniofacial development and growth to ensue. The hindbrain and spinal cord fail to close in *Lp* embryos, NCCs remain attached to the neural folds and so fail to delaminate, indicating the importance of neural fold closure for delamination and initiation of migration. *Lp* is a mutation or series of mutations in genes that interact with members of the Disheveled family of proteins that function in the canonical *Wnt* pathway (see Fig. 2.5).[26]

Splotch

The *Splotch* mutation is more fully understood. *Splotch* results from an inversion of intron 3 of the transcription factor *Pax3*. Normally, *Pax3* downregulates *Msx2* via *Pax3*-binding sites on the *Msx2* promoter, *Msx2*, in turn, regulating the function of Bmps.

   Splotch is characterized by missing or small spinal ganglia and by spina bifida. Ganglionic reduction is independent of the neural tube defects. A transgenic neuroanatomical marker was used by Tremblay *et al.* (1995) to reveal CNC deficiencies, especially of cranial ganglia and cranial nerves. In humans, *Splotch* is manifested as *Waardenburg syndrome* (see Table 9.1). *Splotch* stops NCCs from migrating away from the neural tube, cultured neural tubes from *Sp* or *SP4H* mutant mice exhibit a 24-h delay in the initiation of NCC migration. In part, this delay is a consequence of alterations in N-CAM, an effect also seen in avian embryos treated with retinoic acid—the NCCs retain N-CAM and so cannot migrate normally·

   Migration of murine VNC is especially affected. Transplanting neural tubes between *Splotch* mice and chicken embryos reveals that the migration defect is not intrinsic to the NC but rather arises from abnormal cell interactions.

Mice homozygous for *Splotch* have defective aortic arches, thymic, thyroid, and parathyroid glands and die at or before 13.5 days gestation. Normally, cells from the CarNC migrate through pharyngeal arches 3, 4, and 6 to colonize the outflow tract of the developing heart. *Sp2H*, one of the six mutant alleles of the *Splotch* locus, is associated with persistent truncus arteriosus and related conotruncal heart defects. The transient expression of *Pax3* normally seen in the developing hearts of mice at 10.5 days of gestation is not seen in embryos carrying the *Splotch* mutation. From their studies with *Sp2H* mutant mice, Conway and colleagues conclude that *Pax3* is required for CarNCCs to migrate. Indeed, because of important actions at the condensation stage of organogenesis (see Figs. 3.13 and 7.18), the *Pax* family of genes is a key player in the development of many organs—kidneys, eyes, ears, nose, limbs, vertebrae, and brain.[27]

An interesting addendum to the action of the *Splotch* mutant, one that may apply to many mutants, is the reduction in frequency of neural tube defects in *Splotch* mice exposed to retinoic acid at a specific stage of neural tube development. Thus, injecting 5 mg of retinoic acid/kg maternal body weight at day 9 of gestation stimulates growth of the neural tube with concomitant reduction in neural tube defects, while injection at day 8 is ineffective (Kapron-Bras and Trasler, 1985). Therefore, environmental agents can reduce defects that are normally considered and described as genetically determined.

Other inherited defects have specific effects on particular populations of NCCs:

- *Familial facial osteodysplasia* in humans is an inherited syndrome in which bone with an NC origin is entirely absent but NC-derived cartilages, and bones and cartilages of mesodermal origin, are present and normal.
- *Cleidocranial dysplasia*, an autosomal-dominant skeletal dysplasia in humans, is characterized by short stature, widely separated (patent) sutures between the skull bones, underdevelopment or complete loss (agenesis) of the clavicle, and supernumerary teeth, all defects than can be traced to NCCs (Box 10.3).
- *Lethal Spotted* and *Piebald Lethal* —mouse mutations in the gene for the mitogenic peptide Edn3 and the endothelin receptor b (EdnRb), respectively—are associated with megacolon, which develops because of the lack of enteric ganglia in the terminal 2–3 mm of the bowel (see Box 3.2). NCC migration is normal in both mutants, the defects arising because an abnormal non-NC environment in the bowel will not permit NCCs to colonize the bowel wall, resulting in the development of an aganglionic segment.[28]

Box 10.3 Cleidocranial dysplasia and RUNX2

Cleidocranial dysplasia (also known as cleidocranial dysostosis and Marie–Sainton disease), an autosomal-dominant skeletal dysplasia with associated short stature, specifically affects membrane bones of NC origin. In heterozygotes, the skull sutures remain open, the clavicles fail to develop or are underdeveloped, and supernumerary teeth form.

Cleidocranial dysplasia results from a mutation in the *runt-related transcription factor 2* (*RUNX2*), a gene that is a major regulator of bone cell differentiation. *RUNX2* is a member of the runt-domain family of core-binding transcription factors; **run** from the *Drosophila runt* gene and **x** to denote the mammalian gene. Other synonyms are Osf2 (osteoblast-stimulating factor2) and Cbfa1 (core-binding factor a1).

Two-thirds of the affected humans have unique mutations (70 distinct mutations have been described so far), indicating a complex relationship between gene and phenotype. The most severe cases and 60% of the mutations involve mutations in the *SMAD*-binding domain of *RUNX2*, without affecting the DNA-binding domain. Some 23% of described mutations are of the C terminal of the runt-binding domain (Cunninham *et al.*, 2006).

Targeted disruption of *Runx2* in mice results in failure of formation of all membrane and endochondral bones in homozygotes (*Runx2−/−*) because of arrested osteoblast differentiation. The formation of postcondensation osteoblasts is inhibited completely, although some alkaline phosphatase-positive 'osteoblasts' develop. *Runx2−/−* embryos express considerable Runx2 protein in chondrocytes, especially hypertrophic chondrocytes. *Runx2* is upregulated in the second arch of *Hoxa2* mutants (Box 10.2), consistent with *Hoxa2* inhibiting *Runx2*, and *Hoxa2* inhibiting osteogenesis by downregulating *Runx2*.[a]

---

[a] Literature on this topic is summarized in Hall (2005b*). Functions of members of the Runx family of transcription factors are not conserved across the vertebrates; *Runx1* is required for initial condensation initiating chondrogenesis in mice, but not in zebrafish (Flores *et al.*, 2006). Nor does *Runx2* function in an all or none manner; variation in tandem coding length repeats of *Runx2* alleles are correlated with variation in the shape of the craniofacial skeleton in dogs (Fondon and Garner, 2007).

Continued study of these mutations affecting laboratory animals sheds further light on how defective NCCs or a defective NCC environment produces the mutant phenotype and, *en passant*, increases our understanding of normal NCC development.

Regulation

Even if a mutant gene or teratogen has a deleterious effect on a cell population, embryos may still develop normally. This is partly because of *redundancy and overlapping functions* between members of gene families—mutation in one being compensated for by another family member—and partly because of the developmental phenomenon of **regulation**, the ability of embryos or portions of embryos to compensate for changes that otherwise would result in abnormal development by replacing often quite large groups of cells. You can think of regulation as the embryonic equivalent of regeneration seen in some postnatal organisms. Regeneration replaces already differentiated cells by the dedifferentiation of cells at the wound or from stem cells; regulation replaces undifferentiated cells.

It has long been known that amphibian and avian embryos can compensate for the removal of premigratory NCCs. Avian and mammalian embryos can also regulate for deficiencies that occur during NCC migration.

Exposing 9-day-old rodent embryos to Mitomycin C temporarily blocks cell division, although division starts again after the drug is metabolized. Snow (1981) demonstrated that the number of cells in mouse embryos is reduced by as much as 30% during the 24 h following Mitomycin administration. Such a dramatic reduction in embryo size at such a sensitive stage of development would be expected to kill the embryos or lead to the development of small or abnormal embryos. In an amazing demonstration of regulative ability, however, these embryos have undergone compensatory cell division and growth by 12 or 13 days of gestation, restoring them to the size of normal, untreated littermates. Furthermore, NC derivatives develop normally in these embryos, indicating that NCC regulation must also have occurred. More detailed examination of the regulative capability of the mammalian NC would aid enormously in interpreting the etiology of embryonic defects, possibly provide us with the means to compensate for the abnormalities that otherwise follow exposure to teratogens, and even lead to ways of preventing such defects in future generations.

A study undertaken by Bronner-Fraser (1986) beautifully documents how regulation can compensate for an initial defect in NCC migration. Bronner-Fraser utilized the cell substrate attachment (CSAT) antibody to β1-integrin, a cell surface receptor for fibronectin and laminin, two of the chief ECM components influencing NCC migration. CSAT antibody was injected lateral to the midbrain of chicken embryos immediately before the onset of NCC migration. Twenty-four hours after antibody injection Bronner-Fraser observed:

- a reduction in the number of NCCs, indicative of reduced cell division;
- an aggregation of NCCs within the lumen of the neural tube, indicative of defective initiation of cell migration; and
- NCCs in ectopic locations, indicative either of defective pathways of migration or of abnormal trapping of NCCs.

The only adverse effects seen 36–48 h after antibody injection, however, were ectopic condensations of NCCs and abnormally developed neural tubes. NCCs that had migrated ectopically remained in those ectopic sites. Otherwise, NC derivatives developed normally. Both the defective proliferation and the delayed onset of migration of the majority of the NCCs were compensated for.

Sources of Cells

The source of the cells that replace missing NCCs during regulation depends in part on the region of the neural axis from which NCCS are deleted and in part on how much NC is deleted. Compensation can take the form of

- *migration of NCCs across the midline*, as occurs if the NC are only removed from one side of the neural tube;
- *migration of NCCs from a more cranial or caudal region* of the NC, as occurs if the NC is deleted only from one cranio-caudal region of the neural folds;
- *increased proliferation* of the remaining NCCs, as occurs if the NC are removed only from one region of the neural tube;
- regulation from *neural ectodermal cells* that would not otherwise have formed NCCs; or
- regulation from placodal or epidermal ectoderm.[29]

An approach that may help to increase our understanding of NC regulation is the use of mutations affecting neural tube development.

We saw in the previous chapter that a *TBOX* gene, *TBX1*, is involved in all individuals with DiGeorge syndrome. Another Tbox gene, *Tbx6,* produces a DNA-binding protein. As demonstrated by Chapman and Papaioannou (1998), mutation of *Tbx6* converts murine axial mesoderm into neural tissue, resulting in embryos with three neural tubes: one in the normal axial position and one on either side of the region where paraxial mesoderm would normally develop. Because of the potential to analyze NCC development in the ectopic neural tubes, a mutation that switches mesodermal tissues from somitic to neuronal to produce ectopic neural tubes with normal dorsoventral patterning clearly is of interest. Would such mesoderm–turned-neural tissue be capable of forming NC, given that NC normally arises at the border between neural and epidermal ectoderm? If capable, could ectopic neural tubes regulate to compensate for removal of those NCs?

Completeness of Regulation

As demonstrated for TNCCs in zebrafish embryos by Vaglia and Hall (1999, 2000), regulation, or the degree of regulation, is influenced by:

- the availability of a source of competent cells,
- the time/stage of the embryo when the cells are removed,
- the site from which NC is removed, and
- the amount of crest removed.

For regulation to occur at all after ablating NC from chicken embryos, the neuroepithelium and epidermis must come into contact. Regulation after hindbrain ablation in chicken embryos requires that dorsal midline closure be reestablished, a cellular process associated with modulating expression of *Pax3* and *Snail2*.

Regulation of CNCCs is maximal at the 3- to 7- somite stage. Regulation is most effective when caudal midbrain or hindbrain NC is removed and less effective after ablating caudal forebrain (which does not produce NCCs) or rostral midbrain (which does), demonstrating differential regulative ability as a property of subpopulations of NCCs.[30]

Pharyngeal-Arch Regulation

The pharyngeal-arch skeleton is replaced by regulation after NC is bilaterally extirpated from avian embryos. Regulation from the edges of the neural tube, resulting in a pharyngeal-arch skeleton of normal size and morphology, is preceded by restoration of the *Hox* gene code appropriate to the NC region removed. *Hoxa2* has the most rostral expression of any Hox gene, extending to the boundary between r1 and r2 in neural ectoderm but not in the NC (see Fig. 2.17). Expression of *Hoxa2* in r4 in both neural tube and NC, even if rhombomeres are transposed, indicates intrinsic patterns of expression and independent regulation.

The patterns of migration of crest cells from r3 to r5 are also restricted segmentally. Paul Hunt and Gérald Couly and their colleagues found that surgically removing the rhombencephalic NC from chicken embryos was followed by normal pharyngeal-arch development and normal expression patterns of Hoxa2, *Hoxa3*, and *Hoxb4*. Because of enhanced proliferation and migration, regulation from the neural epithelium on either side of the surgical site compensated for the NCCs removed; that is, there is regulation of *Hox* genes as well as of cells and tissues.

If the mesencephalon and metencephalon are removed from chicken embryos and replaced in inverted orientation, the normal rostrocaudal polarity of *engrailed* is reversed. Within 20 h of the operation, however, a normal *engrailed* pattern is reestablished. Interestingly, although ablation of the dorsal hindbrain alters pathways of NCC migration and patterns of gene expression, changes that would be expected to result in deficiencies of the hyoid arch skeleton, regulation restored normal development of the hyoid arch skeleton.[31]

Dorsal Root Ganglion Regulation

Older embryos also have the ability to regulate regions comprised of NCCs or their derivatives.

The number of sensory neurons in dorsal root ganglia is regulated by interactions between ganglionic neurons and peripheral target tissues. Ablating the branchial NC from H.H. stages 13–14 chicken embryos elicits regulation and the formation of dorsal root ganglia from the remaining NC, although dorsal root ganglia 15, 16, and 17 are small or missing. In addition to this local regulation and, unexpectedly, ganglia remote from the region of the NC removed undergo hypertrophy, some by as much as 220%. Ganglia as far removed as the cervical ganglia show compensatory hypertrophy; indeed, removal of cervical ganglia elicits the greatest hypertrophy.

The dorsal root ganglia along the neural axis in avian embryos differ in size. Somitic mesoderm is not required for precursors of dorsal root ganglia to begin migrating but does regulate their growth, Ganglia adjacent to segments 14 and 15 are 80% larger than those adjacent to segments 5 and 6, a secondary consequence of the local sclerotomal environment, not because of intrinsic properties of the NCCs forming the ganglia.

Quail embryos are smaller than chicken embryos and so have smaller dorsal root ganglia. Regulation of the number of cells contributing to the dorsal root ganglia in avian embryos was demonstrated by transplanting quail NCCs into the equivalent position in chicken embryos—a variation of the classic transplantation experiments undertaken with amphibian embryos by Harrison and Wagner in the first third of the 20th century (see Chapter 7). The size of the ganglia in the chimera was appropriate for the host embryo; local compensatory mechanisms regulated ganglion size.

Quail ganglia transplanted into chicken hosts adapt to the larger size of chicken ganglia. Part of the host environment is the pattern of distribution of the ligand ephrin-B2 and the Eph family of receptor tyrosine kinases and their transmembrane ligands, which mediate interactions between sclerotome and migrating NCCs, confining the latter to specific rostrocaudal territories. Complementary expression of EphA4/EphB1 receptors and the ligand also prevents cells from the second and third arches from intermingling, and so targets third-arch NCCS to their destination (Fig. 10.6 [Color Plate 9]).

Receptor tyrosine kinases (RTKs) may play important and separate roles in regulation, in part because different subpopulations of NCCs have specific RTKs and thus respond differentially to different growth factors: Enteric NCCs, for exam-

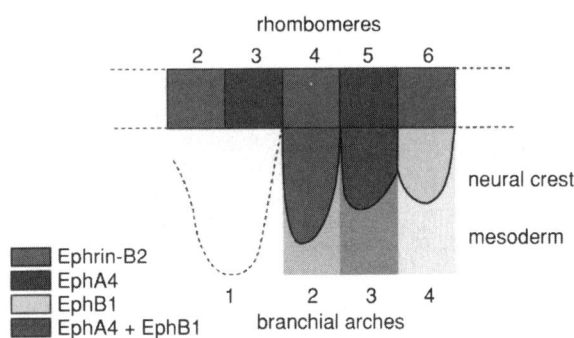

Fig. 10.6 Expression of EphrinB2 (*red*), EphA (*blue*) and EphB1 (*yellow*) in rhombomeres 2–6, in migrating neural crest cells and in cranial mesoderm, is highly regionalized with respect to the four pharyngeal arches. EphrinB2 is expressed in r4 and in the second pharyngeal arch. EphB1 is expressed in r6 and the fourth pharyngeal arch. The boundary of EphrinB2 is shown in *red*, the boundary of cells expressing EphA4 plus EphB1 in *green*. Receptor–ligand interactions at the boundary between pharyngeal arches 2 and 3 restricts cells that are migrating from r5 to the third pharyngeal arch and prevents intermingling of second and third arch cells. Reproduced from A. Smith *et al.* (1997) with the permission of Current Biology Ltd., from a figure kindly supplied by David Wilkinson (see Color Plate 9)

ple, require the Ret RTK receptor for proper migration into the developing gut, while craniofacial mesenchyme in early chicken embryos does not require the RTK receptor, Sprouty, even though Sprouty is required for subsequent growth of craniofacial processes. The RTK receptor, which was cloned from chickens, plays a role in regionalizing the VNC and somites (notably the dermamyotome) and is expressed in placodal ectoderm, where it may play a role in segregation or regionalization.[32]

Regulation of Cardiac Neural Crest

Ablated CNC (especially caudal midbrain–rostral hindbrain) can be replaced completely by regulation, ablated TNC only partially (see Chapter 10). When we consider CarNC, the potential for regulation is even more limited.

Removing CarNC results in cardiovascular anomalies (above), a result that speaks to limited regulation or even lack of regulation. Regulation can occur, however, provided that some CarNCCs remain. Thus, back-transplanting CarNCCs to replace extirpated cardiac crest produces normal cardiac crest and rescues embryos that would otherwise exhibit heart anomalies (Kirby et al., 1993).

Suzuki and Kirby (1997) explicitly tested for the ability of the CarNC of chicken embryos to regulate following bilateral ablation and DiI labeling of the ventral neural tube adjacent to the ablated crest. Ablation before the 5-somite stage evoked only little regulation, while crest ablated after the 6-somite stage was not replaced at all (Fig. 10.7 [Color Plate 9]).[33] This temporal loss of regulatory ability seen in chicken embryos should not automatically be extrapolated to mammals, especially given the recent identification of cardiac stem cells in rodents and their ability to populate chicken hearts after transplantation (see the last section of this chapter on stem cells).

Surprisingly, placodal derivatives provide a source of cells that can replace CarNCCs. The reverse is also true; nodose ganglia can be replaced by regulation from the CarNC (see below). The nodose placode (see Figs. 6.5 and 6.6) can regulate to provide autonomic neurons and mesenchyme to chicken hearts (normally this placode produces neurons but not mesenchyme), but cannot prevent all the cardiac defects that flow from removing the CarNC. Neurons that form from the nodose placode are indistinguishable physiologically from the NC-derived neurons that would have developed without the extirpation.

Removing CarNC and the nodose placode lead to severe cardiovascular abnormalities. Neither midbrain nor TNC transplanted in place of ablated CarNC can compensate for the truncal septation that is a consequence of cardiac crest ablation. Indeed, mesenchyme derived from midbrain NC that has invaded the heart primordium forms ectopic cartilage within the heart (see Chapter 8) and interferes with cardiac mesenchyme, indicating a predetermination of mesenchyme that parallels the predetermination of mandibular-arch skeletal structures reported by Noden in 1983 and discussed in Chapter 7.[34]

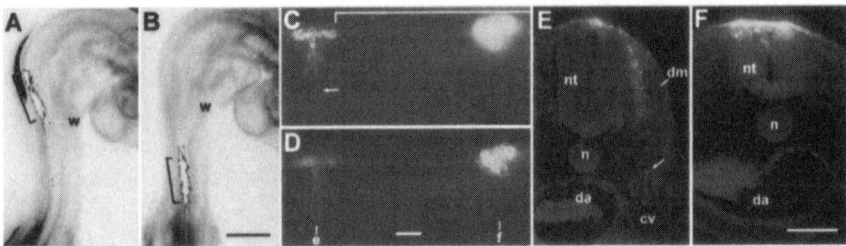

Fig. 10.7 Ablating the neural crest followed by injection of DiI demonstrates absence of regulation; no labeled cells are seen either in the normal migration pathway (illustrated in Fig. 7.16) or at the final site. (**A, B**) Trunk neural crest was ablated (*square bracket*) either adjacent to the most caudal somites (somites 12–17 in **A**) or to the most cranial unsegmented mesoderm (*square bracket* in **B**) from 17- or 20-somite-stage embryos (**A** and **B,** respectively). Multiple sites in the neural tube in the ablated region were then injected with DiI. In neither case can neural crest cells be seen migrating from the ablated region, signifying lack of regulation. w, wing bud. Bar = 1 mm. (**C, D**) This 17-somite embryo is viewed in two focal planes. The *bracket marks* the ablated neural crest. The *yellow* fluorescence marks sites of DiI injection either immediately rostral to the ablated region (the *left* zone of fluorescence, seen in focus in **C**) or within the ablated region (the *right* zone of fluorescence, seen in focus in **D**). Neural crest cells are migrating from the rostral neural crest (*arrow* in **C**) but do not deviate into the ablated area. As expected, no cells migrate from the neural tube in the ablated region. e and f mark the position of the transverse sections shown in **E** and **F**. Bar = 100 μm. (**E**) Transverse section rostral to the ablated region marked e in (**D**). DiI-labeled cells are migrating medial to the dermamyotome (dm) and in the primary sympathetic trunk (*arrow*). cv, cardinal vein; da, descending aorta; n, notochord; nt, neural tube. (**F**) Transverse section through the ablated region marked as f in (**D**). DiI-labeled cells are concentrated adjacent to the dorsal neural tube and not seen in the normal migration pathway; compare with **E**. Abbreviations as in **E**. Bar = 100 μm for **E** and **F**. Reproduced from Suzuki and Kirby (1997) from a figure kindly supplied by Margaret Kirby. Reprinted by permission of Academic Press (see Color Plate 9)

Placodal Regulation from the Cardiac Neural Crest?

In addition to being able to regulate to replace cells within the heart (see Chapter 8), chicken embryo nodose ganglia (which are placodal in origin) can be replaced by regulation. Nodose ganglia arise from placodal ectoderm; *Fgf3* (*interleukin2*), which is produced in rhombomeres of embryonic chicken hindbrains, diffuses to the ectoderm, where it induces nodose ganglia and otocyst.

Early in development, before placodal ectoderm is determined (1.5 days of incubation in avian embryos), trunk nonneurogenic ectoderm transplanted in place of placodal ectoderm can replace placodal ectoderm and form nodose sensory neurons. Ablating the placodal precursor of the nodose ganglion from chicken embryos at H.H. stage 9 elicits partial replacement by 12 days of incubation. Quail/chicken chimeras demonstrate that the cells that regulate for the nodose ganglion arise from CarNC adjacent to the third pair of somite, although this study has not been repeated or confirmed.[35]

Thus, at least in embryonic chickens, it may be that nodose placodes can compensate for the extirpation of CarNC, and vice versa. If such regulation occurs, it would be because of shared properties among NCCs destined to form different cell types; a monoclonal antibody generated against nodose ganglia recognizes peripheral ganglia, Schwann cells, some sensory and some autonomic neurons, as well as 25% of migrating NCCs.

Neural Crest Cells as Stem Cells

Having discussed the regulative ability of various subpopulations of NCCs, an appropriate issue to address is whether any NCCs are **stem cells**. The presence of stem cells in populations of NCCs might reduce the need for regulation from other cells and enhance that regulative potential or extend the ability to regulate into later stages of development, even into adult life. Indeed, the presence of stem cells in adults raises the possibility of repair and/or regeneration of NC derivatives. Consequently, this book concludes with a discussion of stem cells and whether the NC, NCCs, or subpopulations of NCCs at any stages during their development are or contain stem cells. To do so, we must reevaluate the concepts of bi-, tri-, and multipotentiality.

Bi-, Tri-, and Multipotential NCCs

The terms and concepts bi-, tri-, and multipotentiality (see Table 1.2 for definitions) have been discussed in relation to premigratory NCCs and various subpopulations during or after the completion of NCC migration (see Chapter 3).

You will recall that these concepts refer to the potential (hence 'potentiality') of an individual NCC to differentiate along one or two or several directions. Chapter 3 discussed the evidence leading to the conclusions that:

- many subpopulations of NCCs are bipotential before or during migration;
- some remain bipotential even after differentiating part way along one pathway; and
- signals outside the cell determine the pathway of differentiation taken.

The major signals to which NCCs respond—transcription factors, growth factors, and hormones—were discussed in Chapter 2 and in the last section of Chapter 3. An example from Chapter 2 is the *Wnt* signaling pathway, which is involved in specification of cell fate in bipotential cells (Hayward *et al.*, 2008). Entering one differentiation pathway normally closes off other options, although, as we saw in Chapter 3, cells can dedifferentiate and redifferentiate into another cell type, usually within a limited range of options, and usually in association with regeneration, and also sometimes, as discussed above, during regulation.

The development and application of clonal cell culture (see Chapter 3) revealed the bipotentiality of many subpopulations of NCCs before or during migration (see Fig. 3.17). You will recall from Chapter 3 two examples from Japanese quail embryos:

- *premigratory or early migrating CNC* contains two populations of tripotential NCCs (see Fig. 3.17) and
- *premigratory or early migrating CarNC* gives rise to five clonal cell lines, two of which are multipotential, one of which is bipotential (producing chondrocytes and sensory neurons), and two of which are unipotential, producing either pigment cells or smooth muscle cells.

We can conclude that subpopulations of NCCs are lineage restricted before they delaminate and migrate and that lineage restriction reflects identified regions within the NC (cranial and cardiac in this example). We can also conclude that these subpopulations are not stem cells (see below).

Clonal cell culture reveals that populations of tri- and bipotential cells exist if the cells are taken, not from the NC but, from NC derivatives, dorsal root and sympathetic ganglia in this case (see Fig. 3.18). Both cell populations can produce nonganglionic cell types, pigment cells and adrenal medullary cells in this case. We can conclude that bi- and tripotentiality remain in NC derivatives. Can we conclude that the cells isolated from developing ganglia are stem cells? Not necessarily. *A cell can have multiple cell fates (multiple potentials) without being a stem cell.*

What is a Stem Cell?

Although there is some (often considerable) disagreement over how to define stem cells, stem cells may be categorically defined as

  (i) a proliferating and self-renewing population of cells (or a single cell),
 (ii) capable of asymmetrical cell division, and
(iii) the products of which are a stem cell and a cell capable of differentiating in one of many directions.

Some not in agreement with this definition would add the criterion that a stem cell is *totipotent*, that is (and as defined in Table 1.2), able to form any cell of the body. Others would require that a stem cell be pluripotent, meaning able to form many cell types, perhaps all those from a single germ layer (see Table 1.2). The author has hedged with the phrase 'capable of differentiating in one of many directions'. How many is many? This is moot. The essential criteria are self-renewing, asymmetrical cell division, and lack of lineage restriction.

The egg is a totipotent stem cell capable of producing all cells types of the body, which is no mean feat as humans have upward of 400 cell types, half of them neurons (Vickaryous and Hall, 2006). In many species, cells of blastula-stage embryos are stem cells; divide a four-cell stage sea urchin or human embryo into four cells and four sea urchins or quadruplets develop.

If we use echinoderm or amphibian embryos as examples, gastrulation marks the stage during development when embryonic cells are no longer totipotent; this change occurs much earlier in mice, typically at the eight-cell stage, when inner cell mass and trophoblast are determined. Cells of echinoderm or amphibian gastrulae are pluripotent, although that pluripotency becomes progressively restricted as gastrulation continues. Ectoderm from frog early gastrulae can form neural or epidermal ectoderm; ectoderm from late gastrulae is restricted to a neural or epidermal fate (because of inductive interactions during gastrulation; see Chapter 2).

During postgastrula development, the fate of most cells becomes progressively more restricted. Of course, this is not universal. Renewing cell populations are comprised of a proliferating/nondifferentiated and a differentiating/differentiated compartment, the proliferative cells being called into play as differentiated cells are lost. Under normal conditions, a balance is maintained between cell proliferation in the one compartment and cell loss from the other. The proliferative cells are progenitor cells but not stem cells. The most well-known examples are vertebrate blood cells and skin cells.

The experiment that has received much attention in the very recent past, however, is the reprogramming of fibroblasts from the facial skin of a 36-year-old woman into what the authors describe as *induced pluripotent stem cells*, which under *in vitro* and *in vivo* conditions give rise to a large range of cell types, including (in teratomas formed *in vivo*) muscle, keratinized epidermis, cartilage, fat, neural tissue, and intestinal epithelium (K. Takahashi *et al.*, 2007).

As we saw above, cells in the NC and NC derivatives retain the potential to differentiate into more than one cell type. But does the NC itself contain stem cells?

NCCs as Stem Cells

Identification of NCCs as stem cells requires that the cells fulfill minimally two criteria: (i) multipotentiality and (ii) self-renewal. Whether the stem cells are found in the NC, along migratory pathways, in developing NC derivatives within embryos, or in adults, is not material to their identification as stem cells.

Stemple and Anderson (1992) demonstrated that the rat NC contains what they refer to as a stem cell for glia and neurons (see Chapter 6), but they did not show that such cells could produce other cell types or were self-renewing. In a review of NCC-lineage specification published a year later, Stemple and Anderson summarized the evidence of multipotential NCCs and indicated that no studies had demonstrated that these NCCs were self-renewing cell populations.

A decade later and evidence had accumulated. As discussed above, clonal cell culture in the early 1980s showed that premigratory and early migrating NCCs of Japanese quail embryos contain subpopulations of bi-, tri-, and multipotential cells.[36] Trentin and colleagues (2004) used serial subcloning of quail NCCs to demonstrate that two populations of bipotential NCCs (glial/pigment cell and glial–myofibroblast) are self-renewing. But even those may not fit all criteria of stem cells, unless they can be shown to form other cells types. Calling them 'restricted stem cells' would be an inappropriate conflation of two concepts—bipotentiality and stem cell-ness. In an insightful synthesis of the function of Sox10 in NCC development, Kelsh (2006) summarized evidence that led him to the conclusion that, in the context of NCCs as multipotential, (i) expression of specific receptors defines (limits, if you will) the potency of particular NCCs, (ii) receptor expression becomes more restricted as NCCs differentiate, and (iii) exposure to extracellular ligands essentially determines the pathway of differentiation taken by individual NCCs.

Over the past decade, however, stem cells meeting the two criteria of multipotentiality and self-renewal have been identified and isolated, including

(i) from the sciatic nerves of embryonic rats of 14.5 days gestation,
(ii) from the intestines of adult Sprague-Dawley rats, and
(iii) from the whisker pads of adult mice.[37]

We now discuss a fourth example: CarNC stem cells.

CarNC Stem Sells in Adults: Because cardiomyocytes can proliferate in response to damage, humans have some ability to repair damaged hearts. The search for stem cells in adult vertebrates has revealed that the CarNC contributes multipotential stem cells to the hearts of adult rodents, cells that we assume can be activated in situations requiring repair.

Cardiac stem cells also have been isolated from adult humans. They are stem cells because they form clones of self-renewing, multipotential cells that, upon appropriate stimulation, differentiate into cardiac or smooth muscle myocytes or endothelial cells. Cardiac cells isolated from adult humans and transplanted into chicken embryos migrate to the truncus arteriosus and cardiac outflow tract where they differentiate into aortic smooth muscle cells and contribute to dorsal root ganglia and to the glia of spinal nerves.

NC-derived cells have now been identified in the fetal rodent myocardia, where it is assumed they reside into adulthood as dormant multipotential stem cells. It is further assumed that these cells can differentiate into cardiomyocytes (zebrafish CarNCCs form cardiomyocytes) and into the NC derivatives (neurons, glia, smooth muscle) seen after transplanting the stem cells derived from adult hearts.[38]

You will have noted that the criterion of 'capable of differentiating in one of many directions' is used here to apply to cells that in each case produce three cell types: (i) cardiac and smooth muscle myocytes, endothelial cells; (ii) smooth muscle cells, dorsal root ganglia, glia; and (iii) smooth muscle, neurons, glia. This is a far cry from totipotency. Given their self-renewal, ability to undergo asymmetrical cell division and tripotency, these cells are, at the very least, multipotent stem cells. Some might classify them as restricted stem cells, but this is neither a helpful term nor a helpful concept. Multipotent/pluripotent, but not totipotent, stem cells seems more appropriate.

While this may seem an incomplete, even unsatisfactory, way to finish this discussion of the NC and NCC, recall that 120 years ago we would have finished with the uncertainly of the existence of a NC or NCCs, and how such cells could disobey the germ-layer theory. Biologists of the time certainly did not want to disregard the theory. The NC, its cells, and derivatives teach us that we should allow embryos to speak to us. Only in that way will we understand their secrets.

Notes

1. Sarnat and Flores-Sarnat (2004) based a classification of CNS malformations on the integration of morphology and genetics in the context of gradients of gene expression. Tovar (2007) evaluated 195 publications in an overview of the importance of recognizing the NC origin in individuals requiring pediatric surgery.

2. See McLeod *et al.* (1980), Kang and Svoboda (2005), and Kanakubo *et al.* (2006) for defective cell migration as a major factor in craniofacial defects and Newgreen (1995*) for mechanisms underlying early stages of migration.

3. For the induction of birth defects by altering cell migration, see Hassell *et al.* (1977) and Morriss-Kay (1992). For defective ECM as a factor in altered migration see the discussion in Chapter 3.

4. See http://www.genoma.ib.usp.br/TCOF1˙database/ for mutations of *TCOF1* and Dixon *et al.* (2006) for the study in mice.

5. See Hrubec *et al.* (2006) for malformations in infants born to diabetic mothers and Siman *et al.* (2000) for studies in diabetic rats that mirror DiGeorge syndrome.

6. See Suzuki *et al.* (1996) for the studies on high glucose concentrations in rats and Roest *et al.* (2007) for the studies with chicken embryos.

7. For proliferation of migrating NCCs, see Noden (1987*) and Hall (2005b*). See Hall and Miyake (1997) and Hall (2005b*) for sensitive periods and Johnston and Bronsky (1995) for craniofacial defects associated with defective proliferation. See Czeizel *et al.* (2008) for the utility of critical periods over concepts such as 'the first trimester' when assessing congenital anomalies in humans and Hall and Miyake (1997) for an evaluation of how embryos 'count' time.

8. See Winograd *et al.* (1997) for transgenic *Msx2* mice. For Bmp4 and apoptosis of hindbrain NCCs, see Lumsden *et al.* (1991), Thorogood (1993), Lumsden and Graham (1996), and Lumsden and Krumlauf (1996). For similar roles in NC derivatives in pharyngeal arches and neuron precursors, see Barlow and Francis-West (1997), Ekanayake and Hall (1997), and Song *et al.* (1998). See K. Takahashi *et al.* (1998) for *Msx2*-induced apoptosis in even-numbered rhombomeres and Hosokawa *et al.* (2007) for Tgfβ-mediated expression of *Msx2* in the control of the occipital somites that form the caudal region of the skull in mice.

9. See Bennett *et al.* (1995) for the distribution of Bmp in murine orofacial tissues and Le Douarin (1988), Hall and Ekanayake (1991), and Chapter 3 for other growth factors that regulate the differentiation of NCCs.

10. See Dencker *et al.* (1990), and Ruberte *et al.* (1992) for the localization of CRABP to NC and other cell types; Rowe and Brickell (1995) for the retinoid X receptor-γ gene; Ito and Morita (1995) for differential effects of vitamin A on cranial and TNC; and Wendling *et al.* (2000) for arch defects.

11. See Di Renzo *et al.* (2006) for the study with the antifungal agent triadimefon and Cerny (2005), Cerny *et al.* (2004*), and Lee *et al.* (2004) for the reinterpretation of the origins of mandibular and maxillary cartilages.

12. See Hart *et al.* (1990) for the avian model for retinoic acid embryopathy. For effects of vitamin A on NCCs and craniofacial development, see Hassell *et al.* (1977), Wedden (1987), and Maden *et al.* (1996). See Reijntjes *et al.* (2007) for the zebrafish study.

13. See Mallo (1998) and Mallo and Gridley (1996) for ear defects that arise from lack of migration, failure of formation or the respecification of NCCs, or for defective developmental interactions. See Johnston and Bronsky (1995) for animal models for retinoic-acid-based human craniofacial malformations.

14. See Atchley and Hall (1991), Hall (1999b*, 2003a*), and Franz-Odendaal and Hall (2006) for selection operating throughout development and Hall (1999b*) for a general discussion of transitions between classes of control in ontogeny.

15. For defective migration as a major factor in human pharyngeal-arch malformations, see McLeod *et al.* (1980). For the field concept in hemifacial microsomia, mandibulofacial

dysostosis, and embryological interpretations of carcinomas, see Pierce (1985), Hodges and Rowlatt (1994), and Carstens (2002, 2004). For a possible genetic model for hemifacial microsomia, the *far* mutation in mice, see Juriloff *et al.* (1987).

16. See Yip *et al.* (1980) for links between vitamin A and primate craniofacial defects, Morriss-Kay (1992) for direct affects of vitamin A on NCC migration and Dencker *et al.* (1990) for mechanistic studies.

17. See Mahmood *et al.* (1992) for retinoic acid and Tgfβ1 and 2, Mahmood *et al.* (1996) for retinoids and Fgf3, Prince *et al.* (1998) for *valentino* and *kreisler*, and Barrow *et al.* (2000) for the role of *kreisler* in patterning the murine hindbrain. Ruberte *et al.* (1997), who claim that pro-rhombomeres are not functional precursors of rhombomeres, query whether identification of pro-rhombomere is functionally meaningful.

18. See Vandersea *et al.* (1998), Suzuki *et al.* (1998), and Kopinke *et al.* (2006) for studies with fish embryos and see Wendling *et al.* (2000) for mice.

19. See Redline *et al.* (1992) for an early appreciation of the link between *Hox* genes and malformations and Creuzet *et al.* (2002) for the study with *Hoxa2, Hoxa3* and *Hoxb4*.

20. See Y. Chen *et al.* (1995) for the studies on vitamin A and *Msx1* and Tribulo *et al.* (2004) for the role of *Msx* genes. See Koentges and Lumsden (1996) and Miyake *et al.* (1996) for skeletal structures with a composite origin within the mandibular arch (Meckel's cartilage, mandibular membrane bones) or from two pharyngeal arches (middle ear ossicles); the mandibular arch skeleton in chickens has a composite origin from NC, arising from midbrain and r1, r2, and r4 of the hindbrain. For a discussion of slight differences between the fate maps of the hyoid region of chicken embryos produced by Couly *et al.* (1993, 1996) and by Koentges and Lumsden (1996), see Le Douarin and Kalcheim (1999*, pp. 110–115).

21. See Luo *et al.* (2003) for Hoxa2 and *Ap2α*, and Kutejova *et al.* (2008), for *Hoxa1, Six2*, and pharyngeal-arch development in mice.

22. For studies on the disruption of *Hoxa1, Hoxa2, Hoxa3*, see Chisaka and Capecchi (1991), Alexandre *et al.* (1996), and Kanzler *et al.* (1998). For homologous recombination of *Hoxd4* and *Hoxa2*, see Lufkin *et al.* (1992), Gendron-Maguire *et al.* (1993), and Rijli *et al.* (1993). For the demonstration that *Hoxd4* specifies regional identity between pharyngeal arches 1 and 2 in the Japanese flounder, see Suzuki *et al.* (1998).

23. See Mark *et al.* (1993) for the *Hoxa1* mutant mice, Gavalas *et al.* (1998) for *Hoxa1* /*Hoxb1* double mutants, and Miyake *et al.* (1996) for contributions of first- and second-arch elements to mammalian ear development. For an insightful analysis of whether knocking out patterning genes from murine embryos produces evolutionary reversals of the first arch, see K. K. Smith and Schneider (1998).

24. See Gale *et al.* (1996) for the studies on r4, Marshall *et al.* (1992) for transformation of rhombomere identity, Wilkinson (1995) for *Krox20* regulation of *Hoxb2*, and Barrow *et al.* (2000) for a model of how *Hox* genes, *Krox20* and *kreisler* function together to pattern the mouse hindbrain.

25. See Erickson and Weston (1983), Cohen (1989, 1990), and Gorlin *et al.* (2001) for mutation-induced craniofacial defects.

26. See Wilson and Wyatt (1995) for cranial morphogenesis in *Lp* mutant mice. See also Wilson and Wyatt (1988) for failure of the caudal neuropore to close in *vL* mutant mice and the contribution to spina bifida of defective migration of NCCs.

27. See Serbedzija and McMahon (1997) for the basic defects in the *Splotch* mutant, Conway *et al.* (1997) for the *Pax3* studies, Dahl *et al.* (1997) for an overview of *Pax* genes and vertebrate organogenesis, including their involvement in such NC-based defects as Waardenburg syndrome, and Epstein (1996) and Epstein *et al.* (2000) for *Splotch, Pax3*, the NC, and cardiovascular development. Interestingly, if the *Splotch* allele is induced with radiation, neural tube, and tail defects can be induced without affecting the NC.

28. See Rothman *et al.* (1993) and Puffenberger *et al.* (1994) for *Piebald Lethal* and *Lethal Spotted*. Fourteen of some 50 human syndromes involve NCCs demonstrate aganglionic megacolon, of which *Hirschsprung disease* (aganglionic megacolon) resulting from a mutation in *SOX10* (Table 9.1) is perhaps the most common.

29. See the reviews and literature discussed in Hall (1999a*). See Hörstadius (1950, pp. 41–43), Chibon (1974*), Weston (1970*), Hunt *et al.* (1995), Sechrist *et al.* (1995), Raible and Eisen (1996), Buxton *et al.* (1997), Vaglia and Hall (1999*, 2000), and Schilling *et al.* (2001) for mechanisms of regulation of the NC.

30. See Sechrist *et al.* (1995) and Buxton *et al.* (1997) for regulation of the chicken NC.

31. See Hunt *et al.* (1995), Couly *et al.* (1996), Buxton *et al.* (1997), and Saldivar *et al.* (1997) for regulation of the avian NC, Snow and Tam (1979) for regulation of the NC in mouse embryos, Prince and Lumsden (1994) for intrinsic patterning of *Hoxa2*, Sechrist *et al.* (1995*) for the segmental migration of NCCs from r3 and r5, Martinez and Alvarado-Mallart (1990) for regulation of *engrailed* expression in chick/quail hybrids, and Maclean and Hall (1987), Vaglia and Hall (1999, 2000), and Gilbert (2006) for general discussions of regulation.

32. See Asamoto *et al.* (1992) and Gvirtzmann *et al.* (1992) for identification of regulation of the size of dorsal root ganglia, Goldstein *et al.* (1995) for the intrinsic size differences, A. Smith *et al.* (1997) and Krull *et al.* (1997) for the Eph-family of receptors, Robertson and Mason (1995) and Wehrle-Haller and Weston (1997) for tyrosine kinase receptors in chickens, and Asai *et al.* (2006) for the study with enteric NCCs. SNCCs in chicken embryos cannot compensate for the loss of VNC that produce enteric ganglia (Burns *et al.*, 2000).

33. Trunk crest cells that would normally have followed a ventral pathway of migration also were not replaced following ablation, although melanocytes from adjacent regions migrated to fill the void so that pigment patterns were not disturbed despite loss of the NC precursors for pigment cells in a local region.

34. See Kirby (1989) and Hutson and Kirby (2003*) for the ability of NC from other regions to replace CarNC.

35. See Qin and Kirby (1995) for the studies with *Fgf3*, Vogel and Davies (1993) for replacement of placodal ectoderm by trunk non-neurogenic ectoderm, and T. A. Harrison *et al.* (1995) for regulation of nodose ganglia.

36. Two seminal studies on multipotentiality of NCCs from Japanese quail embryos discussed earlier in the text are those by Sieber-Blum and Cohen (1980) and Baroffio *et al.* (1991). See Crane and Trainor (2006) for an analysis of NCC as progenitor and stem cells.

37. See Kruger *et al.* (2002) for NC stem cells in the adult rat gut and for changes in self-renewal, potential and responsiveness to growth factors found in the gut, Fernandes *et al.* (2004) for the whisker pad study and for earlier literature, and Delfino-Machín *et al.* (2007) for an insightful analysis of lineage and hierarchy relationships among and between NC stem cells and for a listing and analysis of the features of 13 NC-derived stem cells.

38. See Sato and Yost (2003) for zebrafish CarNCCs from cardiomyocytes, Tomita *et al.* (2005*) for the transplantation studies, and for an overview of the NC and stem cells and Jones and Trainer for the potential therapeutic use of NCCs as stem cells.

Common Names of Species Discussed$^\oplus$

Alpine (Arctic) newt, *Triturus alpestris*
American alligator, *Alligator mississippiensis*
American opossums, *Didelphis* spp.
Arctic lamprey, *Lethenteron japonicum*$^\otimes$
Atlantic cod, *Gadus morhua*
Atlantic hagfish, *Myxine glutinosa*
Atlantic salmon, *Salmo salar*
Bandicoots, *Perameles* spp.
Bicolor damselfish, *Pomacentrus partitus*
Black-striped frog, *Rana (Sylvirana) nigrovittata*
Bowfin, *Amia calva*
Budgerigar, *Melopsittacus undulatus*
California newt, *Taricha torosa*
Canary, *Serinus canaria*
Cephalochordate, *Asymmetron*
Cephalochordate, *Branchiostoma belcheri*
Cephalochordate (Florida lancelet), *Branchiostoma floridae*
†Cephalochordate, *Cathaymyrus diadexus*
†Cephalochordate, *Cathaymyrus haikoensis*
†Cephalochordate (?)$^\diamond$, *Pikaia gracilens*
†Cephalochordate (?), *Yunnanozoon lividum*
Chicken, *Gallus domesticus*
Clearnose skate, *Raja eglanteria*
Common carp, *Cyprinus carpio*
Common European salamander, *Triturus taeniatus*
Common newt, *Lissotriton (Triturus) vulgaris*
Common skate, *Raja batis*

$^{\oplus\,\dagger}$ designates a fossil (extant) species.

$^\otimes$ Formerly the Japanese lamprey, *Lampetra japonicum*.

$^\diamond$ (?) designates a species of uncertain affinity.

Common torpedo, *Torpedo ocellata*
Convict cichlid, *Cichlasoma nigrofasciatum*
Cotton rat, *Sigmodon fulviventer*
Crested newt, *Triturus cristatus*
Darwin's finches, *Geospiza* spp
Edible frog, *Rana esculenta*
Electric ray, *Torpedo ocellata*
European common frog, *Rana temporaria*
European brook lamprey, *Lampetra (Petromyzon) planeri*
European perch, *Perca fluviatilis*
European river lamprey, *Lampetra fluviatilis*
Florida lancelet (cephalochordate), *Branchiostoma floridae*
Fruit flies, *Drosophila* spp.
Fugu (puffer fish), *Takifugu rubripes*
Glow light tetra, *Danio choprae*
Goldfish, *Carassius auratus*
Green tree frog, *Hyla cinerea*
Groundling (weather loach), *Misgurnus fossilis*
† Hagfish (?), *Haikouichthys ercaicunensis*
† Hagfish (?), *Myllokunmingia fengjiaoa*
Hawaiian acorn worm, *Ptychodera flava*
†Hemichordate (?), *Haikouella jianshanensis*
†Hemichordate (?), *Haikouella lanceolata*
Hexagrammid fishes, *Hexagrammis* spp.
Horseshoe crab, *Limulus polyphemus*
Humans, *Homo sapiens*
Hydroid, *Cordylophora* sp.
Indonesian coelacanth, *Latimeria menadoensis*
Inshore (Japanese) hagfish, *Eptatretus burgeri*
Japanese flounder, *Paralichthys olivaceus*
Japanese (inshore) hagfish, *Eptatretus burgeri*
Japanese lamprey, *Lampetra japonica*⊗
Japanese medaka, *Oryzias latipes*
Japanese Northeast salamander, *Hynobius lichenatus*
Japanese quail, *Coturnix coturnix japonica*
Japanese turtle, *Trionyx sinensis japonicas*
Jefferson's salamander, *Ambystoma jeffersonianum*
Kangaroos, *Macropus* spp.
Kenyan Clawed Frog, *Xenopus borealis*
Larvacean, *Oikopleura dioica*
Lesser-spotted dogfish (small spotted catshark), *Scyliorhinus canicula*
Little skate, *Leucoraja erinacea*

⊗ *Lampetra japonica*, the Japanese lamprey, is now *Lethenteron japonicum*, the Arctic lamprey.

Long-tailed macaque, *Macaca fascicularis*
Mangrove tunicate, *Ecteinascidia turbinate*
Marsupial frog, *Gastrotheca riobambae*
Marsupial 'mouse' (stripe-faced dunnart), *Sminthopsis macroura*
†Mesozoic crown mammal, *Yanoconodon allini*
Mexican axolotl, *Ambystoma mexicanum*
Mexican tetra, *Astyanax mexicanus*
Mississippi paddlefish, *Polyodon spathula*
Mozambique tilapia, *Oreochromis mossambicus*
Mudpuppy, *Necturus maculosus*
Mummichog, *Fundulus heteroclitus*
('Native cat') (northern quoll), *Dasyurus hallucatus*
Nibe croaker, *Nibea mitsukurii*
Nile perch, *Lates niloticus*
Nile tilapia, *Oreochromis niloticus*
North American sea lamprey, *Petromyzon marinus*
North American spotted salamander, *Ambystoma maculatum*
Northern leopard frog, *Rana pipiens*
Northern quoll ('native cat'), *Dasyurus hallucatus*
Oriental fire-bellied toad, *Bombina orientalis*
Pacific sea squirt, *Ciona savignyi*
Painted frog, *Discoglossus pictus*
Palmate newt, *Triturus helveticus*
Pearl danio, *Danio albolineatus*
Pickerel frog, *Rana palustris*
Pig-tailed macaque, *Macaca nemestrina*
Piked (spiny) dogfish, *Squalus acanthias*
Platyfish, *Xiphophorus maculatus*
Puerto Rican coqui, *Eleutherodactylus coqui*
Puffer fish; see Fugu
†Putative chordate, *Pikaia gracilens*
Red-bellied newt, *Taricha rivularis*
Red-eared slider turtle, *Trachemys scripta*
Rock wallabies, *Petrogale* spp.
Rosy barb, *Barbus conchonius*
Sea bass, *Serranus atrarius*
Sea squirt (ascidian), *Halocynthia roretzi*
Sea vase (ascidian), *Ciona intestinalis*
Short-tailed opossum, *Monodelphis domestica*
Siberian newt, *Hynobius keyserlingii*
Small spotted catshark (lesser spotted dogfish), *Scyliorhinus canicula*
Snapping turtle, *Chelydra serpentina*

South African clawed-toed frog, *Xenopus laevis*⊗
South American cichlid, *Apistogramma caetei*
Spanish ribbed newt, *Pleurodeles waltl*
Spanish terrapin, *Mauremys leprosa*
Spiny (piked) dogfish, *Squalus acanthias*
Spotted danio, *Danio nigrofasciatus*
Star ascidian (tunicate), *Botryllus schlosseri*
Stone loach, *Nemacheilus barbatulus*
Stripe-faced dunnart (marsupial 'mouse' of Australia), *Sminthopsis macroura*
Swordtail, *Xiphophorus helleri*
Tenrec, *Hemicentetes* (an insectivore)
Thornback ray, *Raja clavata*
Tiger salamander, *Ambystoma tigrinum*
Tree shrew, *Tupaia belangeri*
Tungara frog, *Physalaemus pustulosus*
Velvet-belly lantern shark, *Etmopterus spinax*
Weather loach (groundling), *Misgurnus fossilis*
Western clawed frog, *Xenopus tropicalis*
Western northern Pacific ascidian, *Halocynthia roretzi*
White sturgeon, *Acipenser transmontanus*
Winter skate, *Leuroraja ocellata*
Wood frog, *Rana sylvatica*
Yellow-bellied toad, *Bombina variegata*
Zebrafish, *Danio rerio*

⊗ Also known by other common names: the African clawed toad; the African clawed frog; the South African clawed toad; the South African clawed frog.

Species (with Common Names) Arranged by Major Groups[⊗]

Ascidians (see Sea Squirts)
Birds

Coturnix coturnix japonica—Japanese quail
Gallus domesticus—domestic chicken
Geospiza spp.—Darwin's finches
Melopsittacus undulatus—budgerigar
Serinus canaria—canary

Bony Fish

Amia calva—bowfin
Apistogramma caetei—South American cichlid
Astyanax mexicanus—Mexican tetra
Barbus conchonius—rosy barb
Carassius auratus—goldfish
Cichlasoma nigrofasciatum—convict cichlid
Cyprinus carpio—common carp
Danio albolineatus—pearl danio
Danio choprae—glow light tetra
Danio feegradei
Danio nigrofasciatus—spotted danio
Danio rerio—zebrafish
Fundulus heteroclitus—mummichog
Hexagrammis spp.—hexagrammid fishes
Lates niloticus—Nile perch
Misgurnus fossilis—groundling (weather loach)
Nemacheilus barbatulus—stone loach
Nibea mitsukurii—Nibe croaker

     Oreochromis mossambicus—Mozambique tilapia
     Oreochromis niloticus—Nile tilapia
     Oryzias latipes—Japanese medaka
     Paralichthys olivaceus—Japanese flounder
     Perca fluviatilis—European perch
     Pomacentrus partitus—bicolor damselfish
     Salmo salar—Atlantic salmon
     Serranus atrarius—sea bass
     Takifugu rubripes—fugu (puffer fish)
     Xiphophorus helleri—swordtail
     Xiphophorus maculatus—platyfish

Cephalochordates[⊗]

     Asymmetron
     Branchiostoma belcheri
     Branchiostoma floridae—Florida lancelet
     [†]*Cathaymyrus diadexus*
     [†]*Cathaymyrus haikoensis*
     [†]*Pikaia gracilens* cephalochordate (?)
     [†]*Yunnanozoon lividum* cephalochordate (?)

Chordate (?)

     [†]*Pikaia gracilens*

Coelacanth

     Latimeria menadoensis—Indonesian coelacanth

Coelenterates

     Cordylophora—hydroid

Fruit flies

     Drosophila spp.

Frogs and Toads

     Bombina orientalis—oriental fire-bellied toad
     Bombina variegata—yellow-bellied toad
     Discoglossus pictus—painted frog
     Eleutherodactylus coqui—Puerto Rican coqui
     Gastrotheca riobambae—marsupial frog
     Hyla cinerea—green tree frog
     Philautus silus—direct-developing Sri-Lankan frog
     Physalaemus pustulosus—Tungara frog
     Rana esculenta—edible frog
     Rana (*Sylvirana*) *nigrovittata*—black-striped frog
     Rana palustris—pickerel frog

[⊗] Amphioxus is the common name for most living cephalochordates.

Rana pipiens—Northern leopard frog
Rana sylvatica—wood frog
Rana temporaria—European common frog
Xenopus borealis—Kenyan clawed Frog
Xenopus laevis—South African clawed-toed toad
Xenopus tropicalis—Western clawed frog

Hagfish

Eptatretus burgeri Japanese (inshore)—hagfish
†*Haikouichthys ercaicunensis*—hagfish (?)
Myxine glutinosa—Atlantic hagfish
†*Myllokunmingia fengjiaoa*—hagfish (?)

Hemichordates

†*Haikouella jianshanensis*—hemichordate (?)
†*Haikouella lanceolata*—hemichordate (?)
Ptychodera flava—Hawaiian acorn worm

Horseshoe crab

Limulus polyphemus

Lampreys

Lampetra fluviatilis—European river lamprey
Lampetra japonica—Japanese lamprey⊗
Lampetra (Petromyzon) planeri—European brook lamprey
Lethenteron japonicum—Arctic lamprey
Petromyzon marinus—North American sea lamprey

Larvaceans

Oikopleura dioica

Marsupial Mammals

Dasyurus hallucatus—northern quoll ('native cat')
Didelphis spp.—American opossums
Macropus spp.—kangaroos
Monodelphis domestica—short-tailed opossum
Perameles spp.—bandicoots
Petrogale spp.—rock wallabies
Sminthopsis macroura—stripe-faced dunnart (marsupial 'mouse')

†Monotremes

†Teinolophos trusleri— *oldest known monotreme (E. Cretaceous)*

Newts and Salamanders

Ambystoma jeffersonianum—Jefferson's salamander
Ambystoma maculatum—North American spotted salamander

⊗ Now *Lethenteron japonicum*, the Arctic lamprey.

Ambystoma mexicanum—Mexican axolotl
Ambystoma tigrinum—tiger salamander
Hynobius keyserlingii—Siberian newt
Hynobius lichenatus—Japanese Northeast salamander
Lissotriton (Triturus) vulgaris—common newt
Necturus maculosus—mudpuppy
Pleurodeles waltl—Spanish ribbed newt
Taricha rivularis—red-bellied newt
Taricha torosa—California newt
Triturus alpestris—alpine (Arctic) newt
Triturus cristatus—crested newt
Triturus helveticus—palmate newt
Triturus taeniatus—common European salamander

Paddlefish

Mississippi paddlefish—*Polyodon spathula*

Placental Mammals

Hemicentetes—tenrec (an insectivore)
Homo sapiens—humans
Macaca fascicularis—long-tailed macaque
Macaca nemestrina—pig-tailed macaque
Sigmodon fulviventer—cotton rat
Tupaia belangeri—tree shrew
†*Yanoconodon allini*—Mesozoic crown mammal

Reptiles

Alligator mississippiensis—American alligator
Chelydra serpentina—snapping turtle
Mauremys leprosa—Spanish terrapin
Trachemys scripta—red-eared slider turtle
Trionyx sinensis japonicas—Japanese turtle

Sea Squirts (Ascidians)

Botryllus schlosseri—star tunicate
Ciona intestinalis—sea vase
Ciona savignyi —Pacific sea squirt
Ecteinascidia turbinata—mangrove tunicate
Halocynthia roretzi—sea squirt

Sharks—Dogfish—Skates—Rays

Etmopterus spinax—velvet belly lantern shark
Leucoraja erinacea—little skate
Leuroraja ocellata—winter skate
Raja batis—common skate
Raja clavata—thornback ray
Raja eglanteria—clearnose skate
Scyliorhinus canicula—small spotted catfish (lesser-spotted dogfish)

Squalus acanthias—piked (spiny) dogfish
Torpedo ocellata—common torpedo (electric ray)

Sturgeon

Acipenser transmontanus—white sturgeon

References

A

Abe, G., Ide, H., and Tamura, K. 2007. Function of Fgf signaling in the developmental process of the median fin fold in zebrafish. *Dev Biol* 304:355–366.

Abu-Elmagd, M., Ishii, Y., Cheung, M., Rex, M., *et al.* 2001. cSox3 expression and neurogenesis in the epibranchial placodes. *Dev Biol* 237:258–269.

Abu-Issa, R., Smyth, G., Smoak, I., Yamamura, K.-I., and Meyers, E. N. 2002. *Fgf8* is required for pharyngeal arch and cardiovascular development in the mouse. *Development* 129:4613–4625.

Abzhanov, A., Protas, M., Grant, B. R., Grant, P. R., and Tabin, C. J. 2004. Bmp4 and morphological variation of beaks in Darwin's finches. *Science* 305:1462–1465.

Akitaya, T., and Bronner-Fraser, M. 1992. Expression of cell adhesion molecules during initiation and cessation of neural crest cell migration. *Dev Dyn* 194:12–20.

Albertson, R. C., Streelman, J. T., Kocher, T. D., and Yelick, P. C. 2005. Integration and evolution of the cichlid mandible: The molecular basis of alternate feeding strategies. *Proc Natl Acad Sci USA* 102:16287–16292.

Albertson, R. C., and Yelick, P. C. 2005. Roles for *fgf8* signaling in left–right patterning of the visceral organs and craniofacial skeleton. *Dev Biol* 283:310–321.

Albertson, R. C., and Yelick, P. C. 2007. *Fgf8* haploinsufficieny results in distinct craniofacial defects in adult zebrafish. *Dev Biol* 306:505–515.

Aldridge, R. J., and Donoghue, P. C. J. 1998. Conodonts: A sister group to Hagfishes? In *The Biology of Hagfishes*, eds. J. M. Jørgensen, J. P. Lomholt, R. E. Weber and H. Malthe, pp, 15–31. London: Chapman & Hall.

Aldridge, R. J., and Purnell, M. A. 1996. The conodont controversies. *Trends Ecol Evol* 11: 463–468.

Alexandre, D., Clark, J. D. W., Oxtoby, E., Yan, Y.-L., *et al.* 1996. Ectopic expression of *Hoxa-1* in the zebrafish alters the fate of the mandibular arch neural crest and phenocopies a retinoic acid-induced phenotype. *Development* 122:735–746.

Alfandari, D., Cousin, H., Gaultier, A., Smith, K., *et al.* 2001. *Xenopus* ADAM 13 is a metalloprotease required for cranial neural crest-cell migration. *Curr Biol* 11:918–930.

Alfandari, D., Wolfsberg, T. G., White, J. M., and DeSimone, D. W. 1997. ADAM 13: A novel ADAM expressed in somitic mesoderm and neural crest cells during *Xenopus laevis* development. *Dev Biol* 182:314–330.

Ali, M. M., Jayabalan, S., Machnicki, M., and Sohal, G. S. 2003. Ventrally emigrating neural tube cells migrate into the developing vestibulocochlear nerve and otic vesicle. *Int J Dev Neurosci* 21:199–208.

Alvarado-Mallart, R.-M. 1993. Fate and potentialities of the avian mesencephalic/metencephalic neuroepithelium. *J Neurobiol* 24:1341–1355.

Andermann, P., Ungos, J., and Raible, D. W. 2002. Neurogenin1 defines zebrafish cranial sensory ganglia precursors. *Dev Biol* 251:45–58.

Anderson, D. J. 1997. Cellular and molecular biology of neural crest cell lineage determination. *Trends Genet* 13:276–280.

Anderson, D. J. 2000. Genes, lineages and the neural crest: A speculative review. *Phil Trans R Soc Lond* B 355:953–964.

Aramaki, M., Kimura, T., Udaka, T., Rika, K., *et al.* 2007. Embryonic expression profile of chicken *Chd7*, the ortholog of the putative gene for CHARGE syndrome. *Birth Defects Res (Part A)* 79:50–57.

Arendt, D., and Nübler-Jung, K. 1996. Common ground plans in early brain development in mice and flies. *BioEssays* 18:255–259.

Artinger, K. B., and Bronner-Fraser, M. 1992. Notochord grafts do not suppress formation of neuronal crest cells or commisural neurons. *Development* 116:877–886.

Asai, N., Fukuda, T., Wu. Z., Enomoto, A., *et al.* 2006. Targeted mutation of serine 697 in the *Ret* tyrosine kinase causes migration defect of enteric neural crest cells. *Development* 133:4507–4516.

Asamoto, K., Nojyo, Y., and Aoyama, H. 1992. Regulation of cell number in formation of the dorsal-root ganglion revealed by transplantation of quail neural crest cells into chick embryos. *Dev Growth* 34:553–560.

Asamoto, K., Nojyo, Y., and Aoyama, H. 1995. Restriction of the fate of early migrating trunk neural crest in gangliogenesis of avian embryos. *Int J Dev Biol* 39:975–984.

Atchley, W. R., and Hall, B. K. 1991. A model for development and evolution of complex morphological structures and its application to the mammalian mandible. *Biol Rev Camb Philos Soc* 66:101–157.

Atsumi, T., Miwa, Y., Kimata, K., and Ikawa, Y. 1990. A chondrogenic cell line derived from a differentiating culture of AT805 teratocarcinoma cells. *Cell Differ Dev* 30:109–116.

Aybar, M. J., Nieto, M. A., and Mayor, R. 2003. Snail precedes Slug in the genetic cascade required for the specification and migration of the *Xenopus* neural crest. *Development* 130:483–494.

B

Bachiller, D., Kungensmith, J., Shneyder, N., Tran, V., *et al.* 2003. The role of chordin/Bmp signaling in mammalian pharyngeal development and DiGeorge syndrome. *Development* 130:3567–3578.

Bagnara, J. T. 1999. The emergence of pigment cell biology; A personal view. *Pigment Cell Res* 12:48–65.

Bagnara, J. T., Matsumoto, J., Ferris, W., Frost, S. K., *et al.* 1979. Common origin of pigment cells. *Science* 203:410–415.

Bagnara, J. T., Taylor, J. D., and Hadley, M. E. 1968. The dermal chromatophore unit. *J Cell Biol* 38:67–79.

Baker, C. V. H., and Bronner-Fraser, M. 1997a. The origins of the neural crest. Part I: Embryonic induction. *Mech Dev* 69:3–11.

Baker, C. V. H., and Bronner-Fraser, M. 1997b. The origins of the neural crest. Part II: An evolutionary perspective. *Mech Dev* 69:13–29.

Baker, C. V. H., and Bronner-Fraser, M. 2000. Establishing neuronal identity in vertebrate neurogenic placodes. *Development* 127:3045–3056.

Baker, C. V. H., and Bronner-Fraser, M. 2001. Vertebrate cranial placodes. I. Embryonic induction. *Dev Biol* 232:1–61.

Baker, C. V. H., Bronner-Fraser, M., Le Douarin, N. M., and Teillet, M.-A. 1997. Early- and late-migrating cranial neural crest cell populations have equivalent developmental potential *in vivo*. *Development* 124:3077–3087.

Baker, J. C., Beddington, R. S. P., and Harland, R. M. 1999. Wnt signaling in *Xenopus* embryos inhibits *Bmp4* expression and activates neural development. *Genes Dev* 13:3149–3159.

Balfour, F. M. 1876. On the development of the spinal nerves in elasmobranch fishes. *Phil Trans R Soc* 166:175–195.

Balfour, F. M. 1878. *A Monograph on the Development of Elasmobranch Fishes*. London: Macmillan & Co.

Balfour, F. M. 1881. *A Treatise on Comparative Embryology*. Volume 2, i–xxii +655 pp. London: Macmillan & Co. (Translated into German by B. Vetter as *Handbuch der Vergleichenden Embryologie*, Verlag von Gustav Fischer, Jena, 1882). Original edition as 13 micro cards, Readex Microprint, New York, 1969.

Balinsky, B. I. 1940. Experiments on total extirpation of the whole entoderm in Triton embryos. *C R Acad Sci URSS* 23:196–198.

Baranski, M., Berdougo, E., Sandler, J. S., Darnell, D. K., and Burrus, L. W. 2000. The dynamic expression pattern of *frzb-1* suggests multiple roles in chick development. *Dev Biol* 217:25–41.

Barembaum, M., and Bronner-Fraser, M. 2005. Early steps in neural crest-specification. *Sem Cell Dev Biol* 16:642–646.

Barembaum, M., and Bronner-Fraser, M. 2007. *Spalt4* mediates invagination and otic placode gene expression in cranial ectoderm. *Development* 134, 3805–3814.

Barlow, A. J., and Francis-West, P. H. 1997. Ectopic application of recombinant BMP-2 and BMP-4 can change patterning of developing chick facial primordia. *Development* 124:391–398.

Barlow, A. J., Wallace, A. S., Thapar, N., and Burns, A. J. 2008. Critical numbers of neural crest cells are required in the pathways from the neural tube to the foregut to ensure complete enteric nervous system formation. *Development* 135:1681–1691.

Barlow, L. A. 1999. A taste for development. *Neuron* 22:209–212.

Barlow, L. A. 2001. Specification of pharyngeal endoderm is dependent on early signals from axial mesoderm. *Development* 128:4573–4583.

Barlow, L. A., and Northcutt, R. G. 1997. Taste buds develop autonomously from endoderm without induction by cephalic neural crest or paraxial mesoderm. *Development* 124:949–957.

Baroffio, A., Dupin, E., and Le Douarin, N. M. 1988. Clone-forming ability and differentiation potential of migratory neural crest cells. *Proc Natl Acad Sci USA* 85:5325–5339.

Baroffio, A., Dupin, E., and Le Douarin, N. M. 1991. Common precursors for neural and mesectodermal derivatives in the cephalic neural crest. *Development* 112:301–305.

Barrow, J. R., Stadler, H. S., and Capecchi, M. R. 2000. Roles of *Hoxa1* and *Hoxa2* in patterning the early hindbrain of the mouse. *Development* 127:933–944.

Bartelmez, G. W. 1960. Neural crest from the forebrain in mammals. *Anat Rec* 138:269–281.

Bartelmez, G. W. 1962. The proliferation of neural crest from forebrain levels in the rat. *Contr Embryol* 37:3–12.

Barth, K. A., Kishimoto, Y., Rohr, K. B., Seydler, C., *et al.* 1999. Bmp activity establishes a gradient of positional information throughout the entire neural plate. *Development* 126:4977–4987.

Basch, M. L., Bronner-Fraser, M., and Garcia-Castro, M. I. 2006. Specification of the neural crest occurs during gastrulation requires Pax7. *Nature* 441:218–222.

Béard, J. 1892. The transient ganglion-cells and their nerves in *Raja batis*. *Anat Anz* 7:191–206.

Béard, J. 1896. The history of transient nervous apparatus in certain Ichthyopsida. An account of the development and degeneration of ganglion cells and nerve fibres. *Zool Jahr Abt Morph* 9:1–106.

Beaudin, A. E., and Stover, P. J. 2007. Folate-mediated one-carbon metabolism and neural tube defects: Balancing genome synthesis and gene expression. *Birth Defects Res (Part C)* 81: 183–203.

Beck, C. W., and Slack, J. M. W. 1998. Analysis of the developing *Xenopus* tail bud reveals separate phases of gene expression during determination and outgrowth. *Mech Dev* 72:41–52.

Beck, C. W., and Slack, J. M. W. 1999. A developmental pathway controlling outgrowth of the *Xenopus* tail bud. *Development* 126:1611–1620.

Beddington, R. S. P., and Robertson, E. J. 1998. Anterior patterning in mouse. *Trends Genet* 14:277–284.

Begbie, J., Ballivet, M., and Graham, A. 2002. Early steps in the production of sensory neurons by the neurogenic placodes. *Mol Cell Neurosci* 21:502–511.

Bennett, J. H., Hunt, P., and Thorogood, P. V. 1995. Bone morphogenetic protein-2 and -4 expression during murine orofacial development. *Archs Oral Biol* 40:847–854.

Berndt, J. D., and Halloran, M. C. 2006. Semaphorin 3d promotes cell proliferation and neural crest cell development downstream of TCF in the zebrafish hindbrain. *Development* 133: 3983–3992.

Bertrand, N., Médevielle, F., and Pituello, F. 2000. FGF signalling controls the timing of *Pax6* activation in the neural tube. *Development* 127:4837–4843.

Besser, L. M., Williams, L. J., and Cragan, J. D. 2007. Interpreting changes in the epidemiology of anencephaly and spina bifida following folic acid fortification of the U.S. grain supply in the setting of long-term trends, Atlanta, Georgia, 1968–2003. *Birth Defects Res (Part A)* 79: 730–736.

Bhattacherjee, V., Mukhopadhyay, P., Singh, S., Johnson, C., *et al.* 2007. Neural crest and mesoderm lineage-dependent gene expression in orofacial development. *Differentiation* 75: 463–477.

Bhushan, A., Chen, Y., and Vale, W. 1998. Smad7 inhibits mesoderm formation and promotes neural cell fate in *Xenopus* embryos. *Dev Biol* 200:260–268.

Biesecker, L. G., Happle, R., Mulliken, J. B., Weksberg, R., *et al.* 1999. Proteus syndrome: Diagnostic criteria, differential diagnosis, and patient evaluation. *Am J Med Genet* 84:389–385.

Billon, N., Iannarelli, P., Monteiro, M. C., Glavieux-Pardanaud, C., *et al.* 2007. The generation of adipocytes by the neural crest. *Development* 134:2283–2292.

Blanco, C., López, D., De Andrés, A. V., Schib, J. L., *et al.* 2001. Cartilage in the bulbus arteriosus of teleostean fishes. *Neth J Zool* 51:361–370.

Blankenship, T. N., Peterson, P. E., and Hendrickx, A. G. 1996. Emigration of neural crest cells from macaque optic vesicles is correlated with discontinuities in its basement membrane. *J Anat* 188:473–483.

Blentic, A., Tandom, P., Payton, S., Walshe, J., *et al.* 2008. The emergence of ectomesenchyme. *Dev Dyn* 237:592–601.

Boisseau, S., and Simonneau, M. 1989. Mammalian neuronal differentiation: Early expression of a neuronal phenotype from mouse neural crest cells in a chemically defined culture medium. *Development* 106:665–674.

Bolande, R. P. 1974. The neurocristopathies: A unifying concept of disease arising in neural crest maldevelopment. *Human Pathol* 5:409–429.

Bolande, R. P. 1981. Neurofibromatosis:-the quintessential neurocristopathy: Pathogenetic concepts and relationships. *Adv Neurol* 29:67–75.

Bolker, J. A. 2004. Embryology. In *Sturgeons and Paddlefish of North America*, eds G. LeBreton, F. Beamish and R. McKinley, pp. 134–146. Dordrecht: Kluwer Academic Publishers.

Boot, M. J., Gittenberger-de Groot, A. C., Poelman, R. E., and Gourdie, R. G. 2006. Connexin43 levels are increased in mouse neural crest cells exposed to homocysteine. *Birth Defects Res (Part A)* 76:133–137.

Boot, M. J., Gittenberger-de Groot, A. C., Van Iperen, L., Hierch, B. P., and Poelman, R. E. 2003. Spatiotemporally separated cardiac neural crest subpopulations that target the outflow tract septum and pharyngeal arch arteries. *Anat Rec Part A* 275A:1009–1018.

Borcea, M. I. 1909. Sur l'origine du coeur, des cellules vasculaires migratrices et des cellules pigmentaires chez les Téléostéens. *C r hebd Séanc Acad Sci Paris* 149:688–689.

Borchers, A., David, R., and Wedlich, D. 2001. *Xenopus* cadherin-11 restrains cranial neural crest migration and influences neural crest specification. *Development* 128:3049–3060.

Bourlat, S. J., Juliusdottir, T., Lowe, C. J., Freeman, R., *et al.* 2006. Deuterostome phylogeny reveals monophyletic chordates and the new phylum Xenoturbellida. *Nature* 444:85–88.

Brannan, C. I., Perkins, A. S., Vogel, K. S., Ratner, N., *et al.* 1994. Targeted disruption of the neurofibromatosis type-1 gene leads to developmental abnormalities in heart and various neural crest-derived tissues. *Genes Dev* 8:1019–1029.

Brauer, P. R., and Markwald, R. R. 1987. Attachment of neural crest cells to endogenous extracellular matrices. *Anat Rec* 219:275–285.

Braun, C. B., and Northcutt, R. G. 1997. The lateral line system of hagfishes (Craniata: Myxinoidea). *Acta Zool (Stockh)* 78:247–268.

Britto, J. M., Tannahill, D., and Keynes, R. J. 2000. Life, death and sonic hedgehog. *BioEssays* 22:499–502.

Britto, J. M., Teillet, M.-A., and Le Douarin, N. M. 2006. An early role for sonic hedgehog from foregut endoderm in jaw development: Ensuring neural crest cell survival. *Proc Acad Natl Sci USA* 103:11607–11612.

Bronner-Fraser, M. 1986. An antibody to a receptor for fibronectin and laminin perturbs cranial neural crest development *in vivo*. *Dev Biol* 117:528–536.

Bronner-Fraser, M. 1987. Perturbation of cranial neural crest migration by the HNK-1 antibody. *Dev Biol* 123:321–331.

Bronner-Fraser, M. 1995. Origins and developmental potential of the neural crest. *Exp Cell Res* 218:405–417.

Bronner-Fraser, M., and Fraser, S. E. 1997. Differentiation of the vertebrate neural tube. *Curr Opin Cell Biol* 7:885–891.

Bronner-Fraser, M., Sieber-Blum, M., and Cohen, A. M. 1980. Clonal analysis of the avian neural crest: Migration and maturation of mixed neural crest clones injected into host chicken embryos. *J Comp Neurol* 193:423–434.

Bronner-Fraser, M., and Stern, C. 1991. Effects of mesodermal tissues on avian neural crest cell migration. *Dev Biol* 143:213–217.

Bronner-Fraser, M., Wolf, J. J., and Murray, B. A. 1992. Effects of antibodies against N-cadherin and N-CAM on the cranial neural crest and neural tube. *Dev Biol* 153:291–301.

Brooks, A. S., Oostra, B. A., and Hofstra, R. M. 2005. Studying the genetics of Hirschsprung's disease: Unraveling an oligogenetic disorder. *Clin Genet* 67:6–14.

Brown, C. B., Feiner, L., Lu, M.-M., Li, J., *et al.* 2001. PlexinA2 and semaphorin signaling during cardiac neural crest development. *Development* 128:3071–3080.

Budi, E. H., Patterson, L. B., and Parichy, D. M. 2008. Embryonic requirements for ErbB signaling in neural crest development and adult pigment formation. *Development* 135:2603–2614.

Burns, A. J., Champeval, D., and Le Douarin, N. M. 2000. Sacral neural crest cells colonise aganglionic hindgut *in vivo* but fail to compensate for lack of enteric ganglia. *Dev Biol* 219:30–43.

Burns, A. J., Delalande, J.-M. M., and Le Douarin, N. M. 2002. In ovo transplantation of enteric nervous system precursors from vagal to sacral neural crest results in extensive hindgut colonization. *Development* 129:2785–2796.

Burstyn-Cohen, T., Stanleigh, J., Sela-Donenfeld, D., and Kalcheim, C. 2004. Canonical Wnt activity regulates trunk neural crest delamination linking BMP/noggin signaling with G1/S transition. *Development* 131:5327–5339.

Buxton, P., Hunt, P., Ferretti, P., and Thorogood, P. 1997. A role for midline closure in the re-establishment of dorsoventral pattern following dorsal hindbrain ablation. *Dev Biol* 183: 150–165.

Buxton, P. G., Hall, B. K., Archer, C. W., and Francis-West, P. 2003. Secondary chondrocyte-derived Ihh stimulates proliferation of periosteal cells during chick cranial development. *Development* 130: 4729–4739.

C

Callery, E. M., Fang, H., and Elinson, R. P. 2001. Frogs without polliwogs: Evolution of anuran direct development. *BioEssays* 23:233–241.

Calloni, G. W., Glavieux-Pardanaud, C., Le Douarin, N. M., and Dupin, E. 2007. Sonic hedgehog promotes the development of multipotent neural crest progenitors endowed with both mesenchymal and neural potentials. *Proc Natl Acad Sci USA* 104:19879–19884.

Camenisch, T. D., Spicer, A. P., Brehm-Gibson, T., Biesterfeldt, J., *et al.* 2000. Disruption of hyaluronan synthase-2 abrogates normal cardiac morphogenesis and hyaluronan-mediated transformation of epithelium to mesenchyme. *J Clin Invest* 106:349–360.

Canan, G., Nebigil, D.-S. C., Dierich, A., Hickel, P., *et al.* 2000. Serotonin 2B receptor is required for heart development. *Proc Natl Acad Sci USA* 97:9508–9513.

Cañestroa, C., and Postlethwait, J. H. 2007. Development of a chordate anterior–posterior axis without classical retinoic acid signaling. *Dev Biol* 305:522–538.

Carl, T. F., Dufton, C., Hanken, J., and Klymkowsky, M. W. 1999. Inhibition of neural crest migration in *Xenopus* using antisense slug RNA. *Dev Biol* 213:101–115.

Carmona-Fontaine, C., Acuna, G., Ellwanger, K., Niehrs, C., and Mayor, R. 2007. Neural crests are actively precluded from the anterior neural fold by a novel inhibitory mechanism dependent on Dickkopf1 secreted by the prechordal mesoderm. *Dev Biol* 309:208–221.

Carney, T. J., Dutton, K. A., Greenhill, E., Delfino-Machín, M., *et al.* 2006. A direct role for Sox10 in specification of neural crest-derived sensory neurons. *Development* 133:4619–4630.

Carrasquillo, M. M., McCallion, A. S., Puffenberger, E. G., Kashuk, C. S., *et al.* 2002. Genome-wide association study and mouse model identify interaction between RET and EDNRB pathways in Hirschsprung disease. *Nature Genet* 32:237–244.

Carroll, S. B., Grenier, J. K., and Weatherbee, S. D. 2005. *From DNA to Diversity. Molecular genetics and the Evolution of Animal Design.* Second Edition. Malden, MA: Blackwell Publishing.

Carstens, M. H. 2002. Development of the facial midline. *J Craniofac Surg* 13:129–187.

Carstens, M. H. 2004. Neural tube programming and craniofacial cleft formation. I. The neuromeric organization of the head and neck. *Eur J Pediat Neurol* 8:181–210.

Cerny, R. 2005. *Embryonic Origin of some Evolutionary Significant Viscerocranial Structures in Amphibians.* Ph.D. Thesis. Department of Zoology, Faculty of Science, Charles University, Prague Czechoslovakia.

Cerny, R., Lwigale, P., Ericksson, R., Meulemans, D., *et al.* 2004. Developmental origins and evolution of jas: New interpretation of "maxillary" and "mandibular." *Dev Biol* 276:225–236.

Chai, Y., Jiang, X., Ito, Y., Bringas, P., Jr., *et al.* 2000. Fate of the mammalian cranial neural crest during tooth and mandibular morphogenesis. *Development* 127:1671–1679.

Chan, W. Y., and Tam, P. P. L. 1986. The histogenetic potential of neural plate cells of early somite-stage mouse embryos. *J Embryol Exp Morphol* 96:183–193.

Chan, W. Y., and Tam, P. P. L. 1988. A morphological and experimental study of the mesencephalic neural crest cells in the mouse embryo using wheat germ agglutinin-gold conjugate as the cell marker. *Development* 102:427–442.

Chapman, D. L., and Papaioannou, V. E. 1998. Three neural tubes in mouse embryos with mutations in the T-box gene *Tbx6. Nature* 391:695–697.

Chen, J.-Y., Dzik, J., Edgecombe, G. D., Ramsköld, L., and Zhou, G.-Q. 1995. A possible early Cambrian chordate. *Nature* 377:720–722.

Chen, J.-Y., Huang, D.-Y., and Li, C.-W. 1999. An early Cambrian craniate-like chordate. *Nature* 402:518–522.

Chibon, P. 1964. Analyse par la méthode de marquage nucléaire à la thymidine tritiée des dérivés de la crête neurale céphalique chez l'Urodèle *Pleurodeles waltlii. C R Acad Sci* 259:3624–3627.

Chibon, P. 1974. Un systeme morphogénètique remarquable: la crête neurale des Vertébrés. *Année Biol* 13:459–480.

Chisaka, O., and Capecchi, M. R. 1991. Regionally restricted developmental defects resulting from targeted disruption of the mouse homeobox gene *Hox*-1.5. *Nature* 350:473–479.

Choi, D. S., Ward, S. J., Messaddeq, N., Launay, J. M., and Maroteaux, L. 1997. 5-HT2B receptor-mediated serotonin morphogenetic functions in mouse cranial neural crest and myocardiac cells. *Development* 124:1745–1755.

Cleves, M. A., Hobbs, C. A., Cleves, P. A., Tilford, J. M., *et al.* 2007. Congenital defects among live born infants with Down syndrome. *Birth Defects Res (Part A)* 79:657–663.

Clouthier, D. E., Hosoda, K., Richardson, J. A., Williams, S. C., *et al.* 1998. Cranial and cardiac neural crest defects in endothelin-A receptor-deficient mice. *Development* 125:613–824.

Clouthier, D. E., and Schilling, T. F. 2004. Understanding endothelin-1 function during craniofacial development in the mouse and zebrafish. *Birth Defects Res Part C, Embryo Today* 72:190–199.

Clouthier, D. E., Williams, S. C., Yanagisawa, H., Wieduqilt, M., *et al.* 2000. Signaling pathways crucial for craniofacial development revealed by endothelin-A receptor-deficient mice. *Dev Biol* 217:10–24.

Cohen, A. M., and Konigsberg, I. R. 1975. A clonal approach to the problem of neural crest determination. *Dev Biol* 46:262–282.

Cohen, M. M. Jr. 1989. Syndromology: An updated conceptual overview I–VI. *Int J Oral Maxillofac Surg* 18:216–228, 281–290, 333–346.

Cohen, M. M. Jr. 1990. Syndromology: An updated conceptual overview VII–X. *Int J Oral Maxillofac Surg* 19:26–37, 81–96.

Cohen, M. M. Jr. 2003. The hedgehog signaling network. *Am J Med Genet* 123A:5–28.

Cohen, M. M. Jr. 2006. Holoprosencephaly: Clinical, anatomic and molecular dimensions. *Birth Defects Res (Part A)* 76:658–673.

Cohen, M. M. Jr., and Baum, B. J. (eds) 1997. *Studies in Stomatology and Craniofacial Biology.* Ohmsha: IOS Press.

Cohn, M. J. 2002. Lamprey *Hox* genes and the origin of jaws. *Nature* 416:386–387.

Colas, J.-F., and Schoenwolf, G. C. 2001. Towards a cellular and molecular understanding of neurulation. *Dev Dyn* 221:117–145.

Cole, A. G., and Hall, B. K. 2004a. Cartilage is a metazoan tissue: integrating data from non-vertebrate sources. *Acta Zool (Stockh)* 85:69–80.

Cole, A. G., and Hall, B. K. 2004b. The nature and significance of invertebrate cartilages revisited: Distribution and histology of cartilage and cartilage-like tissues within the Metazoa. *Zoology* 107:261–274.

Cole, A. G., and Hall, B. K. 2008. Cartilage differentiation in cephalopod molluscs. *Zoology* (in press) doi:10.1016/j.zool.2008.01.003

Cole, L. K., and Ross, L. S. 2001. Apoptosis in the developing zebrafish embryo. *Dev Biol* 240:123–142,

Coles, E. G., Taneyhill, L. A., and Bronner-Fraser, M. 2007. A critical role for Cadherin6B in regulating avian neural crest emigration. *Dev Biol* 312:533–544.

Collazo, A., Bronner-Fraser, M., and Fraser, S. E. 1993. Vital dye labelling of *Xenopus laevis* trunk neural crest reveals multipotency and novel pathways of migration. *Development* 118:363–376.

Collazo, A., Fraser, S. E., and Mabee, P. M. 1994. A dual embryonic origin for vertebrate mechanoreceptors. *Science* 264:426–430.

Conel, J. L. 1931. The genital system of the Myxinoidea: A study based on notes and drawings of these organs in *Bdellostoma* made by Bashford Dean. In *The Bashford Dean Memorial Volume: Archaic Fishes*, ed E. W. Gudger, Article III, pp. 67–102. New York: The American Museum of Natural History.

Conel, J. L. 1942. The origin of the neural crest. *J Comp Neurol* 76:191–215.

Conway, S. J., Bundy, J., Chen, J., Dickman, E., *et al.* 2000. Decreased neural crest stem cell expansion is responsible for the conotruncal heart defects within the *Splotch* (Sp^{2H})/*Pax3* mouse mutant *Cardiovasc Res* 47:314–328.

Conway, S. J., Henderson, D. J., and Copp, A. J. 1997. *Pax3* is required for cardiac neural crest migration in the mouse: Evidence from the *Splotch* (*Sp2H*) mutant. *Development* 124:505–514.

Corbo, J. C., Erives, A., DiGregorio, A., Chang, A., and Levine, M. 1997. Dorsoventral patterning of the vertebrate neural tube is conserved in a protochordate. *Development* 124:2335–2344.

Corcoran, J. 1998. What are the molecular mechanisms of neural tube defects? *BioEssays* 20:6–8.

Cordero, D., Marcucio, R., Hu, D., Gaffield, W., *et al.* 2004. Temporal perturbations in sonic hedgehog signaling elicit the spectrum of holoprosencephaly phenotypes. *J Clin Invest* 114:485–494.

Cornell, R. A., and Eisen, J. S. 2000. Delta signaling mediates segregation of neural crest and spinal sensory neurons from zebrafish lateral neural plate. *Development* 127:2873–2882.

Cornell, R. A., and Eisen, J. S. 2005. Notch in the pathway: The roles of Notch signaling in neural crest development. *Sem Dev Biol* 16:663–672.

Correia, A. C., Costa, M., Moraes, F., Bom. J., *et al.* 2007. *Bmp2* is required for migration but not for induction of neural crest cells in the mouse. *Dev Dyn* 236:2493–2501.

Corsin, J. 1975. Différenciation *in vitro* de cartilage a partir des crêtes neurales céphaliques chez *Pleurodeles waltlii* Michah. *J Embryol Exp Morphol* 33:335–342.

Couly, G. F., Coltey, P. M., and Le Douarin, N. M. 1992. The developmental fate of the cephalic mesoderm in quail chick chimeras. *Development* 114:1–15.

Couly, G. F., Coltey, P. M., and Le Douarin, N. M. 1993. The triple origin of the skull in higher vertebrates. A study in quail chick chimeras. *Development* 117:409–429.

Couly, G. F., Creuzet, S., Bennaceur, S., Vincent, C., and Le Douarin, N. M. 2002. Interactions between Hox-negative cephalic neural crest cells and the foregut endoderm in patterning the facial skeleton in the vertebrate head. *Development* 129:1061–1073.

Couly, G. F., Grapin-Botton, A., Coltey, P., and Le Douarin, N. M. 1996. The regeneration of the cephalic neural crest, a problem revisited: The regenerating cells originate from the contralateral or from the anterior and posterior neural fold. *Development* 122:3393–3407.

Couly, G. F., and Le Douarin, N. M. 1985. Mapping of the neural early primordium in quail-chick chimeras. I. Developmental relationships between placodes, facial ectoderm and prosencephalon. *Dev Biol* 110:422–439.

Couly, G. F., and Le Douarin, N. M. 1987. Mapping of the early neural primordium in quail-chick chimeras. II. The prosencephalic neural plate and neural folds: Implications for the genesis of cephalic human congenital abnormalities. *Dev Biol* 120:198–214.

Couly, G. F., and Le Douarin, N. M. 1990. Head morphogenesis in embryonic avian chimeras: Evidence for a segmental pattern in the ectoderm corresponding to the neuromeres. *Development* 108:543–558.

Cracraft, J. 2005. Phylogeny and evo-devo: Characters, homology, and the historical analysis of the evolution of development. *Zoology* 108:345–356.

Crane, J. F., and Trainor, P. A. 2006. Neural crest stem and progenitor cells. *Annu Rev Cell Dev Biol* 22:267–286.

Creuzet, S. E., Couly, G, Vincent, C., and Le Douarin, N. M. 2002. Negative effect of Hox gene expression on the development of the neural crest-derived facial skeleton. *Development* 129:4301–4313.

Creuzet, S. E., Martinez, S., and Le Douarin, N. M. 2006. The cephalic neural crest exerts a critical effect on forebrain and midbrain development. *Proc Natl Acad Sci USA* 103:14033–14038.

Crump, J. G., Maves, L., Lawson, N. D., Weinstein, B. M., and Kimmel, C. B. 2004. An essential role for Fgfs in endodermal pouch formation influences later craniofacial skeletal patterning. *Development* 131:5703–5716.

Cruz, Y. P., Yousef, A., and Selwood, L. 1996. Fate-map analysis of the epiblast of the dasyurid marsupial *Sminthopsis macroura* (Gould). *Reprod Fertil Dev* 8:779–788.

Cunninham, M. L., Seto, M. L., Hing, A. V., Bull, M. J., *et al.* 2006. Cleidocranial dysplasia with severe parietal bone dysplasia: C-terminal *RUNX2* mutations. *Birth Defects Res (Part A)* 76:78–85.

Cusimano-Carollo, T. 1972. On the mechanism of the formation of the larval mouth in *Discoglossus pictus*. *Acta Embryol Exp* 4:289–332.

Czeizel, A. E., Puhó, E. H., Acs, N., and Bánhidy, F. 2008. Use of specific critical periods of different congenital abnormalities instead of the first trimester concept. *Birth Defects Res (Part A)* 82:139–146.

D

Dahl, E., Koseki, H., and Balling, R. 1997. *Pax* genes and organogenesis. *BioEssays* 19:755–765.

Dai, J., Keller, J., Zhang, J., Lu, Y., *et al.* 2005. Bone morphogenetic protein-6 promotes osteoblastic prostate cancer bone metastases through a dual mechanism. *Cancer Res* 65:8274–8285.

Damas, H. 1951. Observations sur le développement des ganglions crâniens chez *Lampetra fluviatilis* (L). *Archs Biol Paris* 62:55–95.

D'Autréaux, F., Morikawa, Y., Cserjesi, P., and Gershon, M. D. 2007. *Hand2* is necessary for terminal differentiation of enteric neurons from crest-derived precursors but not for their migration into the gut or for formation of glia. *Development* 134:2237–2249.

Davidson, E. H., and Erwin, D. H. 2006. Gene regulatory networks and the evolution of animal body plans. *Science* 311:796–800.

Davis, M. C., Dahn, R. D., and Shubin, N. H. 2007. An autopodial-like pattern of Hox expression in the fins of a basal actinopterygian fish. *Nature* 447:473–476.

Davis, R. L. 2000. The fate of cells in the tailbud of *Xenopus laevis*. *Development* 127:255–267.

Davy, A., and Soriano, P. 2007. Ephrin-B2 forward signaling regulates somite patterning and neural crest cell development. *Dev Biol* 304:182–193.

Dean, B. 1899. On the embryology of *Bdellostoma stouti*. A general account of myxinoid development from the egg and segmentation to hatching. In Festschr. f. C. van Kuppfer, pp. 221–277. Jena: G. Fischer.

de Beer, G. R. 1947. The differentiation of neural crest cells into visceral cartilages and odontoblasts in *Amblystoma*, and a re-examination of the germ-layer theory. *Proc R Soc* B134: 377–398.

de Beer, G. R. 1971. *Homology: An Unsolved Problem*, Oxford Biology Reader No. 11. London: Oxford University Press.

Deardorff, M. A., Tan, C., Saint-Jeannet, J.-P., and Klein, P. S. 2001. A role for frizzled3 in neural crest development. *Development* 128:3655–3663.

Dehal, P., Satou, Y., Campbell, R. K., Chapman, J., *et al.* 2002. The draft genome of *Ciona intestinalis*: Insights into chordate and vertebrate origins. *Science* 298:2157–2167.

Delalande, J.-M., Barlow, A. J., Thomas, A. J., Wallace, A. S., *et al.* 2008. The receptor tyrosine kinase RET regulates hindgut colonization by sacral neural crest cells. *Dev Biol* 313:279–292.

del Barrio, M. G., and Nieto, M. A. 2002. Overexpression of Snail family members highlights their ability to promote chick neural crest formation. *Development* 129:1583–1593.

Delfino-Machín, M., Chipperfield, T. R., Rodrigues, F. S. L. M., and Kelsh, R. N. 2007. The proliferating field of neural crest stem cells. *Dev Dyn* 236:3242–3254.

Del Pino, E. M., and Medina, A. 1998. Neural Development in the marsupial frog, *Gastrotheca riobambae*. *Int J Dev Biol* 42:723–731.

Del Pino, E. M., Venegas-Ferrin, M., Romero-Carvajal, A., Montenegro-Larrea, P., *et al.* 2007. A comparative analysis of frog early development. *Proc Natl Acad Sci USA* 104: 11882–11888.

Delsuc, F., Brinkmann, H., Chourrout, D., and Philippe, H. 2006. Tunicates and not cephalochordates are the closest living relatives of vertebrates. *Nature* 439:965–968.

Dencker, L., Annerwall, E., Busch, C., and Ericksson, U. 1990. Localization of specific retinoid-binding sites and expression of cellular retinoic-acid-binding protein (CRABP) in the early mouse embryo. *Development* 110:343–352.

Devoto, S. H., Stoiber, W., Hammond, C. L., Steinbacher, P., *et al.* 2006. Generality of vertebrate developmental patterns: Evidence for a dermomyotome in fish. *Evol Dev* 8:101–110.

Dickinson, M. E., Sellek, M. A. J., McMahon, A. P., and Bronner-Fraser, M. 1995. Dorsalization of the neural tube by the non-neural ectoderm. *Development* 121:2099–2106.

Di Renzo, F., Broccia, M. L., Giavini, E., and Menegola, E. 2006. Antifungal triazole derivative triadimefon induces ectopic maxillary cartilage by altering the morphogenesis of the first branchial arch. *Birth Defects Res (Part B)* 80:2–11.

Dixon, J., Jones, N. C., Sandell, L. L., Jayasinghe, S. M., *et al.* 2006. *Tcof1*/Treacle is required for neural crest cell formation and proliferation deficiencies that cause craniofacial abnormalities. *Proc Natl Acad Sci USA* 103:13403–13408.

Donoghue, P. C. J. 1998. Growth and patterning in the conodont skeleton. *Phil Trans R Soc Lond B* 353:633–666.

Donoghue, P. C. J. 2002. Evolution of development of the vertebrate dermal and oral skeletons: Unraveling concepts, regulatory theories, and homologies. *Paleobiology* 28:474–507.

Donoghue, P. C. J., Forey, P. L., and Aldridge, R. J. 2000. Conodont affinity and chordate phylogeny. *Biol Rev Camb Philos Soc* 75:191–251.

Dorsky, R. I., Moon, R. T., and Raible, D. W. 2000. Environmental signals and cell fate specification in premigratory neural crest. *BioEssays* 22:708–716.

Dottori, M., Gross, M. K., Labosky, P., and Goulding, M. 2001. The winged-helix transcription factor Foxd3 suppresses interneuron differentiation and promotes neural crest cell fate. *Development* 128:4127–4138.

Drysdale, T. A., and Elinson, R. P. 1993. Inductive events in the patterning of the *Xenopus laevis* hatching and cement glands, two cell types which delimit head boundaries. *Dev Biol* 158: 245–253.

du Brul, E. L. 1964. Evolution of the temporomandibular joint. In *The Temporomandibular Joint* (B. G. Sarnat, ed.), 2nd edition, pp. 3–27. C. C. Thomas, Springfield.

Duband, J.-L., Monier, F., Delannet, M., and Newgreen, D. F. 1995. Epithelium–mesenchyme transition during neural crest development. *Acta Anat* 154:63–78.

Duband, J.-L., Rocher, S., Yamada, K. M., and Thiery, J. P. 1986. Interaction of migrating neural crest cells with fibronectin. *Prog Clin Biol Res* 102:160–178.

Dudas, M., and Kaartinen, V. 2005. Tgf-β superfamily and mouse craniofacial development: Interplay of morphogenetic proteins and receptor signaling controls normal formation of the face. *Curr Topics Dev Biol* 66:65–133.

Dufour, H. D., Chettouh, Z., Deyts, C., de Rosa, R., *et al.* 2006. Precraniate origin of cranial motoneurons. *Proc Natl Acad Sci USA* 103:8727–8732.

Dufresne, C., and Richtsmeier, J. T. 1995. Interaction of craniofacial dysmorphology, growth and prediction of surgical outcome. *J Craniofac Surg* 6:270–281.

Dunlop, L.-L. T., and Hall, B. K. 1995. Relationships between cellular condensation, preosteoblast formation and epithelial–mesenchymal interactions in initiation of osteogenesis. *Int J Dev Biol* 39:357–371.

Dunn, K. J., Williams, B. O., Li, Y., and Pavan, W. J. 2000. Neural crest-directed gene transfer demonstrates Wnt1 role in melanocyte expansion and differentiation during mouse development. *Proc Natl Acad Sci USA* 97:10050–10055.

Duong, T. D., and Erickson, C. A. 2004. MMP-2 plays an essential role in producing epithelial-mesenchymal transformations in the avian embryo. *Dev Dyn* 229:42–53.

Dupin, E., Glavieux, C., Vaigot, P., and Le Douarin, N. M. 2000. Endothelin 3 induces the reversion of melanocytes to glia through a neural crest-derived glial-melanocytic progenitor. *Proc Natl Acad Sci USA* 97:7882–7887.

Dupin, E., Sextier-Sainte-Claire Deville, F., Nataf, V., and Le Douarin, N. M. 1993. The ontogeny of the neural crest. *C R Acad Sci* 316:1072–1081.

DuShane, G. P. 1935. An experimental study of the origin of pigment cells in Amphibia. *J Exp Zool* 72:1–31.

DuShane, G. P. 1943. The embryology of vertebrate pigment cells. Part I. Amphibia. *Q Rev Biol* 18:108–127.

DuShane, G. P. 1944. The embryology of vertebrate pigment cells. Part II. Birds. *Q Rev Biol* 19:98–117.

Dutton, K. A., Pauliny, A., Lopes, S. S., Elworthy, S., *et al.* 2001. Zebrafish colourless encodes Sox10 and specifies non-ectomesenchymal neural crest fates. *Development* 128:4113–4125.

E

Eames, B. F., and Schneider, R. A. 2005. Quail–duck chimeras reveal spatiotemporal plasticity in molecular and histogenic programs of cranial feather development. *Development* 132:1499–1509.

Ebendal, T. 1995. Cell movement in neurogenesis—An interview with Professor Jacobson, Carl Olaf. *Int J Dev Biol* 39:705–711.

Eberhart, J. K., He, X., Swartz, M. E., Yan, Y.-L., *et al.* 2008. MicroRNA Mirn140 modulates Pdgf signaling during palatogenesis. *Nat Genet* 40:290–298.

Eberhart, J. K., Swartz, M. E., Crump, J. G., and Kimmel, C. B. 2006. Early hedgehog signaling from neural to oral epithelium organizes anterior craniofacial development. *Development* 133:1069–1077.

Ekanayake, S., and Hall, B. K. 1994. Formation of cartilaginous nodules and heterogeneity in clones of H.H. 17 mandibular ectomesenchyme from the embryonic chick. *Acta Anat* 151: 171–179.

Ekanayake, S., and Hall, B. K. 1997. The *in vivo* and *in vitro* effects of bone morphogenetic protein-2 on the development of the chick mandible. *Int J Dev Biol* 41:67–81.

Ekker, M., Akimenko, M.-A., Allende, M. L., Smith, R., *et al.* 1997. Relationships among *msx* gene structure and function in zebrafish and other vertebrates. *Mol Biol Evol* 14: 1008–1022.

Elinson, R. P. 1990. Direct development in frogs: Wiping the recapitulationist slate clean. *Sem Dev Biol* 1:263–270.

Ellies, D. L., Langille, R. M., Martin, C. C., Akimenko, M.-A., and Ekker, M. 1997. Specific craniofacial cartilage dysmorphogenesis coincides with a loss of *Dlx* gene expression in retinoic acid-treated zebrafish embryos. *Mech Dev* 61:23–36.

Endo, Y., Osumi, N., and Wakamatsu, Y. 2002. Bimodal functions of Notch-mediated signaling are involved in neural crest formation during avian ectoderm development. *Development* 129: 863–873.

Engleka, K. A., Wu, M., Zhang, M., Antonucci, N. B., and Epstein, J. A. 2007. Menin is required in cranial neural crest for palatogenesis and perinatal viability. *Dev Biol* 311:524–537.

Epperlein, H.-H., Löfberg, J., and Olsson, L. 1996. Neural crest cell migration and pigment pattern formation in urodele amphibians. *Int J Dev Biol* 40:229–238.

Epperlein, H.-H., Meulemans, D., Bronner-Fraser, M., Steinbeisser, H., and Selleck, M. A. J. 2000. Analysis of cranial neural crest migratory pathways in axolotl using cell markers and transplantation. *Development* 127:2751–2781.

Epperlein, H.-H., Radomski, N., Wonka, F., Walther, P., *et al.* 2000. Immunohistochemical demonstration of hyaluronan and its possible involvement in axolotl neural crest cell migration. *J Struct Biol* 132:19–32.

Epperlein, H.-H., Selleck, M. A. J., Meulemans, D., Mchedlishvili, L., *et al.* 2007a. Migratory patterns and developmental potential of trunk neural crest cells in the axolotl embryo. *Dev Dyn* 236:389–403.

Epperlein, H.-H., Vichev, K., Heidrich, F. M., and Kurth, T. 2007b. BMP-4 and noggin signaling modulate dorsal fin and somite development in the axolotl trunk. *Dev Dyn* 236:2462–2472.

Epstein, J. A. 1996. Pax3, neural crest and cardiovascular development. *Trends Cardiovasc Med* 6:255–261.

Epstein, J. A., Li, J., Lang, D., Chen F., *et al.* 2000. Migration of cardiac neural crest cells in *Splotch* embryos. *Development* 127:1869–1878.

Erickson, C. A. 1987. Behavior of neural crest cells on embryonic basal laminae. *Dev Biol* 120: 38–49.

Erickson, C. A. 1993a. Morphogenesis of the avian trunk neural crest—use of morphological techniques in elucidating the process. *Microsc Res Tech* 26:329–351.

Erickson, C. A. 1993b. From the crest to the periphery—control of pigment cell migration and lineage segregation. *Pigment Cell Res* 6:336–347

Erickson, C. A., and Goins, T. L. 1995. Avian neural crest cells can migrate in the dorsolateral path only if they are specified as melanocytes. *Development* 121:915–924.

Erickson, C. A., and Goins, T. L. 2000. Sacral neural crest cell migration to the gut is dependent upon the migratory environment and not cell-autonomous migratory properties. *Dev Biol* 219:79–99.

Erickson, C. A., Loring, J. F., and Lester, S. M. 1989. Migratory pathways of HNK-1 immunoreactive neural crest cells in the rat embryo. *Dev Biol* 134:112–118.

Erickson, C. A., and Perris, R. 1993. The role of cell-cell and cell-matrix interactions in the morphogenesis of the neural crest. *Dev Biol* 159:60–74.

Erickson, C. A., and Weston, J. A. 1983. An SEM analysis of neural crest migration in the mouse. *J Embryol Exp Morphol* 74:97–118.

Erickson, C. A., and Weston, J. A. 1999. VENT cells: A fresh breeze in a stuffy field? *Trends Neurosci* 22:486–488.

Ericsson, R., Cerni, R., Falck, P., and Olsson, L. 2004. Role of cranial neural crest cells in visceral arch muscle positioning and morphogenesis in the Mexican axolotl, *Ambystoma mexicanum*. *Dev Dyn* 231:237–247.

Escriva, H., Holland, N. D., Gronemeyer, H., Laudet, V., and Holland, L. Z. 2002. The retinoic acid signaling pathway regulates anterior/posterior patterning in the nerve cord and pharynx of amphioxus, a chordate lacking neural crest. *Development* 129:2905–2916.

Ewart, J. L., Cohen, M. F., Meyer, R. A., Huang, G. Y., *et al.* 1997. Heart and neural tube defects in transgenic mice overexpressing the Cx43 gap junction gene. *Development* 124: 1281–1292.

F

Fagiani, E., Giardina, G., Luzi, L., Cesaroni, M., *et al.* 2007. RaLP, a new member of the Sc homology and collagen family, regulates cell migration and tumor growth of metastatic melanomas. *Cancer Res* 67:3064–3073.

Falck, P., Hanken, J., and Olsson, L. 2002. Cranial neural crest emergence and migration in the Mexican axolotl (*Ambystoma mexicanum*). *Zoology* 105:195–202.

Fang, H., and Elinson, R. P. 1996. Patterns of Distal-less gene expression and inductive interactions in the head of the direct developing frog *Eleutherodactylus coqui*. *Dev Biol* 179:160–172.

Fang, J., and Hall, B. K. 1997. Chondrogenic cell differentiation from membrane bone periostea. *Anat Embryol* 196:349–362.

Fang, J., and Hall, B. K. 1999. N-CAM is not required for initiation of secondary chondrogenesis: The role of N-CAM in skeletal condensation and differentiation. *Int J Dev Biol* 43:335–342.

Feiner, L., Webber, A. L., Brown, C. B., Lu, M. M., *et al.* 2001. Targeted disruption of semaphorin 3C leads to persistent truncus arteriosus and aortic arc interruption. *Development* 128: 3061–3070.

Ferguson, C. A., and Graham, A. 2004. Redefining the head–trunk interface of the neural crest. *Dev Biol* 269:70–80.

Ferguson, M. W. J. 1985. Reproductive biology and embryology of the crocodilians. In *Biology of the Reptilia*, Volume 14, Development A, eds C. Gans, F. Billett and P. F. A. Maderson, pp. 329–492. New York: John Wiley & Sons.

Fernandes, K. J. L., McKenzie, I. A., Mill, P., Smith, K. M., *et al.* 2004. A dermal niche for multipotent adult skin-derived precursor cells. *Nature Cell Biol* 6:1082–1093.

Fernandes, M., Gutin, G., Alcorn, H., McConnell, S. K., and Hébert, J. M. 2007. Mutation in the BMP pathway in mice support the existence of two molecular classes of holoprosencephaly. *Development* 134:3789–3794.

Fernández-Garre, P., Rodriguez-Gallardo, L., Gallego-Diaz, V., Alvarez, I. S., and Puelles, L. 2002. Fate map of the chicken neural plate at stage 4. *Development* 129:2807–2822.

Finzsch, M., Stolt, C. C., Lommes, P., and Wegner, M. 2008. Sox9 and Sox10 influence survival and migration of oligodendrocyte precursors in the spinal cord by regulating PDGF receptor α expression. *Development* 135:637–646.

Fiorica-Howells, E., Maroteaux, L., and Gershon, M. D. 2000. Serotonin and the 5-HT2B Receptor in the development of enteric neurons. *J Neurosci* 20:294–305.

Flores, M. V., Lam, E. Y. N., Crosier, P., and Crosier, K. 2006. A hierarchy of Runx transcription factors module the onset of chondrogenesis in craniofacial endochondral bones in zebrafish. *Dev Dyn* 235:3166–3176.

Fondon, J. W. III., and Garner, H. R. 2007. Detection of length-dependent effects of tandem repeat alleles by 3D geometric decomposition of craniofacial variation. *Dev Genes Evol* 217:79–85.

Fontaine, J. 1979. Multistep migration of calcitonin cell precursors during ontogeny of the mouse pharynx. *Gen Comp Endocrinol* 37:81–92.

Forey, P. L. 1995. Agnathans recent and fossil and the origin of jawed vertebrates. *Rev Fish Biol Fisheries* 5:267–303.

Forgacs, G., and Newman, S. A. 2005. *Biological Physics of the Developing Embryo*. Cambridge: Cambridge University Press.

Franz-Odendaal, T. A., and Hall, B. K. 2006. Modularity and sense organs in the blind cavefish, *Astyanax mexicanus*. *Evol Dev* 8: 94–100.

Fraser, S. E., and Bronner-Fraser, M. 1991. Migrating neural crest cells in the trunk of the avian embryo are multipotent. *Development* 112:913–920.

Freitas, R., Zhang, G. J., and Cohn, M. J. 2006. Evidence that mechanisms of fin development evolved in the midline of early vertebrates. *Nature* 442:1033–1037.

Frohman, M. A., Boyle, M., and Martin, G. R. 1990. Isolation of the mouse *Hox-2.9* gene; analysis of embryonic expression suggests that positional information along the anterior-posterior axis is specified by mesoderm. *Development* 110:589–607.

Fritzsch, B., and Northcutt, R. G. 1993. Cranial and spinal nerve organization in amphioxus and lampreys: Evidence for an ancestral craniate pattern. *Acta Anat* 148:96–109.

Fukiishi, Y., and Morriss-Kay, G. M. 1992. Migration of cranial neural crest cells to the pharyngeal arches and heart in rat embryos. *Cell Tissue Res* 268:1–8.

G

Gale, E., Prince, U., Lumsden, A., Clarke, J., Holder, N., and Maden, M. 1996. Late effects of retinoic acid on neural crest and aspects of rhombomere identity. *Development* 122:783–793.

Gans, C., and Northcutt, R. G. 1983. Neural crest and the origin of vertebrates: A new head. *Science* 220:268–274.

Gans, C., and Northcutt, R. G. 1985. Neural crest: The implications for comparative anatomy. *Fortschr Zool* 30:507–514.

Garcia-Martinez, V., Alvarez, I. S., and Schoenwolf, G. C. 1993. Locations of the ectodermal and nonectodermal subdivisions of the epiblast at stages 3 and 4 of avian gastrulation and neurulation. *J Exp Zool* 267:431–446.

Garg, V., Yamagishi, C., Hu, T., Kathiriya, I. S., *et al.* 2001. *Tbx1*, a DiGeorge syndrome candidate gene, is regulated by sonic hedgehog during pharyngeal arch development. *Dev Biol* 235: 62–73.

Gass, G., and Hall, B. K. 2007. Collectivity in context: Modularity, cell sociology, and the neural crest. *Biol Theory* 2:1–11.

Gavalas, A., Studer, M., Lumsden, A., Rijli, F. M., *et al.* 1998. *Hoxa1* and *Hoxb1* synergize in patterning the hindbrain, cranial nerves and second pharyngeal arch. *Development* 125: 1123–1136.

Gavalas, A., Trainor, P., Ariza-McNaughton, L., and Krumlauf, R. 2001. Synergy between *Hoxa1* and *Hoxb1*: The relationship between arch patterning and the generation of cranial neural crest. *Development* 128:3017–3027.

Gendron-Maguire, M., Mallo, M., Zhang, M., and Gridley, T. 1993. *Hoxa-2* mutant mice exhibit homeotic transformation of skeletal elements derived from cranial neural crest. *Cell* 75:1317–1331.

George, L., Chaverra, M., Todd, V., Lansford, R., and Lefcort, F. 2007. Nociceptive sensory neurons derive from contralaterally migrating, fate-restricted neural crest cells. *Nature Neurosci* 10:1287–1293.

Gershon, M. D. 1999. Lessons from genetically engineered animal models. II. Disorders of enteric neuronal development: Insights from transgenic mice. *Am J Physiol* 277:G262–G267.

Gessert, S., Maurus, D., Rössner, A., and Kühl, M. 2007. Pescadillo is required for *Xenopus laevis* eye development and neural crest migration. *Dev Biol* 310:99–112.

Ghislain, J., Desmarquet-Trin-Dinh, C., Gilardi-Hebenstreit, P., Charnay, P., and Frain, M. 2003. Neural crest patterning: Autoregulatory and crest-specific elements co-operate for krox-20 transcriptional control. *Development* 130:941–953.

Gilbert, S. F. 2006. *Developmental Biology*. Eighth Edition. New York: Sinauer Inc.

Gilbert, S. F., Bender, G., Betters, E., Yin, M., *et al.* 2007. The contribution of neural crest cells to the nuchal bone and plastron of the turtle shell. *Integ Comp Biol* 47:401–408.

Goding, C. R. 2000. Mitf from neural crest to melanoma: Signal transduction and transcription in the melanocyte lineage. *Genes Dev* 14:1712–1728.

Goh, K. L., Yang, J. T., and Hynes, R. O. 1997. Mesodermal defects and cranial neural crest apoptosis in α5 integrin-null embryos. *Development* 124:4309–4319.

Goldstein, R. S., Avivi, C., and Geffe, R. 1995. Initial axial level-dependent differences in size of avian dorsal root ganglia are imposed by the sclerotome. *Dev Biol* 168:214–222.

Goodrich, E. S. 1930. *Studies on the Structure and Development of Vertebrates.* London: Macmillan & Co. Reprinted, 1958, New York: Dover Publications, Inc.; 1986, Chicago: The University of Chicago Press.

Gorlin, R. J., Cohen, M. M. Jr., and Hennekam, R. C. M. 2001. *Syndromes of the Head and Neck.* Fourth Edition. Oxford and London: Oxford University Press.

Graham, A., and Begbie, J. 2000. Neurogenic placodes: A common front. *Trends Neurosci* 23: 313–316.

Graham, A., Blentic, A., Duque, S., and Begbie, J. 2007. Delamination of cells from neurogenic placodes does not involve an epithelial-to-mesenchymal transition. *Development* 134: 4141–4145.

Graham, A., and Smith, A. 2001. Patterning the pharyngeal arches. *BioEssays* 23:54–61.

Grammatopoulos, G. A., Bell, E., Toole, L., Lumsden, A., and Tucker, A.-S. 2000. Homeotic transformation of branchial arch identity after *Hoxa2* overexpression. *Development* 127: 5355–5365.

Grandel, H., Lun, K., Rauch, G.-J., Rhinn, M., *et al.* 2002. Retinoic acid signalling in the zebrafish embryo is necessary during pre-segmentation stages to pattern the anterior–posterior axis of the CNS and to induce a pectoral fin bud. *Development* 129:2851–2865.

Grant, J. H., Maggioprice, L., Reutebuch, J., and Cunningham, M. L. 1997. Retinoic acid exposure of the mouse on embryonic day 9 selectively spares derivatives of the frontonasal neural crest. *J Craniofac Genet Dev Biol* 17:1–8.

Grant, P. R., and Grant, B, R. 2002. Unpredictable evolution in a 30-year study of Darwin's finches. *Science* 296:707–711.

Graveson, A. C., and Armstrong, J. B. 1996. Premature death (*p*) mutation of *Ambystoma mexicanum* affects the ability of ectoderm to respond to neural induction. *J Exp Zool* 274:248–254.

Graveson, A. C., Hall, B. K., and Armstrong, J. B. 1995. The relationship between migration and chondrogenic potential of trunk neural crest cells in *Ambystoma mexicanum. Roux's Arch Dev Biol* 204:477–483.

Graveson, A. C., Smith, M. M., and Hall, B. K. 1997. Neural crest potential for tooth development in a urodele amphibian: Developmental and evolutionary significance. *Dev Biol* 188:34–42.

Griffith, C. M., Wiley, M. J., and Sanders, E. J. 1992. The vertebrate tail bud: Three germ layers from one tissue. *Anat Embryol* 185:101–113.

Grim, M., and Harata, Z. 2000. Developmental origin of avian Merkel cells. *Anat Embryol* 202:401–410.

Grosse, S. D., and Collins, J. S. 2007. Folic acid supplementation and neural tube defect recurrence prevention. *Birth Defects Res (Part A)* 79:737–742.

Groves, A. K., and Bronner-Fraser, M. 2000. Competence, specification and commitment in otic placode induction. *Development* 127:3489–3499.

Grüneberg, H. 1956. A ventral ectodermal ridge of the tail in mouse embryos. *Nature* 177:787–788.

Gurdon, J. B. 1992. The generation of diversity and pattern in animal development. *Cell* 68:185–187.

Gurjarpadhye, A., Hewett, K. W., Justus, C., Wen X., *et al.* 2007. Cardiac neural crest ablation inhibits compaction and electrical function of conduction system bundles. *Am J Physiol Heart Circ Physiol* 292:H1291–H1300.

Guthrie, S., Muchamore, I., Kuroiwa, A., Marshall, H., *et al.* 1992. Neuroectodermal autonomy of *Hox-2.9* expression revealed by rhombomere transpositions. *Nature* 356:157–159.

Gvirtzmann, G., Goldstein, R. S., and Kalcheim, C. 1992. A positive correlation between permissiveness of mesoderm to neural crest migration and early DRG growth. *J Neurobiol* 23:205–216.

H

Habeck, J. O. 1990. Islands of cartilage and bone at the margins of the carotid bodies in rats. *Anat Anz* 171:277–280.

Hall, B. K. 1980. Tissue interactions and the initiation of osteogenesis and chondrogenesis in the neural crest-derived mandibular skeleton of the embryonic mouse as seen in isolated murine tissues and in recombinations of murine and avian tissues. *J Embryol Exp Morphol* 58: 251–264.

Hall, B. K. 1987. Tissue interactions in the development and evolution of the vertebrate head. In *Developmental and Evolutionary Aspects of the Neural Crest*, ed P. F. A. Maderson, pp. 215–259. New York: Wiley.

Hall, B. K. Evolutionary issues in craniofacial biology. Proceedings of the Symposium on Advances in Craniofacial Developmental Biology and Clinical Implications, San Francisco, CA, April, 1989. *The Cleft Palate J* 27:95–100.

Hall, B. K. (ed) 1994. *Homology: The Hierarchical Basis of Comparative Biology*. Boca Raton: Academic Press.

Hall, B. K. 1995. Homology and embryonic development. *Evol Biol* 28:1–37.

Hall, B. K. 1997. Germ layers and the germ-layer theory revisited: Primary and secondary germ layers, neural crest as a fourth germ layer, homology, demise of the germ-layer theory. *Evol Biol* 30:121–186.

Hall, B. K. 1999a. *The Neural Crest in Development and Evolution*. New York: Springer,

Hall, B. K. 1999b. *Evolutionary Developmental Biology*. Second Edition. Dordrecht, Netherlands: Kluwer Academic Publishers.

Hall, B. K. 2000a. A role for epithelial–mesenchymal interactions in tail growth/morphogenesis and chondrogenesis in embryonic mice. *Cell Tissues Organs* 166:6–14.

Hall, B. K. 2000b. The evolution of the neural crest in vertebrates. In *Regulatory Processes in Development: The Legacy of Sven Hörstadius* Wenner-Gren International Series Volume 76,(eds C.-O. Jacobson, L. Olsson and T. Laurent, pp. 101–113. London: The Portland Press.

Hall, B. K. 2000c. Epithelial-mesenchymal interactions. In: *Methods in Molecular Biology, Vol. 125: Developmental Biology Protocols*, Volume. 3, eds R. S. Tuan and C. W. Lo, pp. 235–243. Totowa, NJ: Humana Press Inc.

Hall, B. K. 2000d. The neural crest as a fourth germ layer and vertebrates as quadroblastic not triploblastic. *Evol Dev* 2:1–3.

Hall, B. K. 2003a. Unlocking the black box between genotype and phenotype: Cell condensations as morphogenetic (modular) units. *Biol & Philos* 18: 219–247.

Hall, B. K. 2003b. Descent with modification: The unity underlying homology and homoplasy as seen through an analysis of development and evolution. *Biol Rev Camb Philos Soc 7* 8:409–433.

Hall, B. K. 2003c. Developmental and cellular origins of the amphibian skeleton. In *Amphibian Biology, Volume 5, Osteology*, eds. H. Heatwole and M. Davies, pp. 1551–1597. Chipping Norton, NSW: Surrey Beatty & Sons.

Hall, B. K. 2005a. Consideration of the neural crest and its skeletal derivatives in the context of novelty/innovations. *J Exp Zool (Mol Dev Evol)* 304B: 548–557.

Hall, B. K. 2005b. *Bone and Cartilage: Developmental and Evolutionary Skeletal Biology*. London: Elsevier Academic Press.

Hall, B. K. (ed) 2007a. *Fins into Limbs. Development, Transformation, and Evolution*. Chicago, Il: The University of Chicago Press.

Hall, B. K. 2007b. Homology and homoplasy. In *Handbook of the Philosophy of Science. Philosophy of Biology*, eds. M. Matthen and C. Stephens, pp. 429–453. Elsevier B. V.

Hall, B. K. 2008. Vertebrate origins: Riding the crest of a new wave, or the wave of a new crest? *Evol Dev* 10:261–263.

Hall, B. K., and Coffin-Collins, P. A. 1990. Reciprocal interactions between epithelium, mesenchyme and epidermal growth factor (EGF) in the regulation of mitotic activity of mandibular epithelium and mesenchyme in the embryonic chick. *J Craniofac Gen Dev Biol* 10:241–261.

Hall, B. K., and Ekanayake, S. 1991. Effects of growth factors on the differentiation of neural crest cells and neural crest cell derivatives. *Int J Dev Biol* 35:367–386.

Hall, B. K., and Hallgrímsson, B. (ed) 2008. *Strickberger's Evolution. The Integration of Genes, Organisms, and Populations.* Fourth Edition. Sudbury, MA: Jones and Bartlett, Publishers.

Hall, B. K., and Hanken, J. 1985. Foreword, In *The Development of the Vertebrate Skull* ed G. R. de Beer. pp. vii–xxviii. Chicago: The University of Chicago Press.

Hall, B. K., and Hörstadius, S. 1988. *The Neural Crest.* Oxford: Oxford Press.

Hall, B. K., and Miyake, T. 1992. The membranous skeleton: The role of cell condensations in vertebrate skeletogenesis. *Anat Embryol* 186:107–124.

Hall, B. K., and Miyake, T. 1997. How do embryos tell time. In *Evolution Through Heterochrony.* ed K. J. McNamara, pp. 1–20. Chichester: John Wiley & Sons.

Hall, B. K., and Miyake, T. 2000. All for one and one for all: Condensations and the initiation of skeletal development. *BioEssays* 22:138–147.

Hall, B. K., and Wake, M. H. (eds) 1999. *The Origin and Evolution of Larval Forms.* San Diego: Academic Press.

Hall, B. K., and Witten, P. E. 2007. The origin and plasticity of skeletal tissues in vertebrate evolution and development. In *Major Transitions in Vertebrate Evolution,* eds J.S. Anderson and H.-D. Sues, pp. 13–57. Bloomington, IN: Indiana University Press..

Hammerschmidt, M., Serbedzija, G. N., and McMahon, A. P. 1996. Genetic analysis of dorsoventral pattern formation in the zebrafish: Requirement of a BMP-like ventralizing activity and its dorsal repressor. *Genes Dev* 10:2452–2461.

Handrigan, G. R. 2003. *Concordia discors*: Duality in the origin of the vertebrate tail. *J Anat* 202:255–267.

Hanken, J., and Hall, B. K. (eds) 1993. *The Vertebrate Skull. Volume 1–3.* Chicago: The University of Chicago Press.

Hanken, K., Jennings, D. H., and Olsson, L. 1997. Mechanistic basis of life-history evolution in anuran amphibians; Direct development. *Amer Zool* 37:160–171.

Harada, Y., Okai, N., Taguchi, S., Tagawa, K., *et al.* 2000. Developmental expression of the hemichordate *otx* gene ortholog. gene. *Mech Dev* 91:337–333.

Harada, Y., Okai, N., Taguchi, S., Shoguchi, E., *et al.* 2001. Embryonic expression of a hemichordate distal-less gene. *Zool Sci* 18:57–61.

Hardisty, M. W. 1979. *Biology of the Cyclostomes.* London: Chapman & Hall.

Harlow, D. E., and Barlow, L. A. 2007. Embryonic origin of gustatory cranial sensory neurons. *Dev Biol* 310:317–328.

Harris, M. J., and Juriloff, D. M. 2007. Mouse mutants with neural tube closure defects and their role in understanding human neural tube defects. *Birth defects Res (Part A): Clin Mol Teratol* 79:187–210.

Harris, M. L., and Erickson, C. A. 2007. Lineage specification in neural crest cell pathfinding. *Dev Dyn* 236:1–19.

Harrison, R. G. 1910. The outgrowth of the nerve fiber as a mode of protoplasmic movement. *J Exp Zool* 9:787–846.

Harrison, R. G. 1929. Correlation in the development and growth of the eye studied by means of heteroplastic transplantation. *W Roux Arch Entwickl Mech Org* 120:1–55.

Harrison, T. A., Stadt, H. A., Kumiski, D., and Kirby, M. L. 1995. Compensatory responses and development of the nodose ganglion following ablation of placodal precursors in the embryonic chick (*Gallus domesticus*). *Cell Tissue Res* 281:379–385.

Hart, R. C., McCue, P. A., Ragland, W. L., Winn, K. J., and Unger, E. R. 1990. Avian model for 13-cis-retinoic acid embryopathy: Demonstration of neural crest related defects. *Teratology* 41:463–472.

Hassell, J. R., Greenberg, J. H., and Johnston, M. C. 1977. Inhibition of cranial neural crest cell development by vitamin A in cultured chick embryo. *J Embryol Exp Morph* 39:267–271.

Hayward, P., Kalmar, T., and Arias, A. M. 2008. Wnt/Notch signalling and information processing during development. *Development* 135:411–424.

Hearn, C., and Newgreen, D. 2000. Lumbo-sacral neural crest contributes to the avian enteric nervous system independently of vagal neural crest. *Dev Dyn* 218:525–530.

Heath, L., Wild, A., and Thorogood, P. V. 1992. Monoclonal antibodies raised against pre-migratory neural crest reveal population heterogeneity during crest development. *Differentiation* 49:151–165.

Heimberg, A. M., Sempere, L. F., Moy, V. N., Donoghue, P. C. J., and Peterson, K. J. 2008. MicroRNAs and the advent of vertebrate morphological complexity. *Proc Natl Acad Sci USA* 105:2946–2950.

Hemmati-Brivanlou, A., and Melton, D. 1997. Vertebrate embryonic cells will become nerve cells unless told otherwise. *Cell* 88:13–17.

Hemmati-Brivanlou, A., Stewart, R. M., and Harland, R. M. 1990. Region-specific neural induction of an engrailed protein by anterior notochord. *Xenopus Sci.* 250:800–802.

Henion, P. D., Garner, A. S., Large, T. H., and Weston, J. A. 1995. trk C-mediated NT-3 signaling is required for the early development of a subpopulation of neurogenic neural crest cells. *Dev Biol* 172:602–613.

Henion, P. D., Raible, D. W., Beattie, C. E., *et al.* 1996. Screen for mutations affecting development of zebrafish neural crest. *Dev Genet* 18:11–17.

Henion, P. D., and Weston, J. A. 1997. Timing and pattern of cell fate restriction in the neural crest lineage. *Development* 124:4351–4359.

Henry, J. J., and Grainger, R. M. 1987. Inductive interactions in the spatial and temporal restriction of lens-forming potential in embryonic ectoderm of *Xenopus laevis*. *Dev Biol* 124: 200–214.

Hertwig, O., and Hertwig, R. 1882. Die Coelomtheorie, Versuch einer Erklärung des mittleren Keimblattes. *Jenaische Zeit* 15:1–150.

Higashiyama, D., Saitsu, H., Komada, M., Takigawa, T., *et al.* 2007. Sequential developmental changes in holoprosencephalic mouse embryos exposed to ethanol during the gastrulation period. *Birth Defects Res (Part A)*, 79:513–523.

High, F. A., and Epstein, J. A. 2008. The multifaceted role of Notch in cardiac development and disease. *Nature Rev Genet* 9:49–61.

Hill, J. P., and Watson, K. M. 1958. The early development of the brain in marsupials. *J Anat* 92:493–497.

Hinrichs, S. H., Nerenberg, M., Reynolds, R. K., Khoury, G., and Jay G. 1987. A transgenic mouse model for human neurofibromatosis. *Science* 237:1340–1343.

Hirano, S. 1986. Observations on the migration and differentiation of neural crest cells in somite extirpated salamander larvae. *Arch Histologicum Japonicum* 49:309–320.

Hirano, S., and Shirai, T. 1984. Morphogenetic studies on the neural crest of *Hynobius* larvae using vital staining and India ink labelling methods. *Arch Histol Japon* 47:57–70.

Hirata, M., Ito, K., and Tsuneki, K. 1997. Migration and colonization patterns of HNK-1-immunoreactive neural crest cells in lamprey and swordtail embryos. *Zool Sci* 14: 305–312.

His, W. 1868. *Untersuchungen über die erste Anlage des Wirbeltierleibes. Die erste Entwicklung des Hühnchens im Ei.* Leipzig: F. C. W. Vogel.

His, W. 1874. *Unserer Körperform und das Physiologische Problem ihrer Entstehung.* Leipzig: Engelmann.

Hodges, G. M., and Rowlatt, C. (eds) 1994. *Developmental Biology and Cancer.* Boca Raton, FL: CRC Press.

Holland, L. Z., and Holland, N. D. 1996. Expression of *AmphiHox-1* and *AmphiPax-1* in amphioxus embryos treated with retinoic acid: Insights into evolution and patterning of the chordate nerve cord and pharynx. *Development* 122:1829–1838.

Holland, L. Z., and Holland, N. D. 2001. Evolution of neural crest and placodes: Amphioxus as a model for the ancestral vertebrate? *J Anat* 199:85–98.

Holland, L. Z., Rached, L. A., Tamme, R., Holland, N. D., *et al.* 2001. Characterization and developmental expression of the amphioxus homolog of *Notch* (*AmphiNotch*): Evolutionary conservation of multiple expression domains in amphioxus and vertebrates. *Dev Biol* 232: 493–507.

Holland, L. Z., Schubert, M., Holland, N. D., and Neuman, T. 2000. Evolutionary conservation of the presumptive neural plate markers *AmphiSox1/2/3* and *AmphiNeurogenin* in the invertebrate chordate amphioxus. *Dev Biol* 226:18–33.

Holland, N. D., Panganiban, G., Henyey, E. L., and Holland, L. Z. 1996. Sequence and developmental expression of *AmphiDll*, an amphioxus *Distal-less* gene transcribed in the ectoderm, epidermis and nervous system: Insights into evolution of craniate forebrain and neural crest. *Development* 122:2911–2920.

Holland, P. W. H. 1996. Molecular biology of lancelets: Insights into development and evolution. *Israel J Zool* 42:S247–S272.

Holland, P. W. H., and Graham, A. 1995. Evolution of regional identity in the vertebrate nervous system. *Persp Dev Neurobiol* 3:17–27.

Holley, S. A. Jackson, P. D., Sasai, Y., Lu, B., *et al.* 1995. A conserved system for dorsal–ventral patterning in insects and vertebrates involving *Sog* and *Chordin*. *Nature* 376:249–253.

Holmdahl, D. E. 1928. Die Enstehung und weitere Entwicklung der Neuralleiste (Ganglienleiste) bei Vogeln und Saugetieren. *Z Mikrosk-Anat Forsch* 14:99–298.

Holmgren, N. 1940. Studies on the head in fishes. Embryological, morphological, and phylogenetic researches. Part I: Development of the skull in sharks and rays. *Acta Zool* (*Stockh*) 21: 51–267.

Holmgren, N. 1943. Studies on the head in fishes. Embryological, morphological, and phylogenetic researches. Part IV: General morphology of the head in fish. *Acta Zool* (*Stockh*) 24:1–188.

Holtfreter, J. 1968. Mesenchyme and epithelia in inductive and morphogenetic processes. In *Epithelial-Mesenchymal Interactions*, eds R. Fleischmajer and R. E. Billingham, pp. 1–30. Baltimore: Williams & Wilkins.

Holzschuh, J., Wada, N., Wada, C., Schaffer, A., *et al.* 2005. Requirements for endoderm and BMP signaling in sensory neurogenesis in zebrafish. *Development* 132:3731–3742.

Hong, C.-S., and Saint-Jeannet, J.-P. 2005. Sox proteins and neural crest development. *Sem Cell Dev Biol* 16:694–703.

Hong, C.-S., and Saint-Jeannet, J.-P. 2007. The activity of Pax3 and Zic3 regulates three distinct cell fates at the neural plate border. *Mol Biol Cell* 18:2192–2202.

Hörstadius, S. 1928. Über die determination des Keimes bei Echinodermen. *Acta Zool* (*Stockh*) 9:1–191.

Hörstadius, S. 1939. The mechanics of sea urchin development studies by operative methods. *Biol Rev Camb Philos Soc* 14:132–179.

Hörstadius, S. 1950. *The Neural Crest: Its Properties and Derivatives in the Light of Experimental Research.* Oxford: Oxford University Press.

Hörstadius, S. 1951. Introduction to Swedish ornithology. In *Proceedings of the Xth International Ornithological Congress,* ed. S. Hörstadius, pp. 75–90. Almqvist & Wiksells: Uppsala.

Hörstadius, S. 1973. *Experimental Embryology of Echinoderms.* Oxford: The Clarendon Press.

Hörstadius, S., and Sellman, S. 1941. Experimental studies on the determination of the chondrocranium in *Amblystoma mexicanum. Ark Zool* 33(13):1–8

Hörstadius, S., and Sellman, S. 1946. Experimentelle untersuchungen über die Determination des Knorpeligen Kopfskelettes bei Urodelen. *Nova Acta R Soc Scient. Upsal Ser* 4(13):1–170.

Hosokawa, R., Urata, M., Han, J., Zehnaly, A., *et al.* 2007. TGF-β mediated *Msx2* expression controls occipital somites-derived caudal region of skull development. *Dev Biol* 310:140–153.

Hou, L., and Takeuchi, T. 1994. Neural crest development in reptilian embryos: Studies with monoclonal antibody, HNK-1. *Zool Sci* 11:423–431.

Hrubec, T. C., Prater, M. R., Toops, K. A., and Holladay, S. D. 2006. Reduction in diabetes-induced craniofacial defects by maternal immune stimulation. *Birth Defects Res* (*Part B*) 77:1–9.

Huang, R., Zhi, Q., Ordahl, C. P., and Christ, B. 1997. The fate of the first somite. *Anat Embryol* 195:435–449.

Hunt, P., Ferretti, P., Krumlauf, R., and Thorogood, P. 1995. Restoration of normal Hox code and branchial arch morphogenesis after extensive deletion of hindbrain neural crest. *Dev Biol* 168:584–597.

Hunt, P., and Krumlauf, R. 1992. *Hox* codes and positional specification in vertebrate embryonic axes. *Ann Rev Cell* 8:227–256.

Hunter, E., Begbie, J., Mason, I., and Graham, A. 2001. Early development of the mesencephalic trigeminal nucleus. *Dev Dyn* 222:484–493.

Hurtado, C., and de Robertis, E. M. 2007. Neural induction in the absence of organizer in salamanders by MAPK. *Dev Biol* 307:282–289.

Huszar, D., Sharpe, A., Hashmi, S., Bouchard, B., *et al.* 1991. Generation of pigmented stripes in albino mice by retroviral marking of neural crest melanoblasts. *Development* 113:653–660.

Hutson, M. R., and Kirby, M. L. 2003. Neural crest and cardiovascular development: A 20-year perspective. *Embryo Today* 69:2–13

Huxley, T. H. 1849. On the anatomy and the affinities of the family of the Medusæ. *Phil Trans R Soc.* 139:413–434.

Huxley, T. H. 1871. *A Manual of the Anatomy of Vertebrated Animals*. London.

I

Ido, A., and Ito, K. 2006. Expression of chondrogenic potential of mouse trunk neural crest cells by FGF2 treatment. *Dev Dyn* 235:363–367.

Ikeya, M., Lee, S. M. K., Johnson, J. E., McMahon, A. P., and Takada, S. 1997. *Wnt* signalling required for expansion of neural crest and CNS progenitors. *Nature* 389:966–970.

Imai, H., Osumi-Yamashita, N., Ninomiya, Y., and Eto, K. 1996. Contribution of early-migrating midbrain crest cells to the dental mesenchyme of mandibular molar teeth in rat embryos. *Dev Biol* 176:151–165.

Imai, K., Takada, N., Satoh, N., and Satou, Y. 2000. β-catenin mediates the specification of endoderm cells in ascidian embryos. *Development* 127:3009–3020.

Ito, K., and Morita, T. 1995. Role of retinoic acid in mouse neural crest cell development *in vitro*. *Dev Dyn* 204:211–218.

Ito, K., Morita, T., and Sieber-Blum, M. 1993b. *In vitro* clonal analysis of mouse neural crest development. *Dev Biol* 157:517–525.

Ito, K., and Sieber-Blum, M. 1993. Pluripotentiality and developmentally restricted neural crest-derived cells in posterior visceral arches. *Dev Biol* 156:191–200.

J

Jablonski, N. G. 2006. *Skin: A Natural History*. Berkeley, CA: University of California Press.

Jackman, W. R., and Kimmel, C. B. 2002. Coincident iterated gene expression in the amphioxus neural tube. *Evol Dev* 4:366–374.

Jacobson, C.-O. 2000. Sven Hörstadius, the man and his work. In *Regulatory Processes in Development*, eds L. Olsson and C.-O. Jacobson, pp. 1–10. London: Portland Press.

Jacobson, M., and Moody, S. A. 1984. Quantitative lineage analysis of the frog's nervous system. I. Lineages of Rohon-Béard neurons and primary motoneurons. *J Neurosci* 4:1361–1369.

Jaenisch, R. 1985. Mammalian neural crest cells participate in normal embryonic development on microinjection into post-implantation mouse embryos. *Nature* 318:181–183.

Janvier, P. 1996. The dawn of the vertebrates: Characters versus common ascent in the rise of current vertebrate phylogenies. *Palaeontology* 39:259–287.

Janvier, P. 2007. Homologies and evolutionary transitions in early vertebrate history. In *Major transitions in Vertebrate Evolution*, eds J. S. Anderson and H.-D. Sues, pp. 57–121. Bloomington and Indianapolis: Indiana University Press.

Järvinen, E., Salazaar-Ciudad, I., Birchmeier, W., Taketo, M. M., *et al.* 2006. Continuous tooth generation in mouse is induced by activated epithelial Wnt/β-catenin signalling. *Proc Natl Acad Sci USA* 103:18627–18632.

Jeffery, W. R. 1997. Evolution of ascidian development. *BioScience* 47:417–425.

Jeffery, W. R. 2007. Chordate ancestry of the neural crest: New insights from ascidians. *Sem Dev Biol* 18:481–491.

Jeffery, W. R., Strickler, A. G., and Yamamoto, Y. 2004. Migratory neural crest-like cells form body pigmentation in a urochordate embryo. *Nature* 431:696–699.

Jesuthasan, S. 1996. Contact inhibition/collapse and path finding of neural crest cells in the zebrafish trunk. *Development* 122:381–389.

Jiang, X., Iseki, S., Maxson, R. E., Sucov, H. M., and Morriss-Kay, G. M. 2002. Tissue Origins and Interactions in the Mammalian Skull Vault. *Devel Biol* 241:106–116.

Jiang, R., Lan, Y., Norton, C. R., Sundberg, J. P., and Gridley, T. 1998. The *Slug* gene is not essential for mesoderm or neural crest development in mice. *Dev Biol* 198:277–285.

Jiang, X., Rowitch, D. H., Soriano, P., McMahon, A. P., and Sucov, H. M. 2000. Fate of the mammalian cardiac neural crest. *Development* 127:1607–1616.

Johnson, D. R. 1986. *The Genetics of the Skeleton. Animal Models of Skeletal Development.* Oxford: The Clarendon Press.

Johnston, M. C. 1966. A radioautographic study of the migration and fate of cranial neural crest cells in the chick embryo. *Anat Rec* 156:143–156.

Johnston, M. C., and Bronsky, P. T. 1995. Prenatal craniofacial development: New insights on normal and abnormal mechanisms. *Crit Rev Oral Biol Med* 6:25–79.

Jones, M. C. 1990. The neurocristopathies: Reinterpretation based upon the mechanism of abnormal morphogenesis. *Cleft Palate J* 27:136–140.

Jones, N. C., and Trainor, P. A. 2005. Role of morphogens in neural crest cell determination. *J Neurobiol* 64:388–404.

Jørgensen, J. M., Lomholt, J. P., Weber, R. E., and Malthe, H. (eds) 1998. *The Biology of Hagfishes.* London: Chapman & Hall.

Juriloff, D. M., and Harris, M. J. 2008. Mouse genetic models of cleft lip with or without cleft palate. *Birth Defects Res (Part A)* 82:63–77.

Juriloff, D. M., Harris, M. J., and Froster-Iskenius, U. 1987. Hemifacial deficiency induced by a shift in dominance of the mouse mutation *far*: A possible genetic model for hemifacial microsomia. *J. Craniofac. Genet Devel Biol* 7:27–44.

K

Kahane, N., and Kalcheim, C. 1994. Expression of trkC receptor mRNA during development of the avian nervous system. *J Neurobiol* 25:571–584.

Kalcheim, C., Carmeli, C., and Rosenthal, A. 1992. Neurotropin 3 is a mitogen for cultured neural crest cells. *Proc Natl Acad Sci USA* 89:1661–1665.

Kalcheim, C., and Le Douarin, N. M. 1986. Requirement of a neural tube signal for the differentiation of neural crest cells into dorsal root ganglia. *Dev Biol* 116:451–466.

Kanakubo, S., Nomura, T., Yamamura, K.-I., Miyazaki, J.-i., *et al.* 2006. Abnormal migration and distribution of neural crest cells in *Pax6* heterozygous mutant eye, a model for human eye diseases. *Genes Cells* 11:919–933.

Kang, P., and Svoboda, K. K. H. 2005. Epithelial–mesenchymal transformation during craniofacial development. *J Dental Res* 84:678–690.

Kanki, J. P., and Ho, R. K. 1997. The development of the posterior body in zebrafish. *Development* 124:881–893.

Kanzler, B., Foreman, R. K., Labosky, P. A., and Mallo, M. 2000. BMP signaling is essential for development of skeletogenic and neurogenic cranial neural crest. *Development* 127:1095–1104.

Kanzler, B., Kuschert, S. J., Liu, Y.-H., and Mallo, M. 1998. *Hoxa-2* restricts the chondrogenic domain and inhibits bone formation during development of the branchial area. *Development* 125:2587–2597.

Kapron-Bras, C. M., and Trasler, D. G. 1985. Reduction in the frequency of neural tube defects in *Splotch* mice by retinoic acid. *Teratology* 32:87–92.

Kardong, K. V. 2006. *Vertebrates: Comparative Anatomy, Function, Evolution*, Fourth Edition. Boston: WCB/McGraw-Hill.

Kasemeier-Kulesa, J. C., Kulesa, P. M., and Lefcort, F. 2005. Imaging neural crest cell dynamics during formation of dorsal root ganglia and sympathetic ganglia. *Development* 132:235–245.

Kasperk, C., Wergedal, J., Strong, D., Farley, J., *et al.* 1995. Human bone cell phenotypes differ depending on their skeletal site of origin. *J Clin Endocrinol Metab* 80:2511–2517.

Kelsh, R. N. 2006. Sorting out *Sox10* function in neural crest development. *BioEssays* 28:788–798.

Kelsh, R. N., Brand, M., Jiang, Y.-J., Heisenberg, C.-P., *et al.* 1996. Zebrafish pigmentation mutations and the processes of neural crest development. *Development* 123:369–389.

Kelsh, R. N., and Eisen, J. S. 2000. The zebrafish colourless gene regulates development of non-ectomesenchymal neural crest derivatives. *Development* 127:515–525.

Kerney, R, Meegaskumbura, M., Manamendra-Arachchi, K., and Hanken, J. 2007. Cranial ontogeny in *Philautus silus* (Anura: Ranidae: Rhacophorinae) reveals few similarities with other direct-developing anurans. *J Morph* 268:715–725.

Kerr, J. G. 1919. *Text-Book of Embryology. Volume II. Vertebrata with the Exception of Mammalia.* London: Macmillan and Co.

Kerr, R. S. E., and Newgreen, D. F. 1997. Isolation and characterization of chondroitin sulfate proteoglycans from embryonic quail that influence neural crest cell behavior. *Dev Biol* 192:108–124.

Kessel, M. 1992. Respecification of vertebral identities by retinoic acid. *Development* 115: 487–501.

Kettunen, P., Spencer-Dene, B., Furmanek, T., Kvinnsland, I. H., *et al.* 2007. Fgfr2b mediated epithelial-mesenchymal interactions coordinate tooth morphogenesis and dental trigeminal axon patterning. *Mech Dev* 124:868–883.

Kimmel, C. B., Miller, C. T., and Keynes, R. J. 2001. Neural crest patterning and the evolution of the jaw. *J Anat* 199:105–120.

Kimura, C., Takeda, N., Suzuki, M., Oshimura, M., *et al.* 1997. *Cis*-acting elements conserved between mouse and pufferfish *Otx2* genes govern the expression in mesencephalic neural crest cells. *Development* 124:3929–3941.

Kinutani, M., Tan, K., Desaki, J., Coltey, M., *et al.* 1989. Avian spinal cord chimeras. Further studies on the neurological syndrome affecting the chimeras after birth. *Cell Differ Dev* 26: 145–162.

Kirby, M. L. 1989. Plasticity and predetermination of mesencephalic and trunk neural crest transplanted into the region of the cardiac neural crest. *Dev Biol* 134:402–412.

Kirby, M. L., Kumiski, D. H., Myers, T., Cerjan, C., and Mishima, N. 1993. Back transplantation of chick cardiac neural crest cells cultured in LIF rescues heart development. *Dev Dyn* 198: 296–311.

Kirchgessner, A. L., Adlersberg, M. A., and Gershon, M. D. 1992. Colonization of the developing pancreas by neural precursors from the bowel. *Dev Dyn* 194:142–154.

Ko, S. O., Chung, I. H., Xu, X., Oka, S., *et al.* 2007. *Smad4* is required to regulate the fate of cranial neural crest cells. *Dev Biol* 312:435–447.

Koentges, G., and Lumsden, A. 1996. Rhombencephalic neural crest segmentation is preserved throughout craniofacial ontogeny. *Development* 122:3229–3242.

Kölliker, A. von. 1884. Die Embryonalen Keimblätter und die Gewebe. *Zeit wiss Zool* 40:179–213.

Koltzoff, N. K. 1901. Entwicklungsgeschichte des Kopfes von *Petromyzon planeri*. *Bull Soc Imp Nat Moscow* 15:259–589.

Kopinke, D., Sasine, J., Swift, J., Stephens, W. Z., and Piotrowski, T. 2006. Retinoic acid is required for endodermal pouch morphogenesis and not for pharyngeal endoderm specification. *Dev Dyn* 235:2696–2709.

Korade, Z., and Frank, E. 1996. Restriction in cell fate of developing spinal cord cells transplanted to neural crest pathways. *J Neurosci* 16:7638–7648.

Kourakis, M. J., and Smith, W. C. 2007. A conserved role for FGF signaling in chordate otic/atrial placode formation. *Dev Biol* 312:245–257.

Kozmik, Z., Holland, N. D., Kalousova, A., Paces, J., *et al.* 1999. Characterization of an amphioxus paired box gene, *AmphiPax2/5/8*: Developmental expression patterns in optic support cells, nephridium, thyroid-like structures and pharyngeal gill slits, but not in the midbrain-hindbrain boundary region. *Development* 126:1295–1304.

Kreiling, J. A., Prabnat, G. W., and Creton, R. 2007. Analysis of Kupffer's vesicle in zebrafish embryos using a cave automated virtual environment. *Dev Dyn* 236:1963–1969.

Krotoski, D., and Bronner-Fraser, M. 1990. Distribution of integrins and their ligands in the trunk of *Xenopus laevis* during neural crest cell migration. *J Exp Zool* 253:139–150.

Krotoski, D. M., Domingo, C., and Bronner-Fraser, M. 1986. Distribution of a putative cell surface receptor for fibronectin and laminin in the avian embryo. *J Cell Biol* 103:1061–1071.

Krotoski, D. M., Fraser, S. E., and Bronner-Fraser, M. 1988. Mapping of neural crest pathways in *Xenopus laevis* using inter- and intra-specific cell markers. *Dev Biol* 127:119–132.

Kruger, G. M., Mosher, J. T., Bixby, S., Joseph, N., *et al.* 2002. Neural crest stem cells persist in the adult gut but undergo changes in self-renewal, neuronal subtype potential, and factor responsiveness. *Neuron* 35:657–669.

Krull, C. E., Lansford, R., Gale, N. W., Collazo, A., *et al.* 1997. Interactions of Eph-related receptors and ligands confer rostrocaudal pattern to trunk neural crest migration. *Curr Biol* 1: 571–580.

Krumlauf, R. 1993. *Hox* genes and pattern formation in the branchial region of the vertebrate head. *Trends Genet* 9:106–112.

Kuhlman, J., and Eisen, J. E. 2007. Genetic screen for mutations affecting development and function of the enteric nervous system. *Dev Dyn* 236:118–127.

Kulesa, P. M., Lub, C. C., and Fraser, S. E. 2005. Time-lapse analysis reveals a series of events by which cranial neural crest cells reroute around physical barriers. *Brain Behav Evol* 66:255–265.

Kulesa, P. M., Teddy, J. M., Stark, D. A., Smith, S. E., and McLennan, R. 2008. Neural crest invasion is a spatially-ordered progression into the head with higher cell proliferation at the migratory front as revealed by the photoactivatable protein KikGR. *Dev Biol* 316:275–287.

Kumano, G., and Nishida, H. 2007. Ascidian embryonic development: An emerging model system for the study of cell fate specification in chordates. *Dev Dyn* 236:1732–1747.

Kupffer, C. von 1906. Die Morphogenie des Centralnervensystems [*Bdellostoma*]. In *Handbuch der Vergleichenden und Experimentellen Entwickelungslehre der Wirbeltiere*, ed O. Hertwig, Band II, Teil 3, pp. 24–38. Jena: Verlag von Gustav Fisher.

Kuraku, S., Takio, Y., Tamura, K., Aono, H., et al. 2008. Noncanonical role of Hox14 revealed by its expression patterns in lamprey and shark. *Proc Natl Acad Sci USA* 105:6679–6683.

Kuratani, S. C. 1991. Alternate expression of the HNK-1 epitope in rhombomeres of the chick embryo. *Dev Biol* 144:215–219.

Kuratani, S. C. 1997. Spatial distribution of postotic crest cells defines the head/trunk interface of the vertebrate body: Embryological interpretation of peripheral nerve morphology and evolution of the vertebrate head. *Anat Embryol* 195:1–13.

Kuratani, S. C. 2004. Evolution of the vertebrate jaw: Comparative embryology and molecular developmental biology reveal the factors behind evolutionary novelty. *J Anat* 205:335–347.

Kuratani, S. C. 2005. Cephalic neural crest cells and the evolution of craniofacial structures in vertebrates: Morphological and embryological significance of the premandibular–mandibular boundary. *Zoology* 108:13–25.

Kuratani, S., and Aizawa, S. 1995. Patterning of the cranial nerves in the chick embryo is dependent on cranial mesoderm and rhombomeric metamerism. *Dev Growth Differ* 37:717–731.

Kuratani, S., and Bockman, D. E. 1991. Capacity of neural crest cells from various axial levels to participate in thymic development. *Cell Tissue Res* 263:99–106.

Kuratani, S., and Eichele, G. 1993. Rhombomere transplantation repatterns the segmental organization of cranial nerves and reveals cell-autonomous expression of a homeodomain protein. *Development* 117:105–117.

Kuratani, S., and Kirby, M. L. 1992. Migration and distribution of circumpharyngeal crest cells in the chick embryo: Formation of the circumpharyngeal ridge and E/c8+ crest cells in the vertebrate head region. *Anat Rec* 234:263–280.

Kuratani, S., Kuraku, S., and Murakami, Y. 2002. Lampreys as an evo–devo model: Lessons from comparative embryology and molecular phylogenetics. *Genesis* 34:175–183.

Kuratani, S., Nobusada, Y., Horigome, N., and Shigetani, Y. 2001. Embryology of the lamprey and evolution of the vertebrate jaw: Insights from molecular and developmental perspectives. *Phil Trans R Soc Lond* B 356:1615–1632.

Kuratani, S., Nobusada, Y., Saito, H., and Shigetani, Y. 2000. Morphological characteristics of the developing cranial nerves and mesoderm head cavities in sturgeon embryos from early pharyngula to late larval stages. *Zool Sci* 17:911–933.

Kuratani, S., Ueki, T., Hirano, S., and Aizawa, S. 1998. Rostral truncation of a cyclostome, *Lampetra japonica*, induced by All-*trans* retinoic acid defines the head/trunk interface of the vertebrate body. *Dev Dyn* 211:35–51.

Kurokawa, D., Sakurai, Y., Inoue, A, Nakayama, R., *et al.* 2006. Evolutionary constraint on *Otx2* neurectoderm enhancers-deep conservation from skate to mouse and unique divergence in teleost. *Proc Natl Acad Sci USA* 103:19350–19355.

Kutejova, E., Engist, B., Self, M., Oliver, G., et al. 2008. P. *Six2* functions redundantly immediately downstream of *Hoxa2*. *Development* 135:1463–1470.

L

La Bonne, C., and Bronner-Fraser, M. 1998. Neural crest induction in *Xenopus*: Evidence for a two-signal model. *Development* 125:2403–2414.

La Bonne, C., and Bronner-Fraser, M. 2000. Snail-related transcriptional repressors are required in *Xenopus* for both the induction of the neural crest and its subsequent migration. *Dev Biol* 221:195–205.

Lacalli, T. C. 2006. Prospective protochordate homologs of vertebrate midbrain and MHB, with some thoughts on MHB origins. *Int J Biol Sci* 2:104–109.

Lacosta, A. M., Canudas, J., González, C., Muniesa, P., *et al.* 2007. Pax7 identifies neural crest, chromatophores lineages and pigment stem cells during zebrafish development. *Int J Dev Biol* 51:327–331.

Lahav, R., Ziller, C., Dupin, E., and Le Douarin, N. M. 1996. Endothelin 3 promotes neural crest cell proliferation and mediates a vast increase in melanocyte number in culture. *Proc Natl Acad Sci USA* 93:3892–3897.

Lallier, T. E., and De Simone, D. W. 2000. Separation of neural induction and neurulation in *Xenopus*. *Dev Biol* 225:135–150.

Lamb, T. M., Knecht, A. K., Smith, W. C., Stachel, S. E., *et al.* 1993. Neural induction by the secreted polypeptide Noggin. *Science* 262:713–718.

Lamborghini, J. E. 1980. Rohon-Beard cells and other large neurons in *Xenopus* originate during gastrulation. *J Comp Neurol* 189:323–333.

Lamers, C. H. J., Rombout, J. W. H. M., and Timmermans, L. P. M. 1981. An experimental study on neural crest migration in *Barbus conchonius* (Cyprinidae, Teleostei) with special reference to the origin of the enteroendocrine cells. *J Embryol Exp Morph* 62:309–323.

Landacre, F. L. 1921. The fate of the neural crest in the head of the Urodeles. *J Comp Neurol* 33:1–43.

Landis, S. C., and Patterson, P. H. 1981. Neural crest cell lineages. *Trends Neurosci* 4:172–174.

Landman, K. A., Simpson, M. J., and Newgreen, D. F. 2007. Mathematical and experimental insights into the development of the enteric nervous system and Hirschsprung's disease. *Dev Growth Differ* 49:277–286.

Langeland, J., Tomsa, J. M., Jackman, W. R. Jr., and Kimmel, C. B. 1998. An amphioxus snail gene: Expression in paraxial mesoderm and neural plate suggests a conserved role in patterning the chordate embryo. *Dev Genes Evol* 208:569–577.

Langille, R. M., and Hall, B. K. 1987. Development of the head skeleton of the Japanese medaka, *Oryzias latipes* (Teleostei). *J Morph* 193:135–158.

Langille, R. M., and Hall, B. K. 1988a. Role of the neural crest in development of the cartilaginous cranial and visceral skeleton of the medaka, *Oryzias latipes* (Teleostei). *Anat Embryol* 177: 297–305.

Langille, R. M., and Hall, B. K. 1988b. Role of the neural crest in development of the trabeculae and branchial arches in embryonic sea lamprey, *Petromyzon marinus* (L). *Development* 102:301–310.

Langille, R. M., and Hall, B. K. 1989. Developmental processes, developmental sequences and early vertebrate phylogeny. *Biol Rev Camb Philo Soc* 64:73–91.

Langille, R. M., and Hall, B. K. 1993. Pattern formation and the neural crest. In *The Skull, Volume 1, Development*, eds J. Hanken and B. K. Hall, pp. 77–111. Chicago: The University of Chicago Press.

Lankester, E. R. 1873. On the primitive cell-layers of the embryo as the basis of genealogical classification of animals, and on the origin of vascular and lymph systems. *Ann Mag nat Hist Series* 4:11:321–338.

Lankester, E. R. 1877. Notes on the embryology and classification of the animal kingdom: Comprising a revision of speculations relative to the origin and significance of the germ layers. *Quart J Microsc Sci* 17:399–454.

Lassiter, R. N. T., Dude, C. M., Reynolds, S. B., Winters, N. I., *et al.* 2007. Canonical Wnt signaling is required for ophthalmic trigeminal placode cell fate determination and maintenance. *Dev Biol* 308:392–406.

Laudel, T. P., and Lim, T.-M. 1993. Development of the dorsal root ganglion in a teleost, *Oreochromis mossambicus* (Peters). *J Comp Neurol* 327:141–150.

Lawson, K. A., Meneses, J. J., and Pedersen, R. A. 1991. Clonal analysis of epiblast fate during germ layer formation in the mouse embryo. *Development* 113:891–911.

Ledent, V. 2002. Postembryonic development of the posterior lateral line in zebrafish. *Development* 129:597–604.

Le Douarin, N. M. 1974. Cell recognition based on natural morphological nuclear markers. *Med Biol* 52:281–319.

Le Douarin, N. M. 1982. *The Neural Crest*. Cambridge: Cambridge University Press.

Le Douarin, N. M. 1986. Cell line segregation during peripheral nervous system ontogeny. *Science* 231:1515–1522.

Le Douarin, N. M. 1988. The Claude Bernard Lecture 1987. Embryonic chimeras: A tool for studying the development of the nervous and immune systems. *Proc R Soc Lond* B 235: 1–17.

Le Douarin, N. M., Brito, J. M., and Creuzet, S. 2007. Role of the neural crest in face and brain development. *Brain Res Revs* 55:237–247.

Le Douarin, N. M., Dupin, E., Baroffio, A., and Dulac, C. 1992. New insights into the development of neural crest derivatives. *Int Rev Cytol* 138:269–314.

Le Douarin, N. M., Fontaine-Perus, J., and Couly, G. 1986. Cephalic ectodermal placodes and neurogenesis. *Trends Neurosci* 9:175–180.

Le Douarin, N. M., Grapin-Botton, A., and Catala, M. 1996. Patterning of the neural primordium in the avian embryo. *Sem Cell Dev Biol* 1:157–167.

Le Douarin, N. M., and Kalcheim, C. 1999. *The Neural Crest*. Second Edition. Cambridge: Cambridge University Press.

Le Douarin, N. M., Ziller, C., and Couly, G. F. 1993. Patterning of neural crest derivatives in the avian embryo: *In vivo* and *in vitro* studies. *Dev Biol* 159:24–49.

Le Lièvre, C. 1971a. Recherches sur l'origine embryologique des arcs viscéraux chez l'embryon d'Oiseau par la méthode des greffes interspécifiques entre Caille et Poulet. *C r Séanc Soc Biol* 165:395–400.

Le Lièvre, C. 1971b. Recherche sur l'origine embryologique du squelette viscéral chez l'embryon d'Oiseau. *C r Ass Anat* 152:575–583.

Le Lièvre, C. 1974. Rôle des cellules mésectodermiques issues des crêtes neurales céphaliques dans la formation des arcs branchiaux et du squelette viscéral. *J Embryol Exp Morphol* 31: 453–477.

Le Lièvre, C. 1976. Contribution des crêtes neurales à la genèse des structures céphaliques et cervicales chez les Oiseaux. Thèse d'Etat, Nantes, France.

Le Lièvre, C. 1978. Participation of neural crest-derived cells in the genesis of the skull in birds. *J Embryol Exp Morphol* 47:17–37.

Le Lièvre, C., and Le Douarin, N. M. 1974. Origine ectodermique du derme de la face et du cou, montrée par des combinaisons interspécifiques chez l'embryon d'Oiseau. *C r hebd Séanc Acad Sci Paris* 278:517–520.

Le Lièvre, C., and Le Douarin, N. M. 1975. Mesenchymal derivatives of the neural crest: Analysis of chimaeric quail and chick embryos. *J Embryol Exp Morphol* 34:125–154.

Lee, H.-O., Levorse, J. M., and Shin, M. K. 2003. The endothelin receptor-B is required for the migration of neural crest-derived melanocyte and enteric neuron precursors. *Dev Biol* 259: 162–175.

Lee, S. H., Bedard, O., Buchtova, M., Fu, K., and Richman, J. M. 2004. A new origin for the maxillary jaw. *Dev Biol* 276:207–224.

Lee, Y. M., Osumi-Yamashita, N., Ninomiya, Y., Moon, C. K., *et al.* 1995. Retinoic acid stage-dependently alters the migration pattern and identity of hindbrain neural crest cells. *Development* 121:825–837.

Li, W., and Cornell, R. A. 2007. Redundant activities of *Tfap2a* and *Tfap2c* are required for neural crest induction and development of other non-neural ectoderm derivatives in zebrafish embryos. *Dev Biol* 304:338–354.

Liao, B.-Y., and Zhang, J. 2008. Null mutations in human and mouse orthologs frequently result in different phenotypes. *Proc Natl Acad Sci USA* 105:6987–6992.

Liedtke, D., and Winkler, C. 2008. Midkine-b regulates cell specification at the neural plate border in zebrafish. *Dev Dyn* 237:62–74.

Liem, K. F., Bemis, W. E., Walker, W. F. Jr., and Grande, L. 2001. *Functional Anatomy of the Vertebrates. An Evolutionary Perspective.* Third Edition. Fort Worth TX.: Harcourt College Publishers.

Liem, K. F., Jr., Tremml, G., Roelink, H., and Jessell, T. M. 1995. Dorsal differentiation of neural plate cells induced by BMP-mediated signals from epidermal ectoderm. *Cell* 82:969–979.

Light, W., Vernon, A. E., Larorella, A., Lavarone, A., and LaBonne, C. 2005. *Xenopus Id3* is required downstream of Myc for the formation of multipotent neural crest progenitor cells. *Development* 132:1831–1841.

Lincoln, J., Kist, R., Scherer, G., and Yutzey, K. E. 2007. Sox9 is required for precursor cell expansion and extracellular matrix organization during mouse heart valve development. *Dev Biol* 305:120–132.

Lister, J. A., Close, J., and Raible, D. W. 2001. Duplicate mitf genes in zebrafish: Complementary expression and conservation of melanogenic potential. *Dev Biol* 237:333–344.

Lister, J. A., Cooper, C., Nguyen, K., Modrell, M., *et al.* 2006. Zebrafish *Foxd3* is required for development of a subset of neural crest derivatives. *Dev Biol* 290:92–104.

Litsiou, A., Hanson, S., and Streit, A. 2005. A balance of FGF, BMP and WNT signalling positions the future placode territory in the head. *Development* 132:4051–4062.

Liu, D., Chu, H., Maves, L., Yan, Y.-L., *et al.* 2003. Fgf3 and Fgf8 dependent and independent transcription factors are required for otic placode specification. *Development* 130:2213–2224.

Lo, L.-C., Birren, S. J., and Anderson, D. J. 1991a. V-myc immortalization of early rat neural crest cells yields a clonal cell line which generates both glial and adrenergic progenitor cells. *Dev Biol* 145:139–153.

Lo, L.-C., Johnson, J. E., Wuenschell, C. W., Saito, T., and Anderson, D. J. 1991b. Mammalian *achaete-scute* homolog 1 is transiently expressed by spatially restricted subsets of early neu-roepithelial and neural crest cells. *Genes Dev* 5:1524–1537.

Loffeld, A., McLellan, N. J., Cole, T., Payne, S. J., *et al.* 2006. Epidermal naevus in Proteus syndrome showing loss of heterozygosity for an inherited PTEN mutation. *Brit J Dermatol* 154:1194–1198.

Logan, C. Y., and Nusse, R. 2004. The Wnt signaling pathway in development and disease. *Annu Rev Cell Dev Biol* 20:781–810.

Lopashov, G. V. 1944. Origins of pigment cells and visceral cartilage in teleosts. *C r Acad Sci USSR* 44:169–172.

López, D., Durán, A. C., de Andrés, A. V., Guerrero, A., *et al.* 2003. Formation of cartilage in the heart of the Spanish terrapin, *Mauremys leprosa* (Reptilia, Chelonia). *J Morphol* 258:97–105.

López, D., Durán, A. C., and Sans-Coma, A. C. 2000. Formation of cartilage in cardiac semilunar valves of chick and quail. *Ann Anat-Anat Anz* 182:349–359.

Loring, J. F., and Erickson, C. A. 1987. Neural crest cell migratory pathways in the trunk of the chick embryo. *Dev Biol* 121:220–236.

Loucks, E. J., Schwend, T., and Ahlgren, S. C. 2007. Molecular changes associated with teratogen-induced cyclopia. *Birth Defects Res (Part A)* 79:642–651.

Lufkin, T., Mark, M., Hart, C. P., Dollé, P., *et al.* 1992. Homeotic transformation of the occipital bones of the skull by ectopic expression of a homeobox gene. *Nature* 359:839–841.

Luider, T. M., Peters-van der Sanden, M. J. H., Molemaar, J. C., Tibboel, D., *et al.* 1992. Charac-terization of HNK-1 antigens during the formation of the avian enteric nervous system. *Devel-opment* 115:561–572.

Lumsden, A. 1988. Spatial organization of the epithelium and the role of neural crest cells in the initiation of the mammalian tooth germ. *Development* 102(Suppl.):155–169.

Lumsden, A., and Graham, A. 1996. Death in the neural crest: Implications for pattern formation. *Sem Cell Dev Biol* 1:169–174.

Lumsden, A., and Krumlauf, R. 1996. Patterning the vertebrate neuraxis. *Science* 274:1109–1115.

Lumsden, A., Sprawson, N., and Graham, A. 1991. Segmental origin and migration of neural crest cells in the hindbrain region of the chick embryo. *Development* 113:1281–1291.

Luo, R., An, M., Arduini, B. L., and Henion, P. D. 2001. Specific pan-neural crest expression of zebrafish *Crestin* throughout embryonic development. *Dev Dyn* 220:169–174.

Luo, T., Lee, Y. H., Saint-Jeannet, J. P., and Sargent, T. D. 2003. Induction of neural crest in *Xenopus* by transcription factor AP2alpha. *Proc Natl Acad Sci USA* 100:532–537.

Luo, T., Xu, Y., Hoffman, T. L., Zhang, T., *et al.* 2007. Inca: A novel p21-activated kinase-associated protein required for cranial neural crest development. *Development* 134:1379–1289.

Luo, Z.-X., Chen, P., Li, G., and Chen, M. 2007. A new eutriconodont mammal and evolutionary development in early mammals. *Nature* 446:288–293.

Lwigale, P. Y., Conrad, G. W., and Bronner-Fraser, M. 2004. Graded potential of neural crest to form cornea, sensory and cartilage along the rostrocaudal axis. *Development* 131:1979–1991.

M

MacDonald, M. E., and Hall, B. K. 2001. Altered timing of the extracellular-matrix-mediated epithelial-mesenchymal interaction that initiates mandibular skeletogenesis in three inbred strains of mice: Development, heterochrony, and evolutionary change in morphology. *J Exp Zool* 291: 258–273.

MacKenzie, A., Ferguson, M. W. J., and Sharpe, P. T. 1991. *Hox-7* expression during murine cran-iofacial development. *Development* 113:601–611.

Maclean, N., and Hall, B. K. 1987. *Cell Commitment and Differentiation*. Cambridge: Cambridge University Press.

Maden, M., Gale, E., Kostetskii, I., and Zile, M. 1996. Vitamin A-deficient quail embryos have half a hindbrain and other neural defects. *Curr Biol* 6:417–426.

Maderson, P. F. A. (ed) 1987. *Developmental and Evolutionary Aspects of the Neural Crest.* New York: John Wiley & Sons.

Mahmood, R., Flanders, K. C., and Morriss-Kay, G. M. 1992. Interactions between retinoids and TGFßs in mouse embryogenesis. *Development* 115:67–74.

Mahmood, R., Mason, I. J., and Morriss-Kay, G. M. 1996. Expression of *Fgf-3* in relation to hindbrain segmentation, otic pit position, and pharyngeal arch morphology in normal and retinoic-acid-exposed mouse embryos. *Anat Embryol* 194:13–22.

Maisey, J. G. 1986. Heads and tails: A chordate phylogeny. *Cladistics* 2:201–256.

Malicki, J., Schier, A. F., Solnicakrezel, L., Stemple, D. L., *et al.* 1996. Mutations affecting development of the zebrafish ear. *Development* 123 (Suppl.):275–283.

Mallatt, J. 1984. Early vertebrate evolution: Pharyngeal structure and the origin of gnathostomes. *J Zool* 204:169–183.

Mallatt, J. 1996. Ventilation and the origin of jawed vertebrates: A new mouth. *Zool J Linn Soc Lond* 117:329–404.

Mallatt, J., and Chen, J.-Y. 2003. Fossil sister group of craniates: Predicted and found. *J Morphol* 258:1–31.

Mallo, M. 1998. Embryological and genetic aspects of middle ear development. *Int J Dev Biol* 42:11–22.

Mallo, M., and Gridley, T. 1996. Development of the mammalian ear: Coordinate regulation of formation of the tympanic ring and the external acoustic meatus. *Development* 122:173–179.

Mancilla, A., and Mayor, R. 1996. Neural crest formation in *Xenopus laevis*: Mechanisms of *Xslug* induction. *Dev Biol* 177:580–589.

Manley, N. R., and Capecchi, M. R. 1995. The role of *Hoxa-3* in mouse thymus and thyroid development. *Development* 121:1989–2003.

Manley, N. R., and Capecchi, M. R. 1997. Hox group 3 paralogous genes act synergistically in the formation of somitic and neural crest-derived structures. *Dev Biol* 192:274–288.

Manley, N. R., and Capecchi, M. R. 1998. Hox group 3 paralogs regulate the development and migration of the thymus, thyroid, and parathyroid glands. *Dev Biol* 195:1–15.

Manni, L., Lane, N. J., Joly, J.-S., Gasparini, F., *et al.* 2004. Neurogenic and non-neurogenic placodes in ascidians. *J Exp Zool (Mol Dev Evol)* 302B:483–504.

Mansouri, A., Stoykova, A., Torres, M., and Gruss, P. 1996. Dysgenesis of cephalic neural crest derivatives in *Pax7⁻/⁻* mutant mice. *Development* 122:831–838.

Manzanares, M., Wada, H., Itaski, N., *et al.* 2001. Conservation and elaboration of *Hox* gene regulation during evolution of the vertebrate head. *Nature* 408:778–779, 781.

Marchant, L., Linker, C., Ruiz, P., Guerrero, N., and Mayor, R. 1998. The inductive properties of mesoderm suggest that the neural crest cells are specified by a BMP gradient. *Dev Biol* 198:319–329.

Marcucio, R. S., Cordero, D. R., Hu, D., and Helms, J. A. 2005. Molecular interactions coordinating the development of the forebrain and face. *Dev Biol* 284:48–61.

Mark, M., Lufkin, T., Vonesch, J.-L., Ruberte, E., *et al.* 1993. Two rhombomeres are altered in *Hoxa-1* mutant mice. *Development* 119:319–338.

Marshall, A. M. 1878. The development of the cranial nerves in the chick. *Quart J Microsc Sci* 18:10–40.

Marshall, A. M. 1879. The morphology of the vertebrate olfactory organ. *Quart J Microsc Sci* 19:300–340.

Marshall, H., Nonchev, S., Sham, M. H., Muchamore, I., *et al.* 1992. Retinoic acid alters hindbrain *Hox* code and induces transformation of rhombomeres 2/3 into a 4/5 identity. *Nature* 360: 737–741.

Martin, K., and Groves, A. K. 2006. Competence of cranial ectoderm to respond to Fgf signaling suggests a two-step model of otic placode induction. *Development* 133:877–887.

Martindale, M. Q., Pang, K., and Finnerty, J. R. 2004. Investigating the origins of triploblasty: 'Mesodermal' gene expression in a diploblastic animal, the sea anemone *Nematostella vectensis* (phylum, Cnidaria; class, Anthozoa). *Development* 131:2463–2474.

Martinez, S., and Alvarado-Mallart, R.-M. 1990. Expression of the homeobox *chick-en* gene in chick/quail chimeras with inverted mes-metencephalic grafts. *Dev Biol* 139:432–436.

Martinez-Morales, J.-R., Henrich, T., Ramialison, M., and Wittbrodt, J. 2007. New genes in the evolution of the neural crest differentiation program. *Genome Biol* 8:R36.1–R36.17.

Martucciello, G., Luinetti, O., Romano, P., and Magrini, U. 2007. Molekularbiologie, Grundlagenforschung und Diagnose des Morbus Hirschsprung. *Pathologe* 28:119–124.

Matt, N., Ghyselinck, N. B., Wendling, O., Chambon, P., and Mark, M. 2003. Retinoic acid-induced developmental defects are mediated by RARβ/RXR heterodimers in the pharyngeal endoderm. *Development* 130:2083–2093.

Matsumoto, J., Wada, K., and Akiyama, T. 1989. Neural crest cell differentiation and carcinogenesis: Capability of goldfish erythrophoroma cells for multiple differentiation and clonal polymorphism in their melanogenic variants. *J Invest Dermat* 92:255S–260S.

Matsuo, I., Kuratani, S., Kimura, C., Takeda, N., and Aizawa, S. 1995. Mouse *Otx2* functions in the formation and patterning of rostral head. *Genes Dev* 9:2646–2658.

Matsuoka, T., Ahlberg, P. E., Kessaris, N., Iannarelli, P., *et al.* 2005. Neural crest origins of the neck and shoulder. *Nature* 436:347–355.

Maxwell, G. D., and Forbes, M. E. 1991. Spectrum of *in vitro* differentiation of quail trunk neural crest cells isolated by cell sorting using the HNK-1 antibody and analysis of the adrenergic development of HNK-1+ sorted subpopulations. *J Neurobiol* 22:276–286.

Mayor, R. 2007. *Dev Biol* 309:208–221.

Mayor, R., Guerrero, N., and Martínez, C. 1997. Role of FGF and *Noggin* in neural crest induction. *Dev Biol* 189:1–12.

Mayor, R., Morgan, R., and Sargent, M. G. 1995. Induction of the prospective neural crest of *Xenopus*. *Development* 121:767–777.

Mazet, F., and Shimeld, S. M. 2005. Molecular evidence from ascidians for the evolutionary origin of vertebrate cranial sensory placodes. *J Exp Zool (Mol Dev Evol)* 304B:340–346.

McCauley, D. W., and Bronner-Fraser, M. 2003. Neural crest contributions to the lamprey head. *Development* 130:2317–2327.

McCauley, D. W., and Bronner-Fraser, M. 2006. Importance of SoxE in neural crest development and the evolution of the pharynx. *Nature* 441:750–752.

McGonnell, I. M., and Graham, A. 2002. Trunk neural crest has skeletogenic potential. *Curr Biol* 12:767–771.

McGonnell, I. M., McKay, I. J., and Graham, A. 2001. A population of caudally migrating cranial neural crest cells: Functional and evolutionary implications. *Dev Biol* 236:354–363.

McLennan, R., and Kulesa, P. M. 2007. In vivo analysis reveals a critical role for neuropilin- in cranial neural crest cell migration in chick. *Dev Biol* 301:227–239.

McLeod, M. J., Harris, M. J., Chernoff, G. F., and Miller, J. R. 1980. First arch malformation: A new craniofacial mutant in the mouse. *J Hered* 71:331–335.

McMahon, A. P., and Bradley, A. 1990. The *Wnt-1 (int-1)* proto-oncogene is required for development of a large region of the mouse brain. *Cell* 62:1073–1085.

Meier, S. 1978. Development of the embryonic chick otic placode. II. Electron microscopic analysis. *Anat Rec* 191:459–478.

Meier, S., and Packard, D. S. Jr. 1984. Morphogenesis of the cranial segments and distribution of neural crest in the embryo of the snapping turtle, *Chelydra serpentina*. *Dev Biol* 102:309–323.

Meinertzhagen, I. A., and Okamura, Y. 2001. The larval ascidian nervous system: The chordate brain from its small beginnings. *Trends Neurosci* 24:401–410.

Mellgren, E. M., and Johnson, S. L., 2002. The evolution of morphological complexity in zebrafish stripes. *Trends Genet* 18:128–134.

Menoud, P. A., Debrot, S., and Schowing, J. 1989. Mouse neural crest cells secrete both urokinase-type and tissue-type plasminogen activators *in vitro*. *Development* 106:685–690.

Menuet, A., Alunni, A., Joly, J.-S., Jeffery, W. R., and Rétaux, S. 2007. Expanded expression of Sonic Hedgehog in *Astyanax* cavefish: Multiple consequences on forebrain development and evolution. *Development* 134:845–855.

Metscher, B. D., Northcutt, R. G., Gardner, D. M., and Bryant, S. V. 1997. Homeobox genes in axolotl lateral line placodes and neuromasts. *Dev Genes Evol* 207:287–295.

Meulemans, D., and Bronner-Fraser, M. 2002. Amphioxus and lamprey Ap-2 genes: Implications for neural crest evolution and migration patterns. *Development* 129:4953–4962.

Meulemans, D., and Bronner-Fraser, M. 2004. Gene-regulatory interactions in neural crest evolution and development. *Dev Cell* 7:291–299.

Meulemans, D., and Bronner-Fraser, M. 2007. Insights from amphioxus into the evolution of vertebrate cartilage. *PLoS One* 2(8):e787.

Mikkelsen, T. S., Wakefield, M. J., Aken, B., Amemiya, C. T., *et al.* 2007. Genome of the marsupial *Monodelphis domestica* reveals innovation in non-coding sequences. *Nature* 447:167–177.

Miller, C. T., Beleza, S., Pollen, A. A., Schluter, D., *et al.* 2007. *cis*-Regulatory changes in *Kit Ligand* expression and parallel evolution of pigmentation in sticklebacks and humans. *Cell* 131:1179–1189.

Miller, C. T., Schilling, T. F., Lee, K.-H., Parker, J., and Kimmel, C. B. 2000. *Sucker* encodes a zebrafish endothelin-1 required for ventral pharyngeal arch development. *Development* 127:3815–3838.

Mills, M. G., Nuckles, R. J., and Parichy, D, M. 2007. Deconstructing evolution of adult phenotypes: Genetic analyses of *kit* reveal homology and evolutionary novelty during adult pigment pattern development of *Danio* fishes. *Development* 134:1081–1090.

Milos, N. C., Meadows, G., Evanson, J. E., Pinchbeck, J. B, *et al.* 1998. Expression of the endogenous galactoside-binding lectin of *Xenopus laevis* during cranial neural crest development: Lectin localization is similar to that of members of the N-CAM and cadherin families of cell adhesion molecules. *J Craniofac Genet Dev Biol* 18:11–29.

Mina, M., Gluhak, J., Upholt, W. B., Kollar, E. J., and Rogers, B. 1995. Experimental analysis of *Msx-1* and *Msx-2* gene expression during chick mandibular morphogenesis. *Dev Dyn* 202: 195–214.

Mitani, S., and Okamoto, H. 1991. Inductive differentiation of two neural lineages reconstituted in a microculture system from *Xenopus* early gastrula cells. *Development* 112:21–31.

Mitgutsch, C., Piekarski, N., Olsson, l., and Haas, A. 2008. Heterochronic shifts during early cranial neural crest cell migration in two ranid frogs. *Acta Zool (Stockh)* 89:69–78.

Mitsiadis, T. A., and Smith, M. M. 2006. How do genes make teeth to order through development. *J Exp Biol (Mol Dev Evol)* 306B:177–182.

Miya, T., Morita, K., Suzuki, A., Ueno, N., and Satoh, N. 1997. Functional analysis of an ascidian homologue of vertebrate BMP-2/BMP-4 suggests its role in the inhibition of neural fate specification. *Development* 124:5149–5159.

Miyake, T., Cameron, A. M., and Hall, B. K. 1996. Stage-specific onset of condensation and matrix deposition for Meckel's and other first arch cartilages in inbred C57BL/6 mice. *J Craniofac Genet Dev Biol* 16:32–47.

Miyake, T., Hunt von Herbing, I., and Hall, B. K. 1997. Neural ectoderm, neural crest, and placodes: Contribution of the otic placode to the ectodermal lining of the embryonic opercular cavity in Atlantic cod (Teleostei). *J Morphol* 231:231–252.

Miyake, T., McEachran, J. D., and Hall, B. K. 1992. Edgeworth's legacy of cranial muscle development with an analysis of the ventral gill arch muscles in batoid fishes (Batoidea: Chondrichthyes). *J Morphol* 212:213–256.

Miyake, T., Vaglia, J. L., Taylor, L. H., and Hall, B. K. 1999. Development of dermal denticles in skates (Chondrichthyes, Batoidea): Patterning and cellular differentiation. *J Morphol* 241:61–81.

Moiseiwitsch, J. R., and Lauder, J. M. 1995. Serotonin regulates mouse cranial neural crest migration. *Proc Natl Acad Sci USA* 92:7182–7186.

Monsoro-Burq, A.-H., Fletcher, R. B., and Harland, R. M. 2003. Neural crest induction by paraxial mesoderm in *Xenopus* embryos requires FGF signals. *Development* 130:3111–3124.

Montero-Balaguer, M., Lang, M. R., Sachdev, S. W., Knappmeyer, C., *et al.* 2006. The mother superior mutation ablates foxd3 activity in neural crest progenitor cells and depletes neural crest derivatives in zebrafish. *Dev Dyn* 235:3199–3212.

Moody, S. L. (ed) 1999. *Cell Lineage and Fate Determination*. New York: Academic Press

Morales, A. V., Barbas, J. A., and Nieto, M. A. 2005. How to become neural crest: From segregation to delamination. *Sem Cell Dev Biol* 16:655–662.

Morales, A. V., Perez-Alcala, S., and Barbas, J. A. 2007. Dynamic Sox5 protein expression during cranial ganglia development. *Dev Dyn* 236:2702–2707.

Mori-Akiyama, Y., Akiyama, H., Rowitch, D. H., and de Crombrugghe, B. 2003. *Sox9* is required for determination of the chondrogenic cell lineage in the cranial neural crest. *Proc Natl Acad Sci USA* 100:9360–9365.

Moro-Balbás, J. A., Gato, A., Alonso, M. I., and Barbosa, E. 1998. Local increase level of chondroitin sulfate induces changes in the rhombencephalic neural crest migration. *Int J Dev Biol* 42:207–216.

Morris-Wiman, J., and Brinkley, L. L. 1990. Changes in mesenchymal cell and hyaluronate distribution correlate with *in vivo* elevation of the mouse mesencephalic neural folds. *Anat Rec* 226:383–395.

Morrison, S. J., Perez, S. E., Qiao, Z., Verdi, J. M., *et al.* 2000. Transient notch activation initiates an irreversible switch from neurogenesis to gliogenesis by neural crest stem cells. *Cell* 101:499–510.

Morrison-Graham, K., Schatteman, G. C., Bork, T., Bowen-Pope, D. F., and Weston, J. A. 1992. A PDGF receptor mutation in the mouse (*Patch*) perturbs the development of a non-neural subset of neural crest-derived cells. *Development* 115:133–142.

Morrison-Graham, K., and Weston, J. A. 1993. Transient *Steel* factor dependence by neural crest-derived melanocyte precursors. *Dev Biol* 159:346–352.

Morriss-Kay, G. M. (ed) 1992. *Retinoids in Normal Development and Teratogenesis*. Oxford: Oxford University Press.

Morriss-Kay, G. M. 1996. Craniofacial defects in AP-2 null mutant mice. *BioEssays* 18:785–788.

Morriss-Kay, G. M., and Tan, S.-S. 1987. Mapping cranial neural crest cell migration pathways in mammalian embryos. *Trends Genet* 3:257–261.

Mosher, J. T., Yeager, K. J., Kruger, G. M., Joseph, N. M., *et al.* 2007. Intrinsic differences among spatially distinct neural crest stem cells in terms of migratory properties, fate determination, and ability to colonize the enteric nervous system. *Dev Biol* 303:1–15.

Mujtaba, T., Mayer-Proschel, M., and Rao, M. S. 1998. A common neural progenitor for the CNS and PNS. *Dev Biol* 200:1–15.

Müller, E., and Ingvar, S. 1921. Über den Ursprung des sympathicus vei den Amphibien. Uppsala *Läkareförenings Förhandlingar Ny Följd* 26:1–15.

Müller, E., and Ingvar, S. 1923. Über den Ursprung des Sympathicus beim Hunchen. *Arch Mikrosk Anat EntwMech* 99:650–671.

Müller, F., and O'Rahilly, R. 2004. The primitive streak, the caudal eminence and related structures in staged human embryos. *Cells Tissues Organs* 177:2–20.

Muneoka, K., Wanek, N., and Bryant, S. V. 1986. Mouse embryos develop normally *exo utero*. *J Exp Zool* 239:289–293.

Murakami, Y., Ogasawara, M., Sugahara, F., Hirano, S., *et al.* 2001. Identification and expression of the lamprey *Pax6* gene: Evolutionary origin of the segmented brain of vertebrates. *Development* 128:3521–3531.

Murphy, M., Reid, K., William, D. E., Lyman, S. D., and Bartlett, P. F. 1992. Steel factor is required for maintenance, but not differentiation, of melanocyte precursors in the neural crest. *Dev Biol* 153:396–401.

Murray, S. A., and Gridley, T. 2006. Snail family genes are required for left-right asymmetry determination, but not neural crest formation in mice. *Proc Natl Acad Sci USA* 103:10300–10304.

Myojin, M., Ueki, T., Sugahara, F., Murakami, Y., *et al.* 2001. Isolation of Dlx and Emx gene cognates in an agnathan species, Lampetra japonica, and their expression patterns during embryonic and larval development: Conserved and diversified regulatory patterns of homeobox genes in vertebrate head evolution. *J Exp Zool (Mol Dev Evol)* 291:68–84.

N

Nagatomo, K.-I., and Hashimoto, C. 2007. *Xenopus hairy2* functions in neural crest formation by maintaining cells in a mitotic and undifferentiated state. *Dev Dyn* 236:1475–1483.

Nair, S., Li, W., Cornell, R., and Schilling, T. F. 2006. Requirements for endothelin type-A receptors and endothelin-1 signaling in the facial ectoderm for the patterning of skeletogenic neural crest cells in zebrafish. *Development* 134:335–345.

Nakagawa, M., Thompson, R. P., Terracio, L., and Borg, T. K. 1993. Developmental anatomy of HNK-1 immunoreactivity in the embryonic rat heart: Co-distribution with early conductive tissue. *Anat Embryol* 187:445–460.

Nakagawa, S., and Takeichi, M. 1998. Neural crest emigration from the neural tube depends on regulated cadherin expression. *Development* 125:2963–2871.

Nakamura, H., and Ayer-Le Lièvre, C. 1982. Mesectodermal capabilities of the trunk neural crest of birds. *J Embryol Exp Morphol* 70:1–18.

Nakata, K., Koyabu, Y., Aruga, J., and Mikoshiba, K. 2000. A novel member of the *Xenopus* Zic family, *Zic5*, mediates neural crest development. *Mech Dev* 99:83–91.

Nakata, K., Nagai, T., Aruga, J., and Mikoshiba, K. 1997. *Xenopus Zic3*, a primary regulator both in neural and neural crest development. *Proc Natl Acad Sci USA* 94:11980–11985.

Narayanan, C. H., and Narayanan, Y. 1978. Determination of the embryonic origin of the mesencephalic nucleus of the trigeminal nerve in birds. *J Embryol Exp Morphol* 43:85–105.

Nechiporuk, A., Linbo, T., Poss, K. D., and Raible, D. W. 2007. Specification of epibranchial placodes in zebrafish. *Development* 134:611–623.

Neidert, A.-H., Virupannavar, V., Hooker, G. W., and Langeland, J. A. 2001. Lamprey *Dlx* genes and early vertebrate evolution. *Proc Natl Acad Sci USA* 98:1665–1670.

Nellemann, C., de Bellard, M. E., Barembaum, M., Laufer, E., and Bronner-Fraser, M. 2001. Excess lunatic fringe causes cranial neural crest over-proliferation. *Devel Biol* 235:121–130.

Neste, D., and Tobin, D. J. 2004. Hair cycle and hair pigmentation: Dynamic interactions and changes associated with aging. *Micron* 35:193–200.

Neuhass, S. C. F., Solnica-Krezel, L., Schier, A. F., Zwartkruis, F., *et al.* 1996. Mutations affecting craniofacial development in zebrafish. *Development* 123:357–367.

Newgreen, D. F. (ed) 1995. Epithelial-Mesenchymal transitions, Part I. *Acta Anat* 154:1–97.

Newgreen, D. F., and Minichiello, J. 1995. Control of epithelio-mesenchymal transformation. 1. Events in the onset of neural crest cell migration are separable and inducible by protein kinase inhibitors. *Dev Biol* 170:91–101.

Newth, D. R. 1950. Fate of the neural crest in lampreys. *Nature* 165:284.

Newth, D. R. 1956. On the neural crest of the lamprey embryo. *J Embryol Exp Morphol* 4:358–375.

Nichols, D. H. 1986. Formation and distribution of neural crest mesenchyme to the first pharyngeal arch region of the mouse embryo. *Am J Anat* 176:221–231.

Nichols, D. H. 1987. Ultrastructure of neural crest formation in the midbrain/rostral hindbrain and preotic hindbrain regions of the mouse embryo. *Am J Anat* 179:143–154.

Nie, X., Deng, C.-x., Wang, Q., and Jiao, K. 2008. Disruption of *Smad4* in neural crest cells leads to mid-gestation death with pharyngeal arch, craniofacial and cardiac defects. *Dev Biol* 316:417–430.

Nie, X., Luuko, K., and Kettunen, P. 2006. Bmp signaling in craniofacial development. *Int J Dev Biol* 50:511–521.

Niederreither, K., Vermot, J., Le Roux, I., Schuhbaur, B., *et al.* 2003. The regional pattern of retinoic acid synthesis by RALDH2 is essential for the development of posterior pharyngeal arches and the enteric nervous system. *Development* 130:2525–2534.

Nieto, M. A., Sargent, M. G., Wilkinson, D. G., and Cooke, J. 1994. Control of cell behavior during vertebrate development by *Slug*, a zinc finger gene. *Science* 164:835–839.

Nieto, M. A., Sechrist, J., Wilkinson, D. G., and Bronner-Fraser, M. 1995. Relationship between spatially restricted *Krox-20* gene expression in branchial neural crest and segmentation in the chick embryo hindbrain. *EMBO J* 14:1697–1710.

Nieuwkoop, P. D., and Albers, B. 1990. The role of competence in the cranio-caudal segregation of the central nervous system. *Dev Growth Differ* 32:23–31.

Nieuwkoop, P. D., Johnen, A. G., and Albers, B. 1985. *The Epigenetic Nature of Early Chordate Development*. Cambridge: Cambridge University Press.

Nishibatake, M., Kirby, M. L., and van Mierop, L. H. S. 1987. Pathogenesis of persistent truncus arteriosus and dextroposed aorta in the chick embryo after neural crest ablation. *Circulation* 75:255–264.

Nishihira, T., Kasai, M., Hayashi, Y., Kimura, M., *et al.* 1981. Experimental studies on differentiation of cells originated from human neural crest tumors *in vitro* and *in vivo*. *Cell Mol Biol* 27:181–196.

Niu, M. C. 1947. The axial organization of the neural crest, studied with particular reference to its pigmentary component. *J Exp Zool* 105:79–114.

Noden, D. M. 1983a. The role of the neural crest in patterning of avian cranial skeletal, connective and muscle tissues. *Dev Biol* 96:144–165.

Noden, D. M. 1983b. The embryonic origins of avian cephalic and cervical muscles and associated connective tissues. *Am J Anat* 168:257–726.

Noden, D. M. 1984a. The use of chimeras in analysis of craniofacial development. In *Chimeras in Developmental Biology*. eds N. M. Le Douarin and A. McLaren, pp. 241–280. Orlando: Academic Press.

Noden, D. M. 1984b. Craniofacial development: New views on old problems. *Anat Rec* 208:1–13.

Noden, D. M. 1987. Interactions between cephalic neural crest and mesodermal populations. In *Developmental and Evolutionary Aspects of the Neural Crest*, ed. P. F. A. Maderson, pp. 89–119. New York: John Wiley & Sons.

Nokhbatolfoghahai, M., and Downie, J. R. 2005. Larval cement glands of frogs: Comparative development and morphology. *J Morphol* 263:270–283.

Northcutt, R. G. 1992. The phylogeny of octavolateralis ontogenies: A reaffirmation of Garstang's phylogenetic hypothesis. In *The Evolutionary Biology of Hearing*, eds A. Popper, D. Webster and R. Fay, pp. 21–47. New York: Springer-Verlag.

Northcutt, R. G. 1995. The forebrain of gnathostomes—in search of a morphotype. *Brain Behav Evol* 46:275–318.

Northcutt, R. G. 1996. The origin of craniates—neural crest, neurogenic placodes, and homeobox genes. *Israel J Zool* 42:S273–S313.

Northcutt, R. G. 2004. Taste buds: development and evolution. *Brain Behav Evol* 64:198–206.

Northcutt, R. G., and Barlow, L. A. 1998. Amphibians provide new insights into taste-bud development. *Trends Neurosci* 21:38–43.

Northcutt, R. G., Brändle, K., and Fritzsch, B. 1995. Electroreceptors and mechanosensory lateral line organs arise from single placodes in axolotls. *Dev Biol* 168:358–373.

Northcutt, R. G., Catania, K. C., and Criley, B. B. 1994. Development of lateral line organs in the axolotl. *J Comp Neurol* 340:480–514.

Northcutt, R. G., and Gans, C. 1983. The genesis of neural crest and epidermal placodes: A reinterpretation of vertebrate origins. *Quart Rev Biol* 58:1–28.

Nozue, T., and Ono, S. 1989. Exposure of newborn mice to adenosine causes neural crest dysplasia and tumor formation. *Neurofibromatosis* 2:261–273.

Nübler-Jung, K., and Arendt, D. 1994. Is ventral in insects dorsal in vertebrates? A history of embryological arguments favouring axis inversion in chordate ancestors. *Wilhelm Roux' Archiv. Dev Biol* 203:357–366.

O

Ogasawara, M., Shigetani, Y., Hirano, S., Satoh, N., and Kuratani, S. 2000. *Pax1/Pax9*-related genes in an agnathan vertebrate, *Lampetra japonica*: Expression pattern of *LjPax9* implies sequential evolutionary events toward the gnathostome body plan. *Dev Biol* 223:399–410.

Ohyama, T., Mohamed, O. A., Taketo, M. M., Dufort, D., and Groves, A. K. 2006. Wnt signals mediate a fate decision between otic placode and epidermis. *Development* 133:865–875.

Olsson, L. 2000. The scientific publications of Sven Hörstadius — a bibliography. In *Regulatory Processes in Development*. ed L. Olsson and C.-O. Jacobson, pp. 11–18. London: Portland Press.

Olsson, L., Ericsson, R., and Cerny, R. 2005. Vertebrate head development: Segmentation, novelties, and homology. *Theory Biosci* 124:145–163.

Olsson, L., Falck, P., Lopez, K., Cobb, J., and Hanken, J. 2001. Cranial neural crest cells contribute to connective tissue in cranial muscles in the anuran amphibian, *Bombina orientalis*. *Dev Biol* 237:354–367.

Olsson, L., and Hanken, J. 1996. Cranial neural-crest migration and chondrogenic fate in the Oriental Fire-Bellied toad, *Bombina orientalis*: Defining the ancestral pattern of head development in anuran amphibians. *J Morph* 229:105–120.

Olsson, L., and Jacobson, C.-O. (eds) 2000. *Regulatory Processes in Development*. Wenner-Gren International Series, Volume 76. London: Portland Press.

Olsson, L., Moury, J. D., Carl, T. F., Håstad, O., and Hanken, J. 2002. Cranial neural crest-cell migration in the direct developing frog, *Eleutherodactylous coqui*: Molecular heterogeneity within and among migratory streams. *Zoology* 105:3–13.

Olsson, L., Svensson, K., and Perris, R. 1996. Effects of extracellular matrix molecules on subepidermal neural crest cell migration in wild type and white mutant axolotl embryos. *Pigment Cell Res.* 9:18–27.

Olsson, L., Stigson, M., Perris, R., Sorrell, J. M., and Löfberg, J. 1996. Distribution of keratan sulfate and chondroitin sulfate in wild type and white mutant axolotl embryos during neural crest cell-migration. *Pigment Cell Res* 9:5–17.

O'Neill, P., McCole, R. B., and Baker, C. V. H. 2007. A molecular analysis of neurogenic placode and cranial sensory ganglion development in the shark, *Scyliorhinus canicula*. *Dev Biol* 304:156–181.

Opitz, J. M., and Clark, E. B. 2000. Heart development: An introduction. *Am J Med Gen* 97:238–247.

Opitz, J. M., and Gorlin, R. J. (eds) 1988. *Neural Crest and Craniofacial Disorders: Genetic Aspects*. New York: Alan R. Liss, Inc.

Oppedal, B. R., Brandtzaeg, P., and Kemshead, J. T. 1987. Immunohistochemical performance testing of monoclonal antibodies to neuroblastoma cells on normal adrenals, spinal and sympathetic ganglia, and neural crest tumours. *Histopathology* 11:351–362.

Oppenheimer, J. M. 1940. The non-specificity of the germ layers. *Quart Rev Biol* 15:1–27.

O'Rahilly, R., and Gardner, E. 1979. The initial development of the human brain. *Acta Anat* 104:123–133.

O'Rahilly, R., and Müller, F. 2006. *The Embryonic Human Brain: An Atlas of Developmental Stages*. Third edition. New York: Wiley-Liss.

O'Rahilly, R., and Müller, F. 2007. The development of the neural crest in the human. *J Anat* 211:335–351.

Orr, H. 1887. Contribution to the morphology of the lizard. *J Morph* 1:311–372.

O'Shea, K. S., and Dixit, V. M. 1988. Unique distribution of the extracellular matrix component thrombospondin in the developing mouse embryo. *J Cell Biol* 107:2723–2748.

Osumi-Yamashita, N., and Eto, K. 1990. Mammalian cranial neural crest cells and facial development. *Dev Growth Differ* 32:451–460.

Osumi-Yamashita, N., Ninomiya, Y., Doi, H., and Eto, K. 1994. The contributions of both forebrain and midbrain crest cells to the mesenchyme in the frontonasal mass of mouse embryos. *Dev Biol* 164:409–419.

Osumi-Yamashita et al. 1996. Rhombomere formation and hindbrain crest cell migration from prorhombomeric origins in mouse embryos. *Dev Growth Differ* 38:107–119.

Ota, K. G., Kuraku, S., and Kuratani, S. 2007. Hagfish embryology with reference to the evolution of the neural crest. *Nature* 446:672–675.

Ota, K. G., and Kuratani, S. 2006. The history of scientific endeavors towards understanding hagfish embryology. *Zool Sci* 23:403–418.

Ota, K. G., and Kuratani, S. 2007. Cyclostome embryology and early evolutionary history of vertebrates. *Int Comp Biol* 47:329–337.

P

Pagon, R. A., Graham, J. M. Jr., Zonana, J., and Yong, S.-L. 1981. Coloboma, congenital heart disease, and choanal atresia with multiple anomalies: CHARGE association. *J Pediat* 99: 223–227.

Pander, C. 1817. *Dissertatio inauguralis, sistens historiam metamorphoseos quam ovum incubatum prioribus quinque diebus subit.* Würzburg.

Papan, C., and Campos-Ortega, J. A. 1994. On the formation of the neural keel and neural tube in the zebrafish, *Brachydanio rerio. W Roux's Arch Dev Biol* 203:178–186.

Paratore, C., Goerich, D. E., Suter, U., Wegner, M., and Sommer, L. 2001. Survival and glial fate acquisition of neural crest cells are regulated by an interplay between the transcription factor Sox10 and extrinsic combinatorial signaling. *Development* 128:3949–3961.

Parichy, D. M. 1996. Pigment patterns of larval salamanders (Ambytomatidae, Salamandridae): The role of the lateral line sensory system and the evolution of pattern-forming mechanisms. *Dev Biol* 175:265–282.

Parichy, D. M. 1998. Experimental analysis of character coupling across a complex life cycle: Pigment pattern metamorphosis in the tiger salamander, *Ambystoma tigrinum tigrinum. J Morph* 237:53–67.

Parichy, D. M. 2007. Homology and the evolution of novelty during Danio adult pigment pattern development. *J Exp Zool (Mol Dev Evol)* 306B:578–590.

Pascualcastroviejo, I. 1990. Tumors of the neural crest. In *Spinal Tumors in Children and Adolescents*, ed I. Pascualcastroviejo, pp. 111–128. New York: Raven Press.

Pasini, A., Amiel, A., Rothbacher, U., Roure, A., *et al.* 2006. Formation of the ascidian epidermal sensory neurons: Insights into the origin of the chordate peripheral nervous system. *PloS Biol* 4:e225.

Pasquale, E. B. 2008. Eph–Ephrin bidirectional signaling in physiology and disease. *Cell* 133:38–52.

Passarge, E, 2002. Dissecting Hirschsprung disease. *Nature Genet* 31:11–12.

Pasterkamp, R. J., and Kolodkin, A. L. 2003. Semaphorin junction: Making tracks toward neural connectivity. *Curr Opin Neurobiol* 13:79–89.

Patterson, P. H. 1990. Control of cell fate in a vertebrate neurogenic lineage. *Cell* 62:1035–1038.

Pearse, A. G. E. 1977. The APUD concept and its implications: Related endocrine peptides in brain, intestine, pituitary, placenta and anuran cutaneous glands. *Med Biol* 55:115–125.

Perris, R., Krotowski, D., Lallier, T., Domingo, C., *et al.* 1991. Spatial and temporal changes in the distribution of proteoglycans during avian neural crest development. *Development* 111: 583–599.

Persaud, T. V. N., Chudley, A. E., and Skalko, R. G. 1985. *Basic Concepts in Teratology.* New York: Alan R. Liss, Inc.

Peters-van der Sanden, M. J. H., Luider, T. M., van der Kamp, A. W. M., Tibboel, D., and Meijers, C. 1993. Regional differences between various axial segments of the avian neural crest regarding the formation of enteric ganglia. *Differentiation* 53:17–24.

Peterson, K. J., and Davidson, E. H. 2000. Regulatory evolution and the origin of the bilaterians. *Proc Natl Acad Sci USA* 97:4430–4433.

Peterson, P. E., Blankenship, T. H., Wilson, D. B., and Hendrickx, A. G. 1996. Analysis of hindbrain neural crest migration in the long-tailed monkey (*Macaca fascicularis*). *Anat Embryol* 194:235–246.

Petrosino, G., Lalatta Costerbosa, G., Barazzoni, A. M., Grandis, A., *et al.* 2003. The mesencephalic trigeminal nucleus of the duck: Development and apoptosis. *Cells Tissues Organs* 175:165–174.

Pettway, Z., Guillory, G., and Bronner-Fraser, M. 1990. Absence of neural crest cells from the region surrounding implanted notochord *in situ. Dev Biol* 142:335–345.

Pierce, G. B. 1985. Carcinoma is to embryology as mutation is to genetics. *Amer Zool* 25:707–712.

Pinco, O., Carmeli, C., Rosenthal, A., and Kalcheim, C. 1993. Neurotrophin-3 affects proliferation and differentiation of distinct neural crest cells and is present in the early neural tube of avian embryos. *J Neurobiol* 24:1626–1641.

Piotrowski, T., and Nüsslein-Volhard, C. 2000. The endoderm plays an important role in patterning the segmental pharyngeal region in zebrafish (*Danio rerio*). *Dev Biol* 225:339–356.

Piotrowski, T., Schilling, T. F., Brand, M., Jiang, Y.-J., *et al.* 1996. Jaw and branchial arch mutants in zebrafish. II: Anterior arches and cartilage differentiation. *Development* 123:345–356.

Platt, J. B. 1893. Ectodermic origin of the cartilages of the head. *Anat Anz* 8:506–509.

Platt, J. B. 1894. Ontogenetic differentiation of the ectoderm in *Necturus.* Second preliminary note. *Arch Mikrosk Anat EntwMech* 43:911–966.

Platt, J. B. 1896. Ontogenetic differentiation of the ectoderm in *Necturus.* II. On the development of the peripheral nervous system. *Quart J Microsc Sci* 38:485–547.

Platt, J. B. 1897. The development of the cartilaginous skull and of the branchial and hypoglossal musculature in *Necturus. Morphol Jb* 25:377–464.

Poelmann, R. E., Gittenberger-de Groot, A. C., Mentink, M. M. T., Delpech, B., *et al.* 1990. The extracellular matrix during neural crest formation and migration in rat embryos. *Anat Embryol* 182:29–39.

Poelmann, R. E., Mikawa, T., and Gittenberger-de Groot, A. C. 1998. Neural crest cells in out-flow tract septation of the embryonic chicken heart: Differentiation and apoptosis. *Dev Dyn* 212:373–384.

Pomeranz, H. D., Rothman, T. P., Chalazonitis, A., Tennyson, V. M., and Gershon, M. D. 1993. Neural crest-derived cells isolated from the gut by immunoselection develop neuronal and glial phenotypes when cultured on laminin. *Dev Biol* 156:341–361.

Ponder, B. 1990. Neurofibromatosis gene cloned. *Nature* 346:703–704.

Porras, D., and Brown, C. B. 2008. Temporal–spatial ablation of neural crest in the mouse results in cardiovascular defects. *Dev Dyn* 237:153–162.

Potterf, S. B., Mollaaghababa, R., Hou, L., Southard-Smith, E. M., *et al.* 2001. Analysis of SOX10 function in neural crest-derived melanocyte development: SOX10-dependend transcriptional control of dopachrome tautomerase. *Dev Biol* 237:245–257.

Prince, V., and Lumsden, A. 1994. *Hoxa-2* expression in normal and transposed rhombomeres: Independent regulation in the neural tube and neural crest. *Development* 120:911–923.

Prince, V. E., Moens, C. B., Kimmel, C. B., and Ho, R. K. 1998. Zebrafish *hox* genes: Expression in the hindbrain region of wild-type and mutants of the segmentation gene, *valentino. Development* 125:393–406.

Protas, M. E., Hersey, C., Kochanek, D., Zhou, Y., *et al.* 2006. Genetic analysis of cavefish reveals molecular convergence in the evolution of albinism. *Nature Gen* 38:107–111.

Puffenberger, E. G., Hosoda, K., Washington, S. S., Nakao, K., *et al.* 1994. A mis-sense mutation of the endothelin-B receptor gene in multigenic Hirschsprung's disease. *Cell* 79:1257–1266.

Pummila, M., Fliniaux, I., Jaatinen, R., James, M. J., *et al.* 2007. Ectodysplasin has a dual role in ectodermal organogenesis: Inhibition of Bmp activity and induction of Shh expression. *Development* 134:117–125.

Putnam, N. H., Srivastava, M., Hellsten, U., Dirks, B., *et al.* 2007. Sea anemone genome reveals ancestral eumetazoan gene repertoires and genomic organization. *Science* 317:86–95.

Q

Qin, F., and Kirby, M. L. 1995. *Int-2* influences the development of the nodose ganglion. *Pediat Res* 38:485–492.

Qiu, M., Bulfone, A., Ghattas, I., Meneses, J. J., *et al.* 1997. Role of the Dlx homeobox genes in proximodistal patterning of the branchial arches: Mutations of *Dlx-1, Dlx-2,* and *Dlx-1* and *-2* alter morphogenesis of proximal skeletal and soft tissue structures derived from the first and second arches. *Dev Biol* 185:165–284.

Quinlan, G. A., Williams, E. A., Tan, S.-S., and Tam, P. P. L. 1995. Neurectodermal fate of epiblast cells in the distal region of the mouse egg cylinder: Implications for body plan organization during early embryogenesis. *Development* 121:87–98.

R

Raible, D. W., and Eisen, J. S. 1994. Restriction of neural crest cell fate in the trunk of the embryonic zebrafish. *Development* 120:495–503.

Raible, D. W., and Eisen, J. S. 1996. Regulative interactions in zebrafish neural crest. *Development* 122:501–507.

Raible, D. W., and Ragland, J. W. 2005. Reiterated Wnt and Bmp signals in neural crest development. *Sem Cell Dev Biol* 16:673–682.

Rao, M. S., and Jacobson, M. (eds) 2005. *Developmental Neurobiology*. Fourth edition. New York: Kluwer Academic/Plenum Publishers.

Rau, M. J., Fischer, S., and Neumann, C. J. 2006. Zebrafish Trap230/Med12 is required as a coactivators for Sox9-dependent neural crest, cartilage and ear development. *Dev Biol* 296:83–93.

Raven, C. P. 1931. Zur Entwicklung der Ganglienleiste. I. Die Kinematik der Ganglienleisten Entwicklung bei den Urodelen. *Wilhelm Roux Arch EntwMech Org* 125:210–293.

Raven, C. P. 1936. Zur Entwicklung der Ganglienleiste. V. Über die Differenzierung des Rumpfganglienleistenmaterials. *Wilhelm Roux Arch EntwMech Org* 134:122–145.

Raven, C. P., and Kloos, J. 1945. Induction by medial and lateral pieces of the archenteron roof with special reference to the determination of the neural crest. *Acta néerl Morphol* 5:348–362.

Rawles, M. E. 1948. Origin of melanophores and their role in development of color patterns in vertebrates. *Physiol Zoöl* 28:383–408.

Rawls, J. F., and Johnson, S. L. 2001. Requirements for the *kit* receptor tyrosine kinase during regeneration of zebrafish fin mesenchyme. *Development* 128:1943–1949.

Redline, R. W., Neish, A., Holmes, L. B., and Collins, T. 1992. Homeobox genes and congenital malformations. *Lab Investig* 66:659–670.

Reedy, M. V., Faraco, C. D., and Erickson, C. A. 1998. The delayed entry of thoracic neural crest cells into the dorsolateral path is a consequence of the late emigration of melanogenic neural crest cells from the neural tube. *Dev Biol* 200:234–246.

Reeves, R. H., Baxter, L. L., and Richtsmeier, J. T. 2001. Too much of a good thing: Mechanisms of gene action in Down syndrome. *Trends Genet* 17:83–88.

Reid, R. G. B. 2007. *Biological Emergences: Evolution by Natural Experiment*. A Bradford Book. Cambridge MA: The MIT Press.

Reijntjes, S., Rodaway, A., and Maden, M. 2007. The retinoic acid metabolizing gene, CYP26B1, patterns the cartilaginous cranial neural crest in zebrafish. *Int J Dev Biol* 51:351–360.

Reissmann, E., Ernsberger, U., Francis-West, P. H., Rueger, D., *et al.* 1996. Involvement of bone morphogenetic protein-4 and bone morphogenetic protein-7 in the differentiation of the adrenergic phenotype in sympathetic neurons. *Development* 122:2079–2088.

Remak, R. 1850–1855. *Untersuchungen über die Entwickelung der Wirbelthiere*. Berlin.

Rhinn, M., Dierich, A., Shawlot, W., Behringer, R. R., *et al.* 1998. Sequential roles for *Otx2* in visceral endoderm and neurectoderm for forebrain and midbrain induction and specification. *Development* 125:845–856.

Riccardi, V. M., and Eichner, J. E. 1986. *Neurofibromatosis: Phenotype, Natural History, and Pathogenesis*. Baltimore: Johns Hopkins University Press.

Riccardi, V. M., and Mulvihill, J. J. 1981. *Neurofibromatosis (von Recklinghausen Disease)*. Genetics, Cell Biology, and Biochemistry. Advances in Neurology. Volume 29. New York: Raven Press.

Rich, T. H., Hopson, J. A., Musser, A. M., Flannery, T. F., and Vickers-Rich, P. 2005. Independent origins of middle ear bones in monotremes and therians. *Science* 307:910–914.

Richardson, M. K., and Hornbruch, A. 1991. Quail neural crest cells cannot read positional values in the dorsal trunk feathers of the chicken embryo. *Dev Genes Evol* 199:397–401.

Richardson, M. K., Hornbruch, A., and Wolpert, L. 1989. Pigment pattern expression in the plumage of the quail embryo and the quail–chick chimaera. *Development* 107:805–818.

Richardson, M. K., and Sieber-Blum, M. 1993. Pluripotent neural crest cells in the developing skin of the quail embryo. *Dev Biol* 157:348–358.

Richarte, A. M., Mead, H. B., and Tallquist, M. D. 2007. Cooperation between the PDGF receptors in cardiac neural crest cell migration. *Dev Biol* 306:785–796.

Rijli, F. M., Mark, M., Lakkaraju, S., Dietrich, A., *et al.* 1993. A homeotic transformation is generated in the rostral branchial region of the head by disruption of *Hoxa-2*, which acts as a selector gene. *Cell* 75:1333–1349.

Rinon, A., Lazar, S., Marshall, H., Büchmann-Møller, S., *et al.* 2007. Cranial neural crest cells regulate head muscle patterning and differentiation during vertebrate embryogenesis. *Development* 134:3065–3075.

Riopelle, R. J., and Riccardi, V. M. 1987. Neuronal growth factors from tumors of Von Recklinghausen Neurofibromatosis. *Can J Neurol Sci* 14:141–144.

Robertson, K., and Mason, I. 1995. Expression of *ret* in the chicken embryo suggests roles in regionalization of the vagal neural tube and somites and in development of multiple neural crest and placodal lineages. *Mech Dev* 53:329–344.

Rodaway, A., and Patient, R. 2001. Mesendoderm: An ancient germ layer? *Cell* 105:169–172.

Rodríguez-Gallardo, L., Climent, V., Garcia-Martinez, V., Schoenwolf, G. C., and Alvarez, I. S. 1997. Targeted over-expression of FGF in chick embryos induces formation of ectopic neural cells. *Int J Dev Biol* 41:715–723.

Roest, P. A. M., van Ipersen, L., Vis, S., Wisse, L. J., *et al.* 2007. Exposure of neural crest cells to elevated glucose leads to congenital heart defects, an effect that can be prevented by N-acetylcysteine. *Birth Defects Res (Part A): Clin Mol Teratol* 79:231–235.

Rogers, S. L., Cutts, J. L., Gegick, P. J., McGuire, P. G., *et al.* 1994. Transforming growth factor-ß1 differentially regulates proliferation, morphology, and extracellular matrix expression by three neural crest-derived neuroblastoma cell lines. *Exp Cell Res* 211:252–262.

Rohon, V. 1884. Zur histogenese des Rückenmarks der Forelle. *Sitzungsberichte Akad Wien* 1884:39–57.

Rollhäuser-ter-Horst, J. 1980. Neural crest replaced by gastrula ectoderm in amphibia. Effect on neurulation, CNS, gills and limbs. *Anat Embryol* 160:203–212.

Romer, A. S. 1972. The vertebrate as a dual animal somatic and visceral. *Evol Biol* 6:121–156.

Roper, R. J., Baxter, L. L., Saran, N., Klinedinst, N., *et al.* 2006. Defective cerebellar response to mitogenic Hedgehog signaling in Down syndrome mice. *Proc Natl Acad Sci USA* 103:1452–1456.

Rosenquist, G. C. 1981. Epiblast origin and early migration of neural crest cells in the chick embryo. *Dev Biol* 87:201–211.

Rosenquist, T. H., Bennett, G. D., Brauer, P. R., Stewart, M. L., *et al.* 2007. Microarray analysis of homocysteine-responsive genes in cardiac neural crest cells *in vitro. Dev Dyn* 236:1044–1054.

Rossi, C. C., Hernandez-Lagunas, L., Zhang, C., Choi, I. F., *et al.* 2008. Rohon-Beard sensory neurons are induced by BMP4 expressing non-neural ectoderm in *Xenopus laevis. Dev Biol* 314:351–361.

Rothman, T. P., Le Douarin, N. M., Fontaine-Perus, J. C., and Gershon, M. D. 1993. Inhibition of migration of neural crest-derived cells by the abnormal mesenchyme of the presumptive bowel of *Ls/Ls* mice. *Dev Dyn* 196:217–233.

Rowe, A., and Brickell, P. M. 1995. Expression of the chicken retinoic X receptor-γ gene in migrating cranial neural crest cells. *Anat Embryol* 192:1–8.

Ruberte, E., Friederich, V., Morriss-Kay, G. M., and Chambon, P. 1992. Differential distribution patterns of CRABP-I and CRABP-II transcripts during mouse embryogenesis. *Development* 115:973–987.

Ruberte, E., Wood, H. B., and Morriss-Kay, G. M. 1997. Prorhombomeric subdivision of the mammalian embryonic hindbrain: Is it functionally meaningful? *Int J Dev Biol* 41:213–222.

Rychel, A. L., Smith, S. E., Shimamoto, H. T., and Swalla, B. J. 2006. Evolution and development of the chordates: Collagen and pharyngeal cartilage. *Mol Biol Evol* 23:541–549.

Rychel, A. L., and Swalla, B. J. 2007. Development and evolution of chordate cartilage. *J Exp Zool* (*Mol Dev Evol*) 308B:325–335.

S

Sadaghiani, B., Crawford, B. J., and Vielkind, J. R. 1994. Changes in the distribution of extracellular matrix components during neural crest development in *Xiphophorus* spp. embryos. *Can J Zool* 72:1340–1353.

Sadaghiani, B., and Thiébaud, C. H. 1987. Neural crest development in the *Xenopus laevis* embryo, studied by interspecific transplantation and scanning electron microscopy. *Dev Biol* 124:91–110.

Saint-Jeannet, J.-P. (ed) 2006. *Neural Crest Induction and Differentiation.* Advances in Experimental Medicine and Biology, Volume 589, 268 pp. New York: Springer.

Saldivar, J. R., Krull, C. E., Krumlauf, R., Ariza-McNaughton, L., and Bronner-Fraser, M. 1996. Rhombomere of origin determines autonomous versus environmentally regulated expression of *Hoxa3* in the avian embryo. *Development* 122:895–904.

Saldivar, J. R., Sechrist, J. W., Krull, C. E., Ruffins, S., and Bronner-Fraser, M. 1997. Dorsal hindbrain ablation results in rerouting of neural crest migration and changes in gene expression, but normal hyoid development. *Development* 124:2729–2739.

Sansom, I. J., Smith, M. M., and Smith, M. P. 1996. Scales of thelodont and shark-like fishes from the Ordovician of Colorado. *Nature* 379:628–630.

Sansom, I. J., Smith, M. P., Armstrong, H. A., and Smith, M. M. 1992. Presence of the earliest vertebrate hard tissues in conodonts. *Science* 256:1308–1311.

Sansom, I. J., Smith, M. P., and Smith, M. M. 1994. Dentine in conodonts. *Nature* 368:591.

Sansom, I. J., Smith, M. P., Smith, M. M., and Turner, P. 1997. *Astraspis*—The anatomy and histology of an Ordovician fish. *Palaeontology* 40:625–643.

Santiago, A., and Erickson, C. A. 2002. Ephrin-B ligands play a dual role in the control of neural crest cell migration. *Development* 129:3621–3632.

Sardet, C., Swalla, B. J., Satoh, N., Sasakura, Y., *et al.* 2008. Euro chordates: Ascidian community swims ahead. The 4th International Tunicate Meeting in Villefranche sur-Mer. *Dev Dyn* 237:1207–1213.

Sarnat, H. B., and Flores-Sarnat, L. 2004. Integrative classification of morphology and molecular genetics in central nervous system malformations. *Am J Med Gen* 126A:386–392.

Sasai, N., Mizuseki, K., and Sasai, Y. 2001. Requirement of *FoxD3*-class signaling for neural crest determination in *Xenopus*. *Development* 128:2525–2536.

Sasai, Y., and de Robertis, E. M. 1997. Ectodermal patterning in vertebrate embryos. *Dev Biol* 182:5–20.

Sato, M., and Yost, H. J. 2003. Cardiac neural crest contributes to cardiomyocytes in zebrafish. *Dev Biol* 257:127–139.

Satoh, N. 1994. *Developmental Biology of Ascidians.* Cambridge: Cambridge University Press.

Saudemont, A., Dray, N., Hudry, B., Le Gouar, M., *et al.* 2008. Complementary striped expression patterns of NK homeobox genes during segment formation in the annelid *Platynereis*. *Dev Biol* 317:430–434.

Sauka-Spengler, T., Meulemans, D., Jones, M., and Bronner-Fraser, M. 2007. Ancient evolutionary origin of the neural crest gene regulatory network. *Dev Cell* 13:405–420.

Savagner, P. 2001. Leaving the neighborhood: Molecular mechanisms involved during epithelial-mesenchymal transition. *BioEssays* 23:912–923.

Schaeffer, B. 1977. The dermal skeleton in fishes. In *Problems in Vertebrate Evolution*, eds S. M. Andrews, R. S. Miles and A. D. Walker, Linnean Society Symposium # 4, pp. 25–52. London: Academic Press.

Schilling, T. F. 1997. Genetic analysis of craniofacial development in the vertebrate embryo. *BioEssays* 19:459–468.

Schilling, T. F., and Kimmel, C. B. 1994. Segment and cell type lineage restrictions during pharyngeal arch development in the zebrafish embryo. *Development* 120:483–494.

Schilling, T. F., Piotrowski, T., Grandel, H., Brand, M., *et al.* 1996b. Jaw and branchial arch mutants in zebrafish 1: Branchial arches. *Development* 123:329–344.

Schilling, T. F., Prince, V., and Ingham, P. W. 2001. Plasticity in zebrafish *hox* expression in the hindbrain and cranial neural crest. *Dev Biol* 231:201–216.

Schilling, T. F., Walker, C., and Kimmel, C. B. 1996a. The *chinless* mutation and neural crest cell interactions in zebrafish jaw development. *Development* 122:1417–1426.

Schlosser, G. 2002a. Development and evolution of lateral line placodes in amphibians. I. Development. *Zoology* 105:119–146.

Schlosser, G. 2002b. Development and evolution of lateral line placodes in amphibians. II. Evolutionary diversification. *Zoology* 105:177–193.

Schlosser, G. 2002c. Modularity and the units of evolution. *Theory Biol* 121:1–80

Schlosser, G. 2006. Induction and specification of cranial placodes. *Dev Biol* 294:303–351.

Schlosser, G. 2007. How old genes make a new head: Redeployment of *Six* and *Eya* genes during the evolution of vertebrate cranial placodes. *Integ Comp Biol* 47:343–359.

Schlosser, G., Kintner, C., and Northcutt, R. G. 1999. Loss of ectodermal competence for lateral line placode formation in the direct developing frog *Eleutherodactylus coqui Dev Biol* 213: 354–369.

Schlosser, G., and Northcutt, R. G. 2000. Development of neurogenic placodes in *Xenopus laevis*. *J Comp Neurol* 418:121–146.

Schlosser, G., and Northcutt, R. G. 2001. Lateral line placodes are induced during neurulation in the axolotl. *Dev Biol* 234:55–71.

Schmidt, C., McGonnell, I. M., Allen, S., Otto, A., and Patel, K. 2007. Wnt6 controls amniote neural crest induction through the non-canonical signaling pathway. *Dev Dyn* 236:2502–2511.

Schmidt-Ullrich, R., and Paus, R. 2005. Molecular principles of hair follicle induction and morphogenesis. *BioEssays* 27:247–261.

Schmitz, X., Papan, C., and Campos-Ortega, J. A. 1993. Neurulation in the anterior trunk region of the zebrafish, *Brachydanio rerio*. *W Roux's Arch Dev Biol* 203:250–259.

Schoenwolf, G. C., and Alvarez, I. S. 1991. Specification of neuroepithelium and surface epithelium in avian transplantation chimeras. *Development* 112:713–722.

Schoenwolf, G. C., Chandler, N. B., and Smith, J. L. 1985. Analysis of the origins and early fates of neural crest cells in caudal regions of avian embryos. *Dev Biol* 110:467–479.

Schoenwolf, G. C., and Nichols, D. H. 1984. Histological and ultrastructural studies on the origin of caudal neural crest cells in mouse embryos. *J Comp Neurol* 222:496–505.

Schubert, W., Coskun V., Tahmina, A., Raoa, M. S., *et al.* 2000. Characterization and distribution of a new cell surface marker of neuronal precursors. *Dev Neurosci* 22:154–166.

Schumacher, A., and Magnuson, T. 1997. Murine *Polycomb-* and *trithorax*-group genes regulate homeotic pathways and beyond. *Trends Genet* 13:167–170.

Schwarz, W., Vieira, J. M., Howard, B., Eickholt, B. J., and Rhurberg, C. 2008. Neuropilin 1 and 2 control cranial gangliogenesis and axon guidance through neural crest cells. *Development* 135:1605–1613.

Sechrist, J., Nieto, M. A., Zamanian, R. T., and Bronner-Fraser, M. 1995. Regulative response of the cranial neural tube after neural fold ablation: Spatiotemporal nature of neural crest regeneration and up-regulation of *Slug*. *Development* 121:4103–4115.

Sedgwick, A. 1894. On the inadequacy of the cellular theory and on the early development of nerves, particularly of the third nerve and of the sympathetic in Elasmobranchii. *Quart J Microsc Sci* 37:87–101.

Sela-Donenfeld, D., and Kalcheim, C. 2000. Inhibition of noggin expression in the dorsal neural tube by somitogenesis: A mechanism for coordinating the timing of neural crest migration. *Development* 127:4845–4854.

Selleck, M. A. J., and Bronner-Fraser, M. 1995. Origins of the avian neural crest: The role of neural plate-epidermal interactions, *Development* 121:525–538.

Selleck, M. A. J., and Stern, C. D. 1991. Fate mapping and cell lineage analysis of Hensen's node in the chick embryo. *Development* 112:615–626.

Sellman, S. 1946. Some experiments on the determination of the larval tooth in *Amblystoma mexicanum*. *Odont Tidskr* 54:1–128.

Serbedzija, G. N., Bronner-Fraser, M., and Fraser, S. E. 1992. Vital dye analysis of cranial neural crest cell migration in the mouse embryo. *Development* 116:297–307.

Serbedzija, G. N., Burgan, S., Fraser, S. E., and Bronner-Fraser, M. 1991. Vital dye labelling demonstrates a sacral neural crest contribution to the enteric nervous system of chick and mouse embryos. *Development* 111:857–866.

Serbedzija, G. N., Chen, J.-N., and Fishman, M. C. 1998. Regulation of the heart field of zebrafish. *Development* 125:1095–1101.

Serbedzija, G. N., and McMahon, A. P. 1997. Analysis of neural crest cell migration in *Splotch* mice using a neural crest-specific LacZ reporter. *Dev Biol* 185:139–147.

Servetnick, M., and Grainger, R. M. 1991. Homeogenetic neural induction in *Xenopus*. *Dev Biol* 147:73–82.

Seufert, D. W., and Hall, B. K. 1990. Tissue interactions involving cranial neural crest in cartilage formation in *Xenopus laevis* (Daudin). *Cell Differ Dev* 32:153–166.

Seufert, D. W., Hanken, J., and Klymkowsky, M. W. 1994. Type II collagen distribution during cranial development in *Xenopus laevis*. *Anat Embryol* 189:81–89.

Shah, S. B., Skromme, I., Hume, C. R., Kessler, D., *et al.* 1997. Misexpression of chick Vg1 in the marginal zone induces primitive streak formation. *Development* 124:5127–5138.

Shankar, K. R., Chuong, C.-M., Jaskoll, T., and Melnick, M. 1994. Effect of *in ovo* retinoic acid exposure on forebrain neural crest: *In vitro* analysis reveals upregulation of N-CAM and loss of mesenchymal phenotype. *Dev Dyn* 200:89–102.

Sharman, A. C., and Holland, P. W. H. 1998. Estimation of *Hox* gene cluster number in lampreys. *Int J Dev Biol* 42:617–620.

Sherman, L., Stocker, K. M., Morrison, R., and Ciment, G. 1993. Basic fibroblast growth factor (bFGF) acts intracellularly to cause the transdifferentiation of avian neural crest-derived Schwann cell precursors into melanocytes. *Development* 118:1313–1326.

Shigetani, Y., Aizawa, S., and Kuratani, S. C. 1995. Overlapping origins of pharyngeal arch crest cells on the postotic hind-brain. *Dev Growth Differ* 37:733–746.

Shimamura, K., Hartigan, D. J., Martinez, S., Puelles, L., and Rubenstein, J. L. R. 1995. Longitudinal organization of the anterior neural plate and neural tube. *Development* 121:3923–3933.

Shimeld, S. M. 1999. The evolution of dorsoventral pattern formation in the chordate neural tube. *Amer Zool* 39:641–649.

Shimeld, S. M. 2008. Peter Holland, homeobox genes and the developmental basis of animal diversity. *Int J Dev Biol* 52:3–7.

Shimeld, S. M., and Holland, P. W. H. 2000. Vertebrate innovations. *Proc Natl Acad Sci USA* 97:4449–4452

Shimeld, S. M., McKay, I. J., and Sharpe, P. T. 1996. The murine homeobox gene *Msx-3* shows highly restricted expression in the developing neural tube. *Mech Dev* 55:201–210.

Shiota, K., Yamada, S., Komada, M., and Ishibashi, M. 2007. Embryogenesis of holoprosencephaly. *Am J Med Gen Part A* 143A:3079–3087.

Shoval, I., Ludwig, L., and Kalcheim, C. 2007. Antagonistic roles of full-length N-cadherin and its soluble BMP cleavage product in neural crest delamination. *Development* 134:491–501.

Shu, D.-G., Conway Morris, S., and Zhang, X.-L. 1996a. A *Pikaia*-like chordate from the Lower Cambrian of China. *Nature* 384:157–158.

Shu, D.-G., Zhang, X.-L., and Chen, L. 1996b. Reinterpretation of *Yunnanozoon* as the earliest known hemichordate. *Nature* 380:428–430.

Shuey, D. L., Sadler, T. W., and Lauder, J. M. 1992. Serotonin as a regulator of craniofacial morphogenesis—site specific malformations following exposure to serotonin uptake inhibitors. *Teratology* 46:367–378.

Sieber-Blum, M. 1990. Mechanisms of neural crest diversification. *Comments Dev Neurobiol* 4:225–249.

Sieber-Blum, M. 1991. Role of the neurotrophic factors BDNF and NGF in the commitment of pluripotent neural crest cells. *Neuron* 6:949–956.

Sieber-Blum, M., and Cohen, A. M. 1980. Clonal analysis of quail neural crest cells: They are pluripotent and differentiate *in vitro* in the absence of noncrest cells. *Dev Biol* 80:96–106.

Sieber-Blum, M., Kumar, S. R., and Riley, D. A. 1988. *In vitro* differentiation of quail neural crest cells into sensory-like neuroblasts. *Dev Brain Res* 39:69–83.

Sieber-Blum, M., and Zhang, J.-M. 1997. Growth factor action in neural crest cell diversification. *J Anat* 191:493–499.

Siman, C. M., Gittenberger-de Groot, A. C., Wisse, B., and Eriksson, U. J. 2000. Malformations in offspring of diabetic rats: Morphometric analysis of neural crest-derived organs and effects of maternal vitamin E treatment. *Teratology* 61:355–367.

Simeone, A., Acampora, D., Gulisano, M., Stornaiuolo, A., and Boncinelli, E. 1992. Nested expression domains of four homeobox genes in developing rostral brain. *Nature* 358:687–690.

Simpson, M. J., Zhang, D. C., Mariani, M., Landman, K. A., and Newgreen, D. F. 2007. Cell proliferation drives neural crest cell invasion of the intestine. *Dev Biol* 302:553–568.

Smith, A., Robinson, V., Patel, K., and Wilkinson, D. G. 1997. The EphA4 and EphB1 receptor tyrosine kinases and ephrin-B2 ligand regulate targeted migration of branchial neural crest cells. *Curr Biol* 7:561–570.

Smith, J. R., Vallier, L., Lupo, G., Alexander, M., *et al.* 2008. Inhibition of Activin/Nodal signaling promotes specification of human embryonic stem cells into neuroectoderm. *Dev Biol* 313:107–117.

Smith, K. K. 2006. Craniofacial development in marsupial mammals: Developmental origins of evolutionary change. *Dev Dyn* 235:1181–1193.

Smith, K. K., and Schneider, R. A. 1998. Have gene knockouts caused evolutionary reversals in the mammalian first arch? *BioEssays* 20:245–255.

Smith, M. M., and Hall, B. K. 1990. Developmental and evolutionary origins of vertebrate skeletogenic and odontogenic tissues. *Biol Rev Camb Philos Soc* 65:277–374.

Smith, M. M., and Hall, B. K. 1993. A developmental model for evolution of the vertebrate exoskeleton and teeth: The role of cranial and trunk neural crest. *Evol Biol* 27:387–448.

Smith, M. M., Hickman, A., Amanze, D., Lumsden, A., and Thorogood, P. 1994. 0neural crest origin of caudal fin mesenchyme in the zebrafish *Brachydanio rerio*. *Proc R Soc Lond* B 256:137–145.

Smith, S. C., Graveson, A. C., and Hall, B. K. 1994. Evidence for a developmental and evolutionary link between placodal ectoderm and neural crest. *J Exp Zool* 270:292–301.

Smith, S. C., Lannoo, M. J., and Armstrong, J. B. 1990. Development of the lateral-line system in the axolotl: Placode specification, guidance of migration, and the origin of polarity. *Anat Embryol* 182:171–180.

Smith Fernandez, A., Pieau, C., Repérant, J., Boncinelli, E., and Wassef, M. 1998. Expression of the *Emx-1* and *Dlx-1* homeobox genes define three molecularly distinct domains in the telencephalon of mouse, chick, turtle and frog embryos: Implications for the evolution of telencephalic subdivisions in amniotes. *Development* 125:2099–2111.

Smith-Thomas, L. C., Davis, J. P., and Epstein, M. L. 1986. The gut supports neurogenic differentiation of periocular mesenchyme, a chondrogenic neural crest-derived cell population. *Dev Biol* 115:293–300.

Smits-van Prooije, A. E., Vermeijs-Keers, C., Poelmann, R. E., Mentink, M. M. T., and Dubbeldam, J. A. 1988. The formation of mesoderm and mesectoderm in 5- to 40-somite rat embryos cultured *in vitro*, using WGA-Au as a marker. *Anat Embryol* 177:245–256.

Snarr, B. S., Wirrig, E. E., Phelps, A. L., Trusk, T. C., and Wessels, A. 2007. A spatiotemporal evaluation of the contribution of the dorsal mesenchymal protrusion to cardiac development. *Dev Dyn* 236:1287–1294.

Snow, M. H. L. 1981. Growth and its control in early mammalian development. *Brit Med Bull* 37:221–226.

Snow, M. H. L., and Tam, P. P. L. 1979. Is compensatory growth a complicating factor in mouse teratology? *Nature* 279:555–557.

Sobkowa, L., Epperlein, H.-H., Herklotza, S., Straubea, W. L., and Tanakaa, E. M. 2006. A germ line GFP transgenic axolotl and its use to track cell fate: Dual origin of the fin mesenchyme during development and the fate of blood cells during regeneration. *Dev Biol* 290:386–397.

Sohal, G. S., Ali, M. M., Ali, A. A., and Dai, D. 1999. Ventrally emigrating neural tube cells contribute to the formation of Meckel's and quadrate cartilage. *Dev Dyn* 216:37–44.

Sohal, G. S., Bockman, D. E., Ali, M. M., and Tsai, N. T. 1996. DiI labeling and homeobox gene Islet-1 expression reveal the contribution of ventral neural tube cells to the formation of the avian trigeminal ganglion. *Int J Dev Neurosci* 14:419–427.

Solomon, K. S., and Fritz, A. 2002. Concerted action of two *dlx* paralogs in sensory placode formation. *Development* 129:3127–3136.

Solomon, K. S., Kudon, T., Dawid, I. B., and Fritz, A. 2003. Zebrafish *foxi1* mediates otic placode formation and jaw development. *Development* 130:929–940.

Song, Q., Mehler, M. F., and Kessler, J. A. 1998. Bone morphogenetic proteins induce apoptosis and growth factor dependence of cultured sympathoadrenal progenitor cells. *Dev Biol* 196:119–127.

Soriano, P. 1997. The PDGFα receptor is required for neural crest cell development and for normal patterning of the somit. *Development* 124:2691–2700.

Spence, S. G., and Poole, T. J. 1994. Developing blood vessels and associated extracellular matrix as substrates for neural crest migration in Japanese quail, *Coturnix coturnix japonica. Int J Dev Biol* 38:85–98.

Sperber, S. M., Saxena, V., Hatch, G., and Ekker, M. 2008. Zebrafish *dlx2a* contributes to hindbrain neural crest survival, is necessary for differentiation of sensory ganglia and functions with *dlxa1* in maturation of the arch cartilage elements. *Dev Biol* 314:59–70.

Spicer, A. P., and Tien, J. Y. L. 2004. Hyaluronan and morphogenesis. *Birth Defects Res (Part C)* 72:89–108.

Spranger, J., Benirschke, K., Hall, J. G., Lenz, W., *et al.* 1982. Errors of morphogenesis: Concepts and terms. *J Pediatr* 100:160–165.

Stach, T. 2000. Microscopic anatomy of developmental stages of Branchiostoma lanceolatum (Cephalochordata, Chordata). *Bonner Zool Monog* #47:1–111.

Stark, M. R., Sechrist, J., Bronner-Fraser, M., and Marcelle, C. 1997. Neural tube–ectoderm interactions are required for trigeminal placode formation. *Development* 124:4287–4295.

Stemple, D. L., and Anderson, D. J. 1992. Isolation of a stem cell for neurons and glia derived form the mammalian neural crest. *Cell* 71:973–985.

Stemple, D. L., and Anderson, D. J. 1993. Lineage diversification of the neural crest: *In vitro* investigations. *Dev Biol* 159:12–23.

Stern, C. D. 1990. The distinct mechanisms of segmentation? *Sem Dev Biol* 1:109–116.

Stern, C. D., Artinger, K. B., and Bronner-Fraser, M. 1991. Tissue interactions affecting the migration and differentiation of neural crest cells in the chick embryo. *Development* 113:207–216.

Stern, C. D., and Keynes, R. J. 1987. Interactions between somite cells: The formation and maintenance of segment boundaries in the chick embryo. *Development* 99:261–272.

Stocker, K. M., Sherman, L., Rees, S., and Ciment, G. 1991. Basic FGF and TGF-β1 influence commitment to melanogenesis in neural crest-derived cells of avian embryos. *Development* 111:635–645.

Stoller, J. Z., and Epstein, J. A. 2005. Cardiac neural crest. *Sem Cell Dev Biol* 16:704–715.

Stone, J. R., and Hall, B. K. 2004. Latent homologues for the neural crest as an evolutionary novelty. *Evol Dev* 6:123–129.

Stone, L. S. 1922. Experiments on the development of the cranial ganglia and the lateral line sense organs in *Amblystoma punctatum. J Exp Zool* 35:421–496.

Stone, L. S. 1926. Further experiments on the extirpation and transplantation of mesectoderm in *Amblystoma punctatum. J Exp Zool* 44:95–131.

Stone, L. S. 1929. Experiments showing the role of migrating neural crest (mesectoderm) in the formation of head skeleton and loose connective tissue in *Rana palustris*. *Wilhelm Roux Arch EntwMech Org* 118:40–77.

Stone, L. S. 1932. Transplantation of the hyobranchial mesentoderm including the right lateral anlage of the second basibranchium in *Ambystoma punctatum. J Exp Zool* 62:109–123.

Streit, A. 2002. Extensive cell movements accompany formation of the otic placode. *Dev Biol* 249:237–254.

Streit, A. 2004. Early development of the cranial sensory nervous system: From a common field to individual placodes. *Dev Biol* 276:1–15.

Sun, S.-K., Dee, C. T., Tripathi, V. B., Rengifo, A., *et al.* 2007. Epibranchial and otic placodes are induced by a common Fgf signal, but their subsequent development is independent. *Dev Biol* 303:675–686.

Suzuki, A., Ueno, N., and Hemmati-Brivanlou, A. 1997. *Xenopus msx-1* mediates epidermal induction and neural inhibition by BMP4. *Development* 124:3037–3044.

Suzuki, H. R., and Kirby, M. L. 1997. Absence of neural crest cell regeneration from the postotic neural tube. *Dev Biol* 184:222–233

Suzuki, N., Svensson, K., and Eriksson, V. J. 1996. High glucose concentration inhibits migration of rat cranial neural crest cells *in vitro. Diabetologia* 39:401–411.

Suzuki, T., Oohara, I., and Kurokawa, T. 1998. *Hoxd-4* expression during pharyngeal arch development in flounder (*Paralichthys olivaceus*) embryos and effects of retinoic acid on expression. *Zool Sci* 15:57–67.

Suzuki, T., Sakai, D., Osumi, N., Wada, H., and Wakamatsu, Y. 2006. *Sox* genes regulate type 2 collagen expression in avian neural crest cells. *Dev Growth Differ* 48:477–486.

Svajger, A., Levak-Svajger, B., Kostovic-Knezevic, L., and Bradamante, Z. 1981. Morphogenetic behaviour of the rat embryonic ectoderm as a renal homograft. *J Embryol Exp Morph.* 65(Suppl.):243–267.

Svoboda, K. R., Linares, A. E., and Ribera, A. B. 2001. Activity regulates programmed cell death of zebrafish Rohon–Beard neurons. *Development* 128:3511–3520.

T

Takahashi, K., Nuckolls, G. H., Tanaka, O., Semba, I., *et al.* 1998. Adenovirus-mediated ectopic expression of *Msx2* in even-numbered rhombomeres induces apoptotic elimination of cranial neural crest cells *in ovo. Development* 125:1627–1635.

Takahashi, K., Tanabe, K., Ohnuki, M., Narita, M., *et al.* 2007. Induction of pluripotent stem cells from adult human fibroblasts by defined factors. *Cell* 131:861–872.

Takahashi, Y., Bontoux, M., and Le Douarin, N. M. 1991. Epithelio-mesenchymal interactions are critical for Quox-7 expression and membrane bone differentiation in the neural crest derived mandibular mesenchyme. *EMBO J* 10:2387–2393.

Takahashi, Y., and Le Douarin, N. 1990. cDNA cloning of a quail homeobox gene and its expression in neural crest-derived mesenchyme and lateral plate mesoderm. *Proc Natl Acad Sci USA* 87:7482–7486.

Takashima, Y., Era, T., Nakao, K., Kondo, S., *et al.* 2007. Neuroepithelial cells supply an initial transient wave of MSC differentiation. *Cell* 129:1377–1388.

Takihara, Y., Tomotsune, D., Shirai, M., Katoh-Fukui, *et al.* 1997. Targeted disruption of the mouse homologue of the *Drosophila polyhomeotic* gene leads to altered anteroposterior patterning and neural crest defects. *Development* 124:3673–3682.

Tallqvist, M. D., and Soriano, P. 2003. Cell autonomous requirement for PDGFRα in populations of cranial and cardiac neural crest cells. *Development* 130:507–518.

Tam, P. P. L. 1989. Regionalization of the mouse embryonic ectoderm: Allocation of prospective ectodermal tissues during gastrulation. *Development* 107:55–68.

Tam, P. P. L., and Quinlan, G. A. 1996. Mapping vertebrate embryos. *Curr Biol* 6:104–106.

Tam, P. P. L., and Selwood, L. 1996. Development of lineages of primary germ layers, extra-embryonic membranes and fetus. *Reprod Fertil Dev* 8:803–805.

Tan, C., Deardorff, M. A., Saint-Jeannet, J.-P., Yang, J., *et al.* 2001. Kermit, a frizzled inter-acting protein, regulates frizzled 3 signaling in neural crest development. *Development* 128: 3665–3674.

Tan, S.-S., and Morriss-Kay, G. M. 1986. Analysis of cranial neural crest cell migration and early fates in postimplantation rat chimaeras. *J Embryol Exp Morphol* 98:21–58.

Taneyhill, L. A., Coles, E. G., and Bronner-Fraser, M. 2007. Snail2 directly represses cadherin6B during epithelial-to-mesenchymal transitions of the neural crest. *Development* 134:1481–1490.

Tardy, M. L., and Webb, J. F. 2003. Development of the supraorbital and mandibular lateral line canals in the cichlid, *Archocentrus nigrofasciatus*. *J Morphol* 255:44–57.

Teillet, M.-A., Kalcheim, C., and Le Douarin, N. M. 1987. Formation of the dorsal root ganglia in the avian embryo: Segmental origin and migratory behavior of neural crest progenitor cells. *Dev Biol* 120:329–347.

Teitelman, G. 1990. Insulin cells of pancreas extend neurites but do not arise from neuroectoderm. *Dev Biol* 142:368–379.

Teng, L., Mundell, N. A., Frist, A. Y., Wang, W., and Labosky, P. A. 2008. Requirement for Foxd3 in the maintenance of neural crest progenitors. *Development* 135:1615–1624.

Testaz, S., and Duband, J.-L. 2001. Central role of the α4β1 integrin in the coordination of avian truncal neural crest cell adhesion, migration, and survival. *Dev Dyn* 222:127–140.

Theiler, K. 1972. *The House Mouse. Development and Normal Stages from Fertilization to 4 Weeks of Age*. Berlin: Springer-Verlag,

Thisse, C., Thisse, B., and Postlethwait, J. H. 1995. Expression of *snail2*, a second member of the zebrafish Snail family, in cephalic mesendoderm and presumptive neural crest of wild-type and *spadetail* mutant embryos. *Dev Biol* 172:86–99.

Thomas, P., and Beddington, R. 1996. Anterior primitive endoderm may be responsible for pat-terning the anterior neural plate in the mouse embryo. *Curr Biol* 6:1487–1496.

Thomas, T., Kurihara, H., Yamagishi, H., Yazaki, Y., *et al.* 1998. A signaling cascade involving endothelin-1, dHAND and Msx1 regulates development of neural-crest-derived branchial arch mesenchyme. *Development* 125:3005–3014.

Thorogood, P. V. 1993. Differentiation and morphogenesis of cranial skeletal tissues. In *The Skull, Volume 1, Development*, eds J. Hanken and B. K. Hall, pp. 112–152. Chicago: The University of Chicago Press.

Toerien, M. J. 1965. Experimental studies on the columella–capsular interrelationships in the turtle, *Chelydra serpentina*. *J Embryol Exp Morphol* 14:265–272.

Tomita, Y., Matsumura, K., Wakamatsu, Y., Shibuya, I., *et al.* 2005. Cardiac neural crest cells contribute to the dormant multipotent stem cell in the mammalian heart. *J Cell Biol* 170: 1135–1146.

Tovar, J. A. 2007. The neural crest in pediatric surgery. *J Pediatric Surg* 42:915–926.

Trainor, P. A. 2005. Specification of neural crest cell formation and migration in mouse embryos. *Sem Cell Dev Biol* 16:683–693.

Trainor, P. A., Ariza-McNaughton, L., and Krumlauf, R. 2002. Role of the isthmus and FGFs in resolving the paradox of neural crest plasticity and prepatterning. *Science* 295:1288–1291.

Trainor, P. A., Sobieszczuk, D., Wilkinson, D., and Krumlauf, R. 2002. Signalling between the hindbrain and paraxial tissues dictates neural crest cell migration pathways. *Development* 129:433–442.

Trainor, P. A., and Tam, P. P. L. 1995. Cranial paraxial mesoderm and neural crest cells of the mouse embryo: Co-distribution in the craniofacial mesenchyme but distinct segregation in branchial arches. *Development* 121:2569–2582.

Tran, S., and Hall, B. K. 1989. Growth of the clavicle and development of clavicular secondary cartilage in the embryonic mouse. *Acta Anat* 135:200–207.

Tremblay, P., Kessel, M., and Gruss, P. 1995. A transgenic neuroanatomical marker identifies cra-nial neural crest deficiencies associated with the *Pax3* mutant *Splotch*. *Dev Biol* 171:317–329.

Trentin, A., Glavieux-Pardanaud, C., Le Douarin, N. M., and Dupin, E. 2004. Self-renewal capacity is a widespread property of various types of neural crest precursor cells. *Proc Natl Acad Sci USA* 101:4495–4500.

Tribulo, C., Aybar, M. J., Sánchez, S. S., and Mayor, R. 2004. A balance between the anti-apoptotic activity of *Slug* and the apoptotic activity of *msx1* is required for the proper development of the neural crest. *Dev Biol* 275:325–342.

Tucker, R. P. 2004. Neural crest cells: A model for invasive behavior. *Int J Biochem Cell Biol* 36:173–177.

Tucker, R. P., Hagios, C., Chiquet-Ehrismann, R., Lawler, J., *et al.* 1999. Thrombospondin-1 and neural crest cell migration. *Dev Dyn* 214:312–322.

Tuckett, F., and Morriss-Kay, G. M. 1986. The distribution of fibronectin, laminin and entactin in the neurulating rat embryo studies by indirect immunofluorescence. *J Embryol Exp Morphol* 94:95–112.

Twitty, V. C., and Bodenstein, D. 1939. Correlated genetic and embryological experiments on *Triturus*. III. Further transplantation experiments on pigment development. IV. The study of pigment cell behavior *in vitro*. *J Exp Zool* 81:357–398.

U

Urbánek, P., Fetka, I., Meisler, M. H., and Busslinger, M. 1997. Cooperation of *Pax2* and *Pax5* in midbrain and cerebellum development. *Proc Natl Acad Sci USA* 94:5703–5708.

V

Vaglia, J. L., and Hall, B. K. 1999. Regulation of neural crest cell populations in vertebrates: Occurrence, distribution and underlying mechanisms. *Int J Dev Biol* 43:95–110.

Vaglia, J. L., and Hall, B. K. 2000. Patterns of migration and regulation of trunk neural crest cells in zebrafish (*Danio rerio*). *Int J Dev Biol* 44:867–881.

Vaglia, J. L., and Smith, K. K. 2003. Early differentiation and migration of cranial neural crest in the opossum, *Monodelphis domestica*. *Evol Dev* 5:121–135.

Vandersea, M. W., McCarthy, R. A., Fleming, P., and Smith, D. 1998. Exogenous retinoic acid during gastrulation induces cartilaginous and other craniofacial defects in *Fundulus heteroclitus*. *Biol Bull* 194:281–296.

Varadkar, P., Kraman, M., Despres, D., Ma, G., *et al.* 2008. Notch2 is required for the proliferation of cardiac neural crest-derived smooth muscle cells. *Dev Dyn* 237:1144–1152.

Vickaryous, M. K., and Hall, B. K. 2006. Human cell type diversity, evolution, development, and classification with special reference to cells derived from the neural crest. *Biol Rev Camb Philos Soc* 81:425–455.

Vielkind, J. R., Tron, V. A., Schmidt, B. M., Dougherty, G. J., *et al.* 1993. A putative marker for human melanoma: A monoclonal antibody derived from the melanoma gene in the *Xiphophorus* melanoma model. *Amer J Pathol* 143:656–662.

Vielle-Grosjean, I., Hunt, P., Gulisano, M., Boncinelli, E., and Thorogood, P. 1997. Branchial Hox gene expression and human craniofacial development. *Dev Biol* 183:49–60.

Vogel, K. S., and Davies, A. M. 1993. Heterotopic transplantation of presumptive placodal ectoderm changes the fate of sensory neuron precursors. *Development* 119:263–276.

Voigt, J., and Papalopulu, N. 2005. A dominant-negative form of the E3 ubiquitin ligase *Cullin-1* disrupts the correct allocation of cell fate in the neural crest lineage. *Development* 133:559–568.

Von Bartheld, C. S., and Baker, C. V. H. 2004. Nervus terminalis derived from the neural crest? A surprising new turn in a century-old debate. *Anat Rec (Part B: New Anat)* 278B:12–13.

Vukicevic, S., Latin, V., Ping, C., Batorsky, R., *et al.* 1994. Localization of osteogenic protein-1 (bone morphogenetic protein-7) during human embryonic development: High affinity binding to basement membranes. *Biochem Biophys Res Commun* 198:693–700.

W

Wada, H., Saiga, H., Satoh, N., and Holland, P. W. H. 1998. Tripartite organization of the ancestral chordate brain and the antiquity of placodes: Insights from ascidian *Pax-2/5/8*, *Hox* and *Otx* genes. *Development* 125:1113–1122.

Wada, N., Javidan, Y., Nelson, S., Carney, T. J., *et al.* 2005. Hedghog signaling is required for cranial neural crest morphogenesis and chondrogenesis at the midline in the zebrafish skull. *Development* 123:3977–3988.

Wagner, G. 1949. Die Bedeutung der Neuralleiste für die Kopfgestaltung der Amphibienlarven. Untersuchungen an Chimaeren von *Triton*. *Rev Suisse Zool* 56:519–620.

Wake, D. B. 1976. On the correct scientific names of Urodeles. *Differentiation* 6:195.

Washausen, S., Obermayer, B., Brunnet, G., Kuhn, H.-J., and Knabe, W. 2005. Apoptosis and proliferation in developing, mature, and regressing epibranchial placodes. *Dev Biol* 278: 86–102.

Watanabe, M., Iwashita, M., Masaru, I., Kurachi, Y., *et al.* 2006. Spot pattern of *leopard Danio* is caused by mutation in the zebrafish *connexin41.8* gene. *EMBO Reports* 7:893–987.

Watanabe Y., and Le Douarin, N. M. 1996. A role for BMP-4 in the development of subcutaneous cartilage. *Mech Dev* 57:69–78.

Webb, J. F., and Noden, D. M. 1993. Ectodermal placodes: Contributions to the development of the vertebrate head. *Amer Zool* 33:434–447.

Webb, J. F., and Northcutt, R. G. 1997. Morphology and distribution of pit organs and canal neuromasts in non-teleost bony fishes. *Brain Behav Evol* 50:139–151.

Webb, J. F., and Shirey, J. E. 2003. Postembryonic development of the cranial lateral line canals and neuromasts in zebrafish. *Dev Dyn* 228:370–385.

Wedden, S. E. 1987. Epithelial–mesenchymal interactions in the development of chick facial primordia and the target of retinoid action. *Development* 99:341–352.

Wehrle-Haller, B., and Weston, J. A. 1995. Soluble and cell membrane forms of steel factor play distinct roles in melanocyte precursor dispersal and survival on the lateral neural crest migration pathway. *Development* 121:731–742.

Wehrle-Haller, B., and Weston, J. A. 1997. Receptor tyrosine kinase-dependent neural crest migration in response to differentially localized growth factors. *BioEssays* 19:337–345.

Weinberg, R. A. 2006. *The Biology of Cancer*. Milton Park. U. K: Garland Science.

Weinstein, D. C., and Hemmati-Brivanlou, A. 1997. Neural induction in *Xenopus laevis*: Evidence for the default model. *Curr Opin Neurobiol* 7:7–12.

Weinstein, D. C., Honoré, E., and Hemmati-Brivanlou, A. 1997. Epidermal induction and inhibition of neural fate by translation initiation factor 4AIII. *Development* 124:4235–4242.

Wendling, O., Dennefeld, C., Chambon, P., and Mark, M. 2000. Retinoid signaling is essential for patterning the endoderm of the third and fourth pharyngeal arches. *Development* 127: 1553–1562.

Weston, A. D., Ozolins, T. R. S., and Brown, N. A. 2007. Thoracic skeletal defects and cardiac malformations: A common epigenetic link? *Birth Defects Res (Part C)* 78:354–370.

Weston, J. A. 1963. A radioautographic analysis of the migration and localization of trunk neural crest cells in the chick. *Dev Biol* 6:279–310.

Weston, J. A. 1970. The migration and differentiation of neural crest cells. *Adv Morphog* 8:41–114.

Weston, J. A. 1991. Sequential segregation and fate of developmentally restricted intermediate cell populations in the neural crest lineage. *Curr Topics Dev Biol* 25:133–153.

Weston, J. A., Yoshida, H., Robinson, V. B., Nishikawa, S., *et al.* 2004. Neural crest and the origin of ectomesenchyme: Neural fold heterogeneity suggests an alternative hypothesis. *Dec Dyn* 229:118–130.

Whitlock, K. E., Wolf, C. D., and Boyce, M. L. 2003. Gonadotropin-releasing hormone (GnRH) cells arise from cranial neural crest and adenohypophyseal regions of the neural plate in the zebrafish, *Danio rerio*. *Dev Biol* 257:140–152.

Whittaker, J. R. 1987. Cell lineages and determinants of cell fate in development. *Amer Zool* 27:607–622.

Wicht, H., and Northcutt, R. G. 1995. Ontogeny of the head of the Pacific hagfish (*Eptatretus stouti*, Myxinoidea): Development of the lateral line system. *Phil Trans R Soc* B349:119–134.

Wildner, H., Gierl, M. S., Strehle, M., Pla, P., and Birchmeier, C. 2008. Insm1 (IA-1) is a crucial component of the transcriptional network that controls differentiation of the sympatho-adrenal lineage. *Development* 135:473–481.

Wilkinson, D. G. 1995. Genetic control of segmentation in the vertebrate hindbrain. *Persp Dev Neurobiol* 3:29–38.

Williams, J. A., Barrios, A., Gatchalian, C., *et al.* 2000. Programmed cell death in zebrafish Rohon Beard neurons is influenced by TrkC1/NT-3 signaling. *Dev Biol* 226:220–230.

Williams, N. A., and Holland, P. W. H. 1992. *Nature* 383:490.

Williams, N. A., and Holland, P. W. H. 1998. Gene and domain duplication in the Chordate *Otx* gene family: Insights from Amphioxus *Otx Mol Biol Evol* 15:600–607.

Williamson, D. A., Parrish, E. P., and Edelman, G. M. 1991. Distribution and expression of two interactive extracellular matrix proteins, cytotactin and cytotactin-binding proteoglycan, during development of *Xenopus laevis*. II. Metamorphosis. *J Morph* 209:203–213.

Wilson, A. L., Shen, Y.-C., Babb-Clendenon, S. G., Rostedt, J., *et al.* 2007. Cadherin-4 plays a role in the development of zebrafish cranial ganglia and lateral line system. *Dev Dyn* 236:893–902.

Wilson, D. B., and Wyatt, D. P. 1988. Closure of the posterior neuropore in the *vL* mutant mouse. *Anat Embryol* 178:559–563.

Wilson, D. B., and Wyatt, D. P. 1995. Alterations in cranial morphogenesis in the *Lp* mutant mouse. *J Craniofac Genet Dev Biol* 15:182–189.

Wilson, P. A., and Hemmati-Brivanlou, A. 1995. Induction of epidermis and inhibition of neural fate by BMP-4. *Nature* 376:331–333.

Wilson, T. J., Davidson, N. J., Boyd, R. L., and Gershwin, M. E. 1992. Phenotypic analysis of the chicken thymic microenvironment during ontogenic development. *Dev Immunol* 2:19–27.

Winograd, J., Reilly, M. P., Roe, R., *et al.* 1997. Perinatal lethality and multiple craniofacial malformations in *Msx2* transgenic mice. *Hum Mol Genet* 6:369–379.

Wise, S. B., and Stock, D. W. 2006. Conservation and divergence of Bmp21, Bmp2b, and Bmp4 expression patterns within and between dentitions of teleost fish. *Evol Dev* 8:511–523.

Woellwarth, C. V. 1961. Die rolle des Neuralleistenmaterials und der Temperatur bei der Determination der Augenlinse. *Embyologia* 6:219–242.

Wonsettler, A. L., and Webb, J. F. 1997. Morphology and development of the multiple lateral line canals on the trunk in two species of *Hexagrammis* (Scorpaeniformes, Hexagrammidae). *J Morph* 233:195–214.

Woo, K., and Fraser, S. E. 1998. Specification of hindbrain fate in the zebrafish. *Dev Biol* 197: 283–296.

Wright, G. M., Keely, F. W., and Robson, P. 2001. The unusual cartilaginous tissues of jawless craniates, cephalochordates and invertebrates. *Cell Tissue Res* 304:165–174.

Wurdak, H., Ittner, L. M., and Sommer, L. 2006. DiGeorge syndrome and pharyngeal apparatus development. *BioEssays* 28:1078–1086.

X

Xu, H, Dude, C. M., and Baker, C. V. H. 2008. Fine-grained fate maps for the ophthalmic and maxillomandibular trigeminal placodes in the chick embryo. *Dev Biol* 317:174–186.

Xu, R.-H., Jaebong, K., Taira, M., Shuning, Z., *et al.* 1995. A dominant negative bone morphogenetic protein 4 receptor causes neuralization of *Xenopus* ectoderm. *Biochem Biophys Res Commun* 212:212–219.

Y

Yavarone, M. S., Shuey, D. L., Tamir, H., Sadler, T. W., and Lauder, J. M. 1993. Serotonin and cardiac morphogenesis in the mouse embryo. *Teratology* 47:573–584.

Yelick, P. C., and Schilling, T. F. 2002. Molecular dissection of craniofacial development using zebrafish. *Crit Rev Oral Biol Med* 13:308–322.

Yeo, W., and Gautier, J. 2004. Early neural cell death: Dying to become neurons. *Dev Biol* 274: 233–244.

Yip, J. E., Kokich, V. H., and Shepard, T. H. 1980. The effect of high doses of retinoic acid on prenatal craniofacial development of *Macaca nemestrina*. *Teratology* 21:29–38.

Yip, J. W. 1986. Migratory patterns of sympathetic ganglioblasts and other neural crest derivatives in chick embryos. *J Neurosci* 6:3465–3473.

Yntema, C. L. 1955. Ear and nose. In *Analysis of Development*, ed B. H. Willier, P. A. Weiss and V. Hamburger, pp. 415–428. Philadelphia: Saunders.

Yntema, C. L., and Hammond, W. S. 1945. Depletions and abnormalities in the cervical sympathetic system of the chick following extirpation of the neural crest. *J Exp Zool* 100:237–263.

Yntema, C. L., and Hammond, W. S. 1947. The development of the autonomic nervous system. *Biol Rev Camb Philos Soc* 22:344–357.

Yu, B. D., Hanson, R. D., Hess, J. L., Horning, S. E., and Korsmeyer, S. J. 1998. MLL, a mammalian *trithorax*-group gene, functions as a transcriptional maintenance factor in morphogenesis. *Proc Natl Acad Sci USA* 95:10632–10636.

Yu, H.-H., and Moens, C. B. 2005. Semaphorin signaling guides cranial neural crest cell migration in zebrafish. *Dev Biol* 280:373–385.

Yu, M., Yue, Z., Wu, P., Wu, D.-Y., *et al.* 2004. The biology of feather follicles. *Int J Dev Biol* 48:181–191.

Z

Zagris, N., and Chung, A. E. 1990. Distribution and functional role of laminin during induction of the embryonic axis in the chick embryo. *Differentiation* 43:81–86.

Zelditch, M. L., and Carmichael, A. C. 1989. Ontogenetic variation in patterns of developmental and functional integration in skulls of *Sigmodon fulviventer*. *Evolution* 43:814–824.

Zhang, G., and Cohn, M. J. 2006. Hagfish and lancelet fibrillar collagens reveal that type II collagen-based cartilage evolved in stem vertebrates. *Proc Natl Acad Sci USA* 103:16829–16833.

Zhang, J., Hagopian-Donaldson, S., Serbedzija, G., Elsemore, J., *et al.* 1996. Neural tube, skeletal and body wall defects in mice lacking transcription factor AP-2. *Nature* 381:238–241.

Zhu, H., Wlodarczyk, B. J., Scott, M., Yu, W., *et al.* 2007. Cardiovascular abnormalities in Folr1 knockout mice and folate rescue. *Birth Defects Res (Part A)* 79:257–268.

Ziller, C., Dupin, E., Brazeau, P., Paulin, D., and Le Douarin, N. M. 1983. Early segregation of a neuronal precursor cell line in the neural crest as revealed by culture in a chemically defined medium. *Cell* 32:627–638.

Zondag, G. C. M., Eversa, E. E., ten Klooster, J. P., Jansssen, L., *et al.* 2000. Oncogenic Ras downregulates Rac activity, which leads to increased Rho activity and epithelial-mesenchymal transition. *J Cell Biol* 149:775–781.

Zottoli, S. J., and Seyfarth, E.-A. 1994. Julia B. Platt (1857–1935): Pioneer comparative embryologist and neuroscientist. *Brain Behav Evol* 43:91–106.

Index

NOTE: The letter 'n' in the index denotes a note number, 'Box' denotes reference given in a box on the relevant pages.

Printed in the United States of America